THEORY

OF

DIFFERENTIAL EQUATIONS.

CAMBRIDGE UNIVERSITY PRESS WAREHOUSE,
C. F. CLAY, MANAGER.
London: FETTER LANE, E.C.
Glasgow: 50, WELLINGTON STREET.

Leipzig: F. A. BROCKHAUS.
New York: G. P. PUTNAM'S SONS.
Bombay and Calcutta: MACMILLAN AND CO., LTD.

THEORY

OF

DIFFERENTIAL EQUATIONS.

PART IV.

PARTIAL DIFFERENTIAL EQUATIONS.

BY

ANDREW RUSSELL FORSYTH,

SC.D., LL.D., MATH.D., F.R.S.,

SADLERIAN PROFESSOR OF PURE MATHEMATICS,

FELLOW OF TRINITY COLLEGE, CAMBRIDGE.

VOL. V.

CAMBRIDGE:

AT THE UNIVERSITY PRESS.

1906

Cambridge:
PRINTED BY JOHN CLAY, M.A.
AT THE UNIVERSITY PRESS.

PREFACE.

THESE two volumes, now published as Part IV of the present work, are my final contribution towards the fulfilment of a promise made twenty-one years ago. They are devoted to the theory of partial differential equations.

Though the work thus is completed, no claim is made that every topic of importance has been discussed. In the earlier volumes, indications of omissions from other portions of the whole subject were given and need not now be repeated: here also, there have been definite omissions. Nothing, for instance, is said concerning the researches of Picard and Dini on the method of successive approximations for the construction of an integral which obeys assigned conditions; these investigations limit the variables to real values, and throughout the treatise I have dealt with variables having complex values. Formal questions, such as those which arise out of the application of the theory of groups, are hardly mentioned; here, as in the preceding volumes, I have concerned myself with organic properties, given by applications of the theory of functions, rather than with formal properties. Again, the subject of boundary problems is not dealt with; it appears to me to belong to the theory of functions in its applications to mathematical physics rather than to the theory of differential equations. In the branches of the subject that are discussed, I have tried to deal as completely as possible with what seems to me to be essential: and I have omitted what are purely formal extensions, to equations of general order, of the properties of equations of the second order when such extensions contain no intrinsic novelty.

In the preparation of the volumes, I have consulted the works of many writers; and references are freely given. My aim has been to make these references relate to the main issues; not a few results, extracted from memoirs, have been used to construct examples; and the name of the author is (I hope) given in every such case. But I have not attempted to select and arrange the references, so that they might make the framework of a history of the subject; had the latter been my purpose, names such as Lagrange, Cauchy, Jacobi, whose work is now the common possession of all writers, would have received more frequent specific references in my pages. It will be seen that Darboux's treatise, *Théorie générale des surfaces*, and Goursat's three volumes, *Leçons sur l'intégration des équations aux dérivées partielles*, have been frequently quoted: I wish to make also a comprehensive acknowledgement of my indebtedness to those works.

The earlier of the two volumes is devoted mainly to equations of the first order. The theory of these equations may be regarded as almost complete, because the actual integration of the equations is made to depend solely upon the solution of difficulties which occur in connection with a system of ordinary equations of the first order.

An introduction to the subject is provided by Cauchy's existence-theorem; it is discussed in the first two chapters. The next chapter is specially concerned with linear equations and linear systems; these admit of a separate and special mode of treatment. The fourth chapter gives an exposition of what, on the whole, I regard as the most effective method of integration for non-linear equations: it contains what is usually called Jacobi's second method, with Mayer's developments. In the succeeding chapter will be found Lagrange's classification of integrals, based upon the process of variation of parameters: but something still remains to be done in this branch of the subject, because even simple examples shew that the customary classes may fail to be entirely comprehensive. The next three chapters are devoted to Cauchy's method of characteristics, alike for two and for any number of independent variables, and to the geometrical associations in the case of two independent variables. Then follows a chapter dealing with Lie's methods, based upon contact-transformations and upon the properties of groups of functions: it was possible to abbreviate this chapter, because Pfaff's problem had already been discussed in the first volume of this work. A chapter has been added dealing with the equations of theoretical dynamics, partly because of their intrinsic connection with partial equations, yet mainly in order to shew the origin of what is usually called Jacobi's first method of integration of partial equations. The concluding chapter of this volume discusses those simultaneous equations of the first order, involving more than one dependent variable, which can be integrated by operations of the same class as those in any of the methods mentioned.

The later of these two volumes is devoted to the consideration of partial equations of the second order and of higher orders, mainly (though not entirely) involving two independent variables. A perusal of the volume will shew that, outside the limits of Cauchy's existence-theorem, knowledge is fragmentary: the inversion of operations of the second order has not yet been discovered and, accordingly, any effective process consists of a succession of operations of the first order.

After a chapter devoted to the discussion of questions connected with the existence of integrals and, in particular, to the discussion of the constitution of a general integral, two chapters are occupied with Laplace's method (and with its developments, due to Darboux) for the integration of the homogeneous linear equations of the second order: the effective success of the method depends upon the vanishing of some invariant, in one or other of two progressively constructed sets of functions involving the coefficients of the original equation. The result raises the question of the form of equations, the primitive of which can be expressed in finite terms: and, to this matter, one chapter is assigned.

In the attempt to integrate any equation of the second order, it is natural to enquire whether an equation of the first order exists which is its complete equivalent: and equations, characterised by this property, will obviously constitute a distinct class. Such, indeed, were the equations of the second order for which integrals (now called intermediate) were first obtained; and one method of their construction is due to Monge. Later, another (and a more direct) method for their construction was given by Boole: but both methods assume that a special form attaches to the intermediate integral, and the assumption demands that a very restricted form shall be possessed by the original equation. Basing his argument entirely upon an assumed type of integral, Ampère devised another process of integration: his method makes no demand for the existence of an intermediate integral: and the result is often effective when no such integral exists. All these three methods, (and another method of some generality, as given), require the construction of integrable combinations of one (and ultimately the same) set of subsidiary equations, when they are applied to the same original equation. But Ampère's method is applicable also to equations of less restricted form.

It may, however, happen that an equation of the second order is not of the restricted form or, being of that form, does not possess an intermediate integral, or is not amenable to Ampère's method. In that case, a method due to Darboux may be applicable, whereby a compatible equation of the second order (or of some higher order) can be constructed; provided only that a compatible equation of finite order can be obtained, a primitive of the original equation can be derived. To these matters, three chapters are given: they explain the working processes that are effective for the determination of an integral in finite terms, whether by a single equation or a set of equations.

One chapter is devoted to the generalisation of integrals which involve some arbitrary parameters, and another to the discussion of characteristics of equations of the second order. The investigations in both of these chapters are clearly incomplete: they could be continued along lines that lead to the complete classification of integrals of equations of the first order.

In the theory of equations of the first order, much information is given by Lie's general theory of contact-transformations: and an obvious investigation is thereby suggested as to whether there is a corresponding theory for equations of order higher than the first. The question has been considered, and partly solved, by Bäcklund and others: one chapter gives an outline of their work: it is clear that much yet remains to be done in this subject.

In the last three chapters of the volume, some of the preceding methods and theories are extended to equations, which are of order higher than the second or which involve more than two independent variables. Only the simplest extensions are discussed: they could be amplified to any extent: but the result would be merely an accumulation of formal theorems possessing neither individuality nor intrinsic value.

From this brief sketch of the contents of these two volumes, it will be manifest that, in the theory of equations of order higher than the first, there are many gaps and that the theory is far from complete: and even a summary perusal of the volumes will give some indication of these gaps. It is my intention to point out, in a presidential address which will be delivered to the London Mathematical Society next month, some of the more obvious and practicable questions which are waiting for solution. Of these, there is no lack: it is only the workers who are wanted.

On not a few occasions, it has been my privilege to acknowledge the help which has been given to me by the Staff of the University Press. Once more, an opportunity comes to me: and I gladly seize it, to express my indebtedness to them all for the care, the attention, and the consideration, by which they have lightened what to me is never an easy or a simple duty.

So I pass from a task, which has filled the greater part of many years of my life, which has broadened in my view as they passed, and which has suffered interruptions that threatened to end it before its completion. Many of its defects are known to me: after it has gone from me, others will become apparent. Nevertheless, my hope is that my work will ease the labour of those who, coming after me, may desire to possess a systematic account of this branch of pure mathematics.

A. R. FORSYTH.

TRINITY COLLEGE, CAMBRIDGE.
October, 1906.

CONTENTS.

CHAPTER I.

INTRODUCTION: TWO EXISTENCE-THEOREMS.

CHAPTER II.

CAUCHY'S THEOREM.

CHAPTER III.

LINEAR EQUATIONS AND COMPLETE LINEAR SYSTEMS.

CHAPTER IV.

NON-LINEAR EQUATIONS: JACOBI'S SECOND METHOD, WITH MAYER'S DEVELOPMENTS.

CHAPTER V.

CLASSES OF INTEGRALS POSSESSED BY EQUATIONS OF THE FIRST ORDER: GENERALISATION OF INTEGRALS.

CHAPTER VI.

THE METHOD OF CHARACTERISTICS FOR EQUATIONS IN TWO INDEPENDENT VARIABLES: GEOMETRICAL RELATIONS OF THE VARIOUS INTEGRALS.

CHAPTER VII.

SINGULAR INTEGRALS AND THEIR GEOMETRICAL PROPERTIES: SINGULARITIES OF THE CHARACTERISTICS.

CHAPTER VIII.

THE METHOD OF CHARACTERISTICS IN ANY NUMBER OF INDEPENDENT VARIABLES.

CHAPTER IX.

LIE'S METHODS APPLIED TO EQUATIONS OF THE FIRST ORDER.

CHAPTER X.

THE EQUATIONS OF THEORETICAL DYNAMICS.

CHAPTER XI.

SIMULTANEOUS EQUATIONS OF THE FIRST ORDER.

CHAPTER I.

INTRODUCTION: TWO EXISTENCE-THEOREMS.

1. THE investigations, which constitute this Part of the present work, are devoted to the consideration of properties of partial differential equations. In text-books which deal with the modes of constructing the integrals of such equations, several processes are given, often with the main purpose of obtaining the integrals in finite terms; but the processes are limited in the scope of their application, because the equations which prove amenable to their action are few in character and not infrequently have been artificially constructed. When these processes either are not applicable or cannot conveniently be completed, no information concerning the solution of the equation would then be obtained; indeed, they offer no guarantee that an integral even exists.

Accordingly, it is desirable to discuss the whole theory of partial differential equations from the foundations and, in the course of that discussion, not only to revise known results but also, so far as may be possible, to place them in their fitting positions in the ordered body of doctrine. Such a discussion was found to be necessary for the proper establishment of results relating to ordinary differential equations. It is even more necessary in the case of partial differential equations, partly because the inversion of simultaneous partial differential operations is more difficult than the inversion of ordinary differential operations, partly because the suggestions as to the character of an integral, as offered by processes of inversion, are less significant for partial equations than for ordinary equations.

2. Two kinds of illustration should suffice for a justification of this last statement.

One mode of attempting to discover the character of the most complete integral of a partial equation would be by generalisation from the case of an ordinary equation.

For an ordinary equation, which has y for its dependent variable and x for its independent variable, the integral is made complete by the assignment of initial values to the variables; that is, y is some function of x and so, when a constant value is assigned to x, the function y and all its derivatives become constants. As the equation is to be satisfied and yet the integral is to be as complete as possible, these constants will be as unrestricted as possible: and therefore it is to be expected that some at least of them will be arbitrary constants. There thus arises a suggestion that the most complete integral will be such that, when some constant value is assigned to x, the function y and some of its derivatives acquire arbitrary constant values. The suggested property has been established under appropriate limitations and conditions.

To extend these results, if possible, to partial differential equations, consider a single partial equation of the first order, having z for its dependent variable and $x_1, \ldots, x_n$ for its independent variables. If an integral exists, that integral must determine z as a function of $x_1, \ldots, x_n$; and so, when an initial value a_n is assigned to x_n, the first derivatives of z with respect to $x_1, \ldots, x_{n-1}$ can be deduced from an assigned expression for z, and then (save in special circumstances) the partial equation determines the first derivative with regard to x_n. By using the equation in combination with the expressions for the first derivatives, the derivatives of higher order can be obtained for the value a_n of x_n; and thus no limitation appears to be imposed on the value of z as an assigned function of $x_1, \ldots, x_{n-1}$, when $x_n = a_n$. If the integral is to be as general as possible, it is reasonable to expect that the assigned function shall be as general as possible. But at this stage, questions arise as to what is the most general function admissible? Is it to be made general by possessing the greatest possible number of arbitrary constants? Can the assigned function be an arbitrary function, subject possibly to limitations imposed by the partial equation? and, if so, must it be explicit or may it be given implicitly, for example, by means of quadratures which cannot be effected in finite terms? Or are all the modes indicated for securing the generality of the integral admissible, so that there are different kinds of general integrals? and if so, are there any relations among the various integrals? To such questions the argument offers no hint of an answer.

Similarly, when a partial equation of the second order is propounded in the same variables z, $x_1, \ldots, x_n$, the extension of the results obtained for ordinary equations suggests that z and $\dfrac{\partial z}{\partial x_n}$ should acquire assigned values as functions of $x_1, \ldots, x_{n-1}$, when $x_n = a_n$. For the values of $\dfrac{\partial z}{\partial x_1}, \ldots, \dfrac{\partial z}{\partial x_{n-1}}$, when $x_n = a_n$, could be deduced from the value of z; and then the values of $\dfrac{\partial^2 z}{\partial x_r \partial x_s}$, for $r = 1, \ldots, n-1$, and $s = 1, \ldots, n$, could be deduced from the values of $\dfrac{\partial z}{\partial x_1}, \ldots, \dfrac{\partial z}{\partial x_n}$ already known; and the partial equation would (save in special circumstances) determine the value of $\dfrac{\partial^2 z}{\partial x_n^2}$. As before, the values thus obtained, when combined with the use of the partial equation, lead to the values of all the derivatives. Thus all the quantities associated with z are known: at the utmost, only special limitations appear to be imposed upon the assigned functions by the process adopted; and therefore it is reasonable to expect that the integral will become the most general possible when the two assigned functions are as general as possible. Again, at this stage, questions arise as to the constitution of the generality of these assigned functions. Is the generality to be secured, by arranging that they shall involve the greatest possible number of arbitrary constants? or by making them independent arbitrary functions of $x_1, \ldots, x_{n-1}$? or by associating them with a possibly even more general function of $x_1, \ldots, x_n$ for the particular value a_n of x_n? If the functions are arbitrary, must they be given explicitly or may they be given implicitly as, for example, by uncompleted quadratures? Again, are all the modes admissible as alternatives, so that they lead to different kinds of general integrals? and if so, what relations (if any) subsist among the integrals? As in the former case, the argument offers no hint of answer to the questions.

3. In both the instances that have been briefly considered, the argument offers suggestions and even stirs expectations: that this is the limit of the attention to be paid to it, can perhaps be most simply seen by a particular case. Applied to a couple of simultaneous partial equations determining a couple of dependent variables, it would lead to a suggestion that the most general integral would involve at least two sets of general elements, whatever be their form; yet the integral of the simultaneous equations

$$\left.\begin{aligned} \frac{\partial^2 z_1}{\partial x^2} - a^2 \frac{\partial^2 z_1}{\partial y^2} + 2a \frac{\partial z_1}{\partial y} + \frac{\partial z_2}{\partial x} - a \frac{\partial z_2}{\partial y} + z_2 &= f(x, y) \\ \frac{\partial z_1}{\partial x} + a \frac{\partial z_1}{\partial y} - z_1 + z_2 &= g(x, y) \end{aligned}\right\}$$

is

$$\left.\begin{aligned} z_1 &= f(x, y) - g(x, y) - \frac{\partial g(x, y)}{\partial x} + a \frac{\partial g(x, y)}{\partial y} \\ z_2 &= f(x, y) - \frac{\partial f(x, y)}{\partial x} - a \frac{\partial f(x, y)}{\partial y} \\ &\quad + 2a \frac{\partial g(x, y)}{\partial y} + \frac{\partial^2 g(x, y)}{\partial x^2} - a^2 \frac{\partial^2 g(x, y)}{\partial y^2} \end{aligned}\right\},$$

a being a constant; manifestly it contains no arbitrary element. In fact, the utmost to be inferred from the argument is that some kinds of equations may possess integrals involving arbitrary elements in their most general forms, and that there may be different kinds of general integrals.

Whether these general integrals include all the integrals of an equation is a matter that demands separate consideration, to be undertaken later in another line of inquiry: and, naturally, a detailed consideration of the generality of integrals must also be undertaken later.

4. Another mode of attempting to discover the character of the most complete integral of a partial equation consists in comparing differential equations, constructed from initial integral equations, with those integral equations: but it is easily seen to be untrustworthy.

Thus if an integral equation

$$\frac{az + b}{cz + d} = \phi(x_1, \ldots, x_n),$$

where $ad - bc = 1$, be propounded, the result of eliminating the constants between the equation and its derivatives leads to the set of partial equations

$$\frac{1}{p_1}\frac{\partial\phi}{\partial x_1} = \frac{1}{p_2}\frac{\partial\phi}{\partial x_2} = \ldots = \frac{1}{p_n}\frac{\partial\phi}{\partial x_n},$$

where

$$p_r = \frac{\partial z}{\partial x_r},$$

for $r = 1, \ldots, n$. The process cannot be reversed, so as to lead to an inference that the most general integral of the set of partial equations contains three arbitrary essential constants: the inference would be incorrect, for the set of equations is satisfied by

$$f(z) = \phi(x_1, \ldots, x_n),$$

where $f(z)$ is any function of z containing any number of arbitrary constants.

Again, if there is given an integral equation

$$f(x_1, \ldots, x_n, z, a_1, \ldots, a_m) = 0,$$

it is possible to construct a set of partial equations with which the integral equation is consistent, by forming the n derived equations which give the values of $\frac{\partial z}{\partial x_1}, \ldots, \frac{\partial z}{\partial x_n}$, and then eliminating the m constants $a_1, \ldots, a_m$. For the present purpose, the m constants may be assumed to be not reducible to a smaller number, and m may be assumed not greater than n; also, when elimination takes place, the number of resulting equations will be not less than $n + 1 - m$. It will be assumed that the number of such equations in the set is actually $n + 1 - m$; each of them is partial, and of the first order.

If m is equal to n, there is a single partial equation: and the argument suggests that a single partial equation of the first order may possess an integral involving n arbitrary constants: it does not prove this result, for there is nothing to shew that the partial equation is not of a special form, arising from the limitation that it has been deduced from an integral equation of specified form. The argument offers no contribution to the question as to whether, if the integral is possessed by the partial equation, it is the most general integral.

If m is less than n, there is a set of simultaneous partial equations of the first order: and the argument might be held to suggest that $n+1-m$ simultaneous partial equations of the first order, involving one dependent variable and n independent variables, may possess a common integral involving m constants. The result is, of course, not proved and it is not true in general fact: for, independently of the impossibility of reversing the process of elimination, the $n+1-m$ equations are affected by the form of the original integral and therefore will have relations with one another, while such a set of partial equations postulated initially need not have any relations with one another. Thus the existence of a common integral is even more doubtful than in the case of a single equation: if it exists, no inference as to its generality can be drawn.

5. After these explanations and criticisms, it is manifest that attempts to obtain information as to the solution of partial equations by vague extensions of the knowledge of the solution of ordinary equations must be abandoned. The constructive process, that will be adopted instead of them, consists in the gradual establishment of results, beginning with the proof of the existence of integrals possessing definite assigned characters. The actual construction of the integrals when their existence has once been established, the discussion of the range of their generality, and the possibility of using them in the derivation of integrals of other kinds, all are matters for subsequent investigation.

It will be assumed that, save in special examples, the number of independent variables is n; and they will usually be denoted by $x_1, \ldots, x_n$. The number of dependent variables may be taken as m, the simplest case arising when $m=1$; they will be denoted by $z_1, \ldots, z_m$; and when there is only one variable, it will be denoted by z. For the present purpose, these dependent variables are to be determined by partial differential equations; let the number of such equations in a given set be s, and suppose that the highest derivatives that occur in them are of order μ.

Let derivatives of each of the equations be constructed, of all orders up to those of order κ inclusive. Then the total number of equations in the amplified set is

$$\begin{aligned} s\{1+n+\tfrac{1}{2}n(n+1)+\ldots \text{ to } (\kappa+1) \text{ terms}\} \\ = s\frac{(n+1)(n+2)\ldots(n+\kappa)}{1.2.\ldots\kappa} \\ = sN, \end{aligned}$$

say; and the total number of dependent quantities, being the dependent variables and their derivatives of all orders up to $\mu+\kappa$ inclusive, is (or can be, for some of the dependent quantities may not occur explicitly)

$$\begin{aligned} m\,\{1+n+\tfrac{1}{2}n\,(n+1)+\ldots \text{ to } (\kappa+\mu+1) \text{ terms}\} \\ = m\,\frac{(n+1)(n+2)\ldots(n+\kappa+\mu)}{1\,.\,2\,.\ldots.\,(\kappa+\mu)} \\ = mNK, \end{aligned}$$

where

$$K=\frac{(n+\kappa+1)\ldots(n+\kappa+\mu)}{(\kappa+1)\ldots(\kappa+\mu)}.$$

The factor K is obviously always greater than unity; and therefore if $s<m$, or if $s=m$, the number sN is less than mNK. The number of equations in the amplified set is less than the number of dependent quantities in the amplified aggregate; and therefore it will generally be impossible to eliminate the dependent quantities from among the equations. Were such elimination possible, the results would take the form of relations between the independent variables: and these, of course, do not occur. There is therefore nothing incompatible with the analytical nature of the case, if $s<m$, or if $s=m$.

Next, consider the possible hypothesis that $s>m$. The factor K is greater than unity; but its value decreases as κ increases, and it tends towards unity with large increase of κ. Let κ_1 be the earliest value of κ for which

$$K<\frac{s}{m};$$

then for the value κ_1, and for every value of κ which is greater than κ_1, we have

$$s>mK,$$

and therefore

$$sN>mNK.$$

For such values of κ, the number of equations in the amplified system is greater than the number of dependent quantities in the amplified aggregate. The dependent quantities could then, in general, be eliminated from the amplified system of equations; the results would take the form of relations among the independent variables alone, and such relations cannot occur. Such a conclusion is, in general, not compatible with the nature of the case: and therefore, in general, s cannot be greater than m. If however the elimination could be performed for any given set of equations, amplified in the manner indicated, the final relations would be evanescent, and the incompatibility would not appear. This last event could occur only if such conditions were satisfied by the original system and consequent conditions were satisfied by the amplified system, as would reduce the number of independent equations in the amplified system so that, at the utmost, it should not be larger than the number of dependent quantities in the amplified aggregate.

Hence, in general, the number of equations in a given system must not be greater than the number of dependent variables involved; but the number of equations may be the greater in particular systems, and the investigation of the necessary and sufficient conditions will be a matter for subsequent discussion.

It is clear without detailed argument that, when s is less than m and when the equations are general, then $m-s$ of the dependent variables can have values assigned (either quite arbitrarily or arbitrarily within proper limits), still leaving as many equations as undetermined dependent variables.

Accordingly, the most general case to be considered for the present is that in which the number of equations is the same as the number of dependent variables.

6. Two properties of such a system of equations may be mentioned; their importance is mainly formal, and only a brief consideration is needed.

The first of the properties can be stated as follows: if a system of m partial equations in m dependent variables involves derivatives of order higher than the first, it can be replaced by an equivalent system of equations containing only derivatives of the first order, the number of independent equations in the new system being the same as the number of dependent variables which it involves.

The property is practically obvious and so hardly requires proof: it can be seen in connection with any particular example. Let there be a single equation, involving derivatives of the second order as the highest: when n new dependent variables are introduced by the equations

$$\frac{\partial z}{\partial x_r} = p_r, \qquad (r = 1, \ldots, n),$$

the given equation can be expressed in a form

$$f\left(x_1, \ldots, x_n, z, p_1, \ldots, p_n, \frac{\partial p_1}{\partial x_1}, \ldots, \frac{\partial p_n}{\partial x_n}\right) = 0,$$

which involves only derivatives of the first order; and the new system now contains $n+1$ equations, involving $n+1$ dependent variables with derivatives of the first order.

It may be added that the main use of the property lies in deducing existence-theorems for equations of order higher than the first from the existence-theorems which soon will be established for systems of equations of the first order.

An extended form of the property enables us, not merely to replace any given system by a system containing only derivatives of the first order, but also to secure that each equation, which in the new system involves derivatives of the first order, is linear in those derivatives. Thus, in the preceding example, additional dependent variables would be introduced by the equations

$$\frac{\partial z}{\partial x_\mu} = p_\mu, \qquad \frac{\partial p_\mu}{\partial x_s} = q_{\mu s},$$

for μ and $s = 1, \ldots, n$: the original equation takes the form of a relation

$$f(x_1, \ldots, x_n, z, p_1, \ldots, p_n, q_{11}, \ldots, q_{nn}) = 0$$

among the variables free from derivatives; and the derived equations

$$\frac{\partial f}{\partial x_s} + p_s \frac{\partial f}{\partial z} + \sum_{\mu=1}^{n} \frac{\partial f}{\partial p_\mu} q_{\mu s} + \sum_{\lambda, \mu}\sum \frac{\partial f}{\partial q_{\lambda\mu}} \frac{\partial q_{\lambda\mu}}{\partial x_s} = 0$$

are formed for $s = 1, \ldots, n$. All the equations are linear in the derivatives which are of the first order; but it should be noted that the number of equations in the modified system is larger than the number of dependent variables, though the conditions for coexistence are satisfied.

When the number of variables is other than very few, the extended form of the property tends to be cumbrous. It is, however, of definite use, as will be seen later (Chap. XVIII), as part of a method for obtaining integrals of equations of order higher than the first when they possess integrals that are expressible in finite terms.

7. The other of the two properties indicated in § 6 bases the solution of a system of m partial equations in m dependent variables upon the solution of one equation (or of more than one equation) in a single dependent variable in association with algebraic processes: the substituted equation or equations usually (but not universally) involve derivatives of order higher than those which occurred in the original system.

Reverting to the method adopted in § 5, and applying it for the purpose of eliminating $z_2, \ldots, z_m$ and all their derivatives, we should have mN equations in the amplified set, while the number of dependent quantities to be eliminated is $(m-1)NK$. Accordingly, let κ_1 be the least value of κ for which

$$K < \frac{m}{m-1},$$

and therefore

$$(m-1)NK < mN.$$

The dependent quantities, composed of $z_2, \ldots, z_m$ and their derivatives, can be eliminated from the amplified set of equations: the results of the elimination will take the form of one equation or more than one equation involving z_1 and its derivatives, the latter being of order higher than those which occur in the original system. Moreover, by the algebraic processes, all the dependent quantities that are eliminated are expressible in terms of those that survive. Accordingly, when the solution of the equation or equations in z_1 is known, the other dependent quantities can be regarded as known: and then the solution of the original system will have been obtained.

It should be added that this property is not of importance in the general theory: its chief value lies in the fact that it provides a method which sometimes is effective in leading to the solution of particular classes of equations.

Preparation for Cauchy's Theorem: the first of the subsidiary Existence-theorems.

8. We proceed now to the establishment of some positive results, in particular, to the establishment of Cauchy's theorem affirming the existence of integrals of a system of partial equations.

The system of equations contains the same number of dependent variables as of equations; in form, it does not include all possible systems of such a character, but it will be found to include a large selection of important and representative systems; and the integrals will be proved to exist, subject to an aggregate of assigned conditions. For the purpose in view, the method devised by Madame Kowalevsky* will be adopted, whereby the main theorem is approached through two existence-theorems belonging to partial equations of comparatively simple type.

The first of these theorems can be stated as follows:—

Let a set of partial equations be given in the form

$$\frac{\partial z_i}{\partial x_1} = \sum_{j=1}^{m} \sum_{r=2}^{n} G_{ijr} \frac{\partial z_j}{\partial x_r},$$

for values $i = 1, \ldots, m$, being m equations in m dependent variables; the coefficients G_{ijr} are functions of $z_1, \ldots, z_m$ alone. Let $c_1, \ldots, c_m$ be a set of values of $z_1, \ldots, z_m$ respectively, in the vicinity of which each of the functions G_{ijr} is regular; and let $\phi_1, \ldots, \phi_m$ be a set of functions of $x_2, \ldots, x_n$, which acquire the values $c_1, \ldots, c_m$ respectively when $x_2 = a_2, \ldots, x_n = a_n$, which are regular in the vicinity of these values of $x_2, \ldots, x_n$, and which otherwise are arbitrary. Then *a system of integrals of the equations can be determined, which are regular functions of* $x_1, \ldots, x_n$ *in the vicinity of the values* $x_1 = a_1, x_2 = a_2, \ldots, x_n = a_n$, *and which acquire the values* $\phi_1, \ldots, \phi_m$ *when* $x_1 = a_1$; *moreover, the system of integrals, determined in accordance with these conditions, is the only system of integrals that can be so determined as regular functions.*

9. It is convenient, for the sake of conciseness in the formulæ, to write

$$x_s - a_s = y_s, \quad z_r - c_r = \zeta_r, \quad \phi_r - c_r = \psi_r,$$

for $s = 1, \ldots, n$, and $r = 1, \ldots, m$: and then we have to deal with quantities in the vicinity of zero values of y and ζ.

As the functions G are regular within this vicinity over some finite region, we select a portion of the region defined by the ranges

$$|\zeta_1| \leqslant R, \quad |\zeta_2| \leqslant R, \ \ldots, \ |\zeta_m| \leqslant R;$$

* *Crelle*, t. LXXX (1875), pp. 1—32.

and we denote by M the greatest value among the quantities $|G_{ijr}|$ within this portion of the region selected, M being finite. The functions G can be expressed as power-series; and if we take

$$G_{ijr} = \Sigma\Sigma \ldots a^{ijr}_{s_1 s_2 \ldots} \zeta_1^{s_1} \zeta_2^{s_2} \ldots,$$

where the multiple summation is for all integer values of $s_1, s_2, \ldots$ from zero upwards, simultaneous zeros being included, then*

$$\left| a^{ijr}_{s_1 s_2 \ldots} \right| \leqslant \frac{M}{R^{s_1+s_2+\ldots}}$$
$$\leqslant \frac{(s_1+s_2+\ldots)!}{s_1!\, s_2! \ldots} \frac{M}{R^{s_1+s_2+\ldots}},$$

a fortiori. Similarly, within a selected region of existence of the functions ψ, defined by the ranges

$$|y_2| \leqslant \rho, \quad |y_3| \leqslant \rho, \ldots, \quad |y_n| \leqslant \rho,$$

we have

$$\psi_\mu = \Sigma\Sigma \ldots c_{\mu\mu_2\mu_3} \ldots y_2^{\mu_2} y_3^{\mu_3} \ldots,$$

where the multiple summation is for all integer values of $\mu_2, \mu_3, \ldots$ from zero upwards, simultaneous zeros being excluded; and then, if N denote the greatest value among the quantities $|\psi_\mu|$ within this region, we have

$$|c_{\mu\mu_2\mu_3\ldots}| \leqslant \frac{N}{\rho^{\mu_2+\mu_3+\ldots}}$$
$$\leqslant \frac{(\mu_2+\mu_3+\ldots)!}{\mu_2!\, \mu_3! \ldots} \frac{N}{\rho^{\mu_2+\mu_3+\ldots}},$$

a fortiori.

If functions ζ exist possessing the character required in the theorem, they can be expanded as series of powers of y_1 in the vicinity of the origin; having regard to the value they must acquire when $y_1 = 0$, we can take them in the form

$$\zeta_\mu = \psi_\mu + y_1 \psi_{\mu 1} + y_1^2 \psi_{\mu 2} + \ldots,$$

where (as the functions ζ are to be regular in all their variables) the coefficients $\psi_{\mu 1}, \psi_{\mu 2}, \ldots$ must be regular functions of $y_2, y_3, \ldots$ within the selected region of existence, and they do not involve y_1. These functions ζ, if they exist, are to satisfy the differential equations: we substitute them therein, and compare the coefficients of the various powers of y_1 and, from the fact that the derivatives

* See my *Theory of Functions*, (Second edition), hereafter quoted as *T. F.*, § 22.

with regard to x_1 (and consequently of y_1) occur on the left-hand sides of the equations only, we find relations of the form

$$p\psi_{\mu p} = \Sigma Z,$$

where the number of terms in the summation on the right-hand side is finite. Each term Z is the product of four factors:

(i) a coefficient a from the expansion of the functions G:

(ii) a product of the functions $\psi_{\kappa\lambda}$, the second subscript index λ being less than p:

(iii) a first derivative of one of the functions $\psi_{\kappa\lambda}$, the second subscript index λ being less than p:

(iv) a positive integer.

Now all the functions $\psi_{\kappa\lambda}$ having their second subscript index zero, are the functions $\psi_1, \ldots, \psi_m$; and their expressions as regular functions of $y_2, \ldots, y_n$ are known. Hence the relations

$$p\psi_{\mu p} = \Sigma Z$$

give a formal determination of the functions $\psi_{\mu p}$ in succession; they appear as power-series in $y_2, \ldots, y_n$, the coefficients in which involve the constants a from the expansion of the functions G, the constants c from the expansion of the functions $\psi_1, \ldots, \psi_m$, and positive numerical factors.

When these values are substituted in the y_1-expansions of $\zeta_1, \ldots, \zeta_m$, expressions result which formally satisfy the differential equations. In order that they may possess functional significance, these multiple series must converge; this necessary convergence can be proved as follows.

10. We consider variation in a more restricted range for the quantities ζ, given by

$$|\zeta_1| < \frac{R}{m}, \quad |\zeta_2| < \frac{R}{m}, \ldots, \quad |\zeta_m| < \frac{R}{m},$$

so that

$$|\zeta_1 + \zeta_2 + \ldots + \zeta_m| < R;$$

and we construct a dominant function* G in the form

$$G = \Sigma\Sigma\ldots \frac{(s_1 + s_2 + \ldots)!}{s_1!\, s_2! \ldots} \frac{M}{R^{s_1+s_2+\ldots}} \zeta_1^{s_1} \zeta_2^{s_2} \ldots,$$

* *T. F.*, § 23.

where the multiple summation is for all integer values of $s_1, s_2, \ldots$ from zero upwards, simultaneous zeros being included. Then the modulus of any term in G is at least as great as the modulus of the corresponding term in G_{ijr} having the same combination of the variables. Moreover, in the region of variation considered, G can be expressed in finite terms: we have

$$G = \frac{M}{1 - \frac{1}{R}(\zeta_1 + \zeta_2 + \ldots + \zeta_m)}.$$

We also consider variation in a more restricted range for the quantities $y_2, \ldots, y_n$, given by

$$|y_2| < \frac{\rho}{n-1}, \ \ldots, \ |y_n| < \frac{\rho}{n-1},$$

so that

$$|y_2 + \ldots + y_n| < \rho;$$

and we construct another dominant function ψ in the form

$$\psi = \Sigma\Sigma\ldots \frac{(\mu_2 + \mu_3 + \ldots)!}{\mu_2!\ \mu_3! \ldots} \frac{N}{\rho^{\mu_2+\mu_3+\ldots}} y_2^{\mu_2} y_3^{\mu_3} \ldots,$$

where the multiple summation is for all integer values of $\mu_2, \mu_3, \ldots$ from zero upwards, simultaneous zeros being excluded. Then the modulus of any term in ψ is at least as great as the modulus of the corresponding term in ψ_μ having the same combination of the variables. Moreover, in the region of variation considered, ψ can be expressed in finite terms; writing

$$y = y_2 + \ldots + y_n,$$

we have

$$\psi = \frac{Ny}{\rho} \cdot \frac{1}{1 - \frac{y}{\rho}}$$

$$= \frac{Ny}{\rho - y},$$

and the range for the variable y is given by

$$|y| < \rho.$$

11. By means of these dominant functions, a dominant system of partial equations

$$\frac{\partial Z_i}{\partial y_1} = \sum_{j=1}^{m} \sum_{r=2}^{n} G \frac{\partial Z_j}{\partial y_r}$$

is constructed; and we assign, as conditions, that each of the quantities Z_i shall acquire the value ψ when $y_1 = 0$. Taking G and ψ in their expanded forms as power-series, and applying to these equations the process applied to the original system, we obtain equations of precisely the same form as before, to determine the successive coefficients in the expressions for the variables Z as power-series in the variables $y_1, y_2, \ldots, y_n$: and the successive operations for the construction of these coefficients are the same as before. In these new operations, all the terms in the expression for any coefficient are positive; the modulus of each term is at least as great as was the modulus of the corresponding term in the former operations; and therefore, if

$$Z_\mu = \psi + y_1\psi'_{\mu 1} + y_1^2\psi'_{\mu 2} + \ldots,$$

we have

$$| p\psi'_{\mu p} | \geqslant | p\psi_{\mu p} |,$$

that is,

$$| \psi_{\mu p} | \leqslant | \psi'_{\mu p} |.$$

Hence the series for ζ_μ will certainly converge if the series for Z_μ converges.

The values of $Z_1, \ldots, Z_m$, and their consequent expressions as converging series, can be otherwise obtained. Returning to the dominant system and using the finite forms for G and ψ, we have to determine values of $Z_1, \ldots, Z_m$, satisfying the equations

$$\frac{\partial Z_i}{\partial y_1} = \frac{M}{1 - \frac{1}{R}(Z_1 + \ldots + Z_m)} \sum_{j=1}^{m} \sum_{r=2}^{n} \frac{\partial Z_j}{\partial y_r},$$

and such that

$$Z_\mu = \frac{Ny}{\rho - y},$$

when $y_1 = 0$. Thus Z_μ, a function of all the variables, is a function of the combination of them represented by y, say a function of y alone, when $y_1 = 0$; and therefore $\frac{\partial Z_\mu}{\partial y_r}$, for $r = 2, \ldots, n$, is also a function of y alone when $y_1 = 0$. The differential equations then shew that $\frac{\partial Z_\mu}{\partial y_1}$ is a function of y alone when $y_1 = 0$. Again, differentiating all the equations with regard to y_1, noting that the quantities Z and all their first derivatives are functions of y alone when $y_1 = 0$, and applying a similar argument, we find that $\frac{\partial^2 Z_\mu}{\partial y_1^2}$

is a function of y alone when $y_1 = 0$. Similarly, for all the derivatives in succession. Inserting their forms in

$$Z_\mu = [Z_\mu]_0 + y_1 \left[\frac{\partial Z_\mu}{\partial y_1}\right]_0 + \frac{y_1^2}{2}\left[\frac{\partial^2 Z_\mu}{\partial y_1^2}\right]_0 + \dots,$$

we see that Z_μ, if it exists, is expressible as a function of y_1 and y.

Now from the equations, we have

$$\frac{\partial Z_1}{\partial y_1} = \frac{\partial Z_2}{\partial y_1} = \dots = \frac{\partial Z_m}{\partial y_1};$$

and therefore, taking account of the conditions that

$$Z_1 = Z_2 = \dots = Z_m = \frac{Ny}{\rho - y},$$

when $y_1 = 0$, we have

$$Z_1 = Z_2 = \dots = Z_m,$$

in general. Denote this common value by Z, which now is a function of y_1 and y; then all the equations in the dominant system are satisfied, provided Z can be determined to satisfy the equation

$$\frac{\partial Z}{\partial y_1} = \frac{M}{1 - m\dfrac{Z}{R}} m(n-1)\frac{\partial Z}{\partial y}$$

and is such that it acquires the value $\dfrac{Ny}{\rho - y}$ when $y_1 = 0$. It is easy to verify that the equation is satisfied by a relation

$$\left(1 - m\frac{Z}{R}\right) y + Mm(n-1)\, y_1 = f(Z),$$

where f is any function whatever of Z: and therefore all the requirements will be met if f can be chosen so as to allow Z to acquire the value $\dfrac{Ny}{\rho - y}$ when $y_1 = 0$. For this purpose, the two equations

$$\left(1 - m\frac{u}{R}\right) y = f(u),$$

$$\frac{Ny}{\rho - y} = u,$$

must be the same. The latter gives

$$y = \frac{\rho u}{N + u},$$

which, substituted in the former, gives

$$f(u) = \left(1 - m\frac{u}{R}\right)\frac{\rho u}{N+u},$$

being the appropriate expression of the function f. Thus all the requirements are met by a value or values of Z determined by the integral relation

$$\left(1 - m\frac{Z}{R}\right) y + Mm(n-1)\, y_1 = \left(1 - m\frac{Z}{R}\right)\frac{\rho Z}{N+Z}.$$

This quadratic equation has two roots. One of the two roots becomes $\frac{R}{m}$ when $y_1 = 0$, and must be discarded as not satisfying the imposed condition when $y_1 = 0$. The other root is given in the form

$$2\frac{m}{R}(\rho - y)Z - \left\{(\rho - y) + m\frac{N}{R}y - Mm(n-1)\, y_1\right\}$$

$$= -\left[\left(\rho - y - m\frac{N}{R}y\right)^2 + M^2m^2(n-1)^2 y_1^2\right.$$

$$\left. - 2Mm(n-1)\, y_1\left\{\rho - y + m\frac{N}{R}(2\rho - y)\right\}\right]^{\frac{1}{2}};$$

it can be expanded in powers of y_1 in the series

$$Z = \frac{Ny}{\rho - y} + Mm(n-1)\, y_1 \frac{N\rho}{\rho - y}\frac{1}{\rho - y - m\frac{N}{R}y}$$

$$+ \text{higher powers of } y_1.$$

It is clear from this expression for Z obtained in finite form that Z can be expanded in a series of powers of y and y_1, which converges in a non-infinitesimal range round $y = 0$ and $y_1 = 0$. When y is replaced by its value $y_2 + y_3 + \dots + y_n$, the modified power-series in $y_1, y_2, \dots, y_n$ still converges in a non-evanescent range round $y_1 = 0$, $y_2 = 0$, ..., $y_n = 0$; consequently, the quantity Z of the required type does exist.

It therefore follows that the quantities $Z_1, \dots, Z_m$ exist as determined by the equations in the dominant system; and therefore integrals of the original equations exist satisfying the prescribed conditions.

12. The preceding investigation establishes the existence of integrals which are regular functions of the variables in the selected region; it is easy to see that they are the only set of integrals which, satisfying the prescribed conditions, are regular functions of the variables. If any other set existed, being regular functions and satisfying the prescribed conditions, they would be expressible in a form

$$\zeta'_\mu = \psi_\mu + y_1 \psi'_{\mu 1} + y_1^2 \psi'_{\mu 2} + \ldots ;$$

when these are substituted in the differential equations, they would lead to relations

$$p\psi'_{\mu p} = \Sigma Z'$$

for the determination of the coefficients, similar to the relations

$$p\psi_{\mu p} = \Sigma Z.$$

When $p = 1$, no double-suffix function occurs in Z', which then is the same as Z, because the term in ζ'_μ independent of y_1 is the same as the corresponding term in ζ_μ: hence

$$\psi'_{\mu 1} = \psi_{\mu 1}.$$

When $p = 2$, the only double-suffix functions that can occur in Z' are the functions $\psi'_{\lambda 1}$, which have been shewn to be the same as $\psi_{\lambda 1}$; hence, for this value, $Z' = Z$, and therefore

$$\psi'_{\mu 2} = \psi_{\mu 2}.$$

Similarly for all the coefficients in succession: we find

$$\zeta'_\mu = \zeta_\mu,$$

for all the values of μ; and therefore the set of regular integrals obtained, subject to the prescribed conditions, are unique regular integrals.

As an example illustrating the general theorem, we require the integrals of the simultaneous equations

$$\frac{\partial u}{\partial x} = u^2 \frac{\partial u}{\partial y},$$

$$\frac{\partial v}{\partial x} = (u^2 - u + v) \frac{\partial u}{\partial y},$$

such that, when $x = 0$,

$$u = y(a_0 + a_1 y + a_2 y^2 + \ldots) = R(y),$$
$$v = y(b_0 + b_1 y + b_2 y^2 + \ldots) = S(y).$$

The equations are amenable to the ordinary practical methods. The most general integral of the first equation is easily found to be

$$u = f(y + u^2 x),$$

where f is arbitrary so far as the equation is concerned. Also

$$\frac{\partial v}{\partial x}-\frac{\partial u}{\partial x}=(v-u)\frac{\partial u}{\partial y}$$

$$=(v-u)\frac{1}{u^2}\frac{\partial u}{\partial x},$$

and therefore

$$v-u=e^{-\frac{1}{u}}g(y),$$

where g is arbitrary so far as the equation is concerned.

To determine these arbitrary functions, the imposed conditions are used. As

$$u=f(y+u^2x),$$

and as $u=R(y)$ when $x=0$, we have

$$R(y)=f(y),$$

and therefore, generally,

$$u=R(y+u^2x).$$

Laplace's theorem in expansion can be used to give the explicit expression for u in terms of x and y: this is

$$u=R(y)+xR^2(y)R'(y)+\frac{x^2}{2}\frac{d}{dy}\{R^4(y)R'(y)\}+\frac{x^3}{6}\frac{d^2}{dy^2}\{R^6(y)R'(y)\}+\dots$$

$$=R(y)\{1+xR(y)T(x, y)\},$$

where $T(x, y)$ is a series of powers of x, the coefficients being functions of y which vanish if $R(y)$ vanishes identically.

Again, as $v=S(y)$ when $x=0$, we have

$$S(y)-R(y)=e^{-\frac{1}{R(y)}}g(y),$$

so that, generally,

$$v-u=e^{\frac{1}{R(y)}-\frac{1}{u}}\{S(y)-R(y)\}.$$

The apparent singularity can be removed; for

$$\frac{1}{R(y)}-\frac{1}{u}=\frac{xT(x, y)}{1+xR(y)T(x, y)},$$

where the function on the right-hand side is regular in the vicinity of $x=0$, $y=0$. Thus the required integrals are

$$u=R(y)\{1+xR(y)T(x, y)\},$$

$$v-u=\{S(y)-R(y)\}e^{\frac{xT(x, y)}{1+xR(y)T(x, y)}}.$$

In particular, if the imposed conditions should be that $u=0$ and $v=S(y)$ when $x=0$, then as $R(y)$ vanishes identically, $T(x, y)$ vanishes. The full expressions for the integrals are

$$u=0, \qquad v=S(y);$$

these are easily obtainable directly from the differential equations but not from the integrals which involve the arbitrary functions f and g.

13. It should be observed that, throughout the foregoing proof, there has been a complete restriction to regular functions. The possibility of non-regular functions, satisfying the equations and obeying the prescribed conditions, has nowhere been taken into account and the proof does not shew that it should be rejected as inadmissible; what is established is that a unique set of regular integrals exists. In the statement of the argument, it is practically assumed that the integrals are regular. Thus z_1 is made to acquire the value of a regular function of $x_2, \ldots, x_n$ when $x_1 = a_1$; and this could be secured when z_1 is a uniform function of x_1, even if x_1 is an essential singularity, provided x_1 then be allowed to approach a_1 by an appropriate path*. If however z_1 is made to acquire its value when $x_1 = a_1$, quite independently of the path by which x_1 approaches the position a_1, and so also for the other dependent variables, then it is not difficult to see that the integrals must be regular under the assigned conditions. For the differential equations then make $\frac{\partial z_1}{\partial x_1}, \ldots, \frac{\partial z_m}{\partial x_1}$ regular functions of $x_2, \ldots, x_n$, whatever be the x_1-path of approach to a_1; when derivatives of the equations are formed and suitably combined it could be inferred that $\frac{\partial^2 z_1}{\partial x_1^2}, \ldots, \frac{\partial^2 z_m}{\partial x_1^2}$, in like circumstances, become regular functions of $x_2, \ldots, x_n$; and so on, for the derivatives in succession. The inference that $z_1, \ldots, z_m$ are regular functions of $x_1, x_2, \ldots, x_n$ is then immediate.

The assumption made by ignoring the path of approach of x_1 to a_1 may fairly be described as a customary assumption: it does, in effect, exclude the consideration of the possibility that a_1 is an essential singularity of a uniform function, and it may exclude the consideration of other possibilities of deviation from regularity. Yet it is not inconceivable that, in particular instances, such as the stability of a system in a critical condition, the excluded possibilities are of importance†: in such an instance, it might be actually the fact that the variable must approach its value by a specific path and is not permitted an unrestricted approach to the value.

* See *T. F.*, p. 57 (second edition), Ex. 4.

† The same considerations occur in connection with the integrals of an ordinary equation of the first order: see §§ 28—34 in volume II of this work, where (§ 34) the condition given for that case by Fuchs is explained.

The second of the subsidiary Existence-theorems.

14. In the preceding theorem, the only variables which occur explicitly in the set of partial equations are the dependent variables: it can, however, be extended so as to allow the explicit occurrence of all the variables. The extended theorem is as follows:—

Let a set of partial equations be given in the form

$$\frac{\partial z_i}{\partial x_1} = \sum_{j=1}^{m} \sum_{r=2}^{n} G_{ijr} \frac{\partial z_j}{\partial x_r} + G_i,$$

for values $i = 1, \ldots, m$, being m equations in m dependent variables; the coefficients G_{ijr} and the quantities G_i are functions of all the variables, dependent and independent. Let $c_1, \ldots, c_m, a_1, \ldots, a_n$ be a set of values of $z_1, \ldots, z_m, x_1, \ldots, x_n$ respectively, in the vicinity of which all the functions G_{ijr} and G_i are regular; and let $\phi_1, \ldots, \phi_m$ be a set of functions of $x_2, \ldots, x_n$, which acquire values $c_1, \ldots, c_m$ respectively when $x_2 = a_2, \ldots, x_n = a_n$, which are regular in the vicinity of these values of $x_2, \ldots, x_n$, and which otherwise are arbitrary. Then *a system of integrals of the equations can be determined, which are regular functions of* $x_1, \ldots, x_n$ *in the vicinity of the values* $x_1 = a_1, x_2 = a_2, \ldots, x_n = a_n$, *and which acquire the values* $\phi_1, \ldots, \phi_m$ *when* $x_1 = a_1$; *moreover, the system of integrals, determined in accordance with these conditions, is the only system of integrals that can be thus determined as regular functions.*

The establishment of this theorem can be derived from the former theorem in a simple manner. Let n new dependent variables $t_1, \ldots, t_n$ be introduced, defined by equations

$$\frac{\partial t_1}{\partial x_1} = \frac{\partial t_2}{\partial x_2}, \quad \frac{\partial t_2}{\partial x_1} = 0, \quad \frac{\partial t_3}{\partial x_1} = 0, \ \ldots, \ \frac{\partial t_n}{\partial x_1} = 0,$$

and by the conditions that, when $x_1 = a_1$,

$$t_1 = a_1, \quad t_2 = x_2, \quad \ldots, \quad t_n = x_n.$$

From the last $n-1$ of these equations, combined with the imposed conditions, it is clear that

$$t_2 = x_2, \quad t_3 = x_3, \quad \ldots, \quad t_n = x_n,$$

in general. Then

$$\frac{\partial t_1}{\partial x_1} = \frac{\partial t_2}{\partial x_2} = 1,$$

so that

$$t_1 = x_1 + \text{function of } x_2, \ldots, x_n$$
$$= x_1,$$

on applying the imposed condition.

We replace $x_1, \ldots, x_n$ in G_{ijr} and in G_i by $t_1, \ldots, t_n$, and denote the functions resulting after the change by H_{ijr} and H_i. Also, noting the fact that $\frac{\partial t_2}{\partial x_2}$ is unity, we take an amplified system of equations

$$\frac{\partial z_i}{\partial x_1} = \sum_{j=1}^{m} \sum_{r=2}^{n} H_{ijr} \frac{\partial z_j}{\partial x_r} + H_i \frac{\partial t_2}{\partial x_2},$$
$$\frac{\partial t_1}{\partial x_1} = \frac{\partial t_2}{\partial x_2},$$
$$\frac{\partial t_\mu}{\partial x_1} = 0,$$

for $i = 1, \ldots, m$, and $\mu = 2, \ldots, n$; and the imposed conditions are that, when $x_1 = a_1$,

$$z_1 = \phi_1, \quad z_2 = \phi_2, \quad \ldots, \quad z_m = \phi_m,$$
$$t_1 = a_1, \quad t_2 = x_2, \quad \ldots, \quad t_n = x_n.$$

The coefficients in the modified system are functions of the dependent variables; the properties of the modified system, when account is taken of the imposed conditions, are the properties of the systems to which the former theorem applies. Hence, by that former theorem, a set of integrals

$$z_i = \psi_i(x_1, x_2, \ldots, x_n),$$
$$t_\mu = x_\mu,$$

for $i = 1, \ldots, m$ and $\mu = 1, \ldots, n$, exists; the functions ψ_i are regular functions of the variables x, and when $x_1 = a_1$, the functions $\psi_1, \ldots, \psi_m$ reduce to $\phi_1, \ldots, \phi_m$ respectively; and these regular integrals are the only set of regular integrals which satisfy the imposed conditions.

When we substitute $t_1 = x_1$, $t_2 = x_2$, $\ldots$, $t_n = x_n$ in the modified system, we return to the original system: the results just obtained constitute the theorem required.

Note. There is the same kind of limitation as in the former case (§ 13): it is possible that, for reasons connected with essential singularities such as particular modes of the approach of x_1 to a_1, there may be non-regular integrals of the equations satisfying the imposed conditions.

Ex. 1. Obtain the integral of the equations

$$\left.\begin{aligned} \frac{\partial u}{\partial x} &= u - 2v - x^2 + 2x + (u - 2v - x^2)\frac{\partial u}{\partial y} + \frac{\partial u}{\partial z} \\ \frac{\partial v}{\partial x} &= \qquad\qquad\qquad\quad (u - 2v - x^2)\frac{\partial v}{\partial y} + \frac{\partial v}{\partial z} \end{aligned}\right\},$$

subject to the initial conditions that, when $x=0$,

$$u = 2e^{(y+z)^2} + ae^{-z}, \qquad v = e^{(y+z)^2},$$

(where a is a constant), in the form

$$u - 2v - x^2 = ae^{-z},$$
$$\log v = \{x + y + z + ae^{-z}(1 - e^{-x})^2\}^2.$$

(Riquier.)

Ex. 2. Integrate similarly the equations

$$\left.\begin{aligned} \frac{\partial u}{\partial x} &= 2x + (u - 2v - x^2)(1 - 2t^2) + 2t(u - 2v - x^2)\frac{\partial u}{\partial y} + \frac{\partial u}{\partial z} \\ \frac{\partial v}{\partial x} &= \qquad\qquad (x^2 + 2v - u)t^2 + 2t(u - 2v - x^2)\frac{\partial v}{\partial y} + \frac{\partial v}{\partial z} \\ \frac{\partial t}{\partial x} &= \qquad\qquad\qquad\qquad\qquad 2t(u - 2v - x^2)\frac{\partial t}{\partial y} + \frac{\partial t}{\partial z} \\ \frac{\partial w}{\partial x} &= \qquad\qquad (x^2 + 2v - u)t^2 + 2t(u - 2v - x^2)\frac{\partial w}{\partial y} + \frac{\partial w}{\partial z} \end{aligned}\right\}$$

subject to the initial conditions that, when $x=0$, the variables u, v, t, w, respectively acquire the values

$$u = 2(y^2 + z^2) + ae^{-z}, \qquad v = y^2 + z^2, \qquad t = 1, \qquad w = 2z.$$

(Riquier.)

Ex. 3. As an example of the general theorem, let it be required to obtain the integral of the equation

$$\frac{\partial z}{\partial x} + \frac{y^2}{y^2 - y + x}\frac{\partial z}{\partial y} = 0,$$

which acquires the value y when $x=1$.

After the explanations that have been given, it will be of the form

$$z = y - (x-1)y_1 + (x-1)^2 y_2 - \dots ;$$

in order that this may satisfy the differential equation, the coefficients $y_1, y_2, \dots$ are determined by the relation

$$\begin{aligned} ny_n = \frac{y^2}{y^2 - y + 1}\frac{\partial y_{n-1}}{\partial y} + \frac{y^2}{(y^2 - y + 1)^2}\frac{\partial y_{n-2}}{\partial y} + \frac{y^2}{(y^2 - y + 1)^3}\frac{\partial y_{n-3}}{\partial y} + \dots \\ \dots + \frac{y^2}{(y^2 - y + 1)^{n-1}}\frac{\partial y_1}{\partial y} + \frac{y^2}{(y^2 - y + 1)^n}, \end{aligned}$$

as may be verified by substituting the expression for z and comparing coefficients of powers of $x-1$. In particular,

$$y_1=\frac{y^2}{y^2-y+1},$$

$$2y_2=\frac{y^2}{y^2-y+1}\frac{d}{dy}\left(\frac{y^2}{y^2-y+1}\right)+\frac{y^2}{(y^2-y+1)^2},$$

and so on.

The equation is amenable to the ordinary practical method. The most general integral is found to be

$$(x-y)\,e^{\frac{1}{y}}=f(z),$$

where f is an arbitrary function, to be rendered definite by means of the assigned condition, which is that z must acquire the value y when $x=1$. Hence

$$(1-y)\,e^{\frac{1}{y}}=f(y),$$

and therefore

$$(1-z)\,e^{\frac{1}{z}}=f(z),$$

so that the required integral is given by the equation

$$(1-z)\,e^{\frac{1}{z}}=(x-y)\,e^{\frac{1}{y}}.$$

But in connection with this equation, it must be specified as that value of z which acquires the value y when $x=1$; it is not enough to take any root of the equation for the integral, because (when $x=1$) there is an infinitude of values of z as functions of y, and only one of these is actually equal to y. In fact, the finite form of the equation, though it includes the required integral, does not give a unique expression for z.

Note. It sometimes is convenient* to associate an ordinary equation

$$\frac{dy}{dx}=f(x,\ y)$$

with a partial differential equation

$$\frac{\partial z}{\partial x}+f(x,\ y)\frac{\partial z}{\partial y}=0.$$

It is known that integrals of the respective equations exist. Taking the equation just discussed, so as to have

$$f(x,\ y)=\frac{y^2}{y^2-y+x},$$

the only regular integral of the ordinary equation, acquiring a value zero when $x=1$, is given by

$$y=0.$$

* Picard, *Traité d'Analyse*, t. II, ch. XI, § 15.

Now taking the partial equation and supposing an integral of the ordinary equation, assumed to exist in the same regular field of variation, to be substituted in z, we have

$$\frac{\partial z}{\partial x}+\frac{\partial z}{\partial y}\frac{dy}{dx}=0,$$

that is, $z=A$, a constant, for that substitution: and the constant A manifestly can be made zero. Conversely,

$$z=0$$

clearly gives the regular integral of the ordinary equation in the form

$$y=0.$$

We can also, by quadratures, obtain the complete primitive of the ordinary equation in the form

$$(x-y)\,e^{\frac{1}{y}}=c,$$

where c is an arbitrary constant. This cannot be obtained from the regular integral of the partial equation, by taking

$$z=A,$$

for any value of A, if we take unlimited variation of y; because $y=0$ is an essential singularity for one equation and an ordinary point for the other. But it can be obtained from the non-regular integral

$$(x-y)\,e^{\frac{1}{y}}=f(z),$$

by taking $z=A$: the appropriate value of c is

$$c=f(A).$$

CHAPTER II.

Cauchy's Theorem.

The main results in this chapter are associated with theorems establishing the existence, under assigned conditions, of integrals of systems of partial equations, the number of equations in a system being the same as the number of dependent variables: they are conveniently described as Cauchy's Theorem, because they have their origin in Cauchy's investigations* on the subject. The method adopted is based upon the memoir of Madame Kowalevsky, quoted in the preceding chapter (p. 11); reference may also be made to the expositions given by Jordan, *Cours d'Analyse*, t. III (1896), ch. III, § 1, and by Goursat, *Leçons sur l'intégration des équations aux derivées partielles du premier ordre*, (1891), ch. I.

15. The existence-theorems established in the preceding chapter can be applied to equations, and to systems of equations, of representative types; and to such applications we now proceed. But some passing remarks must be made upon the limitations that have been imposed. All the equations are linear in the derivatives of the dependent variables: this character, if not initially possessed, frequently (though not universally) can be secured by appropriate transformations. All the coefficients of the derivatives in the equations have been assumed to be regular functions of the independent variables (and, in the case of the earlier theorem, of the dependent variables) within the fields of variation considered: no result as to the character, or even the existence, of integrals has been obtained when there is any deviation from the postulated regularity. The imposed initial conditions are of a similar type, because they require the assumption, as values, of arbitrary functions of a regular character for a

* *Œuvres de Cauchy*, 1re Sér., t. VII, p. 17, and elsewhere. These memoirs were published in the *Comptes Rendus* in 1842; his earliest researches on the subject date back to 1819.

chosen value of a particular variable: but no result is established if these assigned arbitrary functions are not regular, in any form of deviation from regularity. It may be possible (and frequently it is possible) to transform the equations in such a manner that another particular variable may be selected as the pivot of initial conditions, with the appropriate modification as to the arguments of the functions in the assigned initial conditions; and the existence-theorems establish no relation between the integrals proved to exist in the respective cases. Moreover, the integrals considered are functions which are regular within the fields of variation; the limitation to uniformity, instead of to regularity so as to exclude essential singularities, is (for the almost complete part) excluded from discussion in the present state of knowledge.

Even within these restrictions, the existence-theorems already proved have a wide range of important applications; some of these applications will now be taken in succession.

Cauchy's Theorem for Equations of the First Order.

16. We begin with the simplest case, being that of a single equation of the first order in one dependent variable and two independent variables; taking the latter to be x and y, and denoting the first derivatives of z with regard to these variables by p and q respectively, we may consider the equation in the form

$$f(x, y, z, p, q) = 0,$$

where f will be taken to be regular in its arguments: and we shall assume that the equation is irreducible. Let $x = a$, $y = b$, $z = c$, $p = \lambda$, $q = \mu$, be a set of values satisfying the equation $f = 0$; then unless the quantity $\dfrac{\partial f}{\partial p}$ vanishes for these values, the equation can be resolved so as to express p in terms of the remaining arguments in a form

$$p - \lambda = \bar{g}(x - a, y - b, z - c, q - \mu),$$

say

$$p = g(x, y, z, q),$$

where g is a regular analytic function of its arguments. Now, as $\dfrac{\partial f}{\partial p}$ usually involves the variables that occur (or some of them), it usually is possible to choose initial values so that $\frac{\partial f}{\partial p}$ does not vanish: and then the analytic resolution of the original equation is possible. But it may happen that values of z (if any) and of its derivatives which satisfy the original equation $f=0$, that is, which make f vanish consistently with a relation between z, x, y and with derivatives from that integral relation, also make $\frac{\partial f}{\partial p}$ vanish similarly: the suggested resolution of the original equation is then impossible, so that p could not be expressed as a regular analytic function of x, y, z, q.

17. As a first alternative, we assume that the resolution with regard to p is possible in the form

$$p = g(x, y, z, q),$$

where g is regular in the vicinity of $x=a$, $y=b$, $z=c$, $q=\mu$. We can apply the existence-theorems, already established, to prove that *an integral z of the equation exists, having the properties:—*

(i) *it is a regular function of x and y within fields of variation round $x=a$ and $y=b$ given by*

$$|x-a| \leqslant r, \quad |y-b| \leqslant r,$$

where r is not infinitesimal:

(ii) *when $x=a$, the integral z reduces to $\phi(y)$, where $\phi(y)$ is a regular function of y within the field $|y-b| \leqslant r$, acquiring the value c when $y=b$, and otherwise arbitrary;*

(iii) *the integral z, as determined by these conditions, is unique as a regular integral.*

In order to deduce this result from the former theorems, we consider a system

$$\left.\begin{aligned} \frac{\partial z}{\partial x} &= p \\ \frac{\partial q}{\partial x} &= \frac{\partial p}{\partial y} \\ \frac{\partial p}{\partial x} &= \frac{\partial g}{\partial x} + \frac{\partial g}{\partial z} p + \frac{\partial g}{\partial q} \frac{\partial p}{\partial y} \end{aligned}\right\},$$

regarding it as a system in three dependent variables z, p, q. Applying the second of the existence-theorems (§ 14), we infer that integrals of this system of equations exist which are regular in the vicinity of $x = a$, $y = b$, and, when $x = a$, are such that

$$z = \phi(y),$$

$$q = \frac{d\phi(y)}{dy},$$

$$p = g\left\{a,\ y,\ \phi(y),\ \frac{d\phi(y)}{dy}\right\},$$

where $\phi(y)$ is the foregoing regular function of y acquiring the value c when $y = b$, and $\dfrac{d\phi(y)}{dy}$ is therefore also regular, acquiring the value μ when $y = b$; moreover, this set of regular integrals is unique. Let the set of integrals of the system, thus known to exist, be denoted by

$$z = Z(x, y), \quad p = P(x, y), \quad q = Q(x, y);$$

we proceed to prove that $z = Z(x, y)$ satisfies the original equation so that, owing to its other properties, it is the announced integral.

As these quantities Z, P, Q satisfy the amplified system of equations, we have

$$\frac{\partial Z}{\partial x} = P$$

from the first of those equations, so that

$$\frac{\partial z}{\partial x} = p.$$

Again, we have

$$\frac{\partial Q}{\partial x} = \frac{\partial P}{\partial y}$$

from the second of those equations, so that

$$\frac{\partial Q}{\partial x} = \frac{\partial}{\partial y}\left(\frac{\partial Z}{\partial x}\right),$$

and therefore

$$\frac{\partial}{\partial x}\left(\frac{\partial Z}{\partial y} - Q\right) = 0.$$

Hence $\dfrac{\partial Z}{\partial y} - Q$ is a function of y only, and its value is the same whatever value be assigned to x. When $x = a$, we have

$$\frac{\partial Z}{\partial y} = \frac{\partial}{\partial y}\phi(y) = \frac{d\phi(y)}{dy}$$

from the assigned value of Z, and

$$Q=\frac{d\phi(y)}{dy}$$

from the assigned value of Q; hence, when $x=a$, the value of $\frac{\partial Z}{\partial y}-Q$ is zero, and therefore

$$\frac{\partial Z}{\partial y}-Q=0$$

generally, that is,

$$\frac{\partial z}{\partial y}=q.$$

Again, denoting $P-g(x, y, Z, Q)$ by u, we have

$$\begin{aligned}\frac{\partial u}{\partial x}&=\frac{\partial P}{\partial x}-\frac{\partial g}{\partial x}-\frac{\partial g}{\partial Z}\frac{\partial Z}{\partial x}-\frac{\partial g}{\partial Q}\frac{\partial Q}{\partial x}\\&=\frac{\partial P}{\partial x}-\frac{\partial g}{\partial x}-\frac{\partial g}{\partial Z}P-\frac{\partial g}{\partial Q}\frac{\partial P}{\partial y}\\&=0,\end{aligned}$$

by the third equation of the system. Thus u is independent of x; when $x=a$, its value is

$$g\left\{a,\ y,\ \phi(y),\ \frac{d\phi(y)}{dy}\right\}-g\left\{a,\ y,\ \phi(y),\ \frac{d\phi(y)}{dy}\right\}$$

on inserting the values acquired by Z, Q, P when $x=a$: this is zero, and therefore $u=0$ generally, that is,

$$P=g(x, y, Z, Q),$$

and therefore

$$p=g(x, y, z, q).$$

We thus have

$$\frac{\partial z}{\partial x}=p,\quad \frac{\partial z}{\partial y}=q,\quad p=g(x, y, z, q),$$

in association with $z=Z(x, y)$, that is, $z=Z(x, y)$ satisfies the equation

$$\frac{\partial z}{\partial x}=g\left(x, y, z, \frac{\partial z}{\partial y}\right),$$

which is the original equation. Owing to its other properties, by which it obeys the assigned conditions, $z=Z(x, y)$ is the integral required.

18. Passing now to the other alternative, under which the equation $f=0$ cannot be resolved with regard to p because the magnitude $\frac{\partial f}{\partial p}$ vanishes for values of the variables that make f vanish, we consider the resolubility of the equation $f=0$ with regard to q. This resolution will be possible unless the magnitude $\frac{\partial f}{\partial q}$ vanishes for values of the variables that make f vanish; when it is possible, the resolved form will be

$$q = h(x, y, z, p),$$

where h is a regular function of its arguments, in the vicinity of values (say) $x=a$, $y=b$, $z=c$. *An integral of the equation exists having the properties:—*

(i) *it is a regular function of x and y within fields of variation round $x=a$ and $y=b$ given by*

$$|x-a| \leqslant r, \quad |y-b| \leqslant r,$$

where r is not infinitesimal:

(ii) *when $y=b$, the integral z reduces to $\psi(x)$, where $\psi(x)$ is a regular function of x within the field $|x-a| \leqslant r$, acquiring a value c when $y=b$, and otherwise arbitrary:*

(iii) *the integral z, as determined by these conditions, is unique as a regular integral.*

The proof of this proposition is similar to the proof of the proposition in the case when the equation $f=0$ was resolved with regard to p; it will not be set out in detail.

19. Combining these results, it follows that an irreducible equation $f=0$ possesses a regular integral with assigned conditions if it is resoluble with regard to p, that it possesses another regular integral with other assigned conditions if it is resoluble with regard to q, and that each of these integrals is unique under its conditions. These integrals have been obtained from equations

$$p = g(x, y, z, q), \quad q = h(x, y, z, p),$$

respectively, which arise from the resolution of $f=0$ in the respective cases: but they do not generally represent the whole of the equation $f=0$, for if f were of degree m in p and n in q, there would generally be m equations of the former type and n of the latter, distinct from one another in their respective sets. Each such equation determines a unique regular integral under the assigned conditions, which may be made the same for each equation in the set. If for the m equations, the respective regular integrals are

$$z = \theta_1(x, y), \quad z = \theta_2(x, y), \quad \ldots, \quad z = \theta_m(x, y),$$

then clearly the equation

$$\{z - \theta_1(x, y)\}\{z - \theta_2(x, y)\} \ldots \{z - \theta_m(x, y)\} = 0$$

gives the integrals of the equation

$$f(x, y, z, p, q) = 0,$$

supposed of degree m in p and resoluble with regard to p, such that when $x = a$, z assumes the assigned functional value $\phi(y)$. Similarly, if

$$f(x, y, z, p, q) = 0$$

be of degree n in q and be resoluble with regard to q, an equation

$$\{z - \vartheta_1(x, y)\}\{z - \vartheta_2(x, y)\} \ldots \{z - \vartheta_n(x, y)\} = 0$$

gives the integrals of the equation such that, when $y = b$, z assumes the assigned functional value $\psi(x)$.

But it may happen that the equation

$$f(x, y, z, p, q) = 0$$

is not resoluble with regard either to p or to q, that is to say, it may happen that the magnitudes $\frac{\partial f}{\partial p}$ and $\frac{\partial f}{\partial q}$ vanish for values of the variables which make f vanish. The existence-theorem cannot then be applied, and so it provides no information as regards integrals of the equation. We must then investigate independently the character of those integrals (if any) of the equation

$$f(x, y, z, p, q) = 0,$$

which at the same time are such as to satisfy the equations

$$\frac{\partial f}{\partial p} = 0, \quad \frac{\partial f}{\partial q} = 0.$$

This discussion will come later.

20. The initial conditions imposed upon the integrals, in the cases where existence has been established, are associated with particular values of the variable x or of the variable y: a more general form can be given to the theorems, by a change in the independent variables. Let these be changed from x and y to X and Y, where

$$X = X(x, y), \quad Y = Y(x, y),$$

and denote by P and Q the derivatives of z with regard to X and Y respectively. Then if the transformed equation is resoluble with regard to P, it *possesses an integral* z (which therefore is an integral of the original equation) *characterised by the following properties:*

(i) *it is a regular function of x and y within domains that are not infinitesimal:*

(ii) *when $X(x, y) = \alpha$, the integral acquires a value $\theta(x, y)$, which is a regular function within the domains considered, which is not expressible in terms of X alone, and which otherwise is arbitrary:*

(iii) *the regular integral thus determined is unique* for the branch of the equation given by the resolution with regard to P.

Note 1. When the equation $f(x, y, z, p, q) = 0$ is resoluble with regard both to p and to q, regular integrals are obtained each of which is unique under the initial conditions imposed. Such integrals are, in general, independent of one another; if an integral possessed by the equation resolved with respect to p proved to be the same as the integral possessed by the equation resolved with respect to q, there must be relations between the two sets of initial conditions.

Note 2. In each set of initial conditions, a single function occurs which, within certain very broad limitations, is arbitrary: subject to the associated conditions, this arbitrary function determines a regular integral uniquely. We may therefore expect that, when classes of integrals of partial equations of the first order are being discussed, one class will emerge characterised by the occurrence of a single arbitrary function.

This result will be found to be a special case of a more general result.

Note 3. The equation $f(x, y, z, p, q) = 0$ has been described as irreducible; the property has been tacitly used, though explicit reference to the irreducibility has not been made after the first statement.

The reason for assuming the property is practically obvious. If $f(x, y, z, p, q)$ can be expressed as the product of independent regular factors, say F and G, and if, in considering integrals of $f=0$, we begin with the integrals of $F=0$, we have

$$\frac{\partial f}{\partial p} = G\frac{\partial F}{\partial p}, \quad \frac{\partial f}{\partial q} = G\frac{\partial F}{\partial q},$$

so that, as G is not zero, the critical quantities for the resolution of the equation are $\frac{\partial F}{\partial p}$ and $\frac{\partial F}{\partial q}$. We thus, in effect, do consider separately the integrals of $F=0$ and $G=0$; and therefore no generality is lost by assuming the equation as irreducible in this case.

21. We shall frequently have recourse to geometrical illustrations, particularly in the case of equations involving one dependent variable and two independent variables. Such illustrations limit the range of variation of the variables to real quantities; they will, however, be found an occasionally convenient method of statement.

Thus consider the equation $f(x, y, z, p, q)=0$: an integral is a relation between x, y, and z, and this can conveniently be interpreted as the equation of a surface. We have seen that, under conditions which do not need restatement for the present purpose, there is an integral such that, when $x=a$, the integral acquires a value $\phi(y)$. But

$$x=a, \quad z=\phi(y),$$

are the equations of a plane curve, as arbitrary as is the function $\phi(y)$. Hence a surface can be drawn that will satisfy the partial equation and will pass through a plane curve which (within certain large limitations) can be taken arbitrarily.

Similarly, as regards the modified result of § 20: the equations

$$X(x, y)=\alpha, \quad z=\theta(x, y),$$

are the equations of a skew curve; and therefore a surface can be drawn that will satisfy the partial equation and will pass through a skew curve which (within certain large limitations) can be taken arbitrarily.

22. A precisely similar application of the general existence-theorems in the last chapter can be made when the differential equation of the first order involves n independent variables: it will therefore be sufficient to state the results.

Denoting the independent variables by $x_1, \ldots, x_n$, and the first derivatives of z by $p_1, \ldots, p_n$ as usual, we take the equation in the form

$$f(x_1, \ldots, x_n, z, p_1, \ldots, p_n) = 0,$$

and we assume it to be irreducible: and we have the following results.

Except for such values of the variables (if any) as make $\dfrac{\partial f}{\partial p_r}$ vanish at the same time as f, the equation can be resolved with regard to p_r; and if $x_1 = a_1, \ldots, x_n = a_n, z = c, p_1 = \lambda_1, \ldots, p_n = \lambda_n$, be an ordinary set of values for the equation $f = 0$, so that f is regular in their vicinity, then the resolved expression for p_r is regular in the vicinity of those values. Let ϕ_r denote a function of $x_1, \ldots, x_{r-1}, x_{r+1}, \ldots, x_n$, which is regular in the vicinity of $x_1 = a_1, \ldots, x_n = a_n$, which at $a_1, \ldots, a_n$ acquires the value c_r, and which is otherwise arbitrary. Then an integral of the resolved equation exists, determined by the conditions

(i) it is a regular function of the variables within fields of variation given by

$$|x_1 - a_1| \leqslant \rho, \quad \ldots, \quad |x_n - a_n| \leqslant \rho,$$

where ρ is not infinitesimal;

(ii) when $x_r = a_r$, the integral acquires the value ϕ_r.

Moreover, the integral of the resolved equation, as determined by these conditions, is unique.

If the original equation is of degree μ in p_r, there are μ resolved equations equivalent to $f = 0$ save when $\dfrac{\partial f}{\partial p_r}$ vanishes with f; each such resolved equation determines a unique integral, subject to the imposed conditions; if these be $\zeta_1, \ldots, \zeta_\mu$, then the equation

$$(z - \zeta_1) \ldots (z - \zeta_\mu) = 0$$

can be regarded as providing the integral of $f = 0$, subject to the imposed conditions.

The resolution of the equation $f=0$ is possible with regard to each of the n quantities p in turn, except only when $\frac{\partial f}{\partial p}$ vanishes with f; and each such solution leads, under corresponding imposed conditions similar to those used for the resolution with regard to p_r, to similar integrals of the resolved equations and to a corresponding integral of $f=0$, unique under the imposed conditions.

Hence, by resolving with regard to one or other of the derivatives $p_1, \ldots, p_n$, we establish the existence of integrals of the equation, uniquely determined by imposed conditions which, within certain large limitations, involve an arbitrary functional element.

This establishment of the existence of integrals of the equation $f=0$ is effective except in the single conjunction that all the quantities

$$\frac{\partial f}{\partial p_1}, \ \frac{\partial f}{\partial p_2}, \ \ldots, \ \frac{\partial f}{\partial p_n},$$

vanish for values of the variables which make $f=0$: in that conjunction, if it can occur, the existence-theorems cannot be applied. There will therefore remain, as a subject for separate consideration, the discussion of the integrals (if any) of the equation

$$f=0,$$

which simultaneously satisfy the equations

$$\frac{\partial f}{\partial p_1}=0, \ \ldots, \ \frac{\partial f}{\partial p_n}=0.$$

As before, we can deduce the existence of integrals which are such that, when some relation

$$\mu(x_1, \ \ldots, \ x_n)=0$$

is satisfied, z acquires a value $\phi(x_1, \ldots, x_n)$, where μ and ϕ are regular functions, and ϕ is not expressible in terms of μ alone: the general condition, necessary for the existence of the integral, is that the quantity

$$\frac{\partial f}{\partial p_1}\frac{\partial \mu}{\partial x_1}+\frac{\partial f}{\partial p_2}\frac{\partial \mu}{\partial x_2}+\ldots+\frac{\partial f}{\partial p_n}\frac{\partial \mu}{\partial x_n}$$

shall not vanish in virtue of $f=0$.

CAUCHY'S THEOREM FOR EQUATIONS OF THE SECOND ORDER.

23. The equation, next in simplicity, to which the existence-theorems can be applied, is an irreducible equation of the second order in one dependent variable and two independent variables. Denoting the second derivatives of z with regard to x and y by r, s, t, as usual, we may take the equation in the form

$$f(x,\ y,\ z,\ p,\ q,\ r,\ s,\ t) = 0,$$

where f will be assumed to be a regular function of its arguments. Let a, b, c, λ, μ, α, β, γ be a set of values of the arguments of f in the vicinity of which f is regular; then, unless $\dfrac{\partial f}{\partial r}$ vanishes for those values, the equation can be resolved so as to express r in terms of the remaining quantities by an equation

$$r - \alpha = \bar{g}(x - a,\ y - b,\ z - c,\ p - \lambda,\ q - \mu,\ s - \beta,\ t - \gamma),$$

say

$$r = g(x,\ y,\ z,\ p,\ q,\ s,\ t),$$

where g is a regular analytic function. Now, as $\dfrac{\partial f}{\partial r}$ usually involves at least some of the variables, it usually is possible to choose initial values so that $\dfrac{\partial f}{\partial r}$ does not vanish; and then the resolution of the original equation can be effected. But it might happen that values of z (if any) and of its derivatives, which make f vanish, also make $\dfrac{\partial f}{\partial r}$ vanish: the resolution of the original equation with regard to r could not be effected, and we should have to proceed otherwise.

When the resolution is possible, the general theorems can be applied to establish the existence of *an integral z having the properties:*

(i) *it is a regular function of x and y within fields of variation round a and b, given by*

$$|\, x - a \,| \leqslant \rho, \quad |\, y - b \,| \leqslant \rho,$$

where ρ is not infinitesimal;

(ii) *when $x=a$, then z reduces to $\phi_0(y)$ and $\dfrac{\partial z}{\partial x}$ reduces to $\phi_1(y)$, where $\phi_0(y)$ and $\phi_1(y)$ are regular functions of y within the domain $|y-b| \leqslant \rho$, acquiring the values c and λ respectively when $y=b$, and are otherwise arbitrary;*

(iii) *the integral z as determined by these conditions is unique.*

The mode of establishment is similar to that in the case of the equation of the first order, and so the exposition will be brief. We consider a system of equations

$$\frac{\partial z}{\partial x}=p\ ,$$

$$\frac{\partial p}{\partial x}=r\ ,$$

$$\frac{\partial q}{\partial x}=\frac{\partial p}{\partial y},$$

$$\frac{\partial s}{\partial x}=\frac{\partial r}{\partial y},$$

$$\frac{\partial t}{\partial x}=\frac{\partial s}{\partial y},$$

$$\frac{\partial r}{\partial x}=\frac{\partial g}{\partial x}+\frac{\partial g}{\partial z}p+\frac{\partial g}{\partial p}r+\frac{\partial g}{\partial q}\frac{\partial p}{\partial y}+\frac{\partial g}{\partial s}\frac{\partial r}{\partial y}+\frac{\partial g}{\partial t}\frac{\partial s}{\partial y},$$

of the same character as in the general existence-theorems; and we regard the system as involving six dependent variables z, p, q, r, s, t. When the former results are applied, we infer the existence of integrals of this system of equations, characterised by the properties:

(i) they are regular functions of x and y within the fields of variation

$$|x-a| \leqslant \rho, \quad |y-b| \leqslant \rho,$$

(ii) when $x=a$, then

$$z=\phi_0(y),$$

$$p=\phi_1(y),$$

$$q=\frac{d\phi_0(y)}{dy},$$

$$s=\frac{d\phi_1(y)}{dy},$$

$$t = \frac{d^2\phi_0(y)}{dy^2},$$

$$r = g\left\{a, y, \phi_0(y), \phi_1(y), \frac{d\phi_0(y)}{dy}, \frac{d\phi_1(y)}{dy}, \frac{d^2\phi_0(y)}{dy^2}\right\},$$

where $\phi_0(y)$ and $\phi_1(y)$ are the foregoing regular functions:

(iii) the set of integrals determined by these conditions is unique.

Let the set of integrals thus determined be

$$z = Z(x, y), \quad p = P(x, y), \quad q = Q(x, y),$$
$$r = R(x, y), \quad s = S(x, y), \quad t = T(x, y);$$

then $z = Z(x, y)$ *is the announced integral of the original resolved equation.*

The proof is simple, on the same lines as before. From the first of the equations, we have

$$\frac{\partial Z}{\partial x} = P,$$

and therefore

$$\frac{\partial z}{\partial x} = p.$$

Similarly

$$r = \frac{\partial p}{\partial x} = \frac{\partial^2 z}{\partial x^2}.$$

Again, the third equation gives

$$\frac{\partial Q}{\partial x} = \frac{\partial P}{\partial y}$$
$$= \frac{\partial^2 Z}{\partial x \partial y},$$

so that

$$\frac{\partial}{\partial x}\left(\frac{\partial Z}{\partial y} - Q\right) = 0.$$

Thus $\frac{\partial Z}{\partial y} - Q$ is independent of x: inserting the values of Z and Q when $x = a$, we find the value to be zero, so that

$$\frac{\partial Z}{\partial y} = Q,$$

and therefore

$$\frac{\partial z}{\partial y} = q.$$

The fourth equation gives

$$\frac{\partial S}{\partial x} = \frac{\partial R}{\partial y}$$

$$= \frac{\partial}{\partial y}\left(\frac{\partial P}{\partial x}\right),$$

and therefore $S - \frac{\partial P}{\partial y}$ is independent of x: its value is zero when $x = a$, and so

$$S = \frac{\partial P}{\partial y},$$

that is,

$$s = \frac{\partial p}{\partial y}$$

$$= \frac{\partial^2 z}{\partial x \partial y}.$$

Similarly from the fifth equation, we have $T - \frac{\partial^2 Z}{\partial y^2}$ independent of x: its value is zero when $x = a$, so that

$$T = \frac{\partial^2 Z}{\partial y^2},$$

that is,

$$t = \frac{\partial^2 z}{\partial y^2}.$$

Lastly, writing

$$v = R - g(x, y, Z, P, Q, S, T),$$

the sixth equation shews that v is independent of x: its value is zero when $x = a$, and so $v = 0$ generally. Thus

$$R - g(x, y, Z, P, Q, S, T) = 0,$$

that is,

$$r - g(x, y, z, p, q, s, t) = 0,$$

where $p = \frac{\partial z}{\partial x}$, $q = \frac{\partial z}{\partial y}$, $r = \frac{\partial^2 z}{\partial x^2}$, $s = \frac{\partial^2 z}{\partial x \partial y}$, $t = \frac{\partial^2 z}{\partial y^2}$, in association with $z = Z(x, y)$.

Owing to the other properties, by which it obeys the assigned conditions, $z = Z(x, y)$ is the integral of the original differential equation, having the prescribed character.

Thus the existence of an integral of the equation

$$f(x, y, z, p, q, r, s, t) = 0$$

is established, save in the case when $\frac{\partial f}{\partial r}$ vanishes for values (if any) of the variables simultaneously with f. When $\frac{\partial f}{\partial r}$ does not thus vanish, the original equation is analytically resoluble: there is one such integral, subject to the imposed conditions, for each resolved branch of the equation: and if $\zeta_1, \zeta_2, \ldots$ be these integrals, then

$$(z-\zeta_1)(z-\zeta_2)\ldots=0$$

provides an integral of the original equation.

Similarly, if the equation is resolved with regard to t—and this will be possible except for such values (if any) of the variables as make $\frac{\partial f}{\partial t}$ vanish simultaneously with f—and if the resolved form is

$$t=h(x, y, z, p, q, r, s),$$

where h is a regular function of its arguments, then *an integral* z exists, characterised by the properties:

(i) *it is a regular function of x and y within fields of variation round a and b, given by*

$$|x-a| \leqslant \rho, \quad |y-b| \leqslant \rho,$$

where ρ is not infinitesimal;

(ii) *when $y=b$, then z reduces to $\psi_0(x)$ and $\frac{\partial z}{\partial y}$ reduces to $\psi_1(x)$, where $\psi_0(x)$ and $\psi_1(x)$ are regular functions of x within the domain $|x-a| \leqslant \rho$, acquiring the values c and μ when $x=a$, and are otherwise arbitrary;*

(iii) *the integral z, determined by these conditions, is unique.*

The proof is similar to that of the earlier proposition and so need not be expounded.

24. It may happen that there are values of the variables for which the equation is not resoluble with regard either to r or to t, so that $\frac{\partial f}{\partial r}, \frac{\partial f}{\partial t}, f$ vanish simultaneously for such values: yet for these values the equation $f=0$ may be resoluble with regard to s. In that case, we change the independent variables from x and y to

x' and y'; denoting the derivatives with regard to the new variables by p', q', r', s', t', we have

$$\left.\begin{aligned} r &= r'\left(\frac{\partial x'}{\partial x}\right)^2 + 2s'\frac{\partial x'}{\partial x}\frac{\partial y'}{\partial x} + t'\left(\frac{\partial y'}{\partial x}\right)^2 + p'\frac{\partial^2 x'}{\partial x^2} + q'\frac{\partial^2 y'}{\partial x^2} \\ s &= r'\frac{\partial x'}{\partial x}\frac{\partial x'}{\partial y} + \ldots\ldots \\ t &= r'\left(\frac{\partial x'}{\partial y}\right)^2 + \ldots\ldots \end{aligned}\right\}.$$

When these are substituted, the new equation will be resoluble with regard to r' except for values (if any) of the variables which make $\frac{\partial f}{\partial r'}$ vanish, that is, which make

$$\frac{\partial f}{\partial r}\left(\frac{\partial x'}{\partial x}\right)^2 + \frac{\partial f}{\partial s}\frac{\partial x'}{\partial x}\frac{\partial x'}{\partial y} + \frac{\partial f}{\partial t}\left(\frac{\partial x'}{\partial y}\right)^2$$

vanish. In the present case, we can choose x' so that it shall involve x and y; and therefore, even though $\frac{\partial f}{\partial r}$ and $\frac{\partial f}{\partial t}$ vanish, the foregoing quantity will not vanish unless $\frac{\partial f}{\partial s}$ vanishes. When $\frac{\partial f}{\partial s}$ does not vanish, the equation can be resolved with respect to r': and an integral of the equation exists, uniquely determined by conditions similar to those in former cases.

This transformation, moreover, shews that the initial conditions can be modified in all the preceding cases: they can be associated with an initial value $x' = a'$, of course with the appropriate modifications, that is, they can be associated with an initial relation

$$\theta(x, y) = a',$$

where θ is a regular function.

The existence of an integral of the equation

$$f(x, y, z, p, q, r, s, t) = 0$$

is thus established except for such values (if any) of the variables as make the equations

$$\frac{\partial f}{\partial r} = 0, \quad \frac{\partial f}{\partial s} = 0, \quad \frac{\partial f}{\partial t} = 0$$

satisfied simultaneously with $f = 0$. If this be possible, the existence-theorem does not apply: and there must be an independent discussion of these integrals, if any. This discussion will be deferred.

In the case of equations of the first order, say in two independent variables only, we are accustomed to the existence of integrals such that f, $\frac{\partial f}{\partial p}$, $\frac{\partial f}{\partial q}$ vanish simultaneously. Without anticipating the discussion of the corresponding question for equations of the second order, it is to be remarked that the four equations

$$f=0, \quad \frac{\partial f}{\partial r}=0, \quad \frac{\partial f}{\partial s}=0, \quad \frac{\partial f}{\partial t}=0,$$

may coexist without rendering the elimination of r, s, t possible. An example is furnished by the equation

$$(rx^2+2sxy+ty^2-2px-2qy+2z)^2-a^4(r^2+2s^2+t^2)$$
$$=\frac{4a^4}{a^4-(x^2+y^2)^2}(px+qy-z)^2.$$

Cauchy's Theorem in general.

25. We now proceed to apply the existence-theorems, in order to establish the existence of integrals of the system of equations

$$\frac{\partial^{r_1} z_1}{\partial x_1{}^{r_1}}=Z_1, \quad \frac{\partial^{r_2} z_2}{\partial x_1{}^{r_2}}=Z_2, \ \ldots, \ \frac{\partial^{r_m} z_m}{\partial x_1{}^{r_m}}=Z_m,$$

where the quantities $Z_1, Z_2, \ldots, Z_m$ are regular functions of the independent variables $x_1, \ldots, x_n$, of the dependent variables $z_1, \ldots, z_m$, and of the derivatives of the latter of all orders up to (and including) $r_1, \ldots, r_m$ respectively, except only those derivatives which appear on the left-hand sides of the equations.

Let $a_1, \ldots, a_n$ be values of $x_1, \ldots, x_n$ within the field of regular existence of the quantities $Z_1, \ldots, Z_m$; and let a number of functions of $x_2, \ldots, x_n$ be chosen, which are regular in the vicinity of $a_2, \ldots, a_n$, and (subject to certain limitations upon their coefficients about to be stated) which are otherwise arbitrary. These functions will be denoted by $\phi_{\lambda\mu}$, for

$$\lambda=1, \ \mu=0, 1, \ \ldots, \ r_1-1;$$
$$\lambda=2, \ \mu=0, 1, \ \ldots, \ r_2-1;$$
$$\ldots\ldots\ldots\ldots\ldots\ldots\ldots\ldots$$
$$\lambda=m, \ \mu=0, 1, \ \ldots, \ r_m-1.$$

Then *a system of integrals* $z_1, \ldots, z_m$ *of the equations exists, characterised by the properties:*

(i) *they are regular functions of* $x_1, \ldots, x_n$ *in fields of variation given by*

$$|x_1 - a_1| \leqslant \rho, \quad |x_2 - a_2| \leqslant \rho, \quad \ldots, \quad |x_n - a_n| \leqslant \rho,$$

where ρ *is not infinitesimal;*

(ii) *when* $x_1 = a_1$, *the values acquired by the integrals and by their derivatives are given by the relations*

$$\frac{\partial^\mu z_\lambda}{\partial x_1^{\ \mu}} = \phi_{\lambda\mu},$$

for the various values of λ *and* μ, *it being assumed that the values of the functions* $\phi_{\lambda\mu}$, *when* $x_2 = a_2, \ldots, x_n = a_n$, *are values of the derivatives of* $z_1, \ldots, z_m$ *within the field of regular existence of the functions* $Z_1, \ldots, Z_m$;

(iii) *the system of integrals, thus determined, is unique.*

In order to establish this result, we merely generalise the method applied in the preceding special cases of the theorem. We introduce a number of auxiliary variables

$$\frac{\partial^{r+s+t+\ldots} z_\lambda}{\partial x_1^{\ r} \partial x_2^{\ s} \partial x_3^{\ t} \ldots} = p_{\lambda rst\ldots},$$

assigning as initial conditions that, when $x_1 = a_1$, the values they shall assume are given by the relations

$$p_{\lambda r_\lambda 00\ldots} = [Z_\lambda],$$

$$p_{\lambda r00\ldots} = \phi_{\lambda r},$$

when $r < r_\lambda$, and $[Z_\lambda]$ is the value of Z_λ when $x_1 = a_1$; and we construct the system of equations

$$\frac{\partial p_{\lambda r_\lambda 00\ldots}}{\partial x_1} = \frac{\partial Z_\lambda}{\partial x_1} + \sum_{s=1}^{m} \frac{\partial Z_\lambda}{\partial z_s} p_{s10\ldots} + \ldots,$$

$$\frac{\partial p_{\lambda r00\ldots}}{\partial x_1} = p_{\lambda,\ r+1,\ 0,\ 0,\ \ldots},$$

$$\frac{\partial p_{\lambda rst\ldots}}{\partial x_1} = \frac{\partial p_{\lambda,\ r+1,\ s-1,\ t\ldots}}{\partial x_2},$$

these holding for $r < r_\lambda$, $r + s + t + \ldots < r_\lambda$, $s > 0$, and for all values of λ; the right-hand side of the first equation is the complete derivative of Z_λ with respect to x_1, the derivatives of the included arguments being modified by the rest of the equations in this amplified system.

The substituted system of equations conforms to the type specified in the general existence-theorems, which accordingly apply. The system possesses a set of integrals having the properties:

(i) they are regular functions of the variables $x_1, \ldots, x_n$ within the specified fields of variation:

(ii) when $x_1 = a_1$, the various dependent variables acquire the respective assigned values:

(iii) the integrals, thus determined, are unique.

Let the values of z, which occur in this set of integrals, be

$$z_r = \psi_r(x_1, \ldots, x_n),$$

for $r = 1, \ldots, m$: *these values constitute the announced set of integrals of the original system of equations.*

The method of proceeding is the same as for the simple cases. Thus we have

$$\frac{\partial z_\lambda}{\partial x_1} = p_{\lambda 100\ldots},$$

direct from equations of the substituted system. Again, we have

$$\begin{aligned}\frac{\partial p_{\lambda 0100\ldots}}{\partial x_1} &= \frac{\partial p_{\lambda 100\ldots}}{\partial x_2} \\ &= \frac{\partial^2 z_\lambda}{\partial x_1 \partial x_2}.\end{aligned}$$

Thus $p_{\lambda 0100\ldots} - \dfrac{\partial z_\lambda}{\partial x_2}$ is independent of x_1; its value is zero when $x_1 = a_1$, and therefore is zero generally, that is,

$$\frac{\partial z_\lambda}{\partial x_2} = p_{\lambda 0100\ldots}.$$

And so on, step by step: the last step gives

$$\frac{\partial}{\partial x_1}(p_{\lambda r_\lambda 00\ldots} - Z_\lambda) = 0,$$

so that $p_{\lambda r_\lambda 00\ldots} - Z_\lambda$ is independent of x_1: its value is zero when $x_1 = a_1$, owing to the assigned conditions; and therefore the value is zero generally, that is,

$$p_{\lambda r_\lambda 00\ldots} = Z_\lambda,$$

and therefore we have

$$\frac{\partial^{r_\lambda} z_\lambda}{\partial x_1^{r_\lambda}} = Z_\lambda,$$

equations which, for all the values of λ, constitute the original system. These are equations satisfied by

$$z_\lambda = \psi_\lambda (x_1, \ldots, x_n), \qquad (\lambda = 1, \ldots, m),$$

which accordingly are the integrals of that original system. The properties, which they possess as integrals of the substituted system, both as regards regular character, values acquired when $x_1 = a_1$, and uniqueness, shew that they obey the conditions imposed in connection with the original system.

A simple illustration is provided by the differential equation of a vibrating plane membrane, which is

$$h^2 \left(\frac{\partial^2 z}{\partial x^2} + \frac{\partial^2 z}{\partial y^2} \right) = \frac{\partial^2 z}{\partial t^2},$$

where h^2 is a constant : an integral of this equation is uniquely determined by the condition of being a regular function of x, y, t, and by the conditions that, when $t=0$,

$$z = f(x, y), \quad \frac{\partial z}{\partial t} = g(x, y).$$

By the nature of the case, the boundary of the membrane is fixed ; hence, along the boundary, z and $\frac{\partial z}{\partial t}$ are always zero, so that the regular functions $f(x, y)$ and $g(x, y)$ have their otherwise arbitrary character restricted by this general condition attached to the particular problem. But it follows from the general theorem that, if an integral can be obtained, in any manner, satisfying the imposed conditions, it is the unique integral, subject to those conditions.

For example, let the membrane be rectangular in form, having its sides equal to a and b: let the equations of the sides be $y=0$, $y=b$, $x=0$, $x=a$, so that f and g must vanish for any one of these four relations. Now an integral of the equation is given by

$$z = (\alpha \cos ct + \beta \sin ct) \sin \lambda x \sin \mu y,$$

provided

$$h^2 (\lambda^2 + \mu^2) = c^2:$$

and this integral will vanish on the rectangular boundary if

$$\sin \lambda a = 0, \qquad \sin \mu b = 0.$$

The latter will be satisfied by taking

$$\lambda = \frac{l\pi}{a}, \qquad \mu = \frac{m\pi}{b},$$

where l and m are integers; then

$$c^2 = h^2\pi^2\left(\frac{l^2}{a^2} + \frac{m^2}{b^2}\right),$$

and the integral is

$$z = (a\cos ct + \beta\sin ct)\sin\frac{l\pi x}{a}\sin\frac{m\pi y}{b}.$$

Clearly, the sum of any number of such integrals is also an integral: so that we have an integral given by

$$Z = \sum_{l=1}^{\infty}\sum_{m=1}^{\infty}(a_{lm}\cos c_{lm}t + \beta_{lm}\sin c_{lm}t)\sin\frac{l\pi x}{a}\sin\frac{m\pi y}{b}.$$

This quantity Z vanishes on the boundary: if, then, the coefficients a_{lm}, β_{lm} can be determined so as to satisfy the imposed conditions, we shall have the required integral. Now, when $t=0$,

$$Z = \sum_{l=1}\sum_{m=1} a_{lm}\sin\frac{l\pi x}{a}\sin\frac{m\pi y}{b},$$

$$\frac{\partial Z}{\partial t} = \sum_{l=1}\sum_{m=1} c_{lm}\beta_{lm}\sin\frac{l\pi x}{a}\sin\frac{m\pi y}{b};$$

and these should be equal to $f(x, y)$, $g(x, y)$, which accordingly impose limitations upon the character of the regular functions. The conditions will be satisfied if

$$a_{lm} = \frac{4}{ab}\int_0^a\int_0^b f(x, y)\sin\frac{l\pi x}{a}\sin\frac{m\pi y}{b}\,dx\,dy,$$

$$\beta_{lm} = \frac{4}{abc_{lm}}\int_0^a\int_0^b g(x, y)\sin\frac{l\pi x}{a}\sin\frac{m\pi y}{b}\,dx\,dy.$$

The required integral is uniquely given by the expression

$$z = \sum_{l=1}^{\infty}\sum_{m=1}^{\infty}(a_{lm}\cos c_{lm}t + \beta_{lm}\sin c_{lm}t)\sin\frac{l\pi x}{a}\sin\frac{m\pi y}{b},$$

with the foregoing values for the coefficients a and β.

Note. It will be noticed that the existence-theorem provides for the introduction of a number of functions which, within certain very wide limitations, are arbitrary functions of all the variables but one, or are arbitrary functions of all the variables subject to an assigned relation among the variables. In the case of the system of equations considered in this section, the number of these functions is

$$r_1 + r_2 + \ldots + r_m,$$

being the sum of the orders of the highest derivatives that occur.

In particular, if there be only a single dependent variable and a single equation of order r, *the number of arbitrary functions* provided by the theorem for the precise determination of the integral *is* r, *the same as the order of the equation.* Special illustrations have been furnished by an equation of the first order and by an equation of the second order.

26. In the establishment of the theorem as to the existence of integrals of the equations

$$\frac{\partial^{r_1} z_1}{\partial x_1^{r_1}} = Z_1, \ \ldots, \ \frac{\partial^{r_m} z_m}{\partial x_1^{r_m}} = Z_m,$$

it was assumed that no derivatives of z_1 of order higher than r_1 occur, and similarly for the derivatives of the other dependent variables. This limitation is important: it is actually necessary in order that the convergence of the series (and therefore the functional significance of the integrals) may be established.

The importance of the condition may be illustrated by a single example*. Consider the equation

$$\frac{\partial^2 z}{\partial y^2} = \frac{\partial z}{\partial x},$$

which belongs to the system when associated with imposed conditions to be satisfied for an assigned value of y; but the limitation is not obeyed when the imposed conditions are to be satisfied for an assigned value of x. To see the effect of the limitation, let it be required to obtain an integral of the equation which shall acquire a value $P(y)$ when $x=0$, $P(y)$ being an analytic function of y, regular in the vicinity of $y=0$. A formal solution is manifestly given by

$$z = \sum_{n=0} \frac{x^n}{n!} \frac{d^{2n} P}{dy^{2n}}.$$

The convergence of the series cannot be established: indeed, the series in general is not a converging series. To make the series more precise, let

$$P(y) = \frac{1}{1-y},$$

which satisfies all the conditions: then

$$z = \sum_{n=0} \frac{x^n}{(1-y)^{2n+1}} \frac{2n!}{n!}$$

$$= \frac{1}{1-y} \sum_{n=0} \frac{2n!}{n!\,(1-y)^{2n}} x^n.$$

Now it is known† that ρ, the radius of convergence of a converging series $\Sigma a_m x^m$, is given by

$$\frac{1}{\rho} = \operatorname*{Lim}_{m=\infty} |\, a_m^{\frac{1}{m}} \,|\,;$$

* Kowalevsky, *Crelle*, t. LXXX (1875), p. 22.

† *T. F.*, § 26.

hence, if ρ be the radius of convergence of the series for z regarded as a power-series in x, we have

$$\frac{1}{\rho}=\frac{1}{|1-y|^2}\operatorname*{Lim}_{m=\infty}\left(\frac{2m\,!}{m\,!}\right)^{\frac{1}{m}}$$

$$=\frac{1}{|1-y|^2}\operatorname*{Lim}_{m=\infty}\frac{4m}{e}2^{\frac{1}{2m}},$$

approximately, by the use of Stirling's theorem; thus ρ is zero, whatever finite value be possessed by y. In other words, there is no region of convergence for z in the case of the assumed form of $P(y)$.

The question thus suggests itself: what are the limitations upon $P(y)$ that the series for z should converge? To answer it, we take the equation as an instance of the equations in § 23: the theorem shews that a regular integral exists determined uniquely by the conditions that, when $y=0$,

$$z=Q(x)=\sum_{n=0}c_n x^n,$$

$$\frac{\partial z}{\partial y}=R(x)=\sum_{n=0}k_n x^n,$$

where $Q(x)$ and $R(x)$ are regular functions. The formal expression of this integral is easily found to be

$$z=\sum_{m=0}\frac{y^{2m}}{2m\,!}\frac{d^mQ(x)}{dx^m}+\sum_{m=0}\frac{y^{2m+1}}{(2m+1)\,!}\frac{d^mR(x)}{dx^m}.$$

Hence, when $x=0$, the value of z is given by

$$\sum_{m=0}\frac{m\,!}{2m\,!}c_m y^{2m}+\sum_{m=0}\frac{m\,!}{(2m+1)\,!}k_m y^{2m+1};$$

if the integral is to be given by the former process, this must be the value of $P(y)$ in the assigned initial conditions.

Let r denote the radius of convergence of the power-series $Q(x)$ and $R(x)$ simultaneously: then, because $\sum_{n=0}c_n x^n$ and $\sum_{n=0}k_n x^n$ are converging series when $|x|<r$, a finite quantity G exists such that

$$|c_n|<\frac{G}{r^n},\qquad |k_n|<\frac{G}{r^n},$$

so that we may take

$$c_n=\frac{Gu}{r^n},\qquad k_n=\frac{Gv}{r^n},$$

where $|u|<1$, $|v|<1$, while u and v are not zero. If ρ denote the radius of convergence of the series of powers of y, then

$$\frac{1}{\rho}=\operatorname*{Lim}_{m=\infty}\left|\frac{m\,!}{2m\,!}\frac{Gu}{r^m}\right|^{\frac{1}{m}}$$

$$=0,$$

or the power-series must converge over the whole plane. Consequently, the only functions admissible as values of $P(y)$ in the earlier investigation are those which are regular over the whole plane and, when expressed as power-series, converge over the whole plane in a manner comparable with the series

$$\sum_{m=0} \frac{m!}{2m!} y^{2m} + \sum_{m=0} \frac{m!}{(2m+1)!} y^{2m+1}.$$

Thus possible values of $P(y)$ are given by

$$P(y) = e^y,$$
$$P(y) = J_0(y),$$
$$P(y) = J_n(y),$$

when (in the last example) the real part of n is positive.

27. The equations, in § 25, though not of a completely general character, constitute a very extensive class; and they are even more extensive than their explicit form indicates, because of the possibilities of transformation.

Suppose that, in a given system of m equations, the order of the highest derivative of z_λ is r_λ, for $\lambda = 1, \ldots, m$; then, by transformation of the independent variables, it is usually possible to secure the explicit occurrence of the derivatives

$$\frac{\partial^{r_1} z_1}{\partial x_1^{r_1}}, \ \ldots, \ \frac{\partial^{r_m} z_m}{\partial x_1^{r_m}},$$

that is, of the highest derivatives with regard to one and the same variable. If all these occur, no change is needed; if any are absent, we change the variables by relations

$$x_s' = a_{s1} x_1 + a_{s2} x_2 + \ldots + a_{sn} x_n,$$

for $s = 1, \ldots, n$, the constant coefficients a being at our disposal provided their determinant is kept different from zero. Suppose that the required derivative of z_1 has not occurred in the original equations, but that there is a derivative $\dfrac{\partial^{r_1} z_1}{\partial x_1^s \partial x_2^t \partial x_3^u \ldots}$, where $s + t + u + \ldots = r_1$; then, after the transformation, the derivative $\dfrac{\partial^{r_1} z_1}{\partial x_1'^{r_1}}$ will occur unless $a_{11}^s a_{12}^t a_{13}^u \ldots$ vanishes. We can always secure that this negative provision is satisfied; hence the m equations can be transformed so that the required derivatives occur explicitly.

But this result is not sufficient to secure the form of the equations adopted for the existence-theorem; it must further be possible to resolve the m equations with respect to the m selected derivatives. When the resolution is possible, the resolved equations are of the form

$$\frac{\partial^{r_1} z_1}{\partial x_1^{r_1}} = Z_1, \ \ldots, \ \frac{\partial^{r_m} z_m}{\partial x_1^{r_m}} = Z_m.$$

When the resolution with regard to the selected derivatives is not possible, and is equally not possible with regard to every set of similarly selected derivatives, the equations do not belong to the class considered.

As an instance shewing that the form cannot be regarded as one to which all equations of the type considered are reducible, take the equations

$$P_1 \frac{\partial z_1}{\partial x_1} + P_2 \frac{\partial z_2}{\partial x_1} + P_3 \frac{\partial z_3}{\partial x_1} = Z_1,$$

$$Q_1 \frac{\partial z_1}{\partial x_2} + Q_2 \frac{\partial z_2}{\partial x_2} + Q_3 \frac{\partial z_3}{\partial x_2} = Z_2,$$

$$R_1 \frac{\partial z_1}{\partial x_3} + R_2 \frac{\partial z_2}{\partial x_3} + R_3 \frac{\partial z_3}{\partial x_3} = Z_3;$$

when the transformation

$$x_s' = a_{s1} x_1 + a_{s2} x_2 + a_{s3} x_3$$

(for $s = 1, 2, 3$) is effected, they have the form

$$a_{11} \left(P_1 \frac{\partial z_1}{\partial x_1'} + P_2 \frac{\partial z_2}{\partial x_1'} + P_3 \frac{\partial z_3}{\partial x_1'} \right) + \ldots = Z_1,$$

$$a_{12} \left(Q_1 \frac{\partial z_1}{\partial x_1'} + Q_2 \frac{\partial z_2}{\partial x_1'} + Q_3 \frac{\partial z_3}{\partial x_1'} \right) + \ldots = Z_2,$$

$$a_{13} \left(R_1 \frac{\partial z_1}{\partial x_1'} + R_2 \frac{\partial z_2}{\partial x_1'} + R_3 \frac{\partial z_3}{\partial x_1'} \right) + \ldots = Z_3.$$

The equations can be resolved for $\frac{\partial z_1}{\partial x_1'}, \frac{\partial z_2}{\partial x_1'}, \frac{\partial z_3}{\partial x_1'}$ (and therefore would be reducible to the selected general form) if

$$a_{11} a_{12} a_{13} \begin{vmatrix} P_1, & P_2, & P_3 \\ Q_1, & Q_2, & Q_3 \\ R_1, & R_2, & R_3 \end{vmatrix}$$

is not zero. But it might very well happen that the determinant of the coefficients P, Q, R should vanish identically; the resolution would then be impossible. In that case, it is equally impossible to resolve the equations with regard to $\frac{\partial z_1}{\partial x_2'}, \frac{\partial z_2}{\partial x_2'}, \frac{\partial z_3}{\partial x_2'}$, and also with regard to the remaining three derivatives, and so the existence-theorem cannot be applied; but then it is also necessary that the relations

$$\left\|\begin{matrix} P_1, & P_2, & P_3, & Z_1 \\ Q_1, & Q_2, & Q_3, & Z_2 \\ R_1, & R_2, & R_3, & Z_3 \end{matrix}\right\| = 0$$

be satisfied.

Hence the form of equations retained in § 25 is not a completely inclusive normal form; but, as already stated, it includes a very extensive class of equations*.

Ex. Consider the equations

$$\left.\begin{aligned} a\frac{\partial u}{\partial x} + b\frac{\partial v}{\partial x} &= \frac{\partial X}{\partial x} \\ a'\frac{\partial u}{\partial y} + b'\frac{\partial v}{\partial y} &= \frac{\partial Y}{\partial y} \end{aligned}\right\},$$

where a's and b's are constants, X is a function of x alone, Y is a function of y alone.

Effecting the transformations, we easily find that the system can be changed so as to have the normal form selected, provided $ab' - a'b$ is not zero. Assuming this proviso satisfied, the existence-theorem applies and the integrals certainly exist: they are most easily obtainable by quadrature from the original equations in the form

$$\left.\begin{aligned} au + bv &= X + f(y) \\ a'u + b'v &= Y + g(x) \end{aligned}\right\},$$

where f and g are arbitrary functions. To determine f and g in connection with assigned initial conditions, we take the existence-theorem for the transformed equations: it would assign values $\phi(\gamma x + \delta y)$ and $\psi(\gamma x + \delta y)$ to u and v respectively when $\alpha x + \beta y$ is constant, say λ, where $\alpha\delta - \beta\gamma$ is not zero. Thus

$$a\phi\left(\frac{\gamma\lambda}{\alpha} + \frac{\alpha\delta - \beta\gamma}{\alpha}y\right) + b\psi\left(\frac{\gamma\lambda}{\alpha} + \frac{\alpha\delta - \beta\gamma}{\alpha}y\right) = X\left(\frac{\lambda - \beta y}{\alpha}\right) + f(y),$$

$$a'\phi\left(\frac{\delta\lambda}{\beta} - \frac{\alpha\delta - \beta\gamma}{\beta}x\right) + b'\psi\left(\frac{\delta\lambda}{\beta} - \frac{\alpha\delta - \beta\gamma}{\beta}x\right) = Y\left(\frac{\lambda - \alpha x}{\beta}\right) + g(x),$$

which determine f and g.

* For a fuller discussion of this matter, see Bourlet, *Ann. de l'Éc. Norm. Sup.*, 3me Sér., t. VIII (1891), supplément. The example that follows is taken from this memoir.

But if $ab'-a'b=0$, the resolution is not possible: and the existence-theorem does not apply. The quadrature is still possible; and we find

$$au+bv=X+f(y),$$
$$a'u+b'v=Y+g(x),$$

where $f(y)$ is arbitrary so far as the first equation is concerned, and $g(x)$ is arbitrary so far as the second equation is concerned. Assuming for purposes of illustration that no one of the constants a, b, a', b' vanishes, we have

$$b'\{X+f(y)\}=b\{Y+g(x)\};$$

hence, as there is no relation between the variables x and y, we must have

$$g(x)=\frac{b'}{b}X-b'c,$$

$$f(y)=\frac{b}{b'}Y-bc,$$

where c is a constant. The two integral equations are now equivalent to one only; hence they do not precisely determine the two quantities u and v. One of these can be taken at will, say

$$v=\theta(x, y);$$

and then

$$u=-\frac{b}{a}\theta(x, y)+\frac{X}{a}+\frac{Y}{a'}-\frac{bc}{a},$$

which accordingly are integral equations in the case when $ab'-a'b=0$.

Other classes of Equations.

28. The preceding forms of equations are thus not universally inclusive; and, in recent years, investigations have been made on general differential systems, so as to establish the existence of integrals under assigned conditions associated with wider classes of equations. These investigations are mainly due to Méray, Riquier, Bourlet, Tresse, and Delassus*: their formal complication is elaborate. There are two main issues in this development of the theory; one is the construction of canonical forms, the other is the establishment of the existence of integrals of the systems of equations, the expression of which involves arbitrary constants or arbitrary functions. And we have seen, by a particular example,

* Many references will be found in von Weber's article on partial differential equations in the *Encyclopädie der mathematischen Wissenschaften*, vol. II, pp. 299, 300. In addition to these, four memoirs by Riquier may be mentioned; they are to be found in the *Acta Math.*, t. XXIII (1900), pp. 203—332, *ib.*, t. XXV (1902), pp. 297—358, *Ann. de l'Éc. Norm. Sup.*, 3me Sér., t. XVIII (1901), pp. 421—472, *ib.*, t. XX (1903), pp. 27—73.

(§ 3), that cases may occur in which integrals certainly exist and cannot contain any arbitrary element whatever.

For such investigations, we refer to the memoirs of the authors quoted; and we shall therefore enter into no further detail as to the existence of integrals of systems of equations in number equal to the number of dependent variables. There still remain, for our consideration, the discussion of the integrals (if any) of an equation or a system of equations in the vicinity of values of the variables where the functions concerned are not regular, and the discussion of the integrals (if any) of systems of equations in which the number of dependent variables is less than the number of equations. To the former, very little space will be devoted as the subject is hardly begun: it certainly seems to have claimed no attention from investigators. The latter is of the utmost importance, particularly in the case when there is only one dependent variable; it will be undertaken in a succeeding chapter.

CHAPTER III.

Linear Equations and Complete Linear Systems.

For the materials of this chapter, reference may be made to the authorities quoted in Part I, ch. II, of this work, in particular, to Mayer's memoir, *Math. Ann.* t. V (1872), pp. 448—470, and also to chapter II of Goursat's *Leçons sur l'intégration des équations aux dérivées partielles du premier ordre.* The chapter is devoted to linear equations, either single or in simultaneous systems.

Single equations and systems of simultaneous equations, which are homogeneous and linear in the differential elements of the variables, have already been discussed. The discussion of exact equations and exact systems of this type is given in the first two chapters of volume I of this work: the remainder of that volume is devoted to the discussion of inexact equations (Pfaff's problem) and of inexact systems.

The linear equation $\sum_{i=1}^{n} X_i p_i = 0$.

29. We proceed to a more detailed consideration of equations of the first order. Cauchy's theorem establishes the existence of integrals having a considerable degree of generality: but it does not prove that the integrals have the widest degree of generality possible or that they include all integrals by the appropriate specification of the arbitrary elements; and the only method which it provides for the actual construction of the integrals leads to expressions in power-series. It should be added, however, that (save for special classes of equations) the method provided in the proof of the existence-theorem is the only universal mode of constructing the integral: but for those special classes of equations simpler methods can be devised for the construction of the integrals, while further information can be obtained as to their relative generality and their classification. In all that follows, we are concerned rather with the theory in general than with the practical solution of particular equations as expounded in text-books*.

The simplest equations of all are those which are linear in the derivatives; among them, the simplest is the equation

$$X_1 p_1 + X_2 p_2 + \dots + X_n p_n = 0,$$

when the coefficients $X_1, \dots, X_n$ are functions of the variables $x_1, \dots, x_n$ but do not involve the dependent variable z. It will be seen later that every linear equation can be expressed in this form.

As usual, we associate with the partial equation the system of ordinary equations

$$\frac{dx_1}{X_1} = \frac{dx_2}{X_2} = \dots = \frac{dx_n}{X_n};$$

by the theory of such equations, their integral equivalent consists of $n-1$ independent equations in a form

$$u_r(x_1, \dots, x_n) = c_r, \qquad (r = 1, \dots, n-1).$$

Taking any one of these integral equations, we have

$$\frac{\partial u_r}{\partial x_1} dx_1 + \dots + \frac{\partial u_r}{\partial x_n} dx_n = 0,$$

concurrently with the ordinary differential equations; hence

$$X_1 \frac{\partial u_r}{\partial x_1} + \dots + X_n \frac{\partial u_r}{\partial x_n} = 0,$$

again an integral equation. Now there cannot be an integral equation independent of the set

$$u_1 = c_1, \quad u_2 = c_2, \quad \dots, \quad u_{n-1} = c_{n-1};$$

so that the new equation is not independent of this set. But it does not involve any of the quantities c; hence, though the equation holds, it does not hold in virtue of the integral set. It therefore can only be an identity; so that the equation

$$X_1 \frac{\partial u_r}{\partial x_1} + \dots + X_n \frac{\partial u_r}{\partial x_n} = 0$$

is satisfied identically. Consequently, when we put

$$z = u_r$$

* For instance, much of chapter IX in my *Treatise on Differential Equations*, (3d. edn. 1903), will be taken for granted.

in the original partial equation, the latter is satisfied identically: and therefore $z = u_r$ is an integral of the partial equation.

Moreover, this holds for all values of r; and therefore there are $n-1$ functionally distinct integrals of the equation. But there are not more than $n-1$ distinct integrals; that is, every integral can be expressed in terms of these. Let any integral be denoted by

$$z = f(x_1, \ldots, x_n);$$

then the equation

$$X_1 \frac{\partial f}{\partial x_1} + \ldots + X_n \frac{\partial f}{\partial x_n} = 0$$

is identically satisfied. The equations

$$X_1 \frac{\partial u_r}{\partial x_1} + \ldots + X_n \frac{\partial u_r}{\partial x_n} = 0,$$

for $r = 1, \ldots, n-1$, are identically satisfied: and the quantities $X_1, \ldots, X_n$ do not all vanish: hence

$$J = J\left(\frac{f, u_1, \ldots, u_{n-1}}{x_1, x_2, \ldots, x_n}\right) = 0.$$

The quantities u are functionally distinct, so that J does not vanish through an aggregate of vanishing first minors. It cannot vanish in virtue of $z = f$, for it does not involve z. It must therefore vanish identically; and therefore some relation must exist among the quantities $f, u_1, \ldots, u_{n-1}$, the relation involving f because $u_1, \ldots, u_{n-1}$ are functionally distinct: let it be

$$f = \phi(u_1, \ldots, u_{n-1}).$$

Hence *the equation possesses exactly $n-1$ functionally independent integrals.*

If f denote the most general integral of the equation, then ϕ must be the most general function possible: the requirement is satisfied by making ϕ a completely arbitrary function of its arguments. Hence *if $u_1, \ldots, u_{n-1}$ be a set of functionally independent integrals, the most general integral of the equation is given by*

$$z = \phi(u_1, \ldots, u_{n-1}),$$

where ϕ is a completely arbitrary function of its arguments.

The arbitrary function ϕ, and the functionally distinct integrals, can be determined so as to satisfy assigned initial conditions and therefore so as to yield the integral established by Cauchy's theorem. Let $a_1, \dots, a_n$ be values of $x_1, \dots, x_n$ in the vicinity of which all the quantities $X_1, \dots, X_n$ are regular; and suppose that some one of these quantities, say X_1, does not vanish for those values, an assumption that can always be justified by an appropriate choice of $a_1, \dots, a_n$. The general initial conditions will be that the integral z is to acquire a value $f(x_2, \dots, x_n)$, when $x_1 = a_1$, the function f being regular in the fields of variation considered.

The appropriate arguments can easily be constructed. Let an integral v_{r-1} be obtained to satisfy the partial differential equation, subject to the condition that it shall acquire a value x_r, when $x_1 = a_1$; its value is

$$v_{r-1} = x_r + (x_1 - a_1) P_r (x_1 - a_1, x_2 - a_2, \dots, x_n - a_n),$$

where P_r is a regular function of $x_1, \dots, x_n$ in the vicinity of the initial values. Taking this result for $r = 2, \dots, n$, we have $v_1, \dots, v_{n-1}$ as $n-1$ functionally distinct integrals; and then

$$z = f(v_1, \dots, v_{n-1})$$

is clearly the integral of the equation which acquires the assigned value $f(x_2, \dots, x_n)$, when $x_1 = a_1$.

The appropriate arguments can also be constructed from the associated ordinary equations.

COROLLARY. After the preceding analysis, we can state the existence-theorem, in a different but equivalent form, as follows.

If $a_1, \dots, a_n$ are values of $x_1, \dots, x_n$, in the vicinity of which all the coefficients X' in the equation

$$p_1 + X_2' p_2 + X_3' p_3 + \dots + X_n' p_n = 0$$

are regular, the equation possesses $n-1$ functionally distinct integrals, which are regular in the selected region and which reduce to $x_2, \dots, x_n$ respectively, when $x_1 = a_1$; and if these integrals be $z_1, \dots, z_{n-1}$, any integral of the equation can be expressed in the form $z = f(z_1, \dots, z_{n-1})$ by appropriate choice of f.

Ex. 1. Required an integral of the equation

$$x_1 p_1 + x_2 p_2 + x_3 p_3 = 0,$$

which shall acquire the value $\theta(x_2, x_3)$, when $x_1 = a_1$.

To obtain an integral v_1 which shall acquire the value x_2, when $x_1 = a_1$, we take

$$v_1 = x_2 + (x_1 - a_1)\phi_1 + (x_1 - a_1)^2 \phi_2 + \dots;$$

and we find

$$\phi_1 = -\frac{x_2}{a_1}, \qquad \phi_2 = \frac{x_2}{a_1^2}, \dots$$

so that

$$v_1 = x_2 \left\{1 - \frac{x_1 - a_1}{a_1} + \left(\frac{x_1 - a_1}{a_1}\right)^2 - \dots\right\}$$

$$= \frac{a_1 x_2}{x_1}.$$

We proceed similarly to obtain an integral v_2 which shall acquire the value x_3, when $x_1 = a_1$; we find

$$v_2 = \frac{a_1 x_3}{x_1}.$$

The required integral is clearly

$$z = \theta(v_1, v_2)$$

$$= \theta\left(\frac{a_1 x_2}{x_1}, \frac{a_1 x_3}{x_1}\right).$$

If we proceed from the associated ordinary equations, we need two integrals of the equations

$$\frac{dx_1}{x_1} = \frac{dx_2}{x_2} = \frac{dx_3}{x_3}:$$

these can be taken in the form

$$u_1 = \frac{x_2}{x_1}, \qquad u_2 = \frac{x_3}{x_1}.$$

We then require the form of ϕ such that $\phi(u_1, u_2)$ becomes $\theta(x_2, x_3)$, when $x_1 = a_1$: hence

$$\phi\left(\frac{x_2}{a_1}, \frac{x_3}{a_1}\right) = \theta(x_2, x_3),$$

and therefore

$$\phi(u_1, u_2) = \theta(a_1 u_1, a_1 u_2),$$

that is, the integral is

$$z = \phi(u_1, u_2)$$

$$= \theta(a_1 u_1, a_1 u_2),$$

as before.

Ex. 2. Three given functions u, v, w of x, y, z are such that

$$\frac{\partial u}{\partial x} + \frac{\partial v}{\partial y} + \frac{\partial w}{\partial z} = 0;$$

and three other functions ξ, η, ζ of the same variables are defined by the relations

$$u = \frac{\partial \zeta}{\partial y} - \frac{\partial \eta}{\partial z}, \qquad v = \frac{\partial \xi}{\partial z} - \frac{\partial \zeta}{\partial x}, \qquad w = \frac{\partial \eta}{\partial x} - \frac{\partial \xi}{\partial y}.$$

Prove that the most general values of ξ, η, ζ are

$$\xi = \frac{\partial F}{\partial x} + G\frac{\partial H}{\partial x}, \qquad \eta = \frac{\partial F}{\partial y} + G\frac{\partial H}{\partial y}, \qquad \zeta = \frac{\partial F}{\partial z} + G\frac{\partial H}{\partial z},$$

where G and H are integrals of the equation

$$u\frac{\partial \theta}{\partial x} + v\frac{\partial \theta}{\partial y} + w\frac{\partial \theta}{\partial z} = 0,$$

and F is an arbitrary function of x, y, z.

THE LINEAR EQUATION $\sum_{i=1}^{n} X_i p_i = Z$.

30. Next, we consider the linear equation

$$X_1 p_1 + \ldots + X_n p_n = Z,$$

where $X_1, \ldots, X_n, Z$ are functions of the variables $x_1, \ldots, x_n, z$. We shall assume that any factor, which is common to $X_1, \ldots, X_n, Z$, has been removed; it will therefore be unnecessary to take account of a value of z which simultaneously satisfies the equations

$$X_1 = 0, \ \ldots, \ X_n = 0, \quad Z = 0,$$

the differential equation being then satisfied without regard to the derivatives of z.

With the linear equation, we associate the set of ordinary equations

$$\frac{dx_1}{X_1} = \frac{dx_2}{X_2} = \ldots = \frac{dx_n}{X_n} = \frac{dz}{Z}.$$

Now whether $X_1, \ldots, X_n, Z$ be uniform or not, we shall assume that there are values of the variables in the vicinity of which $X_1, \ldots, X_n, Z$ behave regularly; and then, from the theory of ordinary equations, we know that the foregoing set possesses n functionally distinct integrals. Let these be

$$\phi_1(x_1, \ldots, x_n, z) = c_1, \ \ldots, \ \phi_n(x_1, \ldots, x_n, z) = c_n,$$

where $c_1, \ldots, c_n$ are arbitrary constants.

In the first place, *any equation*

$$\phi_r = c_r$$

gives an integral of the original equation if it involves z explicitly. As it is an integral of the ordinary equations, the relation

$$\frac{\partial \phi_r}{\partial x_1} dx_1 + \ldots + \frac{\partial \phi_r}{\partial x_n} dx_n + \frac{\partial \phi_r}{\partial z} dz = 0$$

is consistent with those equations; and therefore

$$X_1 \frac{\partial \phi_r}{\partial x_1} + \ldots + X_n \frac{\partial \phi_r}{\partial x_n} + Z \frac{\partial \phi_r}{\partial z} = 0.$$

Now this is a relation between the variables: it clearly is not satisfied in virtue of $\phi_r = c_r$; and therefore it is satisfied identically.

Taking $\phi_r = c_r$ as an equation giving a value (or values) of z, the derivatives are given by

$$\frac{\partial \phi_r}{\partial x_m} + \frac{\partial \phi_r}{\partial z} p_m = 0.$$

When these values of $\frac{\partial \phi_r}{\partial x_m}$, for $m = 1, \ldots, n$, are substituted in the foregoing equation that is identically satisfied, it becomes

$$\frac{\partial \phi_r}{\partial z}(Z - X_1 p_1 - \ldots - X_n p_n) = 0.$$

Now ϕ_r contains z, so that $\frac{\partial \phi_r}{\partial z}$ is not identically zero: and $\frac{\partial \phi_r}{\partial z}$ does not vanish because of the equation $\phi_r = c_r$, for it does not contain c_r: hence $\frac{\partial \phi_r}{\partial z}$ is different from zero. Accordingly, the equation

$$Z - X_1 p_1 - \ldots - X_n p_n = 0$$

is satisfied: or the equation $\phi_r = c_r$, when ϕ_r involves z explicitly, provides an integral of the partial equation.

The same is true for each of the equations $\phi = c$, provided each particular function ϕ involves z. Now some of the quantities ϕ must involve z, even though each of them may not: for otherwise $\frac{\partial \phi_r}{\partial z}$ would vanish for each value of r, and the equations

$$X_1 \frac{\partial \phi_r}{\partial x_1} + \ldots + X_n \frac{\partial \phi_r}{\partial x_n} = 0$$

would be satisfied identically, for $r = 1, \ldots, n$: we should then have

$$J\left(\frac{\phi_1, \ldots, \phi_n}{x_1, \ldots, x_n}\right) = 0$$

satisfied, but not in virtue of $\phi_1 = c_1, \ldots, \phi_n = c_n$: it must be satisfied identically and therefore, as the functions $\phi_1, \ldots, \phi_n$ do not (under the present hypothesis) involve z, there would be a functional relation between them, contrary to the fact that they are functionally independent. Hence, through the integral system of the ordinary equations, we find an integral or integrals of the partial equation.

In the second place, *let* $f(\phi_1, \ldots, \phi_n)$ *denote any arbitrary function of the quantities* ϕ, *and suppose that the equation*

$$f(\phi_1, \ldots, \phi_n) = 0$$

determines a value or values of z: then $f=0$ provides an integral of the differential equation. For the equations

$$X_1 \frac{\partial \phi_r}{\partial x_1} + \dots + X_n \frac{\partial \phi_r}{\partial x_n} + Z \frac{\partial \phi_r}{\partial z} = 0,$$

for $r=1, \dots, n$, are satisfied identically; as f is arbitrary, not all the quantities $\frac{\partial f}{\partial \phi_1}, \dots, \frac{\partial f}{\partial \phi_n}$ vanish; and therefore, on multiplying by $\frac{\partial f}{\partial \phi_r}$ and adding for all the values of r, the equation

$$X_1 \sum_{r=1}^{n} \frac{\partial f}{\partial \phi_r} \frac{\partial \phi_r}{\partial x_1} + \dots + X_n \sum_{r=1}^{n} \frac{\partial f}{\partial \phi_r} \frac{\partial \phi_r}{\partial x_n} + Z \sum_{r=1}^{n} \frac{\partial f}{\partial \phi_r} \frac{\partial \phi_r}{\partial z} = 0$$

is satisfied identically. Now the derivatives of z, as determined by $f=0$, are given by

$$\sum_{r=1}^{n} \frac{\partial f}{\partial \phi_r} \frac{\partial \phi_r}{\partial x_m} + p_m \sum_{r=1}^{n} \frac{\partial f}{\partial \phi_r} \frac{\partial \phi_r}{\partial z} = 0,$$

for all the values of m: when these are used, the identical equation becomes

$$(Z - X_1 p_1 - \dots - X_n p_n) \sum_{r=1}^{n} \frac{\partial f}{\partial \phi_r} \frac{\partial \phi_r}{\partial z} = 0.$$

Now as f contains z, the quantity

$$\sum_{r=1}^{n} \frac{\partial f}{\partial \phi_r} \frac{\partial \phi_r}{\partial z}$$

does not vanish identically; and it does not vanish in virtue of $f=0$, when f is perfectly arbitrary: hence

$$Z - X_1 p_1 - \dots - X_n p_n = 0,$$

or the equation is satisfied. When the values of $p_1, \dots, p_n$ of z are determined by $f=0$, the equation is seen above to be identically satisfied: hence $f=0$ provides an integral of the equation.

Of course, there may be special forms of f such that the equation $f=0$ does not determine z: and there may be special forms of f, such that $\sum_{r=1}^{n} \frac{\partial f}{\partial \phi_r} \frac{\partial \phi_r}{\partial z}$ vanishes in virtue of $f=0$. In what precedes, we are concerned with quite arbitrary forms of f.

31. The integral thus obtained is clearly of a very general character. A question arises as to whether it is completely inclusive of all integrals: that is, can the functional form f be so chosen as to provide any integral that the partial equation possesses? Let such an integral be known to occur in a form

$$\psi(z, x_1, \ldots, x_n) = 0;$$

and let $z = c$, $x_1 = a_1, \ldots, x_n = a_n$ be a simultaneous set of values of the variables, in the vicinity of which ψ is a regular function. Also suppose that these values are such that, in their vicinity, the functions $\phi_1, \ldots, \phi_n$ are regular, these providing integrals of the partial equation as before explained. Now the n quantities $\phi_1, \ldots,$ ϕ_n are functionally independent of one another; and therefore not all the determinants

$$\left\|\begin{array}{cccc} \dfrac{\partial\phi_1}{\partial z}, & \dfrac{\partial\phi_1}{\partial x_1}, & \ldots, & \dfrac{\partial\phi_1}{\partial x_n} \\ \cdots & \cdots & \cdots & \cdots \\ \dfrac{\partial\phi_n}{\partial z}, & \dfrac{\partial\phi_n}{\partial x_1}, & \ldots, & \dfrac{\partial\phi_n}{\partial x_n} \end{array}\right\|$$

vanish identically. Two typical cases will suffice for the general discussion: for the first, it will be assumed that

$$\frac{\partial(\phi_1, \ \ldots, \ \phi_n)}{\partial(x_1, x_2, \ldots, x_n)}$$

does not vanish identically; for the second, it will be assumed that

$$\frac{\partial(\phi_1, \ \ldots, \ \phi_n)}{\partial(z, x_2, \ldots, x_n)}$$

does not vanish identically.

32. In the first place, when the Jacobian of $\phi_1, \ldots, \phi_n$ with regard to $x_1, \ldots, x_n$ does not vanish identically, it is possible to choose the set of values $c, a_1, \ldots, a_n$ for $z, x_1, \ldots, x_n$, so that the Jacobian does not vanish or become infinite for them, unless the set constitute a singularity or other non-regular place for one or more of the quantities ϕ. Assuming this choice made, we then can resolve the n equations

$$\phi_r = \phi_r(z, x_1, \ldots, x_n), \qquad (r = 1, \ldots, n),$$

so as to express $x_1, \ldots, x_n$ as regular functions of $z, \phi_1, \ldots, \phi_n$, in forms

$$x_m = \xi_m(z, \phi_1, \ldots, \phi_n),$$

for $m=1, \ldots, n$. Let these values for the n variables be substituted in $\psi=0$, so that

$$\psi = 0 = \psi(z, \xi_1, \ldots, \xi_m)$$
$$= \chi(z, \phi_1, \ldots, \phi_n);$$

and now the given integral can be taken in the form $\chi=0$. To obtain the derivatives, we have

$$\frac{\partial \chi}{\partial z} p_m + \sum_{r=1}^{n} \frac{\partial \chi}{\partial \phi_r}\left(\frac{\partial \phi_r}{\partial x_m} + \frac{\partial \phi_r}{\partial z} p_m\right) = 0.$$

Multiply by X_m and add for the values $m=1, \ldots, n$: then

$$\frac{\partial \chi}{\partial z} \sum_{m=1}^{n} X_m p_m + \sum_{r=1}^{n} \frac{\partial \chi}{\partial \phi_r}\left\{\left(\sum_{m=1}^{n} X_m \frac{\partial \phi_r}{\partial x_m}\right) + \frac{\partial \phi_r}{\partial z}\left(\sum_{m=1}^{n} X_m p_m\right)\right\} = 0.$$

Now for the integral under consideration, we have

$$\sum_{m=1}^{n} X_m p_m = Z;$$

and we also know that the equation

$$\sum_{m=1}^{n} X_m \frac{\partial \phi_r}{\partial x_m} + Z \frac{\partial \phi_r}{\partial z} = 0$$

is satisfied identically for all values of r. Moreover, in the vicinities concerned, all the functions are regular, so that the quantities $\frac{\partial \chi}{\partial \phi_r}$ are finite in the fields of variation retained. When these relations are used, the above equation becomes

$$Z \frac{\partial \chi}{\partial z} = 0,$$

and this equation must be satisfied in association with $\chi=0$. This requirement may be met in three ways.

It may happen that $\frac{\partial \chi}{\partial z}$ vanishes identically: then z does not occur explicitly in χ, and the expression of χ then gives

$$\psi = \chi(\phi_1, \ldots, \phi_n),$$

that is, a form of function in the integral $f(\phi_1, \ldots, \phi_n)$ has been obtained so that the general integral becomes the given integral.

It may happen that $\frac{\partial \chi}{\partial z}$ vanishes, not indeed identically but only in virtue of $\chi=0$. Then z occurs explicitly in χ; and the form of the arbitrary function cannot be determined so that the general integral becomes the given integral.

It may happen that $\frac{\partial \chi}{\partial z}$ does not vanish. The condition can only be satisfied, if $Z=0$: and this must hold in association with $\chi=0$. Again, z occurs in χ; thus, once more, the form of the arbitrary function cannot be determined so that the general integral becomes the given integral.

Of these three alternatives, it is clear that the last belongs to a special set: as the integral is given by $Z=0$, we must have

$$\frac{\partial Z}{\partial x_m}+p_m\frac{\partial Z}{\partial z}=0;$$

and then the equation

$$X_1\frac{\partial Z}{\partial x_1}+\ldots+X_m\frac{\partial Z}{\partial x_m}=0$$

must be satisfied, concurrently with $Z=0$. Moreover, as $\phi_1=c_1$, ..., $\phi_n=c_n$ are a set of n independent integrals of the system of ordinary equations

$$\frac{dx_1}{X_1}=\ldots=\frac{dx_n}{X_n}=\frac{dz}{Z},$$

we have

$$\frac{Z}{J\left(\frac{\phi_1,\ \ldots,\ \phi_n}{x_1,\ x_2,\ \ldots,\ x_n}\right)}=\frac{(-1)^{r-1}X_r}{J_r\left(\frac{\phi_1,\ \ldots,\ \phi_n}{z,\ x_1,\ \ldots,\ x_n}\right)},$$

where x_r is omitted from the deriving variables in J_r, and $r=1$, ..., n in turn; hence as $Z=0$ for the integral under consideration, X_r must vanish for the value of z unless J_r should vanish for the value. We have assumed that not all the quantities $Z, X_1, \ldots, X_m$ vanish for the same value of z.

The second alternative may belong to a less special set: it will be illustrated by examples. The first alternative provides the most general case.

Integrals, which arise under the second alternative or under the third alternative, may be called *special* integrals*.

* Sometimes they are called *singular*. This term, however, is better reserved for a class of integrals belonging exceptionally to equations of a degree higher than the first in the derivatives.

33. In the second place, when the Jacobian of $\phi_1, \ldots, \phi_n$ with regard to $z, x_2, \ldots, x_n$ does not vanish identically, to take only a typical case when the Jacobian of those quantities with regard to $x_1, x_2, \ldots, x_n$ does vanish identically, it is possible to choose the set of values $c, a_1, \ldots, a_n$, so that the Jacobian does not vanish or become infinite for them unless they constitute a singularity or other non-regular place of one or more of the quantities ϕ. Assuming this done, we can then resolve the n equations

$$\phi_r = \phi_r(z, x_1, \ldots, x_n)$$

so as to express the variables $z, x_2, \ldots, x_n$ in terms of $x_1, \phi_1, \ldots, \phi_n$ in forms

$$z = \zeta(x_1, \phi_1, \ldots, \phi_n),$$
$$x_r = \eta_r(x_1, \phi_1, \ldots, \phi_n),$$

for $r = 2, \ldots, n$. When these values are substituted for $z, x_2, \ldots, x_n$ in the equation $\psi = 0$ which provides the given integral, it takes the form

$$\psi = 0 = \psi(z, x_1, \ldots, x_n)$$
$$= \theta(x_1, \phi_1, \ldots, \phi_n);$$

and the given integral can now be taken in the form $\theta = 0$. The derivatives of z are given by the n relations

$$\frac{\partial\theta}{\partial x_1} + \sum_{r=1}^{n} \frac{\partial\theta}{\partial\phi_r}\left(\frac{\partial\phi_r}{\partial x_1} + \frac{\partial\phi_r}{\partial z} p_1\right) = 0,$$

$$\sum_{r=1}^{n} \frac{\partial\theta}{\partial\phi_r}\left(\frac{\partial\phi_r}{\partial x_m} + \frac{\partial\phi_r}{\partial z} p_m\right) = 0,$$

for $m = 2, \ldots, n$. Multiplying these by X_1 and by X_m respectively, and adding for the various values of m, we have

$$X_1 \frac{\partial\theta}{\partial x_1} + \sum_{r=1}^{n} \frac{\partial\theta}{\partial\phi_r}\left\{\left(\sum_{m=1}^{n} X_m \frac{\partial\phi_r}{\partial x_m}\right) + \frac{\partial\phi_r}{\partial z} \sum_{m=1}^{n} (X_m p_m)\right\} = 0.$$

For the integral under consideration, we have

$$\sum_{m=1}^{n} X_m p_m = Z;$$

and we know that the relation

$$\sum_{m=1}^{n} X_m \frac{\partial\phi_r}{\partial x_m} + Z \frac{\partial\phi_r}{\partial z} = 0$$

is satisfied identically. Moreover, all the functions are regular in all the vicinities concerned, so that all the quantities $\frac{\partial\theta}{\partial\phi_r}$, for

$r=1, \ldots, n$, are finite in the fields of variation retained. When these equations are used, the above equation becomes

$$X_1 \frac{\partial \theta}{\partial x_1} = 0;$$

and it must be satisfied in association with $\theta = 0$. This requirement can, as in the preceding discussion, be met in three ways.

It may happen that $\frac{\partial \theta}{\partial x_1}$ vanishes identically; then x_1 does not occur explicitly in θ, and the expression of θ gives

$$\psi = \theta(\phi_1, \ldots, \phi_n),$$

that is, a form of function has been obtained for $f(\phi_1, \ldots, \phi_n)$ so that $f = 0$ has become the given integral $\psi = 0$.

Or it may happen that $\frac{\partial \theta}{\partial x_1}$ vanishes, not indeed identically but only in virtue of $\theta = 0$. Then x_1 occurs explicitly in θ; the form of the arbitrary function f in the general integral cannot be determined so as to particularise the general integral into the given integral.

Or it may happen that $\frac{\partial \theta}{\partial x_1}$ does not vanish. The condition can then only be satisfied if $X_1 = 0$; and this must hold in association with $\theta = 0$. Again, the variable x_1 occurs explicitly in θ: thus, once more, the form of the arbitrary function f in the general integral cannot be determined so as to make the general integral become the given integral.

The three alternatives are similar to those in the former discussion; integrals, that arise in connection with the second or the third of the alternatives, will be called *special*, as before.

34. Gathering together these results, we can summarise them as follows:—

Let $\psi(z, x_1, \ldots, x_n) = 0$ *provide an integral of the partial differential equation*

$$X_1 p_1 + \ldots + X_n p_n = Z,$$

and let $f(\phi_1, \ldots, \phi_n) = 0$ *denote its most general integral,* f *being an arbitrary function; then the functional form of* f *can be chosen so that* $f(\phi_1, \ldots, \phi_n)$ *becomes* ψ, *unless* ψ *is of the type of integral called special, or unless the value of* z *provided by* $\psi = 0$ *constitutes a singularity or other non-regular place for one or more of the quantities* ϕ.

It thus appears that the general integral for the linear non-homogeneous equation, in which the dependent variable occurs explicitly, is not so completely inclusive as is the general integral for the linear homogeneous equation, in which the dependent variable does not occur explicitly.

Instances of the principal portion of the theorem are so frequent that none need be adduced here: a few examples will be given to illustrate the special integrals and other exceptions.

Ex. 1. Consider the equation

$$xp+yq=z.$$

Two integrals of the associated equations

$$\frac{dx}{x}=\frac{dy}{y}=\frac{dz}{z}$$

can be taken in the form

$$\phi_1=\frac{z}{x}, \qquad \phi_2=\frac{z}{y};$$

and the most general integral is given by

$$f(\phi_1, \phi_2)=0.$$

It is easy to verify that

$$\psi=z-\frac{x^2}{y}=0$$

provides an integral of the equation. Expressing ψ in terms of ϕ_1, ϕ_2, and z, we find

$$\psi=z-z\frac{\phi_2}{\phi_1^2}=\psi',$$

so that

$$\frac{\partial\psi'}{\partial z}=1-\frac{\phi_2}{\phi_1^2};$$

thus $\frac{\partial\psi'}{\partial z}$ does not vanish identically but only in virtue of $\psi'=0$, and then only in virtue of the factor $1-\frac{\phi_2}{\phi_1^2}$ in ψ'. Thus the integral given by $\psi=0$ is a special integral; for the form of f in $f(\phi_1, \phi_2)$ cannot be chosen so as to make $f(\phi_1, \phi_2)$ become ψ.

It should be noted that $f(\phi_1, \phi_2)$ can be chosen, in a form $\phi_1^2-\phi_2$, so as to vanish for the integral provided by $\psi=0$: but it does not follow (and it is not the fact) that f can be chosen so that $f(\phi_1, \phi_2)$ becomes ψ.

Ex. 2. Consider the equation

$$\left(z-x_1\frac{x_2^2}{x_3}\right)p_1+x_2p_2+x_3p_3=z.$$

Integrals of the associated ordinary equations

$$\frac{dx_1}{z - x_1 \frac{x_2^2}{x_3}} = \frac{dx_2}{x_2} = \frac{dx_3}{x_3} = \frac{dz}{z}$$

may be taken in the form

$$\phi_1 = \left(1 - \frac{x_1 x_2^2}{z x_3}\right) e^{\frac{x_2^2}{x_3}}, \qquad \phi_2 = \frac{x_2}{z}, \qquad \phi_3 = \frac{x_3}{z};$$

and the general integral is

$$f(\phi_1, \phi_2, \phi_3) = 0,$$

where f is an arbitrary function.

It is easy to verify that

$$\psi = z - x_1 \frac{x_2^2}{x_3} = 0$$

provides an integral of the equation; but the functional form f cannot be chosen so that $f(\phi_1, \phi_2, \phi_3)$ becomes ψ. In fact, we have

$$\psi = z\phi_1 e^{-\frac{z\phi_2^2}{\phi_3}} = \psi',$$

so that

$$\frac{\partial \psi'}{\partial z} = \phi_1 e^{-\frac{z\phi_2^2}{\phi_3}} \left(1 - z\frac{\phi_2^2}{\phi_3}\right);$$

and $\frac{\partial \psi'}{\partial z}$ does not vanish identically. Taking the value of z given by ψ and substituting it in ϕ_1, we find $\phi_1 = 0$: so that $\frac{\partial \psi'}{\partial z}$ vanishes in virtue of this result, that is, in virtue of $\psi' = 0$. The integral $\psi = 0$ is a special integral.

If, instead of expressing x_1, x_2, x_3 in terms of the quantities $z, \phi_1, \phi_2, \phi_3$ with a view to the transformation of ψ, we express z, x_2, x_3 in terms of $x_1, \phi_1, \phi_2, \phi_3$, we find

$$\psi = \frac{\phi_3}{\phi_2^2}\left(1 - x_1 \frac{\phi_2^2}{\phi_3}\right) \log\left(\frac{\phi_1\phi_3}{\phi_3 - x_1\phi_2^2}\right) = \psi';$$

and then the requisite condition is

$$X_1 \frac{\partial \psi'}{\partial x_1} = 0,$$

in association with $\psi' = 0$. Now $\frac{\partial \psi'}{\partial x_1}$ does not vanish identically, nor does it vanish in virtue of $\psi' = 0$; we must therefore have $X_1 = 0$ in association with $\psi' = 0$. This is satisfied: and therefore, as before, $\psi = 0$ provides a special integral of the equation.

Ex. 3. Consider the equation

$$xp + 2yq = 2\left(z - \frac{x^2}{y}\right)^2.$$

The associated ordinary equations are

$$\frac{dx}{x}=\frac{dy}{2y}=\frac{dz}{2\left(z-\frac{x^2}{y}\right)^2},$$

of which two independent integrals are given by

$$\phi_1=\frac{x^2}{y}, \qquad \phi_2=ye^{\frac{1}{z-\frac{x^2}{y}}}.$$

The most general integral of the partial equation is

$$f(\phi_1, \phi_2)=0,$$

where f is an arbitrary function.

It is easy to verify that

$$\psi=z-\frac{x^2}{y}=0$$

provides an integral of the equation: but the functional form f cannot be chosen so as to make $f(\phi_1, \phi_2)$ become ψ. Proceeding as in the general exposition, we have

$$\psi=z-\phi_1=\psi',$$

so that $\frac{\partial\psi'}{\partial z}=1$ and cannot vanish, shewing that f cannot be chosen for the purpose. But the quantity Z of the general investigation vanishes for the value of z given by $\psi=0$.

It will be noted that ψ' does not involve ϕ_1: the special integral is a singularity of ϕ_2.

Ex. 4. Consider the equation*

$$\{1+(z-x-y)^{\frac{1}{2}}\}p+q=2.$$

The integrals of the ordinary equations

$$\frac{dx}{1+(z-x-y)^{\frac{1}{2}}}=\frac{dy}{1}=\frac{dz}{2}$$

can be taken in the form

$$\phi_1=2y-z,$$
$$\phi_2=y+2(z-x-y)^{\frac{1}{2}};$$

and the general integral is

$$f(\phi_1, \phi_2)=0.$$

It is easy to verify that

$$\psi=z-x-y=0$$

provides an integral of the equation; it is clear that no form of f can be found which will make the general function $f(\phi_1, \phi_2)$ become ψ. The integral provided by $\psi=0$ is a special integral; and manifestly any set of values, satisfying $\psi=0$ and chosen as initial values, constitute a branch-place of the quantity ϕ_2 and of the coefficient of p in the equation.

As this coefficient is not regular in the vicinity, Cauchy's theorem does not apply.

* This example is given by Chrystal, *Trans. R. S. E.*, t. XXXVI (1892), p. 557.

35. The discussion of the integrals of the equation

$$X_1 p_1 + \ldots + X_n p_n = Z$$

can be associated with the discussion of the integrals of the equation, which is without the quantity Z and any explicit occurrence of z, by means of a simple transformation. Let the integral be given by the equation

$$u = u(z, x_1, \ldots, x_n) = 0,$$

where, in the circumstances, u involves z; then we have

$$\frac{\partial u}{\partial z} p_m + \frac{\partial u}{\partial x_m} = 0.$$

Now $\frac{\partial u}{\partial z}$ does not vanish identically, and we shall assume* that it does not vanish in consequence of $u = 0$; hence we may resolve these equations for $p_1, \ldots, p_n$. Substituting in the original equation, we have

$$X_1 \frac{\partial u}{\partial x_1} + \ldots + X_n \frac{\partial u}{\partial x_n} + Z \frac{\partial u}{\partial z} = 0,$$

and this must be satisfied identically when a value of z given by $u = 0$ is inserted: in other words, the modified equation is satisfied, not identically but only simultaneously with $u = 0$. The modified equation is of the earlier type: the coefficients of the derivatives involve only the independent variables but not the dependent variable u. Of this modified equation, let

$$u = \theta(z, x_1, \ldots, x_n)$$

be an integral; then obviously $u = 0$ will give an integral of the original equation. But the fact that $\theta(z, x_1, \ldots, x_n)$ is an integral of the modified equation means, as was seen before, that when this value of u is substituted the equation is satisfied identically. This limitation is additional to the earlier requirement, which was only that the equation should be satisfied simultaneously with $u = 0$; it was not necessary that the equation should be satisfied identically. We cannot therefore infer from the argument that any integral of the original equation can thus be obtained from an integral of the

* The significance of the assumption, and the limitation which it imposes, would need to be examined if the character of the integrals were being determined solely by the present argument.

modified equation; and it is clear that any integral so obtained is a special case of an integral given by

$$\theta(z, x_1, \ldots, x_n) - a = 0,$$

where a is an arbitrary constant*.

Ex. As an example, consider the equation

$$(x^2+2xy)\frac{\partial z}{\partial x} - z^2\frac{\partial z}{\partial y} = y^2.$$

It clearly is satisfied by a value of z given by the equation

$$x+y+z=0.$$

But effecting the transformation indicated, viz. taking

$$u=u(z, x, y)=0,$$

so that u is a new variable, we have

$$(x^2+2xy)\frac{\partial u}{\partial x} - z^2\frac{\partial u}{\partial y} + y^2\frac{\partial u}{\partial z} = 0.$$

Any integral of this equation, when substituted, is known (by our earlier argument) to make the equation satisfied identically. If we take

$$u=x+y+z,$$

the equation is not satisfied identically; it can only be satisfied for this value of u simultaneously with $u=0$; but $u=x+y+z$ is not an integral of the new equation.

On the other hand, the original equation is satisfied by a value of z given by the equation

$$y^3+z^3=a,$$

where a is a constant: and

$$u=y^3+z^3$$

is an integral of the modified equation. Thus the first integral is not given, the second integral is given, by the method.

The distinction between the two cases can be expressed simply by a reference to the theory of continuous groups. Let

$$X(\theta)=(x^2+2xy)\frac{\partial \theta}{\partial x} - z^2\frac{\partial \theta}{\partial y} + y^2\frac{\partial \theta}{\partial z}$$

be an infinitesimal transformation.

We have

$$X(y^3+z^3)=0;$$

the quantity y^3+z^3 is an *invariant* for the given infinitesimal transformation.

We have

$$X(x+y+z)=(x+y+z)(x+y-z),$$

* The limitation was, I believe, first pointed out by Goursat, in § 16 of the work quoted on p. 55.

so that $x+y+z$ is not an invariant for the infinitesimal transformation: but when we have

$$x+y+z=0,$$

then, in virtue of that equation,

$$X(x+y+z)=0;$$

the equation $x+y+z=0$ is an *invariant equation* for the transformation.

36. It remains to associate Cauchy's theorem with the equation; for this purpose, we have to obtain an integral which, when $x_1=a_1$, reduces to

$$z=g(x_2, \ldots, x_n),$$

where g is a function, which is regular in the domains of the values $x_2=a_2, \ldots, x_n=a_n$, and otherwise is arbitrary.

Choosing a_1 so that X_1 does not vanish there, the integrals of the associated ordinary equations

$$dx_2=\frac{X_2}{X_1}dx_1, \quad dx_3=\frac{X_3}{X_1}dx_1, \quad \ldots, \quad dx_n=\frac{X_n}{X_1}dx_1, \quad dz=\frac{Z}{X_1}dx_1$$

can be obtained, subject to assigned conditions that $x_2=a_2, \ldots, x_n=a_n$, $z=g(a_2, \ldots, a_n)=c$, when $x_1=a_1$; and they have the form

$$u_1=z+(x_1-a_1)v_1=c,$$
$$u_2=x_2+(x_1-a_1)v_2=a_2,$$
$$\ldots\ldots\ldots\ldots\ldots\ldots\ldots\ldots$$
$$u_n=x_n+(x_1-a_1)v_n=a_n,$$

where $v_1, \ldots, v_n$ are regular functions of the variables $x_1, \ldots, x_n, z$. Now the general integral is

$$f(u_1, u_2, \ldots, u_n)=0;$$

or, changing the form of the arbitrary function, we may take

$$u_1=F(u_2, \ldots, u_n)$$

as the integral, where F also is arbitrary. When $x_1=a_1$, this equation becomes

$$z=F(x_2, \ldots, x_n);$$

but the value of z when $x_1=a_1$, is to be $g(x_2, \ldots, x_n)$: and therefore when the arbitrary function is chosen so that

$$F(x_2, \ldots, x_n)=g(x_2, \ldots, x_n),$$

and consequently

$$F(u_2, \ldots, u_n)=g(u_2, \ldots, u_n),$$

we have an integral

$$u_1 = g(u_2, \ldots, u_n),$$

which is the integral in Cauchy's theorem.

Ex. Required the integral of

$$xp + yq = z,$$

which, when $x = a$, is such that $z = \dfrac{y^2}{4c}$.

Two integrals of the ordinary equations

$$\frac{dx}{x} = \frac{dy}{y} = \frac{dz}{z}$$

are taken such that, when $x = a$, we have $y = b$, and $z = \dfrac{b^2}{4c}$; these are easily seen to be

$$u_1 = \frac{az}{x} = \frac{b^2}{4c},$$

$$u_2 = \frac{ay}{x} = b.$$

Thus the general integral of the equation can be taken in the form

$$u_1 = f(u_2),$$

where f is arbitrary. When $x = a$, this equation becomes

$$z = f(y),$$

so that, for the required integral,

$$f(y) = \frac{y^2}{4c};$$

and therefore

$$f(u_2) = \frac{u_2^2}{4c}.$$

Hence the required integral is given by the equation

$$u_1 = \frac{u_2^2}{4c},$$

that is,

$$z = \frac{ay^2}{4cx}.$$

If, instead of taking Cauchy's theorem in its simplest form as associated with an initial value $x_1 = a_1$, we require an integral which, when a relation of the form

$$f(z, x_1, \ldots, x_n) = 0$$

exists among the variables, shall be given by the equation

$$g(z, x_1, \ldots, x_n) = 0,$$

effectively what is required is the determination of the arbitrary functional form F in

$$F(\phi_1, \ldots, \phi_n) = 0,$$

so that the equation may be satisfied without any other relation solely in virtue of $f = 0$, $g = 0$.

As $f = 0$ and $g = 0$ are two relations between $n+1$ quantities, $n-1$ of these can be regarded as independent: or we may regard all the $n+1$ variables as expressible in terms of $n-1$ independent quantities. Taking the latter mode of representing them, let their expressions be

$$z = \psi\ (\xi_1, \ldots, \xi_{n-1}),$$
$$x_r = \psi_r(\xi_1, \ldots, \xi_{n-1}),$$

for $r = 1, \ldots, n$. When these are substituted in the quantities $\phi_1, \ldots, \phi_n$, we have

$$\begin{aligned}\phi_m &= \phi_m(z, x_1, \ldots, x_n)\\ &= \phi_m(\psi, \psi_1, \ldots, \psi_n)\\ &= \bar{\phi}_m(\xi_1, \ldots, \xi_{n-1}) = \bar{\phi}_m \text{ say,}\end{aligned}$$

for $m = 1, \ldots, n$; and these n relations, expressing $\phi_1, \ldots, \phi_n$ in terms of $n-1$ quantities, are satisfied concurrently with the relations $f = 0$, $g = 0$. Among these n relations, let the $n-1$ quantities $\xi_1, \ldots, \xi_{n-1}$ be eliminated, and let the result of the elimination be

$$G(\bar{\phi}_1, \ldots, \bar{\phi}_n) = 0.$$

Now when $f = 0$ and $g = 0$, we have ϕ_m degenerating to $\bar{\phi}_m$; and the general integral becomes

$$F(\bar{\phi}_1, \ldots, \bar{\phi}_n) = 0,$$

which coexists with $f = 0$ and $g = 0$, but, as now it involves only the quantities $\xi_1, \ldots, \xi_{n-1}$, it is satisfied by itself and not in virtue of $f = 0$, $g = 0$. We thus have

$$F(\bar{\phi}_1, \ldots, \bar{\phi}_n) = G(\bar{\phi}_1, \ldots, \bar{\phi}_n),$$

and therefore also

$$F(\phi_1, \ldots, \phi_n) = G(\phi_1, \ldots, \phi_n).$$

Hence the required integral is given by the equation

$$G(\phi_1, \ldots, \phi_n) = 0.$$

Ex. In examples, the details sometimes are developed in a different way. Let it be required to find a surface, satisfying the equation

$$xp+yq=z,$$

and passing through the curve

$$\frac{x^2}{a^2}+\frac{y^2}{b^2}+\frac{z^2}{c^2}=1, \quad lx+my+nz=1.$$

The curve can be expressed in the form

$$x=a\lambda, \quad y=b\mu, \quad z=c\nu,$$

where

$$\lambda^2+\mu^2+\nu^2=1,$$

$$al\lambda+bm\mu+cn\nu=1.$$

Two integrals of the associated ordinary equations are

$$u=\frac{z}{x}, \quad v=\frac{z}{y};$$

hence, along the curve, we have

$$\overline{u}=\frac{c}{a}\frac{\nu}{\lambda}, \quad \overline{v}=\frac{c}{b}\frac{\nu}{\mu},$$

so that

$$\lambda=\frac{c}{a}\frac{\nu}{\overline{u}}, \quad \mu=\frac{c}{b}\frac{\nu}{\overline{v}},$$

whence

$$\nu^2\left(1+\frac{c^2}{a^2}\frac{1}{\overline{u}^2}+\frac{c^2}{b^2}\frac{1}{\overline{v}^2}\right)=1,$$

$$\nu\left(cn+\frac{cl}{\overline{u}}+\frac{cm}{\overline{v}}\right)=1,$$

and therefore

$$1+\frac{c^2}{a^2}\frac{1}{\overline{u}^2}+\frac{c^2}{b^2}\frac{1}{\overline{v}^2}=c^2\left(\frac{l}{\overline{u}}+\frac{m}{\overline{v}}+n\right)^2.$$

This equation corresponds to the equation $G(\overline{\phi}_1, \ldots, \overline{\phi}_n)=0$ in the preceding discussion. In the present case, the required integral is accordingly given by

$$1+\frac{c^2}{a^2u^2}+\frac{c^2}{b^2v^2}=c^2\left(\frac{l}{u}+\frac{m}{v}+n\right)^2;$$

inserting the values of u and v, the equation of the required surface is

$$\frac{x^2}{a^2}+\frac{y^2}{b^2}+\frac{z^2}{c^2}=(lx+my+nz)^2.$$

Complete linear Systems that are homogeneous.

37. Before passing to the discussion of the most general equation of the first order and of degree higher than the first, it is convenient to deal with a system of simultaneous linear equations involving one dependent variable. If the dependent variable occurs explicitly, the equations can be changed, by a transformation as in § 35, so that the new dependent variable does not occur explicitly; the number of independent variables is thus increased by unity. In their transformed expression, the equations are homogeneous in the derivatives; they may be written

$$\left.\begin{array}{l} A_1(u) = a_{11}\dfrac{\partial u}{\partial x_1} + a_{21}\dfrac{\partial u}{\partial x_2} + \ldots + a_{s1}\dfrac{\partial u}{\partial x_s} = 0 \\ A_2(u) = a_{12}\dfrac{\partial u}{\partial x_1} + a_{22}\dfrac{\partial u}{\partial x_2} + \ldots + a_{s2}\dfrac{\partial u}{\partial x_s} = 0 \\ \ldots\ldots\ldots\ldots\ldots\ldots\ldots\ldots\ldots\ldots\ldots\ldots \\ A_\mu(u) = a_{1\mu}\dfrac{\partial u}{\partial x_1} + a_{2\mu}\dfrac{\partial u}{\partial x_2} + \ldots + a_{s\mu}\dfrac{\partial u}{\partial x_s} = 0 \end{array}\right\},$$

where the coefficients a_{mn}, for m and $n = 1, \ldots, s$, are functions of the independent variables $x_1, \ldots, x_s$ alone. We have to investigate the conditions under which an integral (if any) can be possessed by the system; we have also to devise means for the construction of an integral when it is possessed.

Equations of this type have already, in Part I (§§ 38—41) of this work, received some consideration; but there they arose as a class, associated with equations linear and homogeneous in differential elements in the variables, and the limitations imposed upon them were derived from the originating equations. Their importance, not least owing to their frequent occurrence in various theories, justifies an independent treatment; the earlier discussion will render some abbreviation possible.

The μ equations in the propounded system are said to be independent, when no linear relation of the form

$$\xi_1 A_1(u) + \ldots + \xi_\mu A_\mu(u) = 0$$

exists, the quantities $\xi_1, \ldots, \xi_\mu$ being functions of $x_1, \ldots, x_s$, and the quantities $A_1(u), \ldots, A_\mu(u)$ being merely the linear combinations of the derivatives of u. If, however, such a relation or relations should exist, then one or more than one of the equations $A(u) = 0$ would be dependent on the others: an integral of those others would satisfy the dependent equation or equations; and so the dependent equations could be ignored for the present purpose. Accordingly, we shall assume that the equations in the system are independent.

It is clear that, if a linear system in s independent variables contains s equations or more than s equations, the equations can only be satisfied by having

$$\frac{\partial u}{\partial x_1}=0,\quad \frac{\partial u}{\partial x_2}=0,\quad \ldots,\quad \frac{\partial u}{\partial x_s}=0;$$

and then

$$u = \text{constant}$$

is obviously the only integral of the system. Having disposed of systems for which $\mu \geqslant s$, we shall now assume $\mu < s$.

The μ equations are independent; but it may be necessary to associate other equations with them, arising as consequences of their coexistence or as conditions of their coexistence. It is clear that, if the equations

$$A_m(u)=0,\quad A_n(u)=0$$

possess a common integral, it makes the left-hand sides vanish identically; and therefore the equations

$$A_m(A_n u)=0,\quad A_n(A_m u)=0,$$

and so also

$$A_m(A_n u)-A_n(A_m u)=0,$$

are satisfied for that common integral; that is, the last equation coexists with $A_m(u)=0$ and $A_n(u)=0$, when the two latter are members of a linear system. But the new equation is found also to be linear in the first derivatives of u: for the coefficient of $\dfrac{\partial^2 u}{\partial x_k\,\partial x_l}$ in $A_m(A_n u)$ is $a_{km}a_{ln}+a_{lm}a_{kn}$, when k and l are different, and is $a_{lm}a_{ln}$, when k is the same as l; and the coefficient of $\dfrac{\partial^2 u}{\partial x_k\,\partial x_l}$ in $A_n(A_m u)$ is $a_{ln}a_{km}+a_{kn}a_{lm}$, when k and l are different, and is $a_{ln}a_{lm}$, when k is the same as l: thus the derivatives of u of the second order disappear, and only derivatives of the first order remain. The equation is

$$\begin{aligned}0&=A_m(A_n u)-A_n(A_m u)\\&=\sum_{r=1}^{s}\{A_m(a_{rn})-A_n(a_{rm})\}\frac{\partial u}{\partial x_r}.\end{aligned}$$

Now this equation may be evanescent, because the coefficient of each of the derivatives of u vanishes. Or it may be satisfied in virtue of the original set, as a linear combination of them; it then is not a new independent equation, and consequently it need not be taken into further account. Or it may be not evanescent, and not a linear combination of the original equations, and yet it must be satisfied; then it is a new equation, and it must be associated with the system.

Similarly for any pair of equations in the system. Suppose that, by taking all possible pairs, r new equations are obtained so that there is a system

$$A_1(u)=0,\ \ldots,\ A_\mu(u)=0,\quad A_{\mu+1}(u)=0,\ \ldots,\ A_{\mu+r}(u)=0.$$

Again we must take all possible pairs; clearly it will be sufficient to take each of the first μ with each of the last r, and all possible pairs of the last r; all new equations are to be retained. And so on, until the process either provides no new equation or until the number of equations has come to be s. The latter case has been dealt with. When the former case occurs, the number of equations being less than s, the system at that stage is called a *complete linear system.* Manifestly, when there is only one dependent variable and there are several linear equations, we have to deal with complete linear systems. Moreover, the only systems of this type that require consideration are those in which the number of independent equations is less than the number of independent variables.

38. Two properties, possessed by complete linear systems, lead to simplification in the analysis: they must be established.

In the first place, *when a complete system is replaced by another, which is its algebraic equivalent, the new system is complete.* Let a system

$$A_1(u)=0,\ \ldots,\ A_\mu(u)=0,$$

supposed complete, be replaced by a system

$$B_1(u)=0,\ \ldots,\ B_\mu(u)=0,$$

where

$$B_m(u)=\sum_{n=1}^{\mu}\xi_{mn}A_n(u),$$

for $m=1,\ \ldots,\ \mu$, and the quantities ξ_{mn} are functions of the variables $x_1,\ \ldots,\ x_s$ such that their determinant does not vanish.

It is clear that the quantities $A_n(u)$ are expressible as linear combinations of the quantities $B_m(u)$; so that, algebraically, the two systems of equations are equivalent to one another.

To decide whether the new system is complete or not, we construct the quantities $B_m(B_n u) - B_n(B_m u)$; and we have

$$B_m(B_n u) - B_n(B_m u)$$

$$= \sum_{r=1}^{\mu} \left[\xi_{mr} A_r \left\{ \sum_{i=1}^{\mu} \xi_{ni} A_i(u) \right\} \right] - \sum_{i=1}^{\mu} \left[\xi_{ni} A_i \left\{ \sum_{r=1}^{\mu} \xi_{mr} A_r(u) \right\} \right]$$

$$= \sum_{r=1}^{\mu} \sum_{i=1}^{\mu} \xi_{mr} \xi_{ni} A_r(A_i u) + \sum_{r=1}^{\mu} \sum_{i=1}^{\mu} \xi_{mr} A_r(\xi_{ni}) A_i(u)$$

$$- \sum_{r=1}^{\mu} \sum_{i=1}^{\mu} \xi_{mr} \xi_{ni} A_i(A_r u) - \sum_{r=1}^{\mu} \sum_{i=1}^{\mu} \xi_{ni} A_i(\xi_{mr}) A_r(u).$$

Combining the first summations in the two lines, we have $A_i(A_r u) - A_r(A_i u)$ as the coefficient of $\xi_{mr} \xi_{ni}$; this quantity is a linear combination of the quantities $A_1(u), \ldots, A_\mu(u)$, because the system is complete: hence these two summations give a linear combination of the quantities $A(u)$. Each of the other two summations is actually a linear combination of these quantities; hence the whole expression for $B_m(B_n u) - B_n(B_m u)$ is a linear combination of the quantities $A(u)$. Each of the quantities $A(u)$ is a linear combination of the quantities $B(u)$; when the values are substituted, we find that $B_m(B_n u) - B_n(B_m u)$ is a linear combination of the quantities $B(u)$. As this holds for all values of m and n, it follows that the system of equations $B_1(u) = 0, \ldots, B_\mu(u) = 0$ is complete.

In the second place, *a complete system remains complete for any transformation of the independent variables.* Let these variables be transformed by the relations

$$x_r' = f_r(x_1, \ldots, x_s),$$

for $r = 1, \ldots, s$, the functions $f_1, \ldots, f_s$ being independent of one another. Then

$$\frac{\partial u}{\partial x_r} = \frac{\partial u}{\partial x_1'} \frac{\partial f_1}{\partial x_r} + \frac{\partial u}{\partial x_2'} \frac{\partial f_2}{\partial x_r} + \ldots + \frac{\partial u}{\partial x_s'} \frac{\partial f_s}{\partial x_r},$$

for all values of r; substituting in $A_n(u)$ for the quantities $\dfrac{\partial u}{\partial x_r}$, we have

$$A_n(u) = A_n'(u),$$

and $A_n'(u)$ is homogeneous and linear in the derivatives

$$\frac{\partial u}{\partial x_1'}, \ldots, \frac{\partial u}{\partial x_s'}.$$

As there is no linear relation among the quantities $A_n(u)$, there can be none among the quantities $A_n'(u)$: the equations $A'(u)=0$ are independent. Further, the operation A_n is replaced by A_n', having the modified coefficients: thus

$$A_m(A_n u) = A_m(A_n' u) = A_m'(A_n' u),$$
$$A_n(A_m u) = A_n(A_m' u) = A_n'(A_m' u),$$

and therefore

$$\begin{aligned} A_m'(A_n'u) - A_n'(A_m'u) &= A_m(A_n u) - A_n(A_m u) \\ &= \text{linear combination of } A_1(u), \ldots, A_\mu(u), \\ &= \ldots\ldots\ldots\ldots\ldots\ldots\ldots\ldots \; A_1'(u), \ldots, A_\mu'(u), \end{aligned}$$

for all values of m and n. Hence the system of equations $A_1'(u)=0, \ldots, A_\mu'(u)=0$ is complete.

39. The first of these properties is used to express a complete linear system in a canonical form: the second of them will be used in the establishment of the existence-theorem.

As regards the expression in a canonical form, let a complete linear system of m equations be given, involving one dependent variable u implicitly through its derivatives and $m+n$ independent variables $x_1, \ldots, x_{m+n}$. As the m equations are independent of one another, they can be resolved algebraically so as to express m of the derivatives of u, say $\frac{\partial u}{\partial x_1}, \ldots, \frac{\partial u}{\partial x_m}$, linearly in terms of the remainder; let their expression be

$$B_t(u) = \frac{\partial u}{\partial x_t} + \sum_{s=m+1}^{m+n} U_{st} \frac{\partial u}{\partial x_s} = 0,$$

for $t = 1, \ldots, m$.

The system was complete in its earlier expression: hence, by the preceding property, it remains complete in the changed expression; consequently

$$B_i(B_j u) - B_j(B_i u) = \sum_{k=1}^{m} \xi_k B_k(u),$$

where the quantities ξ do not involve u or its derivatives. The left-hand side of this relation is

$$\sum_{s=m+1}^{m+n} \{B_i(U_{sj}) - B_j(U_{si})\} \frac{\partial u}{\partial x_s},$$

and it does not contain any of the m derivatives $\frac{\partial u}{\partial x_1}, \ldots, \frac{\partial u}{\partial x_m}$: whereas the right-hand side does contain a derivative $\frac{\partial u}{\partial x_t}$ unless ξ_t is zero. Hence, in order that the relation may be satisfied, each of the quantities $\xi_1, \ldots, \xi_m$ is zero; and it then becomes

$$\sum_{s=m+1}^{m+n} \{B_i(U_{sj}) - B_j(U_{si})\} \frac{\partial u}{\partial x_s} = 0.$$

Now the system is complete, so that no equation of this type is to be associated with it which is not satisfied in virtue of $B_1(u) = 0$, $\ldots$, $B_m(u) = 0$; consequently, this equation must be evanescent for all values of i and j, and therefore

$$B_i(U_{sj}) - B_j(U_{si}) = 0.$$

This relation involves the independent variables only; hence it must be satisfied identically, for all values of i, j, and s.

Conversely, if this relation be satisfied for all values of i, j, and s, then we have

$$B_i(B_j u) - B_j(B_i u) = 0;$$

and the system of equations $B_1(u) = 0, \ldots, B_m(u) = 0$ is evidently complete. Hence we have the formal result:—

A complete linear system of m equations, involving one dependent variable u and $m+n$ independent variables $x_1, \ldots, x_{m+n}$, and such that only derivatives of u occur, the equations being homogeneous in those derivatives, can be expressed in the form

$$B_t(u) = \frac{\partial u}{\partial x_t} + \sum_{s=m+1}^{m+n} U_{st} \frac{\partial u}{\partial x_s} = 0,$$

for $t = 1, \ldots, m$; and the conditions, necessary and sufficient to secure that the system should be complete, are the aggregate of the $\frac{1}{2}m(m-1)n$ relations

$$B_i(U_{sj}) - B_j(U_{si}) = 0,$$

for the values of s, and for the combinations of i and j: each of these relations must be satisfied identically.

In consequence of the conditions, the equation

$$B_i(B_j u) - B_j(B_i u) = 0$$

is satisfied identically, for all values of i and j. A set of equations possessing this property is frequently said to be *in involution.*

A complete linear system, expressed in the above form, is sometimes called a *Jacobian system.*

40. The preceding investigation gives the formal conditions for the coexistence of the equations: it gives no information as to the integral or integrals (if any) of those equations. An existence-theorem, similar to those in the preceding chapters, is as follows:

Let $a_1, \ldots, a_{m+n}$ be a set of values of the independent variables in the vicinity of which all the coefficients U, in the complete Jacobian system

$$B_t(u) = \frac{\partial u}{\partial x_t} + \sum_{s=m+1}^{m+n} U_{st} \frac{\partial u}{\partial x_s} = 0, \qquad (t = 1, \ldots, m),$$

are regular functions; then the system possesses n functionally distinct integrals, which are regular functions in the vicinity of the selected values and which reduce respectively to values $x_{m+1}, \ldots, x_{m+n}$, when $x_1 = a_1, x_2 = a_2, \ldots, x_m = a_m$.

The theorem has been established* when $m = 1$. The inductive method will be used for the general case; and we shall prove that it is true for a Jacobian system of m equations in $m + n$ independent variables, if it is true for a Jacobian system of $m - 1$ equations in $m + n - 1$ independent variables.

Accordingly, we make the latter supposition that the theorem is true for a complete Jacobian system of $m - 1$ equations in $m + n - 1$ variables. For brevity, we make $a_1 = 0, \ldots, a_{m+n} = 0$: all that would be necessary to secure this result would be to take $y_\mu = x_\mu - a_\mu$.

The equation

$$B_1(u) = \frac{\partial u}{\partial x_1} + \sum_{s=m+1}^{m+n} U_{s1} \frac{\partial u}{\partial x_s} = 0$$

possesses $m + n - 1$ functionally independent integrals, which are regular functions of the variables in finite fields of variation round $0, \ldots, 0$, and which acquire values $x_2, x_3, \ldots, x_{m+n}$ respectively at that place; this is a theorem already proved (§ 29). Of these integrals, $m - 1$ clearly are given by

$$u = x_2, x_3, \ldots, x_m$$

respectively; let the remainder be denoted by $u = y_{m+1}, \ldots, y_{m+n}$ respectively, where

$$y_{m+s} = x_{m+s} + x_1 R_{m+s}, \qquad (s = 1, \ldots, n),$$

* In § 29, Corollary.

R_{m+s} denoting a regular function of the variables $x_1, \ldots, x_{m+n}$ in the assigned vicinity. Reversing these equations so as to express $x_{m+1}, \ldots, x_{m+n}$, we have

$$x_{m+s} = y_{m+s} + x_1 P_{m+s},$$

where P_{m+s} is a regular function of the variables $x_1, \ldots, x_m, y_{m+1}, \ldots, y_{m+n}$ in the vicinity of $0, \ldots, 0$.

Now let the independent variables be changed from $x_1, x_2, \ldots, x_{m+n}$ to $x_1, x_2, \ldots, x_m, y_{m+1}, \ldots, y_{m+n}$; we know, from the property established in § 38, that the new system of equations is complete. Also let the result of the transformation on any integral u be denoted by v. The effect of the transformation upon $B_1(u) = 0$ can be obtained at once: as its $m + n - 1$ functionally independent integrals now are $x_2, \ldots, x_m, y_{m+1}, \ldots, y_{m+n}$, which are the aggregate of independent variables other than x_1, we have

$$B_1(v) = \frac{\partial v}{\partial x_1} = 0.$$

For the other equations, we have

$$\frac{\partial u}{\partial x_i} = \frac{\partial v}{\partial x_i} + \sum_{s=1}^{n} \frac{\partial v}{\partial y_{m+s}} \frac{\partial y_{m+s}}{\partial x_i},$$

for $i = 2, \ldots, m$, and

$$\frac{\partial u}{\partial x_{m+j}} = \sum_{s=1}^{n} \frac{\partial v}{\partial y_{m+s}} \frac{\partial y_{m+s}}{\partial x_{m+j}},$$

for $j = 1, \ldots, n$; hence the equation $B_t(u) = 0$ becomes

$$B_t(v) = \frac{\partial v}{\partial x_t} + \sum_{s=m+1}^{m+n} V_{st} \frac{\partial v}{\partial y_s} = 0,$$

with new coefficients V_{st}.

The properties of these coefficients could be deduced from those of the coefficients U_{st}: they are most simply deduced by the use of the known property that the new system $B_1(v) = 0, \ldots, B_m(v) = 0$ is complete. On account of this property possessed by a Jacobian system (it will be noticed that the new system has the form of a Jacobian system), we have

$$B_1(V_{sj}) - B_j(V_{s1}) = 0,$$

for all values of s and j. Now all the coefficients V_{s1} are zero, and B_1 is $\frac{\partial}{\partial x_1}$; hence the foregoing condition is

$$\frac{\partial V_{sj}}{\partial x_1} = 0,$$

that is, the coefficients V_{sj} do not involve x_1. We also have

$$B_i(V_{sj}) - B_j(V_{si}) = 0,$$

for all values of i and j in pairs combined from $2, \ldots, m$, and for $s = m+1, \ldots, m+n$. The modified Jacobian system is

$$B_1(v) = \frac{\partial v}{\partial x_1} = 0,$$

$$B_t(v) = \frac{\partial v}{\partial x_t} + \sum_{s=m+1}^{n} V_{st} \frac{\partial v}{\partial y_s} = 0,$$

for $t = 2, \ldots, m$.

Now the last $m-1$ equations constitute a complete Jacobian system, for the necessary and sufficient conditions

$$B_i(V_{sj}) - B_j(V_{si}) = 0$$

are satisfied; and they are a system in $m+n-1$ independent variables $x_2, \ldots, x_m, y_{m+1}, \ldots, y_{m+n}$, the variable x_1 not occurring. Owing to $B_1(v) = 0$, it follows that an integral of the system of m equations cannot involve x_1 in the modified set of variables: consequently, every integral of the system of m equations in the $m+n$ independent variables is an integral of the system of $m-1$ equations in $m+n-1$ independent variables, and conversely.

The coefficients V_{st} in the Jacobian system of $m-1$ equations are regular functions of the variables in the vicinity of $x_2, \ldots, x_m, y_{m+1}, \ldots, y_{m+n} = 0, \ldots, 0$; for they are polynomial combinations of the coefficients U_{st} and of the derivatives of $y_{m+1}, \ldots, y_{m+n}$ with respect to the original variables, all of which are regular in the assigned vicinity. By the hypothesis adopted for the systems of $m-1$ equations, the Jacobian system of $m-1$ equations in the $m+n-1$ variables possesses n functionally independent integrals which are regular functions of the variables in the domain considered and which reduce respectively to $y_{m+1}, \ldots, y_{m+n}$, when $x_2 = 0, \ldots, x_m = 0$; let these integrals be

$$v_s = y_{m+s} + \phi_{m+s}, \qquad (s = 1, \ldots, n),$$

where ϕ_{m+s} is a regular function of the variables which vanishes, when $x_2 = 0, \ldots, x_m = 0$. It is clear that no one of the quantities $v_1, \ldots, v_n$ contains x_1, so that each of them satisfies

$$B_1(v) = \frac{\partial v}{\partial x_1} = 0.$$

Consequently, they are integrals of the Jacobian system of m equations.

Moreover, these integrals satisfy the assigned conditions; for we have

$$v_s = y_{m+s} + \phi_{m+s}$$
$$= x_{m+s} + x_1 R_{m+s} + \phi_{m+s},$$

so that as ϕ_{m+s} is still a regular function vanishing when $x_2 = 0$, ..., $x_m = 0$, the integral v_s reduces to x_{m+s}, when we revert to the original variables and we make $x_1 = 0$, $x_2 = 0$, ..., $x_m = 0$.

The theorem is thus true for a complete Jacobian system of m equations in $m+n$ variables, if it is true for such a system of $m-1$ equations in $m+n-1$ variables. It is known to be true for a single equation in any number of variables: hence it is true generally.

The existence of n functionally independent integrals has thus been established. When $m=1$, it is known that an equation in $n+1$ independent variables possesses n, and not more than n, such integrals; the course of the preceding argument then shews that *a complete Jacobian system of m equations in $m+n$ variables possesses n, and not more than n, functionally independent integrals.*

41. The set of integrals, determined in association with the assigned conditions of § 40 and reducing to $x_{m+1}, \ldots, x_{m+n}$ for assigned values of $x_1, \ldots, x_m$, is sometimes called a *fundamental system* for the assigned vicinity.

As in the case of a single equation, it can be proved that any integral can be expressed in terms of any set of n functionally independent integrals: and, in particular, the expression in terms of the members of a fundamental system is simple.

To prove the first of these statements, let $u_1, \ldots, u_n$ denote a set of functionally independent integrals of a Jacobian system of m equations in $m+n$ independent variables; so that, with the preceding notation for the system, the equations

$$B_t(u_r) = \frac{\partial u_r}{\partial x_t} + \sum_{s=m+1}^{m+n} U_{st} \frac{\partial u_r}{\partial x_s} = 0,$$

for $r=1, \ldots, n$, and $t=1, \ldots, m$, are satisfied. Moreover, they are satisfied identically, because the quantities $u_1, \ldots, u_n$ do not occur explicitly.

Now, let

$$u = \psi(x_1, \ldots, x_{m+n})$$

denote any integral of the Jacobian system: then the equations

$$\frac{\partial\psi}{\partial x_t} + \sum_{s=m+1}^{m+n} U_{st}\frac{\partial\psi}{\partial x_s} = 0,$$

for $t = 1, \ldots, m$, are satisfied; and they are satisfied identically, because ψ does not occur explicitly. Hence all the determinants

$$\left\|\begin{matrix} \dfrac{\partial\psi}{\partial x_1}, \ldots, & \dfrac{\partial\psi}{\partial x_m}, & \dfrac{\partial\psi}{\partial x_{m+1}}, \ldots, & \dfrac{\partial\psi}{\partial x_{m+n}} \\ \dfrac{\partial u_1}{\partial x_1}, \ldots, & \dfrac{\partial u_1}{\partial x_m}, & \dfrac{\partial u_1}{\partial x_{m+1}}, \ldots, & \dfrac{\partial u_1}{\partial x_{m+n}} \\ \cdots & \cdots & \cdots & \cdots \\ \dfrac{\partial u_n}{\partial x_1}, \ldots, & \dfrac{\partial u_n}{\partial x_m}, & \dfrac{\partial u_n}{\partial x_{m+1}}, \ldots, & \dfrac{\partial u_n}{\partial x_{m+n}} \end{matrix}\right\|$$

vanish, in association with the integral

$$u = \psi(x_1, \ldots, x_{m+n});$$

but u does not occur explicitly so that, as the determinants cannot vanish in virtue of that integral equation, they must vanish identically. Accordingly, some functional relation must exist among the quantities $\psi, u_1, \ldots, u_n$; it cannot exist among $u_1, \ldots, u_n$ alone, for these are independent; and therefore it must involve ψ. Hence it may be taken in the form

$$\psi = f(u_1, \ldots, u_n),$$

where f is some function of its arguments.

We thus have a verification of the proposition that the number of functionally independent integrals in a Jacobian system of m equations in $m + n$ variables is exactly n.

The form of the function f is not difficult to obtain when ψ is given, if we take a fundamental system of integrals for the set $u_1, \ldots, u_n$. In particular, let the latter be $v_1, \ldots, v_n$, where v_s (for $s = 1, \ldots, n$) assumes the value x_{m+s}, when $x_1 = a_1, \ldots, x_m = a_m$; it is required to determine the functional form g, such that

$$\psi(x_1, \ldots, x_m, x_{m+1}, \ldots, x_{m+n}) = g(v_1, \ldots, v_n).$$

Let $x_1 = a_1, \ldots, x_m = a_m$; this equation becomes

$$\psi(a_1, \ldots, a_m, x_{m+1}, \ldots, x_{m+n}) = g(x_{m+1}, \ldots, x_{m+n}).$$

This is true for all variables independent of one another; and therefore

$$\psi(a_1, \ldots, a_m, v_1, \ldots, v_n) = g(v_1, \ldots, v_n),$$

so that

$$\psi(x_1, \ldots, x_m, x_{m+1}, \ldots, x_{m+n}) = \psi(a_1, \ldots, a_m, v_1, \ldots, v_n),$$

being the required expression in terms of a fundamental set of integrals.

Corollary. When the preceding results are combined, the following existence-theorem is obvious:—

Let $a_1, \ldots, a_{m+n}$ be a set of values of $x_1, \ldots, x_{m+n}$ such that, in their vicinity, all the coefficients U in the complete Jacobian system

$$\frac{\partial u}{\partial x_t} + \sum_{s=m+1}^{m+n} U_{st} \frac{\partial u}{\partial x_s} = 0,$$

for $t = 1, \ldots, m$, are regular functions of the variables; and let $h(x_{m+1}, \ldots, x_{m+n})$ denote any regular function of its arguments in the assigned region of variation, which (except for the requirement of being regular) is arbitrary. Then an integral of the Jacobian system exists, which is a regular function of the variables in the vicinity of $a_1, \ldots, a_{m+n}$, and which acquires the value $h(x_{m+1}, \ldots, x_{m+n})$, when $x_1 = a_1, \ldots, x_m = a_m$.

42. It is a part of Cauchy's existence-theorem that an integral satisfying the conditions:

(i) that it is a regular function of the variables within the domain of a set of values where all coefficients in the above linear equation are regular,

(ii) that it acquires the value of an assigned regular function for an initial value of one of the variables,

is a unique integral so determined. Hence the fundamental system of integrals of the equation

$$p_1 + X_2 p_2 + \ldots + X_{n+1} p_{n+1} = 0,$$

required to acquire values $x_2, \ldots, x_{n+1}$ respectively when $x_1 = a_1$, and to be regular functions of the variables, is unique as a set of integrals.

The inductive proof of the establishment of integrals of a Jacobian system shews that, if a set of integrals satisfying the assigned conditions be unique for a Jacobian system of $m-1$ equations, a set of integrals satisfying the assigned conditions is unique for a Jacobian system of m equations. The proposition just quoted indicates that a fundamental system is unique when there is a single equation: hence a fundamental set of integrals is unique for a Jacobian system.

Similarly, the integral at the end of § 41, defined as an integral of the Jacobian system

$$\frac{\partial u}{\partial x_t} + \sum_{s=m+1}^{m+n} U_{st} \frac{\partial u}{\partial x_s} = 0,$$

for $t = 1, \ldots, m$, which is a regular function of the variables and acquires the value of an assigned regular function of $x_{m+1}, \ldots, x_{m+n}$ for initial values of $x_1, \ldots, x_m$, is easily seen to be a unique integral determined by those conditions.

The property of uniqueness of the integrals is thus established in connection with the various existence-theorems belonging to the Jacobian systems. But it must be remembered that the selected initial values of the variables are such that all the coefficients U_{st} are regular in their vicinity: and only on this hypothesis have the theorems been established. Separate investigation is necessary for the consideration of integrals (if any) of the system in the vicinity of a set of selected initial values of the variables, which constitute a singularity or other non-regular place of any of the coefficients.

Two Methods of Integration of complete linear systems.

43. Now that the existence of integrals of a Jacobian system has been established and that the character of the conditions which limit an integral has been indicated, it is desirable to have some means of actually constructing the integral, more especially if there should be an integral which is expressible in finite terms. Two methods seem more direct for this purpose than others: one of these is due to Mayer, the other is based upon the actual stages in the establishment of the existence of the integrals.

Mayer's method has already* been expounded: consequently the discussion need not be repeated, but the results will be restated for convenience. It is as follows:—

* Vol. I of this work, §§ 41, 42.

To obtain a set of n independent integrals of the complete Jacobian system

$$\frac{\partial u}{\partial x_t} + \sum_{s=m+1}^{m+n} U_{st} \frac{\partial u}{\partial x_s} = 0,$$

for $t = 1, \ldots, m$, *we transform the variables* $x_1, \ldots, x_m$ *by the substitutions*

$$x_t = \alpha_t + (y_1 - \theta) f_t(y_1, \ldots, y_m),$$

and construct the equation

$$\frac{\partial u}{\partial y_1} + \sum_{s=m+1}^{m+n} Y_s \frac{\partial u}{\partial x_s} = 0.$$

The equations subsidiary to this single equation, viz.

$$dy_1 = \frac{dx_{m+1}}{Y_{m+1}} = \ldots = \frac{dx_{m+n}}{Y_{m+n}},$$

are to be integrated, keeping $y_2, \ldots, y_m$ *as invariable quantities: let the n integrals be*

$$\phi_p(x_{m+1}, \ldots, x_{m+n}, y_1, \ldots, y_m) = \text{constant},$$

for $p = 1, \ldots, n$. *Then the set of n independent integrals of the Jacobian system are given by the following process: in the equations*

$$\phi_p(x_{m+1}, \ldots, x_{m+n}, y_1, y_2, \ldots, y_m) = \phi_p(c_1, \ldots, c_n, \theta, y_2, \ldots, y_m),$$

the variables $y_1, \ldots, y_m$ *are to be replaced by their values in terms of the variables* $x_1, \ldots, x_m$, *and if, in any of the equations,* $\phi_p(c_1, \ldots, c_n, \theta, y_2, \ldots, y_m)$ *should be a pure constant, the changed equation is*

$$\phi_p(x_{m+1}, \ldots, x_{m+n}, x_1, \ldots, x_m) = \phi_p(c_1, \ldots, c_n, \alpha_1, \ldots, \alpha_m).$$

These n equations are resolved so as to give $c_1, \ldots, c_n$ *(or n independent functional combinations of them): let the result of the resolution be*

$$u_\mu(x_1, \ldots, x_m, x_{m+1}, \ldots, x_{m+n}) = C_\mu, \qquad (\mu = 1, \ldots, n),$$

where $C_1, \ldots, C_n$ *are n independent functions of* $c_1, \ldots, c_n$: *the n integrals of the original system are*

$$u = u_\mu(x_1, \ldots, x_m, x_{m+1}, \ldots, x_{m+n}), \qquad (\mu = 1, \ldots, n).$$

Note 1. The simplest substitutions for the transformation of the variables appear to be

$$x_1 = y_1,$$
$$x_t = \alpha_t + (y_1 - \alpha_1) y_t,$$

for $t=2, \ldots, m$; the quantities $Y_{m+1}, \ldots, Y_{m+n}$ in the subsidiary equations are given by

$$Y_{m+s} = U_{s1} + \sum_{t=2}^{m} U_{st}\, y_t.$$

Note 2. If only a single integral of the Jacobian system is wanted and not a full set, it can certainly be obtained from any one integral of the subsidiary system.

Note 3. If any integrals of a Jacobian system are known, they can be used to modify the system, so as to reduce the amount of integration necessary to complete the set. Thus let

$$u = y_1, \quad u = y_2, \quad \ldots, \quad u = y_p,$$

be known integrals independent of one another, where $p < n$; and use these p quantities to change the variables from $x_1, \ldots, x_{m+n}$ to (say) $x_1, \ldots, x_{m+n-p}, y_1, \ldots, y_p$. Then as y_1 is an integral of the system, the term $\frac{\partial u}{\partial y_1}$ must be absent from each equation of the modified system: its coefficient must vanish in order that the equation may be satisfied. Similarly for $\frac{\partial u}{\partial y_2}, \ldots, \frac{\partial u}{\partial y_p}$. Thus there will be a modified system of m equations: the variables $y_1, \ldots, y_p$ are of the nature of parameters: it involves $m+n-p$ variables $x_1, \ldots, x_{m+n-p}$; and it still is complete. It therefore possesses $n-p$ integrals; and these can be obtained, as in Mayer's method, by the integration of the $n-p$ subsidiary ordinary equations.

44. In outline, and as regards the theoretic amount of inverse operations (such as integration) that are required, Mayer's method for the integration of complete linear systems is the simplest and the briefest: but occasionally, for particular systems, the detailed operations can be complicated. An alternative method of proceeding is provided by an adaptation of Jacobi's method of integrating partial differential equations; the details of the adaptation are almost dictated by the course of the proof of the existence-theorem. In details, it frequently is simpler than Mayer's method, though the number of inverse operations is greater: but the mere number of such operations, without regard paid to their intrinsic difficulty, is not the only trustworthy criterion of practical simplicity.

The method may be described as that of successive reduction. Let the system be taken in its canonical form, the first equation being

$$B_1(u) = \frac{\partial u}{\partial x_1} + \sum_{s=m+1}^{m+n} U_{s1} \frac{\partial u}{\partial x_s} = 0.$$

This equation in $m+n$ variables has $m+n-1$ functionally independent integrals; of these, $m-1$ are evidently given by $x_2, \ldots, x_m$, and the remaining n are provided by integrals of the subsidiary equations

$$dx_1 = \frac{dx_{m+1}}{U_{m+1,1}} = \ldots = \frac{dx_{m+n}}{U_{m+n,1}},$$

the quantities $x_2, \ldots, x_m$ being regarded as parametric. If the integrals of these subsidiary equations are

$$u_\mu(x_1, \ldots, x_{m+n}) = \text{constant}, \qquad (\mu = 1, \ldots, n),$$

then the n remaining integrals of $B_1(u) = 0$ are given by

$$u = u_\mu(x_1, \ldots, x_{m+n}) = u_\mu.$$

Every integral of $B_1(u) = 0$ is a functional combination of $x_2, \ldots, x_m, u_1, \ldots, u_n$; the appropriate functional combinations must be such as to satisfy the remaining equations of the system.

We accordingly make $x_2, \ldots, x_m, u_1, \ldots, u_n$ the independent variables for the equation $B_2(u) = 0$. If any integral of this equation be taken in the form

$$u = f(x_2, \ldots, x_m, u_1, \ldots, u_n),$$

which is also an integral of $B_1(u) = 0$, we have

$$B_2(u) = \frac{\partial f}{\partial x_2} + \sum_{r=1}^{n} \frac{\partial f}{\partial u_r} B_2(u_r) = 0.$$

Because the system is complete, we have

$$B_1(B_2 u_r) - B_2(B_1 u_r) = 0;$$

but $B_1(u_r)$ vanishes identically, so that $B_2(B_1 u_r) = 0$, and therefore

$$B_1(B_2 u_r) = 0.$$

Hence $B_2(u_r)$ satisfies the equation $B_1(u) = 0$; it may be zero, or it may be a pure constant: if it is neither of these but is variable, it is an integral of $B_1(u) = 0$, and therefore can be expressed in terms of $x_2, \ldots, x_m, u_1, \ldots, u_n$. Thus all the coefficients in the transformed expression of $B_2(u) = 0$ are functions of the $m+n-1$

new variables alone. The equation in this form has $m+n-2$ functionally independent integrals; of these, $m-2$ are given by $x_3, \ldots, x_m$; and the remaining n are provided by integrals of the subsidiary equations

$$dx_2 = \frac{du_1}{B_2(u_1)} = \ldots = \frac{du_n}{B_2(u_n)},$$

the quantities $x_3, \ldots, x_m$ being regarded as parametric. All the denominators, if not zero or pure constants, are functions of $x_2, \ldots, x_m, u_1, \ldots, u_n$; let the integrals of this set be

$$v_\rho(x_2, \ldots, x_m, u_1, \ldots, u_n) = \text{constant}, \qquad (\rho = 1, \ldots, n);$$

then the n remaining integrals of $B_2(u) = 0$ are

$$u = v_\rho(x_2, \ldots, x_m, u_1, \ldots, u_n) = v_\rho.$$

Each of these, as a functional combination of $x_2, \ldots, x_m, u_1, \ldots, u_n$, is an integral of $B_1(u) = 0$; and every integral, common to $B_1(u) = 0$ and $B_2(u) = 0$, is a functional combination of $x_3, \ldots, x_m, v_1, \ldots, v_n$. The appropriate functional combinations must be chosen so as to satisfy the remaining equations of the Jacobian system.

We now proceed as before: and for the third equation, we make $x_3, \ldots, x_m, v_1, \ldots, v_n$ the independent variables. If any integral of the equation $B_3(u) = 0$ be

$$u = \phi(x_3, \ldots, x_m, v_1, \ldots, v_n),$$

we have

$$B_3(u) = \frac{\partial\phi}{\partial x_3} + \sum_{r=1}^{n} \frac{\partial\phi}{\partial v_r} B_3(v_r) = 0.$$

But, as the system is complete, we have

$$B_1(B_3 v_r) = B_3(B_1 v_r) = 0,$$
$$B_2(B_3 v_r) = B_3(B_2 v_r) = 0,$$

because $B_1(v_r)$ and $B_2(v_r)$ vanish identically; therefore $B_3(v_r)$ is a simultaneous integral of $B_1(u) = 0$ and $B_2(u) = 0$. Consequently $B_3(v_r)$ is either zero, or a pure constant, or a function of $x_3, \ldots, x_m, v_1, \ldots, v_n$; and the coefficients in the modified form of $B_3(u) = 0$ involve only the variables which occur in the derivatives of ϕ. The position is now the same as in the preceding stage, except that the number of variables has been decreased by unity.

We pass thus from stage to stage: the integrals at the last stage are n functionally independent integrals of the system.

Note. At first sight, it would appear as though the number of quadratures of ordinary equations, required to make the process effective, is mn, being n for each stage. But the number can be reduced, often very substantially, except at the last stage when n such quadratures are then certainly required. For example, let

$$u'(x_1, \ldots, x_{m+n}) = \text{constant}$$

be any integral of the subsidiary system of $B_1(u)=0$: then $u = u'(x_1, \ldots, x_{m+n}) = u'$ is an integral of $B_1(u)=0$. Now for any value of p, we have

$$\begin{aligned} B_1(B_p u') &= B_p(B_1 u') \\ &= 0, \end{aligned}$$

because $B_1(u')$ vanishes identically: hence $B_p(u')$ satisfies $B_1(u)=0$. If $B_p(u')$ is not zero and is not a pure constant, it is an integral of $B_1(u)=0$; if it is functionally independent of u', we may write

$$u'' = B_p(u');$$

and we thus obtain a new integral of $B_1(u)=0$ without any further quadrature, in the case of each operator B_p that leads to a result of this type.

Again, each new integral so obtained may be similarly treated, until possibly an adequate number of integrals has been obtained at the stage. The reduction in the number of quadratures may thus be made by means of the operators in the remaining equations of the system at any stage: it clearly cannot be made at the last stage when no further operator remains for consideration.

Further, if $B_p(u')$ is zero, then u' is an integral common to $B_1(u)=0$ and $B_p(u)=0$; when retained as a new independent variable under transformation of the variables, the integration of $B_p(u)=0$ will be thereby simplified.

Again, if $B_p(u')$ be a pure constant, $=a$ say, and if $B_p(v')$, derived from a functionally distinct integral of $B_1(u)=0$, be also a pure constant, $=b$ say, then

$$B_p(bu' - av') = 0,$$

that is, $bu' - av'$ is an integral common to $B_1(u)=0$ and $B_p(u)=0$; it can be used to simplify the integration of $B_p(u)=0$ at the appropriate stage.

Thus the number of quadratures necessary for the method may be considerably reduced: but even in the most favourable circumstances, their number is greater than the number in Mayer's method.

Ex. 1. As an example, which will be integrated by both methods, consider the system

$$X_1(z)=x_1p_1-x_2p_2+x_3p_3-x_4p_4=0,$$
$$X_2(z)=x_3p_1-x_1p_3=0,$$
$$X_3(z)=x_4p_2-x_2p_4=0,$$

where $p_\mu=\frac{\partial z}{\partial x_\mu}$, for $\mu=1, 2, 3, 4$. We have

$$X_1(X_2 z)-X_2(X_1 z)=0,$$
$$X_1(X_3 z)-X_3(X_1 z)=0,$$
$$X_2(X_3 z)-X_3(X_2 z)=0,$$

so that the system is a complete linear system, being a system in involution. When expressed in a canonical form, it is

$$\xi_1(z)=p_1-\frac{x_1}{x_4}\frac{x_2^2+x_4^2}{x_1^2+x_3^2}p_4=0,$$

$$\xi_2(z)=p_2-\frac{x_2}{x_4}p_4=0,$$

$$\xi_3(z)=p_3-\frac{x_3}{x_4}\frac{x_2^2+x_4^2}{x_1^2+x_3^2}p_4=0.$$

Adopting Mayer's method of integration, we make the transformations

$$x_1=y_1, \quad x_2=a_2+y_1y_2, \quad x_3=a_3+y_1y_3;$$

and then the single equation to be considered is

$$\frac{\partial z}{\partial y_1}-Y\frac{\partial z}{\partial x_4}=0,$$

where

$$Y=\frac{y_1+y_3(a_3+y_1y_3)}{x_4}\frac{(a_2+y_1y_2)^2+x_4^2}{y_1^2+(a_3+y_1y_3)^2}+y_2\frac{a_2+y_1y_2}{x_4}.$$

The subsidiary equation is

$$dy_1+\frac{dx_4}{Y}=0;$$

and an integral is found to be

$$\{y_1^2+(a_3+y_1y_3)^2\}\{(a_2+y_1y_2)^2+x_4^2\}=\text{constant}.$$

Accordingly, by the theorem quoted in § 43, we construct the equation

$$\{y_1^2+(a_3+y_1y_3)^2\}\{(a_2+y_1y_2)^2+x_4^2\}=a_2^2a_3^2,$$

the right-hand side being obtained by putting y_1 equal to zero in the left; and then, replacing the variables x_1, x_2, x_3, we have

$$(x_1^2+x_3^2)(x_2^2+x_4^2)=a_2^2a_3^2,$$

that is, by the theorem, a common integral is

$$z=(x_1^2+x_3^2)(x_2^2+x_4^2).$$

A more general common integral is

$$z=F\{(x_1^2+x_3^2)(x_2^2+x_4^2)\},$$

where F is any arbitrary function.

Proceeding by the other method of integration, we obtain an integral, other than x_2 and x_3, of

$$p_1-\frac{x_1}{x_4}\frac{x_2^2+x_4^2}{x_1^2+x_3^2}p_4=0:$$

the subsidiary system is

$$dx_1+\frac{dx_4}{\dfrac{x_1}{x_4}\dfrac{x_2^2+x_4^2}{x_1^2+x_3^2}}=0,$$

an integral of which is

$$(x_2^2+x_4^2)(x_1^2+x_3^2)=\text{constant}.$$

We take x_2, x_3, v as the independent variables, where

$$v=(x_2^2+x_4^2)(x_1^2+x_3^2).$$

But $\xi_2(v)=0$, so that the second equation becomes

$$\frac{\partial z}{\partial x_2}=0:$$

and any integral common to the first two equations is a function of x_3 and v.

We take x_3 and v as the independent variables for the third equation. But $\xi_3(v)=0$, so that the third equation becomes

$$\frac{\partial z}{\partial x_3}=0.$$

The integral is thus a function of v; a common integral of the system is, as before,

$$z=v=(x_1^2+x_3^2)(x_2^2+x_4^2).$$

If an integral is required to attain an assigned value $g(x_4)$, when $x_1=a$, $x_2=b$, $x_3=c$, it is easily seen to be

$$z=g\left\{\left(\frac{v}{a^2+c^2}-b^2\right)^{\frac{1}{2}}\right\}.$$

Ex. 2. Prove that the system

$$0=p_1+2x_1p_2+3x_2p_3+p_5+x_4p_6,$$

$$0=x_1p_1+2x_2p_2+3x_3p_3+x_5p_5+x_6p_6,$$

$$0=(3x_1^2-2x_2)p_1+(3x_1x_2-x_3)p_2+3x_1x_3p_3+(x_4x_5-x_6)p_4+x_5^2p_5+x_5x_6p_6,$$

is complete: and find a system of three common integrals.

Complete linear Systems that are not homogeneous.

45. The complete linear systems that have been considered are homogeneous in the derivatives of z: and the dependent variable does not explicitly occur. But it is possible to have complete linear systems which are not homogeneous in the derivatives and in which the dependent variable does occur explicitly. This class of equations is a very special example of a system of simultaneous equations and can be treated by the general method devised for general systems: the equations can, however, be more simply treated by being included under the class already considered. We take a new dependent variable u such that

$$u = u(z, x_1, \ldots, x_n),$$

and we transform the equations by means of relations

$$\frac{\partial u}{\partial z} p_m + \frac{\partial u}{\partial x_m} = 0:$$

the transformed equations are homogeneous and u does not occur explicitly. These are amenable to the method already explained: the conditions of coexistence are at once obtainable; and integrals will be given by equations

$$u = \text{constant},$$

provided u involves z.

Note. The same warning must be applied about linear non-homogeneous systems as was applied to a single non-homogeneous equation (§ 35). The method does not necessarily give all the integrals of such a system, for it may fail to give those which belong to the residuary class called *special*.

46. The conditions of the coexistence and the completeness of the system can be easily obtained from the transformed system. Thus let a given linear system be expressed in the form

$$E_1(z) = p_1 + \sum_{s=m+1}^{m+n} a_{s1} p_s \;= Z_1,$$

$$E_2(z) = p_2 + \sum_{s=m+1}^{m+n} a_{s2} p_s \;= Z_2,$$

$$\cdots\cdots\cdots\cdots\cdots\cdots\cdots\cdots\cdots\cdots$$

$$E_m(z) = p_m + \sum_{s=m+1}^{m+n} a_{sm} p_s = Z_m;$$

and consider the system

$$B_1(u) = \frac{\partial u}{\partial x_1} + \sum_{s=m+1}^{m+n} a_{s1} \frac{\partial u}{\partial x_s} + Z_1 \frac{\partial u}{\partial z} = 0,$$

$$B_2(u) = \frac{\partial u}{\partial x_2} + \sum_{s=m+1}^{m+n} a_{s2} \frac{\partial u}{\partial x_s} + Z_2 \frac{\partial u}{\partial z} = 0,$$

....

$$B_m(u) = \frac{\partial u}{\partial x_m} + \sum_{s=m+1}^{m+n} a_{sm} \frac{\partial u}{\partial x_s} + Z_m \frac{\partial u}{\partial z} = 0,$$

the quantities $a_{\lambda\mu}$ and Z being functions of $z, x_1, \ldots, x_n$. The conditions of completeness of the latter system are

$$B_i(a_{sj}) = B_j(a_{si}),$$

$$B_i(Z_j) = B_j(Z_i),$$

for all pairs of values $i, j = 1, \ldots, m$, and for $s = m+1, \ldots, m+n$. But for any value of r, we have

$$B_r = E_r + Z_r \frac{\partial}{\partial z},$$

as a relation between the operators; thus the above conditions become

$$E_i(a_{sj}) + Z_i \frac{\partial a_{sj}}{\partial z} = E_j(a_{si}) + Z_j \frac{\partial a_{si}}{\partial z},$$

$$E_i(Z_j) + Z_i \frac{\partial Z_j}{\partial z} = E_j(Z_i) + Z_j \frac{\partial Z_i}{\partial z},$$

for all pairs of values $i, j = 1, \ldots, m$, and for $s = m+1, \ldots, m+n$. These are the conditions, necessary and sufficient to secure the coexistence and completeness of the system.

We know that the system of equations $B_1(u) = 0, \ldots, B_m(u) = 0$, being a complete linear system of m equations in $m+n+1$ variables, possesses $n+1$ functionally distinct integrals. Let a set of these be taken in the form $u_1, \ldots, u_{n+1}$, some of which will certainly involve z; then any integral of the transformed system can be expressed in a form

$$u = f(u_1, \ldots, u_{n+1}),$$

and its most general integral will be obtained by taking f as a completely arbitrary function. It is also obvious that an integral of the original system will be provided by an equation

$$u = 0,$$

if u is an integral of the transformed system which involves z; hence a very general integral of the original system will be provided by the equation

$$f(u_1, \ldots, u_{n+1}) = 0.$$

But for reasons similar to those adduced for a single equation in § 35, we are not in a position to declare (and it is not, in fact, true) that every integral of the original system of equations is included in the equation $f = 0$ for an appropriate form of f.

As the quantities $u_1, \ldots, u_{n+1}$, necessary for the construction of the function f, are simultaneous integrals of the system $B_1(u) = 0$, $\ldots$, $B_m(u) = 0$, it is clear that either of the two methods (in §§ 43, 44) effective for the construction of $u_1, \ldots, u_{n+1}$ can be adopted.

Ex. Let it be required to find whether the equations

$$x_1 p_1 + x_2 p_2 - x_3 p_3 + z = 0,$$
$$x_2 p_1 - x_1 p_2 + z p_3 + x_3 = 0,$$

have any common integral.

Expressing these equations in the form

$$p_1 + \frac{x_2 z - x_1 x_3}{x_1^2 + x_2^2} p_3 = -\frac{x_1 z + x_2 x_3}{x_1^2 + x_2^2},$$

$$p_2 - \frac{x_1 z + x_2 x_3}{x_1^2 + x_2^2} p_3 = -\frac{x_2 z - x_1 x_3}{x_1^2 + x_2^2},$$

and applying the conditions of the text, we find them satisfied : hence the equations coexist, and they form a complete system.

To obtain the general common integral, we construct the equations

$$\frac{\partial u}{\partial x_1} + \frac{x_2 z - x_1 x_3}{x_1^2 + x_2^2} \frac{\partial u}{\partial x_3} - \frac{x_1 z + x_2 x_3}{x_1^2 + x_2^2} \frac{\partial u}{\partial z} = 0,$$

$$\frac{\partial u}{\partial x_2} - \frac{x_1 z + x_2 x_3}{x_1^2 + x_2^2} \frac{\partial u}{\partial x_3} - \frac{x_2 z - x_1 x_3}{x_1^2 + x_2^2} \frac{\partial u}{\partial z} = 0,$$

which are a Jacobian system. It possesses two functionally distinct integrals: these are found, by the processes previously explained, to be

$$u = u_1 = x_1 x_3 + x_2 z, \qquad u = u_2 = x_2 x_3 - x_1 z.$$

A general integral, common to the two original equations, is given by

$$x_1 x_3 + x_2 z = \phi(x_2 x_3 - x_1 z),$$

where ϕ is an arbitrary functional form.

CHAPTER IV.

NON-LINEAR EQUATIONS: JACOBI'S SECOND METHOD, WITH MAYER'S DEVELOPMENTS.

FOR the material of the present chapter, reference may be made to Jacobi's posthumous memoir, "Nova methodus......integrandi," *Crelle*, t. LX (1862), pp. 1—181, *Ges. Werke*, t. V, pp. 1—189; to Mayer's memoir, "Ueber unbeschränkt......Differentialgleichungen," *Math. Ann.*, t. V (1872), pp. 448—470; and to Imschenetsky's memoir "Sur l'intégration......premier ordre," *Grunert's Archiv*, t. L (1869), pp. 278—474. Mention should also be made of Mansion's treatise "Théorie des équations aux derivées partielles du premier ordre" (1875), Book II; and of the exposition given in chapters VI and VII of Goursat's treatise, already (p. 55) quoted.

47. We now proceed to deal with single equations, and with systems of consistent equations, of the first order and of general degree in the derivatives: clearly no generality is lost by assuming that the equations are irreducible. It will be sufficiently obvious from the discussion in the last chapter that the construction of an integral of the equation or of the system of equations is a process of several stages, differing in this respect from the usual construction of an integral of an ordinary equation; and the difficulty, in general, is the discovery of the effective inverse operations that lead from stage to stage.

Now, whatever equation or equations may be assigned for the determination of the value or for the limitation of the form of a dependent variable, one permanent relation subsists between a number of independent variables $x_1, \ldots, x_n$, a dependent variable z, and the derivatives $p_1, \ldots, p_n$ of the latter: the relation is

$$dz = p_1 dx_1 + \ldots + p_n dx_n.$$

The quantities $p_1, \ldots, p_n$ are themselves dependent variables and consequently are functions of $x_1, \ldots, x_n$: but it frequently happens that they arise as functions of $x_1, \ldots, x_n, z$, the last variable not being explicitly known in terms of the independent variables. Also, there must be only a single functional relation between $z, x_1, \ldots, x_n$, so that the integral equivalent of the preceding differential relation is effectively a single equation among the variables: consequently, the differential relation must be an exact equation. The conditions necessary and sufficient to secure this result are known*: in the present case, they are

$$\frac{\partial p_\mu}{\partial x_m} - \frac{\partial p_m}{\partial x_\mu} + p_m \frac{\partial p_\mu}{\partial z} - p_\mu \frac{\partial p_m}{\partial z} = 0,$$

for the $\frac{1}{2}n(n-1)$ pairs of values of m and μ from the set $1, \ldots, n$. Let

$$\frac{d}{dx_\rho} = \frac{\partial}{\partial x_\rho} + p_\rho \frac{\partial}{\partial z},$$

so that $\dfrac{d}{dx_\rho}$ is the complete derivative with regard to x_ρ, account being taken of the explicit occurrence of x_ρ as well as of its implicit occurrence through z: the necessary and sufficient conditions become

$$\frac{dp_\mu}{dx_m} = \frac{dp_m}{dx_\mu};$$

and these conditions apply, whatever be the quantity z and however its derivatives $p_1, \ldots, p_n$ may be determined. When they are satisfied, the relation

$$dz = p_1 dx_1 + \ldots + p_n dx_n$$

is exact: when an integral equivalent is obtained by the recognised processes of quadrature, that equivalent is an integral relation between $z, x_1, \ldots, x_n$.

Now there are n of these derivatives of z: when regarded for the purpose of quadrature, they will most generally be determined by n equations

$$F_1 = 0, \; F_2 = 0, \; \ldots, \; F_n = 0,$$

where $F_1, \ldots, F_n$ will be assumed to be n regular functions of $x_1, \ldots, x_n, z, p_1, \ldots, p_n$ which, so far as they involve $p_1, \ldots, p_n$, are functionally distinct; consequently the Jacobian

$$J\left(\frac{F_1, \ldots, F_n}{p_1, \ldots, p_n}\right), = J \text{ say},$$

* Part I of this work, § 11.

does not vanish identically. In that case, the n equations can be resolved so as to give expressions for $p_1, \ldots, p_n$ as regular functions of $x_1, \ldots, x_n, z$: when these expressions are substituted in the equations, the latter become identities. Taking two of these equations, say $F_r = 0$ and $F_s = 0$, thus turned into identities, we have, on differentiating with regard to x_m,

$$\frac{dF_r}{dx_m} + \sum_{\mu=1}^{n} \frac{\partial F_r}{\partial p_\mu} \frac{dp_\mu}{dx_m} = 0,$$

$$\frac{dF_s}{dx_m} + \sum_{\mu=1}^{n} \frac{\partial F_s}{\partial p_\mu} \frac{dp_\mu}{dx_m} = 0,$$

and therefore, on the elimination of $\frac{dp_m}{dx_m}$,

$$\frac{dF_r}{dx_m}\frac{\partial F_s}{\partial p_m} - \frac{dF_s}{dx_m}\frac{\partial F_r}{\partial p_m} + \sum_{\mu=1}^{n} \frac{\partial (F_r, F_s)}{\partial (p_\mu, p_m)} \frac{dp_\mu}{dx_m} = 0.$$

This holds for each value of m; taking it then in succession for each value of m, and adding all the left-hand sides together, we have

$$\sum_{m=1}^{n} \left(\frac{dF_r}{dx_m}\frac{\partial F_s}{\partial p_m} - \frac{dF_s}{dx_m}\frac{\partial F_r}{\partial p_m}\right) + \sum_{m=1}^{n} \sum_{\mu=1}^{n} \frac{\partial (F_r, F_s)}{\partial (p_\mu, p_m)} \frac{dp_\mu}{dx_m} = 0.$$

The last double summation can be modified: the terms for which $\mu = m$ do not occur: taking a pair of values for μ and m from the set $1, \ldots, n$, and combining them, the summation may be written

$$\sum_{\mu, m}^{n}{}' \frac{\partial (F_r, F_s)}{\partial (p_\mu, p_m)} \left(\frac{dp_\mu}{dx_m} - \frac{dp_m}{dx_\mu}\right).$$

Moreover, it is convenient to use a symbol to denote the first summation: we write

$$[F_r, F_s] = \sum_{m=1}^{n} \left(\frac{dF_r}{dx_m}\frac{\partial F_s}{\partial p_m} - \frac{dF_s}{dx_m}\frac{\partial F_r}{\partial p_m}\right);$$

and the equation now becomes

$$[F_r, F_s] + \sum_{\mu, m}^{n}{}' \frac{\partial (F_r, F_s)}{\partial (p_\mu, p_m)} \left(\frac{dp_\mu}{dx_m} - \frac{dp_m}{dx_\mu}\right) = 0.$$

This holds for all combinations of r and s from the set $1, \ldots, n$.

SIGNIFICANCE OF THE JACOBIAN RELATIONS.

48. Two inferences can be drawn from this aggregate of relations.

In the first place, the quantities $p_1, \ldots, p_n$, as determined by the equations $F_1 = 0, \ldots, F_n = 0$, have thus far merely been regarded as variable magnitudes: but, in addition, they are to be derivatives of z. The conditions, necessary and sufficient to secure this last property, are

$$\frac{dp_\mu}{dx_m} - \frac{dp_m}{dx_\mu} = 0 ;$$

hence we have

$$[F_r, F_s] = 0,$$

for all values of r and s. These equations, $\frac{1}{2}n(n-1)$ in number, are thus a necessary consequence of the hypothesis that the quantities p are the derivatives of z.

In the second place, if these $\frac{1}{2}n(n-1)$ equations are satisfied, then the quantities $p_1, \ldots, p_n$, determined by the equations $F_1 = 0, \ldots, F_n = 0$, are the n first derivatives of z with regard to $x_1, \ldots, x_n$. Assuming the equations to be satisfied, the foregoing aggregate of relations becomes

$$\sum_{\mu, m}^{n}{}' \frac{\partial (F_r, F_s)}{\partial (p_\mu, p_m)} \left(\frac{dp_\mu}{dx_m} - \frac{dp_m}{dx_\mu} \right) = 0,$$

for all combinations of r and s. We thus have $\frac{1}{2}n(n-1)$ equations, homogeneous and linear in the $\frac{1}{2}n(n-1)$ quantities

$$\frac{dp_\mu}{dx_m} - \frac{dp_m}{dx_\mu}.$$

Taking the equations in the form

$$\sum_{\mu=1}^{n} \sum_{m=1}^{n} \frac{\partial F_r}{\partial p_\mu} \frac{\partial F_s}{\partial p_m} \left(\frac{dp_\mu}{dx_m} - \frac{dp_m}{dx_\mu} \right) = 0,$$

which is only a rearrangement, and writing

$$\sum_{m=1}^{n} \frac{\partial F_s}{\partial p_m} \left(\frac{dp_\mu}{dx_m} - \frac{dp_m}{dx_\mu} \right) = u_{\mu s},$$

we have

$$\sum_{\mu=1}^{n} \frac{\partial F_r}{\partial p_\mu} u_{\mu s} = 0.$$

This equation holds for all values of r and of s: taking one value of s, and the n values of r in turn, we have n equations which are linear and homogeneous in $u_{1s}, u_{2s}, \ldots, u_{ns}$. The determinant of the coefficients does not vanish, for it is the Jacobian of the functions $F_1, \ldots, F_n$ with regard to $p_1, \ldots, p_n$; hence

$$u_{\mu s} = 0,$$

for all values of μ and s, that is,

$$\sum_{m=1}^{n} \frac{\partial F_s}{\partial p_m}\left(\frac{dp_\mu}{dx_m} - \frac{dp_m}{dx_\mu}\right) = 0.$$

Taking the n values of s in turn, we have n equations which are linear and homogeneous in the quantities

$$\frac{dp_\mu}{dx_1} - \frac{dp_1}{dx_\mu}, \ldots, \frac{dp_\mu}{dx_n} - \frac{dp_n}{dx_\mu};$$

the determinant of their coefficients does not vanish, for again it is the Jacobian of $F_1, \ldots, F_n$ with regard to $p_1, \ldots, p_n$: hence

$$\frac{dp_\mu}{dx_m} - \frac{dp_m}{dx_\mu} = 0,$$

for all values of m and μ. These conditions have been proved necessary and sufficient to secure that the quantities p are derivatives of z; and they are a necessary consequence of the equations

$$[F_r, F_s] = 0,$$

which therefore are sufficient to secure that the quantities p are derivatives of z and that, when their values given by $F_1 = 0, \ldots, F_n = 0$ are substituted in the equation

$$dz = p_1 dx_1 + \ldots + p_n dx_n,$$

this equation is exact.

If, in particular, z does not occur explicitly in any of the equations $F = 0$, then

$$\frac{dF_\mu}{dx_m} = \frac{\partial F_\mu}{\partial x_m}:$$

the equations become

$$\sum_{m=1}^{n} \left(\frac{\partial F_r}{\partial x_m}\frac{\partial F_s}{\partial p_m} - \frac{\partial F_r}{\partial p_m}\frac{\partial F_s}{\partial x_m}\right) = 0,$$

and these are frequently represented by the form

$$(F_r, F_s) = 0.$$

Note. All these conditions will equally be required if the equations determining $p_1, \ldots, p_n$ occur in the form

$$F_1 = a_1, \ldots, F_n = a_n,$$

where $a_1, \ldots, a_n$ are constants.

49. Again, suppose that $n+1$ equations, involving $p_1, \ldots, p_n$, $z, x_1, \ldots, x_n$, are given. Let them be

$$G_1 = 0, \ldots, G_{n+1} = 0;$$

they can be regarded as determining $n+1$ quantities $z, p_1, \ldots, p_n$ in terms of $x_1, \ldots, x_n$. We proceed to shew that the conditions

$$[G_r, G_s] = 0,$$

for all combinations of r and s from the set $1, \ldots, n+1$, must be satisfied, if quantities $z, p_1, \ldots, p_n$ are so related that

$$\frac{\partial z}{\partial x_m} = p_m, \quad \frac{\partial p_m}{\partial x_\mu} = \frac{\partial p_\mu}{\partial x_m};$$

also, that the conditions specified suffice to secure these relations.

When the values of $z, p_1, \ldots, p_n$ are substituted in the equations $G = 0$, each of them becomes an identity; and therefore we have, from any equation $G_r = 0$ after the substitution,

$$\frac{\partial G_r}{\partial x_m} + \frac{\partial G_r}{\partial z}\frac{\partial z}{\partial x_m} + \sum_{\mu=1}^{n} \frac{\partial G_r}{\partial p_\mu}\frac{\partial p_\mu}{\partial x_m} = 0,$$

so that

$$\frac{\partial G_r}{\partial x_m} + p_m \frac{\partial G_r}{\partial z} + \frac{\partial G_r}{\partial z}\left(\frac{\partial z}{\partial x_m} - p_m\right) + \sum_{\mu=1}^{n} \frac{\partial G_r}{\partial p_\mu}\frac{\partial p_\mu}{\partial x_m} = 0,$$

or writing

$$\frac{\partial}{\partial x_m} + p_m \frac{\partial}{\partial z} = \frac{d}{dx_m},$$

for all values of m, we have

$$\frac{dG_r}{dx_m} + \frac{\partial G_r}{\partial z}\left(\frac{\partial z}{\partial x_m} - p_m\right) + \sum_{\mu=1}^{n} \frac{\partial G_r}{\partial p_\mu}\frac{\partial p_\mu}{\partial x_m} = 0.$$

Similarly

$$\frac{dG_s}{dx_m} + \frac{\partial G_s}{\partial z}\left(\frac{\partial z}{\partial x_m} - p_m\right) + \sum_{\mu=1}^{n} \frac{\partial G_s}{\partial p_\mu}\frac{\partial p_\mu}{\partial x_m} = 0.$$

Multiplying the former by $\frac{\partial G_s}{\partial p_m}$, the latter by $\frac{\partial G_r}{\partial p_m}$, and subtracting we have

$$\frac{dG_r}{dx_m}\frac{\partial G_s}{\partial p_m} - \frac{dG_s}{dx_m}\frac{\partial G_r}{\partial p_m} + \left(\frac{\partial G_r}{\partial z}\frac{\partial G_s}{\partial p_m} - \frac{\partial G_s}{\partial z}\frac{\partial G_r}{\partial p_m}\right)\left(\frac{\partial z}{\partial x_m} - p_m\right)$$
$$+ \sum_{\mu=1}^{n} \frac{\partial (G_r, G_s)}{\partial (p_\mu, p_m)}\frac{\partial p_\mu}{\partial x_m} = 0.$$

Summing the left-hand sides of this equation, taken for all the n values of m in succession, we have

$$\sum_{m=1}^{n}\left(\frac{dG_r}{dx_m}\frac{\partial G_s}{\partial p_m}-\frac{dG_s}{dx_m}\frac{\partial G_r}{\partial p_m}\right)$$
$$+\sum_{m=1}^{n}\frac{\partial(G_r,\,G_s)}{\partial(z,\,p_m)}\left(\frac{\partial z}{\partial x_m}-p_m\right)+\sum_{m=1}^{n}\sum_{\mu=1}^{n}\frac{\partial(G_r,\,G_s)}{\partial(p_\mu,\,p_m)}\frac{\partial p_\mu}{\partial x_m}=0,$$

or, again using $[G_r,\, G_s]$ to denote the first summation, we have

$$[G_r,\, G_s]+\sum_{m=1}^{n}\frac{\partial(G_r,\,G_s)}{\partial(z,\,p_m)}\left(\frac{\partial z}{\partial x_m}-p_m\right)+\sum_{m=1}^{n}\sum_{\mu=1}^{n}\frac{\partial(G_r,\,G_s)}{\partial(p_\mu,\,p_m)}\frac{\partial p_\mu}{\partial x_m}=0.$$

The last summation can also be written

$$\sum_{m=1}^{n}\sum_{\mu=1}^{n}\frac{\partial G_r}{\partial p_\mu}\frac{\partial G_s}{\partial p_m}\left(\frac{\partial p_\mu}{\partial x_m}-\frac{\partial p_m}{\partial x_\mu}\right);$$

and therefore we have

$$[G_r,\, G_s]+\sum_{m=1}^{n}\frac{\partial(G_r,\,G_s)}{\partial(z,\,p_m)}\left(\frac{\partial z}{\partial x_m}-p_m\right)$$
$$+\sum_{m=1}^{n}\sum_{\mu=1}^{n}\frac{\partial G_r}{\partial p_\mu}\frac{\partial G_s}{\partial p_m}\left(\frac{\partial p_\mu}{\partial x_m}-\frac{\partial p_m}{\partial x_\mu}\right)=0.$$

If then the quantities $z,\, p_1,\, \ldots,\, p_n$, determined as functions of $x_1,\, \ldots,\, x_n$ by the $n+1$ equations $G=0$, be such that their values satisfy the relations

$$\frac{\partial z}{\partial x_m}=p_m,\quad \frac{\partial p_\mu}{\partial x_m}=\frac{\partial p_m}{\partial x_\mu},$$

for all values of m and μ, then we must have

$$[G_r,\, G_s]=0;$$

and this holds for all combinations of r and s.

Conversely, if the relation holds for all the values of r and s, then the values of $z,\, p_1,\, \ldots,\, p_n$, as given by the equations $G=0$, are such that the quantities p are equal to the derivatives of z and satisfy the necessary relations of the foregoing type. When $[G_r,\, G_s]=0$, the equation becomes

$$\sum_{m=1}^{n}\frac{\partial(G_r,\,G_s)}{\partial(z,\,p_m)}\left(\frac{\partial z}{\partial x_m}-p_m\right)+\sum_{m=1}^{n}\sum_{\mu=1}^{n}\frac{\partial G_r}{\partial p_\mu}\frac{\partial G_s}{\partial p_m}\left(\frac{\partial p_\mu}{\partial x_m}-\frac{\partial p_m}{\partial x_\mu}\right)=0;$$

or, if we write

$$u_{rm}=\frac{\partial G_r}{\partial z}\left(\frac{\partial z}{\partial x_m}-p_m\right)+\sum_{\mu=1}^{n}\frac{\partial G_r}{\partial p_\mu}\left(\frac{\partial p_\mu}{\partial x_m}-\frac{\partial p_m}{\partial x_\mu}\right),$$
$$v_r=\sum_{m=1}^{n}\frac{\partial G_r}{\partial p_m}\left(\frac{\partial z}{\partial x_m}-p_m\right),$$

we have

$$-v_r\frac{\partial G_s}{\partial z}+\sum_{m=1}^{n}\frac{\partial G_s}{\partial p_m}u_{rm}=0,$$

for all values of r and s. Taking the $n+1$ values of s in succession, and keeping one and the same value of r, we have $n+1$ equations, homogeneous and linear in the $n+1$ magnitudes v_r, $u_{r1}, \ldots, u_{rn}$. The determinant of the coefficients of these magnitudes is

$$J\left(\frac{G_1, \ldots, G_{n+1}}{z, p_1, \ldots, p_n}\right),$$

which does not vanish, because the $n+1$ equations $G=0$ are presumed to determine z, $p_1, \ldots, p_n$ as functions of the other variables; hence all the magnitudes v_r, $u_{r1}, \ldots, u_{rn}$ vanish, that is,

$$\sum_{m=1}^{n}\frac{\partial G_r}{\partial p_m}\left(\frac{\partial z}{\partial x_m}-p_m\right)=0,$$

$$\frac{\partial G_r}{\partial z}\left(\frac{\partial z}{\partial x_m}-p_m\right)+\sum_{\mu=1}^{n}\frac{\partial G_r}{\partial p_\mu}\left(\frac{\partial p_\mu}{\partial x_m}-\frac{\partial p_m}{\partial x_\mu}\right)=0,$$

the former holding for all values of r, the latter for all values of r and of m. Taking the latter for a single value of m and for all the values of r, we again have $n+1$ equations, homogeneous and linear in the $n+1$ magnitudes

$$\frac{\partial z}{\partial x_m}-p_m,\ \frac{\partial p_1}{\partial x_m}-\frac{\partial p_m}{\partial x_1},\ \ldots,\ \frac{\partial p_n}{\partial x_m}-\frac{\partial p_m}{\partial x_n},$$

one of which is identically zero; the determinant of the coefficients again is the Jacobian of $G_1, \ldots, G_{n+1}$ with regard to z, $p_1, \ldots, p_n$, and so does not vanish: hence the $n+1$ magnitudes are zero, that is,

$$\frac{\partial z}{\partial x_m}-p_m=0,\quad \frac{\partial p_\mu}{\partial x_m}-\frac{\partial p_m}{\partial x_\mu}=0.$$

Next, taking all the values of m in turn, the other set of equations

$$\sum_{m=1}^{n}\frac{\partial G_r}{\partial p_m}\left(\frac{\partial z}{\partial x_m}-p_m\right)=0$$

is satisfied without providing any new condition: accordingly, we have the relations

$$p_m=\frac{\partial z}{\partial x_m},\quad \frac{\partial p_\mu}{\partial x_m}=\frac{\partial p_m}{\partial x_\mu},$$

as a necessary consequence of the equations $[G_r, G_s]=0$.

Note. It will be noticed that, if the n equations

$$F_1 = 0, \ldots, F_n = 0,$$

satisfy all the conditions $[F_r, F_s] = 0$ necessary for coexistence, the determination of a value of z which satisfies them all requires resolution of the equations and a quadrature: while, if the $n+1$ equations

$$G_1 = 0, \ldots, G_{n+1} = 0,$$

satisfy all the conditions $[G_r, G_s] = 0$ necessary for coexistence, the determination of a value of z which satisfies them all requires resolution of the equations only. The former case could be changed into the latter by the provision of an additional appropriate equation: this appropriate equation is actually provided as the result of the necessary quadrature, with the added advantage that it gives a relation between z and the variables $x_1, \ldots, x_n$, free from the quantities $p_1, \ldots, p_n$.

50. These two theorems lead to various issues, as regards the solution of a single equation and of a system of compatible equations. We shall deal with the latter first.

Accordingly, we suppose that several equations

$$F_1 = 0, \ldots, F_s = 0$$

are given: after the foregoing explanations, we can suppose that s is less than n. It may also be assumed that these equations are algebraically independent of one another and, at this stage, that they involve all the variables concerned: also, that it is not possible to eliminate $z, p_1, \ldots, p_n$ from among them, so as to lead to a relation among the independent variables alone. In order that the given equations may coexist, it is clear from the preceding analysis that the further equations

$$[F_i, F_j] = 0$$

must be satisfied, for all combinations of i and j from the set $1, \ldots, s$.

One, or more than one, of these further equations may be impossible: the original equations cannot then coexist as determining a function z of $x_1, \ldots, x_n$ which satisfies $F_1 = 0, \ldots, F_s = 0$ simultaneously. The case requires no further consideration.

One, or more than one, of the further equations may be satisfied identically: no condition is thereby imposed upon the system.

Similarly, no condition is imposed upon the system when any one of the further equations is satisfied in virtue of the original equations.

But it may happen that one of the further equations is not satisfied, either identically or in virtue of the original equations, and yet it must be satisfied: it is a new equation, which must be associated with the original system. Each such further equation, not satisfied either identically or in virtue of the original equations or in virtue of the newly associated equations, must be associated with the system: let the additional aggregate thus provided be

$$F_{s+1}=0,\ \ldots,\ F_t=0,$$

each of which, as representing a relation $[F_i, F_j]=0$, is an equation of the first order.

In order that these may coexist with the original system and with one another, each of them must be combined with every other and with every member of the original system in the relation $[F_i, F_j]=0$. Any new equation thus arising is associated with the increased system: and the process is repeated until the system is so amplified that the relation is satisfied either identically or in virtue of the equations in the amplified system. Such a system, on the analogy of the earlier and simpler case in Chapter III, is called *complete*: if it be denoted by

$$F_1=0,\ \ldots,\ F_m=0,$$

the relation $[F_i, F_j]=0$ is satisfied for every combination of i and j from the set $1, \ldots, m$, either identically or in virtue of the members of the complete system.

If the original system should be such that z does not occur explicitly in any equation, the relation $[F_i, F_j]=0$ becomes $(F_i, F_j)=0$; and then the complete system is

$$F_1=0,\ \ldots,\ F_m=0,$$

being such that the relation $(F_i, F_j)=0$ is satisfied for every combination of i and j from the set $1, \ldots, m$, either identically or in virtue of the members of the complete system. Moreover, z does not occur explicitly in any member of the complete system: for it is not introduced by any of the relations $(F_i, F_j)=0$.

51. In § 22, it was established that the equation

$$f(x_1, \ldots, x_n, z, p_1, \ldots, p_n) = 0$$

possesses an integral with one or other of the assigned initial conditions for one of the values $x_1 = a_1, \ldots, x_n = a_n$, except for such values (if any) of the variables as satisfy $f = 0$ and also

$$\frac{\partial f}{\partial p_1} = 0, \ldots, \frac{\partial f}{\partial p_n} = 0.$$

If these equations can coexist, we must have

$$\left[f, \frac{\partial f}{\partial p_\mu}\right] = 0, \quad \left[\frac{\partial f}{\partial p_r}, \frac{\partial f}{\partial p_s}\right] = 0,$$

for all values of μ, r, s, in connection with the $n+1$ equations.

The former condition is

$$\sum_{m=1}^{n} \left\{\frac{df}{dx_m} \frac{\partial}{\partial p_m}\left(\frac{\partial f}{\partial p_\mu}\right) - \frac{\partial f}{\partial p_m} \frac{d}{dx}\left(\frac{\partial f}{\partial p_\mu}\right)\right\} = 0;$$

but $\dfrac{\partial f}{\partial p_m} = 0$ for all values of m, and thus the condition is

$$\sum_{m=1}^{n} \frac{df}{dx_m} \frac{\partial^2 f}{\partial p_m \partial p_\mu} = 0,$$

for all values of μ, so that we have n relations, homogeneous and linear in the n quantities $\dfrac{df}{dx_1}, \ldots, \dfrac{df}{dx_n}$. Suppose now that, if the equations $\dfrac{\partial f}{\partial p_1} = 0, \ldots, \dfrac{\partial f}{\partial p_n} = 0$ can coexist with $f = 0$, they determine $p_1, \ldots, p_n$, so that

$$J\left(\frac{\dfrac{\partial f}{\partial p_1}, \ldots, \dfrac{\partial f}{\partial p_n}}{p_1, \ldots, p_n}\right)$$

is not zero; then the preceding n relations can only be satisfied by

$$\frac{df}{dx_m} = 0,$$

that is, by

$$\frac{\partial f}{\partial x_m} + p_m \frac{\partial f}{\partial z} = 0,$$

for $m = 1, \ldots, n$. These are n additional relations: they must be satisfied by the values of $p_1, \ldots, p_n, z$, provided by the $n+1$ equations.

It is easy to see that these relations must be satisfied whether $\frac{\partial f}{\partial p_1}=0, \dots, \frac{\partial f}{\partial p_n}=0$ are independent of one another or not, quà equations in $p_1, \dots, p_n$. For the value of z, and the values of $p_1, \dots, p_n$ deduced from it, must make $f=0$ satisfied identically when they are substituted: hence

$$\frac{\partial f}{\partial x_m}+p_m\frac{\partial f}{\partial z}+\sum_{\mu=1}^{n}\frac{\partial f}{\partial p_\mu}\frac{\partial p_\mu}{\partial x_m}=0,$$

that is,

$$\frac{\partial f}{\partial x_m}+p_m\frac{\partial f}{\partial z}=0,$$

for all the values of m, which are the conditions in question.

The other set of conditions is

$$\sum_{m=1}^{n}\left[\left\{\frac{d}{dx_m}\left(\frac{\partial f}{\partial p_r}\right)\right\}\frac{\partial^2 f}{\partial p_m\partial p_s}-\left\{\frac{d}{dx_m}\left(\frac{\partial f}{\partial p_s}\right)\right\}\frac{\partial^2 f}{\partial p_m\partial p_r}\right]=0,$$

for all values of r and s.

When all these conditions are satisfied, and when the $n+1$ equations $f=0$, $\frac{\partial f}{\partial p_1}=0, \dots, \frac{\partial f}{\partial p_n}=0$, determine $p_1, \dots, p_n, z$ as functions of $x_1, \dots, x_n$, the value of z is certainly an integral of the original equation $f=0$. Clearly, it then is not capable of obeying assigned initial conditions: for it possesses no arbitrary element which is at our disposal.

Such integrals are of the class usually called *singular*: we shall recur to them later. When they exist, they result from the elimination of $p_1, \dots, p_n$ between

$$f=0,\ \frac{\partial f}{\partial p_1}=0, \dots, \frac{\partial f}{\partial p_n}=0.$$

Ex. 1. As an example, we may take

$$f=-z+x_1p_1+\dots+x_np_n+g(p_1, \dots, p_n)=0;$$

the additional equations are

$$\frac{\partial f}{\partial p_m}=x_m+\frac{\partial g}{\partial p_m}=0,$$

so that evidently g must involve all the quantities $p_1, \dots, p_m$. The relations

$$\frac{\partial f}{\partial x_m}+p_m\frac{\partial f}{\partial z}=0$$

are satisfied identically; likewise the other set of relations. Moreover, the form of g is known: hence, eliminating $p_1, \ldots, p_n$ from the equations

$$f=0, \quad x_m+\frac{\partial g}{\partial p_m}=0, \qquad (m=1, \ldots, n),$$

the value of z given by the resulting equation is an integral of the original equation. But it contains no arbitrary element.

Ex. 2. It must not be supposed that elimination of $p_1, \ldots, p_n$ is always possible among the $n+1$ equations. Taking $n=2$, a simple instance is provided by the equation

$$f=(px+qy-z)^2-p^2-q^2+\frac{z^2}{x^2+y^2-1}=0.$$

All the equations

$$f=0, \quad \frac{\partial f}{\partial p}=0, \quad \frac{\partial f}{\partial q}=0,$$

and all the relations

$$\frac{\partial f}{\partial x}+p\frac{\partial f}{\partial z}=0, \quad \frac{\partial f}{\partial y}+q\frac{\partial f}{\partial z}=0,$$

$$\frac{d}{dx}\left(\frac{\partial f}{\partial p}\right)\frac{\partial}{\partial p}\left(\frac{\partial f}{\partial q}\right)-\frac{d}{dx}\left(\frac{\partial f}{\partial q}\right)\frac{\partial}{\partial p}\left(\frac{\partial f}{\partial p}\right)$$
$$+\frac{d}{dy}\left(\frac{\partial f}{\partial p}\right)\frac{\partial}{\partial q}\left(\frac{\partial f}{\partial q}\right)-\frac{d}{dy}\left(\frac{\partial f}{\partial q}\right)\frac{\partial}{\partial q}\left(\frac{\partial f}{\partial p}\right)=0,$$

are satisfied by the two equations

$$\frac{p}{x}=\frac{q}{y}=\frac{z}{x^2+y^2-1};$$

attempted elimination gives no further equation. An integral for the particular example is clearly given by

$$z^2=a(x^2+y^2-1),$$

where a is an arbitrary constant.

To equations having integrals of this kind, we shall recur later.

The Combinants (F, G), $[F, G]$: Some properties.

52. Before passing to the further consideration of a complete system of given equations, it is convenient to note a property of the operation, represented by (F, G) when F and G do not involve z, and by $[F, G]$ when F and G do involve z.

Let F, G, H be any three independent functions of $2n$ quantities $x_1, \ldots, x_n, p_1, \ldots, p_n$; then it is not difficult to verify that the equation

$$((F, G)H)+((G, H)F)+((H, F)G)=0$$

is satisfied identically. From this identical relation, one inference can be made at once. Let $u = \phi$, and $u = \psi$, where ϕ and ψ are functions of $x_1, \ldots, x_n, p_1, \ldots, p_n$, be two independent integrals of

$$(F, u) = 0,$$

which is a homogeneous linear equation in u; then

$$(F, \phi) = 0, \quad (F, \psi) = 0,$$

both equations being satisfied identically. Hence

$$((F, \phi)\psi) = 0, \quad ((\psi, F)\phi) = -((F, \psi)\phi) = 0,$$

and therefore

$$((\phi, \psi)F) = 0,$$

that is,

$$(F(\phi, \psi)) = 0.$$

Thus, taking

$$u = (\phi, \psi),$$

we have

$$(F, u) = 0.$$

Now (ϕ, ψ) may be zero identically, or it may be a pure constant: in either case, the equation $u = (\phi, \psi)$ gives a trivial (and negligible) integral of the differential equation. But if (ϕ, ψ) be a variable quantity, then $u = (\phi, \psi)$ gives an integral of $(F, u) = 0$; and if it be distinct from ϕ and from ψ, it is a new integral. We therefore have the theorem*:

If $u = \phi$ and $u = \psi$ are integrals of the equation

$$(F, u) = 0,$$

then $u = (\phi, \psi)$ also satisfies the equation: and if (ϕ, ψ) be a variable quantity distinct from ϕ and from ψ, then $u = (\phi, \psi)$ is a new integral of the equation.

Another mode of stating this result is as follows: *Let $f_1 = 0, \ldots, f_m = 0$ be a complete system of equations which do not involve z explicitly, so that the relation*

$$(f_r, f_s) = 0,$$

for all values of r and s from the set $1, \ldots, m$, is satisfied; and let $u = \theta$ be an integral of the homogeneous linear equation $(f_1, u) = 0$. Then if the relation $(f_1, f_r) = 0$ is satisfied identically, the quantity

* The theorem is customarily associated with Poisson's name. It was used by Jacobi without explicit indication of the limitations, though he uses it only generally, not universally: see Jacobi, *Ges. Werke*, t. v, pp. 49, 50.

$u=(f_r,\ \theta)$ also satisfies the equation $(f_1,\ u)=0$; and it is a new integral of that equation if, being a variable quantity, it is functionally independent of θ.

The result is derived, on the lines of the earlier explanation, from the relation

$$((f_1,\ f_r)\,\theta)+((f_r,\ \theta)f_1)+((\theta,\ f_1)f_r)=0.$$

For $(f_1,\ \theta)=0$ identically, and $(f_1,\ f_r)=0$ identically, so that $((f_1,\ f_r)\,\theta)=0$ and $((\theta,\ f_1)f_r)=0$: thus $u=(f_r,\ \theta)$ satisfies the equation $(f_1,\ u)=0$.

The relation $(f_1, f_r)=0$ must always be satisfied; but if it is satisfied only in virtue of equations of the system, the inference as to the significance of $(f_r,\ \theta)$ cannot be drawn.

53. Next, let f, g, h be any three independent functions of $2n+1$ magnitudes $x_1,\ \ldots,\ x_n,\ z,\ p_1,\ \ldots,\ p_n$; then it is not difficult to verify that the equation

$$[[f,\ g]\,h]+[[g,\ h]f]+[[h,\ f]\,g]=-\frac{\partial f}{\partial z}[g,\ h]-\frac{\partial g}{\partial z}[h,\ f]-\frac{\partial h}{\partial z}[f,\ g]$$

is satisfied identically.

A corresponding inference can be drawn from this identity: but it is not so completely useful as in the former case. Let $u=\phi$ and $u=\psi$, where ϕ and ψ are functions of $x_1,\ \ldots,\ x_n,\ z,\ p_1,\ \ldots,\ p_n$, be two independent integrals of

$$[f,\ u]=0,$$

which is a homogeneous linear equation in u: then

$$[f,\ \phi]=0,\quad [f,\ \psi]=0$$

are satisfied identically. Hence also

$$[[f,\ \phi]\,\psi]=0,\quad [[\psi,\ f]\,\phi]=-[[f,\ \psi]\,\phi]=0;$$

therefore, taking $\phi=g$ and $\psi=h$ in the above identity, and using these relations, we have

$$[[\phi,\ \psi]f]=-\frac{\partial f}{\partial z}[\phi,\ \psi],$$

or, writing

$$v=[\phi,\ \psi],$$

we have

$$[f,\ v]=v\frac{\partial f}{\partial z}.$$

Thus, in general, $u = v = [\phi, \psi]$ does not satisfy the equation

$$[f, u] = 0;$$

and thus, in general, a new integral of that equation is not obtained. If, however, f does not explicitly involve z, so that $\frac{\partial f}{\partial z}$ is zero, then

$$[f, v] = 0;$$

hence $u = v$ satisfies the equation. As before, $[\phi, \psi]$ may be zero identically, or it may be a pure constant: then $u = v$ gives a trivial (and negligible) integral of the equation. But if $[\phi, \psi]$ is a variable quantity, then $u = [\phi, \psi]$ is an integral of the equation: and it is a new integral if distinct from ϕ and from ψ, that is, if it is not expressible in terms of ϕ and of ψ alone. Hence we have the theorem*:—

If $u = \phi$ and $u = \psi$ are integrals of the equation

$$[f, u] = 0,$$

then $u = [\phi, \psi]$ is a new integral of that equation, only if f does not explicitly involve z and if $[\phi, \psi]$ is a variable quantity not expressible in terms of ϕ and ψ alone.

Another mode of stating the result is as follows: *Let $f_1 = 0, \ldots, f_m = 0$ be a complete system of equations, some at least of which involve z explicitly, so that the relation*

$$[f_r, f_s] = 0,$$

for all values of r and s from the set $1, \ldots, m$, is satisfied; and let $u = \vartheta$ be an integral of the homogeneous linear equation $[f, u] = 0$. Then if the relation $[f_1, f_r] = 0$ is satisfied identically and if the equation $f_1 = 0$ does not involve z explicitly, the quantity $u = [f_r, \vartheta]$ also satisfies the equation $[f_1, u] = 0$; and it is a new integral of that equation if, being a variable quantity, it is functionally independent of ϑ.

The result is derived from the same identity as before. Taking $f = f_1$, $g = f_r$, $h = \vartheta$, we have $[f_1, \vartheta] = 0$ identically and $[f_1, f_r] = 0$ also identically, so that

$$[[f_1, f_r]\vartheta] = 0, \quad [[f_1, \vartheta]f_r] = 0:$$

* The correct statement of the theorem appears to have been given first by Mayer, *Math. Ann.*, t. IX (1876), p. 370.

also $\frac{\partial f_1}{\partial z} = 0$, by the hypothesis adopted: thus

$$[[f_r, \vartheta] f_1] = 0,$$

and so $u = [f_r, \vartheta]$ satisfies the equation $[f_1, u] = 0$.

The relation $[f_1, f_r] = 0$ must always be satisfied: if it is satisfied only in virtue of the equations of the system, the inference as to the significance of $[f_r, \vartheta]$ cannot be drawn.

54. Now the ultimate object of investigations, connected with a single equation or with complete systems of equations, is either the construction of the most general integral that is possessed or the formation of processes effective for such construction. Moreover, speaking generally, such processes will be made simpler by every reduction in the number of inverse operations to be performed and by every increase in the number of direct operations.

It is clear, from the two preceding sets of results, that a direct operation for the construction of a new integral of $(F_1, u) = 0$ will more frequently be effective than a direct operation for the construction of a new integral of $[f_1, u] = 0$: indeed, the latter is effective only when the equation $f_1 = 0$ is more limited than is generally permitted to the system of equations in which it is included.

We know that it is always possible, by means of a transformation

$$u = u(z, x_1, \ldots, x_n) = 0,$$

to remove the dependent variable from explicit occurrence in an equation, or in a system of equations, involving only one dependent variable: the number of independent variables is, however, thereby increased by unity. When the integral u of the transformed system has been obtained in the most general form, which comprehends all its integrals, a general integral of the original system is at once deduced from the equation $u = 0$; but this general integral is not completely comprehensive, for it need not include special integrals if any such exist. But as has been seen in the case of a single equation, that is non-homogeneous and of the first order, the processes adopted for the untransformed equation do not lead to the special integrals, if any.

Thus there would appear to be no real loss of generality and no real diminution in the number of integrals obtainable, if we pass to a transformed system in which the dependent variable does not explicitly occur. On the other hand, there is an added element of effectiveness, because the quantity $u = (F_r, \theta)$ is often an integral of $(F_1, u) = 0$, whereas the quantity $u = [f_r, \vartheta]$ requires compliance with an additional condition in order that it may be an integral of $[f_1, u] = 0$.

Accordingly, for the immediate present, it will be assumed that the dependent variable does not occur explicitly: hence we have to deal with a system of equations

$$F_1 = 0, \ \ldots, \ F_m = 0,$$

involving the quantities $x_1, \ldots, x_n, p_1, \ldots, p_n$. We may further assume that the equations are linearly independent of one another, so that no one of the quantities F can be expressed as a linear combination of the remainder with coefficients whether variable or constant. And after the discussion in § 50, we shall assume that the system is complete, so that the relation

$$(F_r, F_s) = 0$$

is satisfied, for all values of r and s from the set $1, \ldots, m$, either identically or in virtue of the equations of the system.

Moreover, after the same discussion, it will be assumed that $m < n$. What is required is a value of z satisfying all the equations of the system: in order to proceed by quadratures, other $n - m$ compatible and independent equations are needed.

Mayer's development of Jacobi's second method.

55. There are various ways of deducing the further $n - m$ equations that are requisite: one of the simplest of these ways is Mayer's development of what is often called Jacobi's second method.

The m equations in the complete system

$$F_1 = 0, \ \ldots, \ F_m = 0$$

are linearly independent of one another, in the sense that no one of the quantities F can be expressed as a linear combination of the remainder: thus there can be no effective functional relation among the quantities F. Consequently, the m equations can be resolved so as to express m of the involved variables in terms of the rest.

In the first instance, let it be supposed that the equations can be resolved so as to express m of the variables $p_1, \ldots, p_n$, say to express $p_1, \ldots, p_m$, in terms of all the other quantities involved; and let the result of the resolution be denoted by

$$p_i - \phi_i(p_{m+1}, \ldots, p_n, x_1, \ldots, x_n) = 0,$$

or by

$$p_i - \phi_i = 0,$$

for $i = 1, \ldots, m$. We prove, as follows, that the resolved system of equations is complete, the original system being complete: that is to say, the relation

$$(p_r - \phi_r, p_s - \phi_s) = 0$$

is satisfied, for all values of r and of s from the set $1, \ldots, m$.

When the values $\phi_1, \ldots, \phi_m$ for $p_1, \ldots, p_m$ respectively are substituted in all the equations of the original system, each of the latter becomes an identity. Therefore

$$\frac{\partial F_r}{\partial x_i} + \sum_{k=1}^{m} \frac{\partial F_r}{\partial p_k} \frac{\partial \phi_k}{\partial x_i} = 0,$$

for all values of $i = 1, \ldots, n$: that is,

$$\frac{\partial F_r}{\partial x_i} = \sum_{k=1}^{m} \frac{\partial F_r}{\partial p_k} \frac{\partial (p_k - \phi_k)}{\partial x_i},$$

for all these values. Similarly, we have

$$\frac{\partial F_r}{\partial p_j} + \sum_{k=1}^{m} \frac{\partial F_r}{\partial p_k} \frac{\partial \phi_k}{\partial p_j} = 0,$$

for the values $j = m + 1, \ldots, n$; that is,

$$\frac{\partial F_r}{\partial p_j} = \sum_{k=1}^{m} \frac{\partial F_r}{\partial p_k} \frac{\partial (p_k - \phi_k)}{\partial p_j},$$

for these values. Also

$$\frac{\partial F_r}{\partial p_{j'}} = \sum_{k=1}^{m} \frac{\partial F_r}{\partial p_k} \frac{\partial (p_k - \phi_k)}{\partial p_{j'}},$$

for the values $j' = 1, \ldots, m$, because for each of these values only a single term on the right-hand side occurs: hence

$$\frac{\partial F_r}{\partial p_i} = \sum_{k=1}^{m} \frac{\partial F_r}{\partial p_k} \frac{\partial (p_k - \phi_k)}{\partial p_i},$$

for the values $i = 1, \ldots, n$. And these results hold for all the values $r = 1, \ldots, m$.

Substituting in (F_r, F_s), we find

$$(F_r, F_s) = \sum_{i=1}^{n} \left(\frac{\partial F_r}{\partial x_i} \frac{\partial F_s}{\partial p_i} - \frac{\partial F_r}{\partial p_i} \frac{\partial F_s}{\partial x_i} \right)$$

$$= \sum_{i=1}^{n} \sum_{k=1}^{m} \sum_{l=1}^{m} \frac{\partial F_r}{\partial p_k} \frac{\partial F_s}{\partial p_l} \left\{ \frac{\partial (p_k - \phi_k)}{\partial x_i} \frac{\partial (p_l - \phi_l)}{\partial p_i} - \frac{\partial (p_k - \phi_k)}{\partial p_i} \frac{\partial (p_l - \phi_l)}{\partial x_i} \right\}$$

$$= \sum_{k=1}^{m} \sum_{l=1}^{m} \frac{\partial F_r}{\partial p_k} \frac{\partial F_s}{\partial p_l} (p_k - \phi_k, \, p_l - \phi_l).$$

Now $(F_r, F_s) = 0$, for all values of r and of s from the set $1, \ldots, m$: hence

$$\sum_{k=1}^{m} \sum_{l=1}^{m} \frac{\partial F_r}{\partial p_k} \frac{\partial F_s}{\partial p_l} (p_k - \phi_k, \, p_l - \phi_l) = 0,$$

for all these combinations of values. Taking the relation for one value of s and for all the values of r, we have m equations, homogeneous and linear in the m quantities

$$\sum_{l=1}^{m} \frac{\partial F_s}{\partial p_l} (p_k - \phi_k, \, p_l - \phi_l),$$

for $k = 1, \ldots, m$. The determinant of the coefficients of these quantities in the m equations is

$$J\left(\frac{F_1, \ldots, F_m}{p_1, \ldots, p_m}\right),$$

which, by hypothesis, does not vanish: consequently,

$$\sum_{l=1}^{m} \frac{\partial F_s}{\partial p_l} (p_k - \phi_k, \, p_l - \phi_l) = 0,$$

for all the values of s and k. Taking this relation for all the m values of s, we again have m equations, homogeneous and linear in the m quantities

$$(p_k - \phi_k, \, p_l - \phi_l),$$

for $l = 1, \ldots, m$; and the determinant of the coefficients of these quantities in the m equations is again

$$J\left(\frac{F_1, \ldots, F_m}{p_1, \ldots, p_m}\right),$$

which does not vanish: consequently,

$$(p_k - \phi_k, \, p_l - \phi_l) = 0,$$

for k and $l=1, \ldots, m$. Hence the system of equations

$$p_i - \phi_i = 0, \qquad (i=1, \ldots, m),$$

is complete.

It is easy to see that the completeness of the system of equations

$$p_i - \phi_i = 0,$$

for $i=1, \ldots, m$, is of a special kind. The relation

$$(p_k - \phi_k, \; p_l - \phi_l) = 0,$$

k and l having any values from the set $1, \ldots, m$, is

$$0 = -\frac{\partial \phi_l}{\partial x_k} + \frac{\partial \phi_k}{\partial x_l} + \sum_{j=m+1}^{n} \left(\frac{\partial \phi_k}{\partial x_j} \frac{\partial \phi_l}{\partial p_j} - \frac{\partial \phi_k}{\partial p_j} \frac{\partial \phi_l}{\partial x_j} \right).$$

This relation is to be satisfied, and it clearly is not satisfied in virtue of the equations

$$p_i - \phi_i = 0,$$

for $i=1, \ldots, m$, because it does not involve any of the quantities $p_1, \ldots, p_m$; hence the relations for the modified system are satisfied identically.

When the relations, necessary and sufficient to secure the completeness of a system, are satisfied identically, the system is said to be *in involution*. The resolved system of complete equations is a system in involution, because each of the relations

$$(p_k - \phi_k, \; p_l - \phi_l) = 0$$

is satisfied identically*.

Note. Even when a complete system of equations $F_1=0, \ldots, F_m=0$, is such that z occurs explicitly, a corresponding result is obtainable. Suppose that the m equations can be resolved with regard to z and to $m-1$ of the quantities p, say $p_1, \ldots, p_{m-1}$, in the form

$$z - \psi = 0, \quad p_1 - \psi_1 = 0, \quad \ldots, \quad p_{m-1} - \psi_{m-1} = 0,$$

where $\psi, \psi_1, \ldots, \psi_{m-1}$ do not involve $z, p_1, \ldots, p_{m-1}$; then the relations

$$[p_i - \psi_i, \; p_j - \psi_j] = 0,$$

$$[z - \psi - x_1(p_1 - \psi_1) - \ldots - x_{m-1}(p_{m-1} - \psi_{m-1}), \; p_i - \psi_i] = 0,$$

* Sometimes, for convenience, the unresolved complete system is said to be *in involution*.

are satisfied in virtue of the relations

$$[F_r, F_s] = 0,$$

for all values of i and j. When the quantities

$$[p_i - \psi_i, p_j - \psi_j]$$

are expressed in full, they contain none of the variables $z, p_1, \ldots, p_{m-1}$; they cannot vanish, therefore, in virtue of the resolved equations: hence they must vanish identically. Similarly, the quantity

$$[z - \psi - x_1(p_1 - \psi_1) - \ldots - x_{m-1}(p_{m-1} - \psi_{m-1}), p_i - \psi_i]$$

vanishes identically.

The resolved system can be regarded as a system in involution.

56. In order to obtain common integrals of the system, a satisfactory method will be devised if, by its means, other $n-m$ equations are associated with the m equations in the system: and the remaining stage will be a quadrature with reference to the variables $x_1, \ldots, x_n$, if the Jacobian of

$$p_1 - \phi_1, \ldots, p_m - \phi_m, \quad u_{m+1}, \ldots, u_n,$$

(where $u_{m+1} = \text{constant}, \ldots, u_n = \text{constant}$, are the additional $n-m$ equations) with regard to $p_1, \ldots, p_n$ does not vanish identically, that is, if

$$J\left(\frac{u_{m+1}, \ldots, u_n}{p_{m+1}, \ldots, p_n}\right)$$

does not vanish identically. Accordingly, this method requires the determination of $n-m$ equations $u = \text{constant}$.

Each such equation, as it is to coexist with the equations of the given system, must satisfy the conditions which are necessary and sufficient to secure the coexistence: that is, it must satisfy the relations

$$(p_1 - \phi_1, u) = 0, \ \ldots, \ (p_m - \phi_m, u) = 0,$$

which, in effect, are m equations for the determination of u. Now these equations constitute a complete system of the type considered in the last chapter. We have

$$((p_r - \phi_r, u), p_s - \phi_s) - ((p_s - \phi_s, u), p_r - \phi_r) + ((p_s - \phi_s, p_r - \phi_r), u) = 0$$

identically; also $(p_s - \phi_s, p_r - \phi_r) = 0$ identically for all values of r and s, so that $((p_s - \phi_s, p_r - \phi_r), u) = 0$: hence

$$(p_r - \phi_r, (p_s - \phi_s, u)) = (p_s - \phi_s, (p_r - \phi_r, u)).$$

Let

$$(p_i - \phi_i, u) = A_i(u),$$

where A_i is a linear operator: the foregoing relation becomes

$$A_r(A_s u) = A_s(A_r u),$$

for all values of r and s. This aggregate of relations is necessary and sufficient to secure that the system of equations $A_1(u) = 0, \ldots, A_m(u) = 0$, is complete.

Written in full, the equations are

$$\frac{\partial u}{\partial x_1} + \sum_{i=m+1}^{n} \left(\frac{\partial u}{\partial p_i}\frac{\partial \phi_1}{\partial x_i} - \frac{\partial u}{\partial x_i}\frac{\partial \phi_1}{\partial p_i}\right) + \sum_{j=1}^{m} \frac{\partial u}{\partial p_j}\frac{\partial \phi_1}{\partial x_j} = 0,$$

....

$$\frac{\partial u}{\partial x_m} + \sum_{i=m+1}^{n} \left(\frac{\partial u}{\partial p_i}\frac{\partial \phi_m}{\partial x_i} - \frac{\partial u}{\partial x_i}\frac{\partial \phi_m}{\partial p_i}\right) + \sum_{j=1}^{m} \frac{\partial u}{\partial p_j}\frac{\partial \phi_m}{\partial x_j} = 0,$$

a system of m equations in the $2n$ variables $x_1, \ldots, x_n, p_1, \ldots, p_n$. When any integral of the system has been obtained involving any of the variables $p_1, \ldots, p_m$, the relations $p_1 = \phi_1, \ldots, p_m = \phi_m$, can be used (without affecting its value or its significance) so as to remove these variables. In the transformed expression for u, we have

$$\frac{\partial u}{\partial p_1} = 0, \ \ldots, \ \frac{\partial u}{\partial p_m} = 0,$$

that is, we may take the m equations in the form

$$\frac{\partial u}{\partial x_1} + \sum_{i=m+1}^{n} \left(\frac{\partial u}{\partial p_i}\frac{\partial \phi_1}{\partial x_i} - \frac{\partial u}{\partial x_i}\frac{\partial \phi_1}{\partial p_i}\right) = 0,$$

....

$$\frac{\partial u}{\partial x_m} + \sum_{i=m+1}^{n} \left(\frac{\partial u}{\partial p_i}\frac{\partial \phi_m}{\partial x_i} - \frac{\partial u}{\partial x_i}\frac{\partial \phi_m}{\partial p_i}\right) = 0.$$

The original system of m equations was complete: the transformed system, with the condition that the integrals u do not involve $p_1, \ldots, p_m$, is also complete. The number of variables involved is $2n - m$, being $x_1, \ldots, x_n, p_{m+1}, \ldots, p_n$: so that, as the system for the quantity u is now a complete Jacobian system, it possesses $2n - 2m$ integrals*, which are functionally independent of one another.

* It is easy to see that this result comes also from the earlier form of the equations for u: that form involves $2n$ variables, and so there are $2n - m$ integrals. But $u = p_1 - \phi_1, \ldots, u = p_m - \phi_m$ are seen to be m integrals: and therefore there are other $2n - 2m$ integrals.

Two limitations are, however, imposed upon these integrals, so that not all of them can be retained for our purpose. In the first place, the aggregate of $n-m$ equations required must be such that

$$J\left(\frac{u_{m+1}, \ldots, u_n}{p_{m+1}, \ldots, p_n}\right)$$

does not vanish identically. In the second place, let $u_{m+1}=a_{m+1}$ be an integral of the system: it must involve one or more of the quantities $p_{m+1}, \ldots, p_n$, and it must be resoluble with regard to one of them, because otherwise all the derivatives $\frac{\partial u}{\partial p_{m+1}}, \ldots, \frac{\partial u}{\partial p_n}$ would vanish: let the resolved form be

$$p_{m+1}-\phi_{m+1}=0,$$

where ϕ_{m+1} involves a_{m+1}. Any other of the $2n-2m-1$ remaining integrals, say v, undoubtedly satisfies

$$(p_1-\phi_1, v)=0, \ \ldots, \ (p_m-\phi_m, v)=0;$$

but, for our purpose of proceeding to the determination of the n quantities p, it must also satisfy the relation

$$(p_{m+1}-\phi_{m+1}, v)=0;$$

and this relation will not, in general, be satisfied for any one of the $2n-2m-1$ integrals, selected at random. Accordingly, it is not necessary to obtain all the $2n-2m$ integrals of the system though, if they are known, they can be used in the construction of v as an appropriate functional combination of the $2n-2m-1$ integrals other than u_{m+1}: it is sufficient at this stage to obtain a single integral of the transformed system. Denoting this single integral by u_{m+1}, we resolve the equation $u_{m+1}=a_{m+1}$ with regard (say) to p_{m+1}, in the form $p_{m+1}=\phi_{m+1}$; and then any other equation $v=$ constant, that can coexist with the original system and with $p_{m+1}=\phi_{m+1}$, must satisfy the necessary and sufficient conditions

$$(p_r-\phi_r, v)=0,$$

for $r=1, \ldots, m+1$. As before, v may be assumed not to contain $p_1, \ldots, p_m$: and for reasons similar to those adduced before, it may be assumed not to contain p_{m+1}, so that

$$\frac{\partial v}{\partial p_1}=0, \ \ldots, \ \frac{\partial v}{\partial p_{m+1}}=0;$$

and then the system of equations for v is

$$\frac{\partial v}{\partial x_r} + \sum_{i=m+2}^{n} \left(\frac{\partial v}{\partial p_i} \frac{\partial \phi_r}{\partial x_i} - \frac{\partial v}{\partial x_i} \frac{\partial \phi_r}{\partial p_i} \right) = 0,$$

for $r = 1, \ldots, m+1$. The system of $m+1$ equations is complete: it involves the $2n-m-1$ variables $x_1, \ldots, x_n, p_{m+2}, \ldots, p_n$; and so it possesses $2n-2m-2$ functionally independent integrals.

57. At each stage, we have a complete Jacobian system for the determination of a quantity u, such that an equation $u=a$ can be associated with the system of equations for the variable z. The theory of these Jacobian systems, as explained in the preceding chapter, shews that they do possess a number of integrals; and therefore quantities u of the appropriate type do exist, so that we require only their explicit expressions in order to formulate the successive equations $u=a$.

We thus may pass from stage to stage: at each step, an integral of a number of simultaneous equations, forming a complete Jacobian system, is required: and as, at any stage, the number of equations has become greater while the number of variables has become less than at the preceding stage, the construction of the integrals in succession is successively simpler.

At each stage, what is required is a single integral belonging to the complete Jacobian system then framed: this integral must involve one of the variables p still surviving in the system*. For this purpose, we may use either Mayer's method or the amplified Jacobian method devised for complete linear systems; but it is not necessary to work either method to the complete issue, because all that is wanted is a single integral of the simultaneous system, not the aggregate of functionally independent integrals of the system.

Each new equation of the type $u =$ constant, associated with the system in its amplified condition before the derivation of the particular u, introduces an arbitrary constant. Thus, at the end of the series of operations which result in giving n equations, the number of arbitrary constants introduced is $n-m$; when the

* In case, at any stage, an appropriate integral of this type may not conveniently be obtainable, while an integral involving the variables x and some of the old variables may be forthcoming, a transformation similar to that adopted for a corresponding difficulty, hereafter (§§ 58, 59) discussed, will be effective.

n equations are fully resolved for $p_1, \ldots, p_n$, the expressions for these quantities involve $x_1, \ldots, x_n, a_1, \ldots, a_{n-m}$. When these values are introduced into the equation

$$dz = p_1 dx_1 + \ldots + p_n dx_n,$$

the right-hand side is an exact differential; when the quadrature of this exact differential is effected, we have

$$z = \phi(x_1, \ldots, x_n, a_1, \ldots, a_{n-m}) + b,$$

where b is an arbitrary constant. This equation gives the required value of z as an integral common to the system of equations: its expression contains $n - m + 1$ constants.

Ex. Obtain a common integral (if it exist) of the simultaneous equations

$$\left.\begin{aligned} F_1 = p_1 p_2 - x_3 x_4 = 0 \\ F_2 = p_3 p_4 - x_1 x_2 = 0 \end{aligned}\right\}.$$

We have

$$(F_1, F_2) = p_1 x_1 + p_2 x_2 - p_3 x_3 - p_4 x_4;$$

the right-hand side must vanish, and it clearly does not vanish in virtue of $F_1 = 0$, $F_2 = 0$; hence we have a new equation to be associated with the first two, and we write

$$F_3 = p_1 x_1 + p_2 x_2 - p_3 x_3 - p_4 x_4 = 0.$$

We now have

$$\begin{aligned} (F_1, F_2) &= F_3, \\ (F_1, F_3) &= -2F_1, \\ (F_2, F_3) &= 2F_2; \end{aligned}$$

all the quantities of the type (F_r, F_s) vanish in virtue of the three equations; hence these equations are a complete system.

Resolving the three equations $F_1 = 0$, $F_2 = 0$, $F_3 = 0$, so as to express p_1, p_2, p_3 in terms of the other variables that occur, we find two systems, viz.

$$\text{(i)} \qquad p_1 = \frac{x_2 x_3}{p_4}, \quad p_2 = \frac{x_4 p_4}{x_2}, \quad p_3 = \frac{x_1 x_2}{p_4};$$

$$\text{(ii)} \qquad p_1 = \frac{x_4 p_4}{x_1}, \quad p_2 = \frac{x_1 x_3}{p_4}, \quad p_3 = \frac{x_1 x_2}{p_4}.$$

The second of these two sets is derivable from the first by interchanging the variables x_1 and x_2; hence its integral must be similarly derivable from the integral of the first.

To obtain this integral, we need an equation $u = a$, where a is a constant and u must involve p_4; and u is determined by the equations

$$\left(p_1 - \frac{x_2 x_3}{p_4}, u\right) = 0, \quad \left(p_2 - \frac{x_4 p_4}{x_2}, u\right) = 0, \quad \left(p_3 - \frac{x_1 x_2}{p_4}, u\right) = 0,$$

together with the justifiable assumption that u is explicitly independent of p_1, p_2, p_3. These equations are

$$0=\frac{\partial u}{\partial x_1}+\frac{x_2x_3}{p_4^2}\frac{\partial u}{\partial x_4},$$

$$0=\frac{\partial u}{\partial x_2}-\frac{x_4}{x_2}\frac{\partial u}{\partial x_4}+\frac{p_4}{x_2}\frac{\partial u}{\partial p_4},$$

$$0=\frac{\partial u}{\partial x_3}+\frac{x_1x_2}{p_4^2}\frac{\partial u}{\partial x_4};$$

and the fact that they are complete can easily be verified.

The Mayer solution of these equations is as follows. We transform the variables by the relations

$$x_1=y_1,$$
$$x_2=x_2+(y_1-a_1)\,y_2,$$
$$x_3=x_3+(y_1-a_1)\,y_3;$$

and we form the single equation

$$\frac{\partial u}{\partial y_1}+Y_1\frac{\partial u}{\partial x_4}+Y_2\frac{\partial u}{\partial p_4}=0,$$

where

$$Y_1=\frac{x_2x_3}{p_4^2}-\frac{x_4}{x_2}y_2+\frac{x_1x_2}{p_4^2}y_3,$$

$$Y_2=\frac{p_4}{x_2}y_2.$$

An integral of the subsidiary system

$$dy_1=\frac{dx_4}{Y_1}=\frac{dp_4}{Y_2},$$

where y_2 and y_3 are arbitrary parameters, is required involving p_4: one such integral is clearly derivable from

$$dy_1=\frac{dp_4}{Y_2}=\frac{dp_4}{p_4}\,\frac{a_2+(y_1-a_1)\,y_2}{y_2},$$

in the form

$$\frac{p_4}{a_2+(y_1-a_1)\,y_2}=\text{constant}.$$

Then an integral of the original system is given by

$$\frac{p_4}{a_2+(y_1-a_1)\,y_2}=\frac{c}{a_2},$$

that is, by

$$\frac{p_4}{x_2}=\frac{c}{a_2}.$$

Hence $u=\dfrac{p_4}{x_2}$ is an integral of the three equations; as it involves p_4, it is of the required type.

To deduce the value of z, we take

$$\frac{p_4}{x_2}=a,$$

and then the values of p_1, p_2, p_3, p_4 are

$$p_1=\frac{x_3}{a}, \quad p_2=ax_4, \quad p_3=\frac{x_1}{a}, \quad p_4=ax_2:$$

inserting these values in

$$dz=p_1dx_1+p_2dx_2+p_3dx_3+p_4dx_4,$$

we find that

$$z=\frac{x_1x_3}{a}+ax_2x_4+b$$

is an integral of the first set of equations derived from the resolution of $F_1=0$, $F_2=0$, $F_3=0$, and therefore is an integral of the original equations.

Effecting upon this integral the interchange of variable whereby the first and the second resolved sets are interchanged, we find that

$$z=\frac{x_2x_3}{a}+ax_1x_4+b$$

is also an integral of the original equations $F_1=0$, $F_2=0$

We thus have two distinct integrals, each involving two arbitrary constants: and they are the only integrals that are thus obtainable. Their relation (if any) to one another, and the derivation of other integrals (if any) from them, belong to a range of subsequent investigation.

The amplified Jacobian method of solution is simple in the present case and leads very directly to the integral

$$u=\frac{p_4}{x_2}:$$

the rest of the analysis is the same as before.

58. The preceding investigation has rested on the two assumptions: (i) that the equations $F_1=0, \ldots, F_m=0$ of the complete system can be resolved with regard to m of the variables $p_1, \ldots, p_n$: (ii) that the equations of the complete amplified system $F_1=0, \ldots, F_m=0$, $u_{m+1}=a_{m+1}, \ldots, u_n=a_n$, can be resolved with regard to $p_1, \ldots, p_n$, so that the Jacobian

$$J\begin{pmatrix}F_1, \ldots, F_m, u_{m+1}, \ldots, u_n \\ p_1, \ldots\ldots\ldots\ldots\ldots\ldots, p_n\end{pmatrix}$$

does not vanish identically. The latter assumption is, however, unnecessary: and, as has been proved by Mayer*, it is sufficient that the n functionally independent equations $F_1=0, \ldots, F_m=0$, $u_{m+1}=a_{m+1}, \ldots, u_n=a_n$ should be resoluble with regard to n of the quantities which they involve.

* *Math. Ann.*, t. VIII (1875), p. 313.

Still retaining the first assumption, let the m equations $F_1=0$, ..., $F_m=0$ be resoluble with respect to p_1, ..., p_m; and let the resolved set be

$$p_1-\phi_1=0, \ldots, p_m-\phi_m=0.$$

Let X be a function of all the variables such that

$$(F_r, X)=0,$$

for $r=1$, ..., m; and let ξ denote the value of X which results from substituting ϕ_1, ..., ϕ_m as the values of p_1, ..., p_m in X; then

$$(p_r-\phi_r, \xi)=0.$$

For

$$\frac{\partial \xi}{\partial x_i}=\frac{\partial X}{\partial x_i}+\sum_{k=1}^{m}\frac{\partial X}{\partial p_k}\frac{\partial \phi_k}{\partial x_i},$$

so that

$$\frac{\partial X}{\partial x_i}=\frac{\partial \xi}{\partial x_i}+\sum_{k=1}^{m}\frac{\partial X}{\partial p_k}\frac{\partial (p_k-\phi_k)}{\partial x_i},$$

for $i=1$, ..., m. Similarly

$$\frac{\partial X}{\partial p_j}=\frac{\partial \xi}{\partial p_j}+\sum_{k=1}^{m}\frac{\partial X}{\partial p_k}\frac{\partial (p_k-\phi_k)}{\partial p_j},$$

for $j=m+1$, ..., n; and this last relation is identically true for $j=1$, ..., m, because neither ξ nor any one of the quantities ϕ_1, ..., ϕ_m, involves p_1, ..., p_m; that is, the relation is true for $j=1$, ..., n. Also, when the values of p_1, ..., p_m are substituted in the equations $F_1=0$, ..., $F_m=0$, these become identities: hence

$$\frac{\partial F_r}{\partial x_i}=\sum_{h=1}^{m}\frac{\partial F_r}{\partial p_h}\frac{\partial (p_h-\phi_h)}{\partial x_i},$$

for values of $i=1$, ..., n; and (as above)

$$\frac{\partial F_r}{\partial p_j}=\sum_{h=1}^{m}\frac{\partial F_r}{\partial p_h}\frac{\partial (p_h-\phi_h)}{\partial p_j},$$

first for $j=m+1$, ..., n, from the identical equation, and obviously identically for $j=1$, ..., m, that is, for values of $j=1$, ..., n. Hence

$$(F_r, X)=\sum_{h=1}^{m}\frac{\partial F_r}{\partial p_h}(p_h-\phi_h, \xi)+\sum_{h=1}^{m}\sum_{k=1}^{m}\frac{\partial F_r}{\partial p_h}\frac{\partial X}{\partial p_k}(p_h-\phi_h, p_k-\phi_k).$$

Now we have

$$(p_h-\phi_h, p_k-\phi_k)=0,$$

and by hypothesis,

$$(F_r, X) = 0,$$

for all values of r; hence

$$\sum_{h=1}^{m} \frac{\partial F_r}{\partial p_h} (p_h - \phi_h, \xi) = 0,$$

holding for $r = 1, \ldots, m$. But

$$J\left(\frac{F_1, \ldots, F_m}{p_1, \ldots, p_m}\right)$$

is not zero, because of the assumed resolubility of the equations $F_1 = 0, \ldots, F_m = 0$ with respect to $p_1, \ldots, p_m$: hence the preceding m relations can only be satisfied by

$$(p_h - \phi_h, \xi) = 0;$$

which was to be proved.

Next, suppose that (by some method or other) we possess $n - m$ equations

$$u_{m+1} = a_{m+1}, \ldots, u_n = a_n,$$

which coexist with $F_1 = 0, \ldots, F_m = 0$, and with one another; the n equations in the aggregate being functionally independent of one another. The original system of m equations is certainly resoluble with regard to $p_1, \ldots, p_m$; the amplified system of n equations is resoluble with regard to n of the variables, which can certainly be chosen so as to include $p_1, \ldots, p_m$ and may include others of the quantities p though perhaps not all of them. Suppose, then, that the variables chosen for resolution include $p_1, \ldots, p_\mu$, where $\mu \geqslant m$, but not more than μ of the quantities p; the resolved equations will be equivalent to μ equations of the amplified system, say to

$$F_1 = 0, \ldots, F_m = 0, \quad u_{m+1} = a_{m+1}, \ldots, u_\mu = a_\mu.$$

In the remaining equations of the system, let the values of $p_1, \ldots, p_\mu$ be substituted, and suppose that they become

$$v_{\mu+1} = a_{\mu+1}, \ldots, v_n = a_n,$$

these equations not being resoluble for any of the quantities $p_{\mu+1}, \ldots, p_n$, and consequently not involving any of these quantities. Then, exactly as in the preceding case, we have

$$(p_h - \phi_h, v_k) = 0,$$

for $h=1, \ldots, \mu$, and $k=\mu+1, \ldots, n$: taking account of the fact that v_k involves none of the variables $p_1, \ldots, p_n$, we may write this set of equations in the form

$$\frac{\partial v_k}{\partial x_h} - \sum_{i=\mu+1}^{n} \frac{\partial v_k}{\partial x_i} \frac{\partial \phi_h}{\partial p_i} = 0,$$

for all the values of h and k.

The quantities $v_{\mu+1}, \ldots, v_n$ involve the variables $x_1, \ldots, x_n$: we prove, as follows, that they are functionally independent combinations of $x_{\mu+1}, \ldots, x_n$. Otherwise, there would be some relation

$$f(x_1, \ldots, x_\mu, v_{\mu+1}, \ldots, v_n) = 0,$$

which would have to be satisfied identically when the values of $v_{\mu+1}, \ldots, v_n$ in terms of the variables $x_1, \ldots, x_n$ are substituted; and as it would involve one or more of the quantities v, it could be resolved with regard (say) to v_n in the form

$$v_n = g(x_1, \ldots, x_\mu, v_{\mu+1}, \ldots, v_{n-1}).$$

This relation would also be an identity when the values of $v_{\mu+1}, \ldots, v_n$ in terms of $x_1, \ldots, x_n$ are substituted. Now

$$\frac{\partial v_n}{\partial x_h} - \sum_{i=\mu+1}^{n} \frac{\partial v_n}{\partial x_i} \frac{\partial \phi_h}{\partial p_i} = 0;$$

substituting $g(x_1, \ldots, x_\mu, v_{\mu+1}, \ldots, v_{n-1})$ for v_n, this gives

$$\left(\frac{\partial g}{\partial x_h} + \sum_{\lambda=\mu+1}^{n-1} \frac{\partial g}{\partial v_\lambda} \frac{\partial v_\lambda}{\partial x_h}\right) - \sum_{\lambda=\mu+1}^{n-1} \frac{\partial g}{\partial v_\lambda} \sum_{i=\mu+1}^{n} \frac{\partial v_\lambda}{\partial x_i} \frac{\partial \phi_h}{\partial p_i} = 0;$$

but

$$\frac{\partial v_\lambda}{\partial x_h} - \sum_{i=\mu+1}^{n} \frac{\partial v_\lambda}{\partial x_i} \frac{\partial \phi_h}{\partial p_i} = 0,$$

for $\lambda = \mu+1, \ldots, n-1$, and therefore

$$\frac{\partial g}{\partial x_h} = 0,$$

for all the values of h. Thus the above expression for v_n would give

$$v_n = g(v_{\mu+1}, \ldots, v_{n-1}),$$

and the quantities $v_{\mu+1}, \ldots, v_n$ would not be functionally independent of one another, contrary to the construction of the quantities u. Hence $v_{\mu+1} = a_{\mu+1}, \ldots, v_n = a_n$, are functionally independent combinations of $x_{\mu+1}, \ldots, x_n$.

The equations $v_{\mu+1} = a_{\mu+1}, \ldots, v_n = a_n$ can therefore be resolved with regard to $x_{\mu+1}, \ldots, x_n$; and consequently the system

$$F_1 = 0, \ldots, F_m = 0, \quad u_{m+1} = a_{m+1}, \ldots, u_n = a_n$$

can be resolved with regard to $p_1, \ldots, p_\mu, x_{\mu+1}, \ldots, x_n$. Let a resolved set be

$$p_h = \psi_h(x_1, \ldots, x_\mu, p_{\mu+1}, \ldots, p_n),$$

$$x_k = \theta_k(x_1, \ldots, x_\mu, p_{\mu+1}, \ldots, p_n),$$

for $h = 1, \ldots, \mu$, and $k = \mu + 1, \ldots, n$, the functions θ and ψ involving the arbitrary constants.

59. Now take a new dependent variable Z, defined by the contact-transformation*

$$Z = z - p_{\mu+1}x_{\mu+1} - \ldots - p_n x_n,$$

being a transformation of a type first used by Lagrange†; then

$$dZ = p_1 dx_1 + \ldots + p_\mu dx_\mu - x_{\mu+1}\, dp_{\mu+1} - \ldots - x_n dp_n.$$

We write

$$x_1, \ldots, x_\mu, -p_{\mu+1}, \ldots, -p_n = y_1, \ldots, y_\mu, y_{\mu+1}, \ldots, y_n,$$

respectively, and regard $y_1, \ldots, y_n$ as new independent variables: then, denoting by $q_1, \ldots, q_n$ the derivatives of the new dependent variable with regard to the new independent variables, we have

$$p_h = q_h, \quad x_k = q_k,$$

for $h = 1, \ldots, \mu$, and $k = \mu + 1, \ldots, n$. Let

$$F_r(x_1, \ldots, x_n, p_1, \ldots, p_n) = G_r(y_1, \ldots, y_n, q_1, \ldots, q_n),$$

$$u_s(x_1, \ldots, x_n, p_1, \ldots, p_n) = w_s(y_1, \ldots, y_n, q_1, \ldots, q_n),$$

on effecting these changes: then as the equations

$$F_1 = 0, \ldots, F_m = 0, \quad u_{m+1} = a_{m+1}, \ldots, u_n = a_n,$$

are resoluble with regard to $p_1, \ldots, p_\mu, x_{\mu+1}, \ldots, x_n$, the equations

$$G_1 = 0, \ldots, G_m = 0, \quad w_{m+1} = a_{m+1}, \ldots, w_n = a_n,$$

are resoluble with regard to $q_1, \ldots, q_n$. Moreover, the equations

$$F_1 = 0, \ldots, F_m = 0, \quad u_{m+1} = a_{m+1}, \ldots, u_n = a_n,$$

* Called *tangential transformation* in Part I of this work: the phrase *contact-transformation* is now customary, and it will be herein adopted whenever reference to it is required.

† *Œuvres complètes*, t. IV, p. 84.

satisfy the relations

$$(F_r, F_s)=0, \quad (F_r, u_i)=0, \quad (u_i, u_j)=0,$$

for $r, s=1, \ldots, m$, and $i, j=m+1, \ldots, n$. Now

$$\frac{\partial F_r}{\partial x_h}=\frac{\partial G_r}{\partial y_h}, \quad \frac{\partial F_r}{\partial p_h}=\frac{\partial G_r}{\partial q_h},$$

$$\frac{\partial F_r}{\partial x_k}=\frac{\partial G_r}{\partial q_k}, \quad \frac{\partial F_r}{\partial p_k}=-\frac{\partial G_r}{\partial y_k},$$

for $h=1, \ldots, \mu$, and $k=\mu+1, \ldots, n$; and similarly for relations between derivatives of u and w. Thus

$$\begin{aligned}(F_r, F_s) &= \sum_{h=1}^{\mu}\left(\frac{\partial G_r}{\partial y_h}\frac{\partial G_s}{\partial q_h}-\frac{\partial G_r}{\partial q_h}\frac{\partial G_s}{\partial y_h}\right) \\ &\qquad + \sum_{k=\mu+1}^{n}\left\{\frac{\partial G_r}{\partial q_k}\left(-\frac{\partial G_s}{\partial y_k}\right)-\left(-\frac{\partial G_r}{\partial y_k}\right)\frac{\partial G_s}{\partial q_k}\right\} \\ &= \sum_{i=1}^{n}\left(\frac{\partial G_r}{\partial y_i}\frac{\partial G_s}{\partial q_i}-\frac{\partial G_r}{\partial q_i}\frac{\partial G_s}{\partial y_i}\right) \\ &= (G_r, G_s);\end{aligned}$$

and similarly

$$(F_r, u_i)=(G_r, w_i), \quad (u_i, u_j)=(w_i, w_j).$$

Consequently

$$(G_r, G_s)=0, \quad (G_r, w_i)=0, \quad (w_i, w_j)=0;$$

and the equations

$$G_1=0, \ldots, G_m=0, \quad w_{m+1}=a_{m+1}, \ldots, w_n=a_n$$

are resoluble with regard to $q_1, \ldots, q_n$, expressing these quantities in terms of $y_1, \ldots, y_n, a_{m+1}, \ldots, a_n$. Moreover, the earlier results shew that the values of $q_1, \ldots, q_n$ thus given make

$$dZ=q_1 dy_1+\ldots+q_n dy_n$$

an exact equation: when the quadrature is effected, the result will be of the form

$$Z=\psi(y_1, \ldots, y_n, a_{m+1}, \ldots, a_n)+b,$$

where b is an arbitrary constant. To obtain the value of z, we note that

$$x_k=\frac{\partial Z}{\partial y_k}=\frac{\partial \psi}{\partial y_k},$$

for $k = \mu + 1, \ldots, n$; and

$$Z = z - x_{\mu+1}p_{\mu+1} - \ldots - x_n p_n$$
$$= z + x_{\mu+1}y_{\mu+1} + \ldots + x_n y_n;$$

hence

$$z + x_{\mu+1}y_{\mu+1} + \ldots + x_n y_n$$
$$= \psi(y_1, \ldots, y_n, a_{m+1}, \ldots, a_n) + b$$
$$= \psi(x_1, \ldots, x_\mu, y_{\mu+1}, \ldots, y_n, a_{m+1}, \ldots, a_n) + b,$$
$$x_k = \frac{\partial \psi}{\partial y_k},$$

for $k = \mu + 1, \ldots, n$. Eliminating $y_{\mu+1}, \ldots, y_n$ among these $n - \mu + 1$ relations, and resolving the eliminant with regard to z, we have a relation

$$z = I(x_1, \ldots, x_n, a_{m+1}, \ldots, a_n) + b,$$

for z occurs in the elimination only in the combination $z - b$; this relation is the integral of the original system of equations, and it involves $n - m + 1$ constants.

One such integral will arise for each resolved set of equations arising out of the resolution of the equations

$$F_1 = 0, \ldots, F_m = 0, \quad u_{m+1} = a_{m+1}, \ldots, \quad u_n = a_n;$$

the aggregate of these integrals includes all the integrals that are thus obtainable. But other integrals may be deduced by other processes, which will form the subject of subsequent explanations.

Ex. Consider an equation

$$f = 3px + qy + q^3x^2 = 0.$$

An equation $u = a$ is required, which may coexist with $f = 0$: it is given by

$$(f, u) = 0,$$

an equation that is homogeneous and linear in u: and an integral is required which involves either p or q or both. The subsidiary equations are

$$\frac{dx}{3x} = \frac{dp}{-3p - 2q^3x^2} = \frac{dy}{y + 3q^2x^2} = \frac{dq}{-q};$$

one integral of these equations is

$$q^3x = \text{constant};$$

and another integral is

$$\frac{y}{xq^2} - x = \text{constant}.$$

Taking the former integral, we have to resolve the equations

$$f = 0, \quad q^3x = c^3,$$

where c is a constant. Resolution with respect to p and q is simple, giving

$$q=cx^{-\frac{1}{3}},\quad p=-\tfrac{1}{3}c^3-\tfrac{1}{3}cyx^{-\frac{4}{3}};$$

substituting in

$$dz=p\,dx+q\,dy,$$

and effecting the quadrature, we find

$$z=A-\tfrac{1}{3}c^3x+cx^{-\frac{1}{3}}y,$$

which is an integral involving two arbitrary constants.

Taking the other integral of the subsidiary equations, we have to resolve the equations

$$f=0,\quad \frac{y}{xq^2}-x=2a,$$

where a is a constant. Resolution with regard to p and q is possible: it is simpler with regard to p and y, and with respect to these variables gives the relations

$$F=y-q^2(x^2+2ax)=0,\quad G=p+\tfrac{2}{3}q^3(x+a)=0,$$

which satisfy the relation $(F, G)=0$ identically. After the investigations above, we take q and x as the new independent variables and Z as the new dependent variable, where

$$Z=z-qy.$$

Thus

$$\begin{aligned}dZ&=p\,dx-y\,dq\\&=-\tfrac{2}{3}q^3(x+a)\,dx-q^2(x^2+2ax)\,dq,\end{aligned}$$

so that

$$Z=B-\tfrac{1}{3}q^3(x^2+2ax).$$

Now

$$y=-\frac{\partial Z}{\partial q}=q^2(x^2+2ax),$$

so that we have to eliminate q between the equations

$$y=q^2(x^2+2ax),\quad z-qy=B-\tfrac{1}{3}q^3(x^2+2ax).$$

The result is

$$z-B=\tfrac{2}{3}y^{\frac{3}{2}}(x^2+2ax)^{-\frac{1}{2}},$$

another integral involving two arbitrary constants.

Later, the relation between different integrals will be considered.

60. Reasons were adduced in § 54 for discussing equations in a form which does not explicitly contain the dependent variable; but it should be added that the preceding method can be applied also when the dependent variable does occur explicitly. In that case, the investigation follows the same lines as before, but the analysis is rather more complicated on account of the occurrence of z: it will be sufficient to give merely an outline.

Let $f_1=0, \ldots, f_m=0$ be a complete system of equations in the n independent variables, involving the dependent variable and its first derivatives: then the relation

$$[f_r, f_s]=0$$

is satisfied for all values of r and s, either explicitly or in virtue of the equations of the system. Any other equation, coexisting with the equations of the system in a form $u=$ constant, must be such that

$$[f_i, u]=0,$$

for all values $i=1, \ldots, m$.

Suppose that the system of equations $f_1=0, \ldots, f_m=0$ is resolved with regard to z and $m-1$ of the variables p, say $p_1, \ldots, p_{m-1}$, in the form

$$z-\psi=0, \quad p_1-\psi_1=0, \ldots, p_{m-1}-\psi_{m-1}=0;$$

the resolved system is in involution, for (§ 55, Note) the relations

$$[p_i-\psi_i, p_j-\psi_j]=0,$$

$$[z-\psi-x_1(p_1-\psi_1)-\ldots-x_{m-1}(p_{m-1}-\psi_{m-1}), p_i-\psi_i]=0,$$

for all values of i and j from the set $1, \ldots, m-1$, are satisfied identically. Let these values of $z, p_1, \ldots, p_{m-1}$, in terms of $x_1, \ldots, x_n, p_m, \ldots, p_n$, be substituted in u and let the resulting value be denoted by w; then the equations

$$[z-\psi-x_1(p_1-\psi_1)-\ldots-x_{m-1}(p_{m-1}-\psi_{m-1}), w]=0,$$

$$[p_i-\psi_i, w]=0,$$

are satisfied in virtue of $[f_i, u]=0$, and conversely.

Moreover, the system of equations determining w is a complete system. For if

$$g_1=0, \ldots, g_m=0$$

is a complete system in involution, then the identical relation

$$[[g_r, g_s]w]+[[g_s, w]g_r]+[[w, g_r]g_s]$$
$$=-\frac{\partial w}{\partial z}[g_r, g_s]-\frac{\partial g_r}{\partial z}[g_s, w]-\frac{\partial g_s}{\partial z}[w, g_r]$$

becomes

$$[g_r, [g_s, w]]-[g_s, [g_r, w]]=\frac{\partial g_r}{\partial z}[g_s, w]-\frac{\partial g_s}{\partial z}[g_r, w],$$

because $[g_r, g_s]$ vanishes identically: hence, denoting $[g_i, w]$ by $B_i(w)$, we have

$$B_r(B_s w)-B_s(B_r w)=\frac{\partial g_r}{\partial z}B_s(w)-\frac{\partial g_s}{\partial z}B_r(w)$$
$$=0,$$

in virtue of $B_r(w)=0$, $B_s(w)=0$, which is the test of a complete system.

As the equations

$$[z-\psi-x_1(p_1-\psi_1)-\ldots-x_{m-1}(p_{m-1}-\psi_{m-1}),\ w]=0,$$
$$[p_i-\psi_i,\ w]=0,$$

are a complete system, they possess a simultaneous set of integrals: let one such integral involving some one of the variables $p_m, \ldots, p_n$ be obtainable in the form

$$w=w(x_1, \ldots, x_n, p_m, \ldots, p_n);$$

then the equation

$$w(x_1, \ldots, x_n, p_m, \ldots, p_n)=a,$$

where a is an arbitrary constant, coexists with

$$z-\psi=0, \quad p_1-\psi_1=0, \ldots, p_{m-1}-\psi_{m-1}=0.$$

Let it be resolved so as to give (say) p_m in terms of the other variables it contains, and denote the result by

$$p_m=\chi_m;$$

and let this value be inserted in the other equations so that they take the form

$$z-\chi=0, \quad p_1-\chi_1=0, \ldots, p_{m-1}-\chi_{m-1}=0.$$

Then for the next stage, we proceed from the $m+1$ equations

$$z-\chi=0, \quad p_1-\chi_1=0, \ldots, p_m-\chi_m=0,$$

as in this stage from the m equations.

When $n+1$ equations have been obtained, the first of them has a form

$$z-\theta=0,$$

where θ involves $n-m+1$ constants; $z=\theta$ is an integral of the original system.

Jacobi's Second Method, when z does not occur.

61. The preceding investigation has been carried out after an initial assumption that the m equations in the given complete system are resoluble with regard to m of the variables $p_1, \ldots, p_n$: the selection of $p_1, \ldots, p_m$ was merely typical. This assumption is not any real limitation: for if the m equations

$$F_1=0, \ldots, F_m=0$$

are not theoretically resoluble with regard to any m of the variables $p_1, \ldots, p_n$, so that all the determinants

$$\begin{vmatrix} \frac{\partial F_1}{\partial p_1}, & \ldots, & \frac{\partial F_1}{\partial p_n} \\ \cdots & \cdots & \cdots \\ \frac{\partial F_m}{\partial p_1}, & \ldots, & \frac{\partial F_m}{\partial p_n} \end{vmatrix}$$

vanish, then $p_1, \ldots, p_n$ can be eliminated among the m equations: as the m equations are functionally distinct, the eliminant cannot vanish identically and so would take a form

$$\Theta(x_1, \ldots, x_n) = 0,$$

a relation among the independent variables alone. Such a result is excluded: and so the m equations are resoluble with regard to some selection of m variables from the set $p_1, \ldots, p_n$.

The forms of the resolved equations may, however, be complicated: and then it might be desirable to proceed from the unresolved equations. Such a process was given by Jacobi, and it is sometimes called his *second method*; naturally, it is less simple than the method that has just been expounded, for it deals with equations of a less simple form than those to which Mayer's method is applied. Indeed, the preceding process is really a form of Jacobi's method: but it has been simplified and shortened by the improvements and the developments due to Lie and to Mayer.

Thus far in the range of these discussions, we have been considering m equations: and though there is no intrinsic element in the analysis which makes m greater than unity, all the superficial appearance suggests that m is not unity. For variety, we shall now deal with the integration of a single equation: and it will be found that, in general, the process leads to the issue through the integration of systems. For this purpose, we shall use Jacobi's method: a sufficient indication of its detailed working, whether for single equations or for detailed systems, will thus be provided.

As already hinted, Jacobi's method of integration (without the modifications and amplifications introduced by the investigations of Lie and Mayer) appears to be most useful when, from whatever cause, the equation or equations are not resolved with regard to one or more of the derivatives. We begin with a single irreducible equation, unresolved with regard to any of the variables p and not explicitly containing the dependent variable: it may be taken in the form

$$f = f(x_1, \ldots, x_n, p_1, \ldots, p_n) = 0.$$

By the process adopted, other $n-1$ equations are required which, speaking generally*, would suffice for the expression of $p_1, \ldots, p_n$ in terms of $x_1, \ldots, x_n$.

If $u =$ constant be such an equation, then the relation

$$(f, u) = 0$$

must be satisfied; any integral of this equation, distinct from f (which manifestly is an integral) and involving some of the variables $p_1, \ldots, p_n$, will suffice for the purpose. The system of ordinary equations, subsidiary to the construction of this integral, is

$$\frac{dx_1}{-\frac{\partial f}{\partial p_1}} = \ldots = \frac{dx_n}{-\frac{\partial f}{\partial p_n}} = \frac{dp_1}{\frac{\partial f}{\partial x_1}} = \ldots = \frac{dp_n}{\frac{\partial f}{\partial x_n}};$$

let $f_1 =$ constant be one integral of the system, where f_1 involves one at least of the quantities $p_1, \ldots, p_n$: then we may take

$$u = f_1.$$

The relation

$$(f, f_1) = 0$$

is satisfied identically: and the two equations

$$f_1 = a, \quad f = 0,$$

where a is an arbitrary constant, satisfy the conditions of coexistence.

62. We now proceed to obtain another equation, involving some of the variables p and coexisting with the two equations; if it be $v =$ constant, then the relations

$$(f, v) = 0, \quad (f_1, v) = 0,$$

must be satisfied. These effectively are two equations for the determination of v; any common integral of the appropriate form and functionally distinct from f and f_1 (both of which manifestly are integrals) will suffice. Now the equation $(f, v) = 0$ is the

* That is to say, omitting from consideration the alternative already discussed in §§ 58, 59.

same as that for the determination of u, so that the subsidiary ordinary equations are the same as before: let

$$\phi = \phi(x_1, \ldots, x_n, p_1, \ldots, p_n) = \text{constant},$$

be an integral, which involves some of the variables $p_1, \ldots, p_n$ and is functionally distinct from f and f_1; then the equation

$$(f, \phi) = 0$$

is satisfied identically.

If ϕ is such that $(f_1, \phi) = 0$, then we may take

$$v = \phi$$

as a common integral of the two equations.

If ϕ is such that (f_1, ϕ) does not vanish, then (f_1, ϕ) is either a constant, say c, or is a variable quantity, say ϕ_1. In the latter case, ϕ_1 is an integral of the equation $(f, u) = 0$, by Poisson's theorem (§ 52); and it is a new integral, if it is functionally distinct from f, f_1, ϕ.

Similarly, if ϕ_1 is a new integral of $(f, u) = 0$, we may have $(f_1, \phi_1) = 0$, in which case we may take

$$v = \phi_1$$

as a common integral of the two equations; or if (f_1, ϕ_1) is not zero, it is either a constant, say c', or is a variable quantity, say ϕ_2. As before, Poisson's theorem shews that ϕ_2 is an integral of the equation $(f, u) = 0$: it is a new integral, if it is functionally distinct from f, f_1, ϕ, ϕ_1.

Proceeding in this sequence, we have a number of functions $\phi, \phi_1, \phi_2, \ldots$; and provided (f_1, ϕ_μ) is a variable quantity, it is a new integral of the equation $(f, u) = 0$ if it is functionally distinct from $f, f_1, \phi, \phi_1, \ldots, \phi_\mu$. Now the number of functionally distinct integrals of $(f, u) = 0$ is not greater than $2n - 1$; hence, if the series of functions either should not cease, by the occurrence of a zero-value for (f, ϕ_r), or should not give a constant non-zero value for (f, ϕ_r), then we must sooner or later obtain a function ϕ_r which is expressible in terms of those already found. Let ϕ_i be the first function in the sequence which either is zero, or is a pure constant different from zero, or is expressible in terms of the preceding functions.

Then no new distinct integrals will arise from continuing the construction of the functions (f_1, ϕ). For in the first alternative and in the second alternative, we have $(f_1, \phi_i) = 0$: and if, in the third alternative

$$\phi_i = \theta(f, f_1, \phi, \phi_1, \ldots, \phi_{i-1}),$$

then

$$(f_1, \phi_i) = (f_1, f)\frac{\partial\theta}{\partial f} + (f_1, f_1)\frac{\partial\theta}{\partial f_1} + (f_1, \phi)\frac{\partial\theta}{\partial\phi} + \ldots + (f_1, \phi_{i-1})\frac{\partial\theta}{\partial\phi_{i-1}}$$

$$= \phi_1\frac{\partial\theta}{\partial\phi} + \ldots + \phi_i\frac{\partial\theta}{\partial\phi_{i-1}},$$

which is expressible in terms of the functions anterior to ϕ_i; and so for each succeeding function.

Accordingly, consider a functional combination of $\phi, \phi_1, \ldots, \phi_{i-1}$ represented by

$$v = g(\phi, \phi_1, \ldots, \phi_{i-1});$$

then

$$(f, v) = 0,$$

whatever be the form of the function g. Also, as above,

$$(f_1, v) = \phi_1\frac{\partial g}{\partial\phi} + \phi_2\frac{\partial g}{\partial\phi_1} + \ldots + \phi_i\frac{\partial g}{\partial\phi_{i-1}};$$

hence, if g can be determined so that the right-hand side vanishes, we shall have $(f_1, v) = 0$. In order to determine g from the relation

$$\phi_1\frac{\partial g}{\partial\phi} + \phi_2\frac{\partial g}{\partial\phi_1} + \ldots + \phi_i\frac{\partial g}{\partial\phi_{i-1}} = 0,$$

we consider the system of $i-1$ ordinary equations

$$\frac{d\phi}{\phi_1} = \frac{d\phi_1}{\phi_2} = \ldots = \frac{d\phi_{i-1}}{\phi_i};$$

their integral equivalent consists of $i-1$ distinct integral equations of the form

$$h_r(\phi, \phi_1, \phi_2, \ldots, \phi_{i-1}) = \text{constant}, \qquad (r = 1, \ldots, i-1),$$

whether ϕ_i be zero, or a constant, or be the foregoing quantity θ; and each of these functions h_r is such that

$$\phi_1\frac{\partial h_r}{\partial\phi} + \phi_2\frac{\partial h_r}{\partial\phi_1} + \ldots + \phi_i\frac{\partial h_r}{\partial\phi_{i-1}} = 0.$$

Hence, taking

$$v = h_r(\phi, \phi_1, \phi_2, \ldots, \phi_{i-1}),$$

we have

$$(f, v)=0, \quad (f_1, v)=0;$$

and thus we have $i-1$ distinct integrals common to the two equations.

If ϕ_i is zero, the simplest of these integrals is

$$v=\phi_{i-1};$$

even so, it is only one of $i-1$ distinct integrals common to the two equations.

Also i is greater than zero, because we have assumed that (f_1, ϕ), which is ϕ_1, does not vanish. Hence, if i is greater than unity, a common integral has been obtained; in that case, indeed, we have obtained $i-1$ common integrals of $(f, u)=0$, $(f_1, u)=0$, distinct from f and f_1. Consequently, this stage is completed except only when (f_1, ϕ), though not zero, either is a constant or is not functionally independent of f, f_1, ϕ: that is, in the case when $i=1$.

In the case when $i=1$ in connection with a quantity ϕ, we return to the equations subsidiary to $(f, u)=0$: and we determine another integral of them in the form

$$\psi=\psi(x_1, \ldots, x_n, p_1, \ldots, p_n)=\text{constant},$$

where ψ is functionally distinct from f, f_1, ϕ. We proceed with ψ in the same way as with ϕ, by forming the functions

$$(f_1, \psi)=\psi_1, \quad (f_1, \psi_1)=\psi_2, \ldots,$$

in succession; and, as before, we obtain an integral or a number of integrals common to the two equations

$$(f, u)=0, \quad (f_1, u)=0,$$

save only in the case where ψ_1, though not zero, is either a constant or is not functionally independent of f, f_1, ψ.

Even if the integral required is not provided because of the double lapse of the process into this exceptional stage, an integral as required can be obtained by a combination of the two integrals ϕ and ψ. Take any function $g(\phi, \psi, f_1)$: owing to the origin of ϕ and ψ, we have

$$(f, g)=0;$$

and

$$(f_1, g) = \frac{\partial g}{\partial \phi}(f_1, \phi) + \frac{\partial g}{\partial \psi}(f_1, \psi) + \frac{\partial g}{\partial f_1}(f_1, f_1)$$
$$= \frac{\partial g}{\partial \phi}\phi_1 + \frac{\partial g}{\partial \psi}\psi_1.$$

We form the equations

$$\frac{d\phi}{\phi_1} = \frac{d\psi}{\psi_1} = \frac{df_1}{0}:$$

in these equations, ϕ_1 either is a constant or is a functional combination of f, f_1, ϕ, say $\Phi(f, f_1, \phi)$; and likewise for ψ_1, which either is a constant or is a functional combination of f, f_1, ψ, say $\Psi(f, f_1, \psi)$. For our purposes, f is zero: one integral of the two ordinary equations is $f_1 = a_1$, where a_1 is an arbitrary constant; another integral is given by integrating

$$\frac{d\phi}{\Phi(0, a_1, \phi)} = \frac{d\psi}{\Psi(0, a_1, \psi)}.$$

Let an integral equivalent of this be

$$u(a_1, \phi, \psi) = \text{constant},$$

or say

$$u(f_1, \phi, \psi) = \text{constant}:$$

then if we take

$$g(\phi, \psi, f_1) = u(f_1, \phi, \psi),$$

we have

$$(f_1, u) = 0.$$

In other words, $u = u(f_1, \phi, \psi)$ is an integral common to the two equations $(f, u) = 0$, $(f_1, u) = 0$.

The simplest instance occurs when $\phi_1 = c$, $\psi_1 = c'$, where c and c' are constants: then

$$u = c'\phi - c\psi.$$

In every case, an integral common to the two equations $(f, u) = 0$, $(f_1, u) = 0$ has been obtained. It has required the assignment of certainly one integral of the equations subsidiary to $(f, u) = 0$: even when the functions (f_1, ϕ_r) have to be formed, each of them gives an integral of that subsidiary system, and so does each combination of the type $h_1(\phi, \phi_1, \ldots)$, $h_2(\phi, \phi_1, \ldots)$, ...; and only one of these combinations is assigned. The most unfavourable association is that in which the ϕ-series ends with ϕ_1 and a ψ-series ends with ψ_1; and then the two integrals ϕ and ψ

of the subsidiary system of $(f, u)=0$ must be assigned for the construction of an integral common to $(f, u)=0$, $(f_1, u)=0$.

Now the subsidiary system consists of $2n-1$ ordinary equations; its integral equivalent must consist of $2n-1$ independent equations. One of these is $f=0$, and another consists of $f_1=a_1$; hence there are other $2n-3$ independent equations, which may be denoted by

$$\phi=\text{constant},\quad \psi=\text{constant},\quad \chi=\text{constant},\quad \vartheta=\text{constant},\quad \ldots.$$

If $(f_1, \phi)=0$, then $u=\phi$ is the quantity desired. If (f_1, ϕ) is neither zero, nor a constant, nor a functional combination of f, f_1, ϕ, then there is a ϕ-series: and a single combination of the members of the series, (which must also, in the circumstances, be a combination of some of the quantities $\phi, \psi, \chi, \vartheta, \ldots$), will give a quantity u as required. The most unfavourable set of results possible is that in which the ϕ-series terminates with (f_1, ϕ), the ψ-series terminates with (f_1, ψ), and so on, no one of these quantities vanishing: then each of the quantities

$$\int\frac{d\phi}{(f_1, \phi)}-\int\frac{d\psi}{(f_1, \psi)},$$

$$\int\frac{d\phi}{(f_1, \phi)}-\int\frac{d\chi}{(f_1, \chi)},$$

$$\int\frac{d\phi}{(f_1, \phi)}-\int\frac{d\vartheta}{(f_1, \vartheta)},$$

$$\vdots$$

is an integral common to $(f, u)=0$, $(f_1, u)=0$. As there are $2n-3$ quantities $\phi, \psi, \chi, \vartheta, \ldots$, it follows that, even with the most unfavourable set of results, the two equations $(f, u)=0$ and $(f_1, u)=0$ possess $2n-4$ integrals in common, independent of f, of f_1, and of one another, and obtainable in this manner.

Let $u=f_2$ be one of these integrals: then the equation

$$f_2=a_2,$$

where a_2 is an arbitrary constant, associates itself with

$$f=0,\quad f_1=a_1.$$

We thus have succeeded in associating two new equations $f_1=a_1$, and $f_2=a_2$, with f and with one another.

63. The next stage is the determination of a new equation $u=\text{constant}$, consistent with

$$f=0,\quad f_1=a_1,\quad f_2=a_2;$$

the necessary and sufficient conditions for coexistence are

$$(f, u)=0, \quad (f_1, u)=0, \quad (f_2, u)=0.$$

Let $u=\lambda$ be an integral, common to $(f, u)=0$, $(f_1, u)=0$, and functionally distinct from f, f_1, f_2, where

$$\lambda=\lambda(x_1, \ldots, x_n, p_1, \ldots, p_n):$$

it may be taken as one of the $2n-5$ common integrals, other than f, f_1, f_2. We proceed as before, and form a series of functions

$$(f_2, \lambda)=\lambda_1, \quad (f_2, \lambda_1)=\lambda_2, \ldots.$$

Each of these quantities is a common integral of $(f, u)=0$, $(f_1, u)=0$. For

$$(f(f_2, \theta))+(f_2(\theta, f))+(\theta(f, f_2))=0,$$
$$(f_1(f_2, \theta))+(f_2(\theta, f_1))+(\theta(f_1, f_2))=0;$$

and $(f, f_2)=0$, $(f_1, f_2)=0$, both identically, so that

$$(\theta(f, f_2))=0, \quad (\theta(f_1, f_2))=0;$$

and therefore

$$(f(f_2, \theta))=(f_2(f, \theta)),$$
$$(f_1(f_2, \theta))=(f_2(f_1, \theta)).$$

Let $\theta=\lambda$; these results give

$$(f, \lambda_1)=(f_2(f, \lambda))=0,$$
$$(f_1, \lambda_1)=(f_2(f_1, \lambda))=0,$$

because $(f, \lambda)=0$, $(f_1, \lambda)=0$, both identically satisfied; thus λ_1 is an integral common to $(f, u)=0$, $(f_1, u)=0$. Let $\theta=\lambda_1$; then the two relations give

$$(f, \lambda_2)=(f_2(f, \lambda_1))=0,$$
$$(f_1, \lambda_2)=(f_2(f_1, \lambda_1))=0,$$

as before: that is, λ_2 is an integral common to $(f, u)=0$, $(f_1, u)=0$. And so for all the functions λ in succession.

The number of independent integrals is limited: and thus the λ-series will terminate either in a zero, or in a pure constant, or in a function expressible in terms of the anterior functions. Proceeding as before, we obtain some λ-function, or some combination of λ-functions, say Λ, such that

$$(f_2, \Lambda)=0,$$

save only in the case when (f_2, λ) is either a constant (not zero) or is not distinct from f, f_1, f_2, λ.

In the latter circumstance, we take another integral μ, common to $(f, u)=0$, $(f_1, u)=0$, and distinct from f, f_1, f_2, λ. Proceeding in the same way, we obtain some μ-function or some combination of μ-functions, say M, such that

$$(f_2, M)=0,$$

save only in the case when (f_2, μ) either is a constant (not zero) or is not distinct from f, f_1, f_2, μ.

And should the latter happen, then if

$$N=\int \frac{d\lambda}{(f_2, \lambda)}-\int \frac{d\mu}{(f_2, \mu)},$$

we have

$$(f_2, N)=0.$$

Thus in every case we obtain an integral common to the three equations

$$(f, u)=0, \quad (f_1, u)=0, \quad (f_2, u)=0:$$

and in the least favourable combination of circumstances, there are $2n-6$ such integrals, independent of f, f_1, f_2, and of one another.

Let f_3 be one of those integrals; then the equation

$$f_3=a_3,$$

where a_3 is an arbitrary constant, associates itself with

$$f=0, \quad f_1=a_1, \quad f_2=a_2.$$

64. We proceed in this way from stage to stage, obtaining equations $f_4=a_4, \ldots$ in succession which are associated with all the equations that precede them. The last stage of all is the construction of an equation $f_{n-1}=a_{n-1}$. Our earlier results shew that, when the equations

$$f=0, \quad f_1=a_1, \ \ldots, f_{n-1}=a_{n-1}$$

are resolved for $p_1, \ldots, p_n$ in terms of $x_1, \ldots, x_n$, the values thus obtained are such as to make

$$p_1 dx_1+\ldots+p_n dx_n$$

an exact differential; after quadrature, an integral of the original equation $f=0$ is given by

$$z-a_n=\int(p_1 dx_1+\ldots+p_n dx_n),$$

involving n arbitrary constants.

If it is not possible or not convenient to resolve the equations $f = 0, \ldots, f_{n-1} = a_{n-1}$ with regard to $p_1, \ldots, p_n$, we choose another set of the variables involved and, resolving with regard to these, adopt the process explained in §§ 58, 59.

Jacobi's Second Method when z does occur.

65. In the preceding account of Jacobi's method of solving an equation $f = 0$, the dependent variable z has been supposed not to occur explicitly. If it should occur explicitly, we have already seen that there is a mode of proceeding by a change of dependent variable, associated with a unit increase in the number of independent variables. This mode of proceeding may be cumbrous: and in any case, it is desirable (if possible) to have a direct method for constructing an integral.

Accordingly, let

$$f = f(x_1, \ldots, x_n, z, p_1, \ldots, p_n) = 0$$

be an irreducible equation which involves z explicitly: if

$$u = u(x_1, \ldots, x_n, z, p_1, \ldots, p_n) = \text{constant}$$

be an equation which can coexist with $f = 0$, it is necessary and sufficient that the relation

$$[f, u] = 0$$

should be satisfied. This equation is homogeneous and linear in the derivatives of u; written in full, it is

$$\sum_{i=1}^{n} \left\{ \left(\frac{\partial f}{\partial x_i} + p_i \frac{\partial f}{\partial z} \right) \frac{\partial u}{\partial p_i} - \frac{\partial u}{\partial x_i} \frac{\partial f}{\partial p_i} - \frac{\partial u}{\partial z} p_i \frac{\partial f}{\partial p_i} \right\} = 0.$$

To obtain a value of u, we construct the system of subsidiary equations

$$\frac{dx_1}{-\dfrac{\partial f}{\partial p_1}} = \ldots = \frac{dx_n}{-\dfrac{\partial f}{\partial p_n}} = \frac{dp_1}{\dfrac{\partial f}{\partial x_1} + p_1 \dfrac{\partial f}{\partial z}} = \ldots = \frac{dp_n}{\dfrac{\partial f}{\partial x_n} + p_n \dfrac{\partial f}{\partial z}}$$

$$= \frac{dz}{-\left(p_1 \dfrac{\partial f}{\partial p_1} + \ldots + p_n \dfrac{\partial f}{\partial p_n} \right)},$$

which (for reasons that will appear hereafter) are called the *equations of the characteristics*; and we take an integral of these equations, choosing by preference one that is not free from $z, p_1, \ldots, p_n$, if any such exist. Let such an integral be

$$g(x_1, \ldots, x_n, z, p_1, \ldots, p_n) = \text{constant};$$

then the equation

$$u = g(x_1, \ldots, x_n, z, p_1, \ldots, p_n)$$

gives a value of u as required; and the relation

$$[f, g] = 0$$

is satisfied identically, so far as concerns $g = \text{constant}$, but not necessarily identically, so far as concerns $f = 0$: indeed, it may be satisfied only in virtue of $f = 0$.

Ex. The characteristics of the equation

$$pxz + qyz - xy = 0$$

are given by

$$\frac{dx}{-xz} = \frac{dy}{-yz} = \frac{dp}{pz - y + p(px + qy)} = \frac{dq}{qz - x + q(qx + py)} = \frac{dz}{-pxz - qyz}.$$

An integral, as required, is given by

$$z^2 - xy = \text{constant};$$

the relation

$$[pxz + qyz - xy,\ z^2 - xy] = 0$$

is satisfied only in virtue of $f = 0$. Another integral, as required, is given by

$$\frac{pz - \frac{1}{2}y}{qz - \frac{1}{2}x} = \text{constant};$$

the relation

$$\left[pxz + qyz - xy,\ \frac{pz - \frac{1}{2}y}{qz - \frac{1}{2}x}\right] = 0$$

is satisfied identically.

66. Accordingly, at this stage it is convenient, for the sake of very substantial simplification of the analysis, to resolve the two equations

$$f = 0, \quad g = a,$$

for z and one of the variables p, chosen so as to give the simplest resolution: let the selected variable be p_1, and let the result of the resolution be denoted by

$$z - \psi = 0, \quad p_1 - \psi_1 = 0,$$

where ψ and ψ_1 are functions of $x_1, \ldots, x_n, p_2, \ldots, p_n$. Then, after the explanations in § 55, Note, and § 60, we take these two equations in the form

$$z - \psi - x_1(p_1 - \psi_1) = 0, \quad p_1 - \psi_1 = 0;$$

any equation $w = c$ that can coexist with them must satisfy the equations

$$[z - \psi - x_1(p_1 - \psi_1), w] = 0,$$
$$[p_1 - \psi_1, w] = 0.$$

These two equations to determine w are, by § 60, a complete system.

As any integral of these two equations is to furnish an equation $w =$ constant, which shall coexist with

$$z - \psi = 0, \quad p_1 - \psi_1 = 0,$$

it can be transformed so that, if z and p_1 do occur, they are replaced by ψ and ψ_1 respectively: that is, without loss of generality, w may be assumed not to involve either z or p_1 explicitly. Let

$$w = \phi = \phi(x_1, \ldots, x_n, p_2, \ldots, p_n)$$

be an integral of the equation

$$[z - \psi - x_1(p_1 - \psi_1), w] = 0;$$

then, as $[z - \psi - x_1(p_1 - \psi_1), \phi]$ does not contain z or p_1, so that it cannot vanish in virtue of $z - \psi = 0$ or $p_1 - \psi_1 = 0$, and as it must vanish, it vanishes identically. Construct the function $[p_1 - \psi_1, \phi]$, $= \phi_1$ say. If ϕ_1 vanishes identically, this last condition is satisfied: also $[p_1 - \psi_1, \phi] = 0$; and therefore $w = \phi$ is a common integral of the two equations. In that case, the equation

$$\phi = a_1,$$

where a_1 is an arbitrary constant, can be associated with

$$z - \psi = 0, \quad p_1 - \psi_1 = 0.$$

Suppose, on the other hand, that ϕ_1 does not vanish identically; then, as

$$[[\zeta, \pi]\phi] + [[\pi, \phi]\zeta] + [[\phi, \zeta]\pi]$$
$$= -\frac{\partial \phi}{\partial z}[\zeta, \pi] - \frac{\partial \zeta}{\partial z}[\pi, \phi] - \frac{\partial \pi}{\partial z}[\phi, \zeta]$$

identically, we have, on writing

$$\zeta = z - \psi - x_1(p_1 - \psi_1), \quad \pi = p_1 - \psi_1,$$

the relation

$$[[p_1 - \psi_1, \phi]\zeta] = -[p_1 - \psi_1, \phi],$$

that is,

$$[\phi_1, \zeta] = -\phi_1.$$

Thus ϕ_1 is not an integral of

$$[z - \psi - x_1 (p_1 - \psi_1), w] = 0;$$

and as $[\phi_1, \zeta]$ involves derivatives of ϕ_1, it is clear that ϕ_1 cannot be a pure constant.

In that case, let

$$w = \chi = \chi(x_1, \ldots, x_n, p_2, \ldots, p_n)$$

be another integral of the equation

$$[z - \psi - x_1 (p_1 - \psi_1), w] = 0,$$

functionally distinct from $w = \phi$; and construct the function $[p_1 - \psi_1, \chi], = \chi_1$ say. If χ_1 vanishes identically, it follows that $w = \chi$ is an integral of the two equations determining w; and then the equation

$$\chi = c_1,$$

where c_1 is an arbitrary constant, can be associated with

$$z - \psi = 0, \quad p_1 - \psi_1 = 0.$$

But if χ_1 does not vanish identically, then we have

$$[\chi_1, \zeta] = -\chi_1,$$

as before; and χ_1 cannot be a constant. Also

$$[\phi_1, \zeta] = -\phi_1,$$

so that

$$\left[\frac{\chi_1}{\phi_1}, \zeta\right] = -\frac{1}{\phi_1}\chi_1 - \frac{\chi_1}{\phi_1^2}(-\phi_1) = 0;$$

and therefore

$$w_1 = \frac{\chi_1}{\phi_1}$$

is an integral of the equation

$$[z - \psi - x_1 (p_1 - \psi_1), w] = 0.$$

Now both ϕ_1 and χ_1 are variable: but $\dfrac{\chi_1}{\phi_1}$ may be a constant, say c. Then

$$\begin{aligned}[p_1 - \psi_1, \chi] &= \chi_1 \\ &= c\phi_1 \\ &= [p_1 - \psi_1, c\phi],\end{aligned}$$

and therefore

$$[p_1 - \psi_1, \chi - c\phi] = 0;$$

hence as

$$[z - \psi - x_1 (p_1 - \psi_1), \chi - c\phi] = 0,$$

so that, under the particular hypothesis, $w = \chi - c\phi$ is an integral common to the two equations. But, in general, w_1 will be a variable quantity.

Assuming w_1 now not to be a constant, construct the function $[p_1 - \psi_1, w_1], = \chi_2$ say. If χ_2 vanishes identically, it follows that $w = w_1$ is an integral common to the two equations for the determination of w; and then the equation

$$w_1 = c_1,$$

where c_1 is an arbitrary constant, can be associated with

$$z - \psi = 0, \quad p_1 - \psi_1 = 0.$$

If χ_2 does not vanish identically, then

$$w_2 = \frac{\chi_2}{\phi_1}$$

is an integral of the equation

$$[z - \psi - x_1 (p_1 - \psi_1), w] = 0.$$

If w_2 be constant and equal to α, then

$$w = w_1 - \alpha\phi$$

is an integral common to the two equations. But, in general, w_2 will be a variable quantity.

Assuming that w_2 is variable, we construct the function $[p_1 - \psi_1, w_2], = \chi_3$ say. As before, if χ_3 vanishes identically, we have an integral $w = w_2$ common to the two equations. If $\chi_3 = \beta\phi_1$, where β is a constant, then $w = w_2 - \beta\phi$ is an integral common to the two equations. If $\dfrac{\chi_3}{\phi_1}$ is not zero nor a constant, then

$$w_3 = \frac{\chi_3}{\phi_1}$$

is an integral of the equation

$$[z - \psi - x_1 (p_1 - \psi_1), w] = 0.$$

Proceeding in this way, either we shall at some stage obtain an integral common to the two equations, or we shall obtain an integral of the equation

$$[z - \psi - x_1 (p_1 - \psi_1), w] = 0$$

which is expressible in terms of the preceding integrals; for the number of functionally distinct integrals of that equation is limited.

When the last alternative occurs, all succeeding integrals are also so expressible; for if

$$w_m = f(w_{m-1}, \ldots, \chi, \phi),$$

then, as

$$[p_1 - \psi_1, w_\mu] = w_{\mu+1}\phi_1,$$

we have

$$w_{m+1} = \frac{1}{\phi_1}[p_1 - \psi_1, w_m]$$

$$= \frac{\partial f}{\partial w_{m-1}} w_m + \frac{\partial f}{\partial w_{m-2}} w_{m-1} + \ldots + \frac{\partial f}{\partial \chi} w_1 + \frac{\partial f}{\partial \phi},$$

shewing that w_{m+1} is expressible in terms of the earlier integrals: and so for all succeeding integrals. Now take some functional combination of ϕ, χ, w_1, ..., w_{m-1}, say

$$g = g(\phi, \chi, w_1, \ldots, w_{m-1});$$

then

$$[p_1 - \psi_1, g] = \phi_1 \left\{ \frac{\partial g}{\partial \phi} + \frac{\partial g}{\partial \chi} w_1 + \frac{\partial g}{\partial w_1} w_2 + \ldots + \frac{\partial g}{\partial w_{m-1}} f \right\}:$$

if g can be chosen so that the right-hand side vanishes, then $[p_1 - \psi_1, g] = 0$, and we shall have an integral common to our two equations. Let any integral of the system of ordinary equations

$$\frac{d\phi}{1} = \frac{d\chi}{w_1} = \frac{dw_1}{w_2} = \ldots = \frac{dw_{m-1}}{f(w_{m-1}, \ldots, \chi, \phi)}$$

be

$$g_1(\phi, \chi, \ldots, w_{m-1}) = \text{constant};$$

then taking

$$g = g_1(\phi, \chi, \ldots, w_{m-1}),$$

we have

$$[p_1 - \psi_1, g] = 0.$$

Moreover, there are m functionally distinct integrals of the system of ordinary equations: hence there are m distinct integrals common to the two equations

$$[z - \psi - x_1(p_1 - \psi_1), w] = 0, \quad [p_1 - \psi_1, w] = 0;$$

and these are constructed out of $m + 1$ distinct integrals of the first equation*.

* The simplest case occurs when w_1 is not functionally distinct from the integrals that precede it, viz. from ϕ and χ, so that we then have

$$w_1 = f(\chi, \phi);$$

if the integral of the equation

$$d\phi = \frac{d\chi}{f(\chi, \phi)}$$

be $g(\phi, \chi) = \text{constant}$, we take $g(\phi, \chi)$ as a common integral of the two equations.

Let one of these integrals be $u(x_1, \ldots, x_n, p_2, \ldots, p_n)$: then the equation

$$u(x_1, \ldots, x_n, p_2, \ldots, p_n) = a_2$$

coexists with the equations

$$z - \psi = 0, \quad p_1 - \psi_1 = 0.$$

In every case, therefore, an equation has been constructed which coexists with the equations already obtained.

67. To proceed to the next stage, we resolve the equation

$$u(x_1, \ldots, x_n, p_2, \ldots, p_n) = a_2$$

with regard to one of the variables p which it contains: let the resolved form be

$$p_2 = \chi_2,$$

where χ_2 involves $a_2, x_1, \ldots, x_n, p_3, \ldots, p_m$. Let this value of p_2 be inserted in ψ and ψ_1, and let the resulting expressions be χ and χ_1; then we have the simultaneous equations

$$z - \chi = 0, \quad p_1 - \chi_1 = 0, \quad p_2 - \chi_2 = 0.$$

Now χ, χ_1, χ_2 do not involve z: hence, writing

$$\zeta = z - \chi - x_1(p_1 - \chi_1) - x_2(p_2 - \chi_2), \quad \pi_1 = p_1 - \chi_1, \quad \pi_2 = p_2 - \chi_2,$$

and denoting by θ any quantity which does not involve z, we have

$$[[\pi_1, \theta]\, \zeta] + [[\theta, \zeta]\, \pi_1] + [[\zeta, \pi_1]\, \theta] = -[\pi_1, \theta],$$
$$[[\pi_2, \theta]\, \zeta] + [[\theta, \zeta]\, \pi_2] + [[\zeta, \pi_2]\, \theta] = -[\pi_2, \theta],$$
$$[[\pi_1, \theta]\, \pi_2] + [[\theta, \pi_2]\, \pi_1] + [[\pi_2, \pi_1]\, \theta] = 0;$$

also we have

$$[\zeta, \pi_1] = 0, \quad [\zeta, \pi_2] = 0, \quad [\pi_2, \pi_1] = 0,$$

identically.

Let σ and ρ be integrals common to the two equations

$$[\zeta, v] = 0, \quad [\pi_1, v] = 0,$$

obtained as in the preceding sections, and limited so that they do not involve z and that they are functionally distinct from π_2 and from one another; and let

$$[\pi_2, \sigma] = \sigma_1, \quad [\pi_2, \rho] = \rho_1.$$

If either σ_1 or ρ_1 vanishes, then we have a common integral of the three equations

$$[\zeta, v] = 0, \quad [\pi_1, v] = 0, \quad [\pi_2, v] = 0.$$

If neither of them vanishes, we make θ equal to σ and then to ρ in succession in the above identities. The first of the identities gives no condition; the second gives

$$[\zeta, \sigma_1] = \sigma_1, \quad [\zeta, \rho_1] = \rho_1;$$

and the third gives

$$[\pi_1, \sigma_1] = 0, \quad [\pi_1, \rho_1] = 0.$$

Hence $\frac{\rho_1}{\sigma_1}$ is an integral common to the two equations

$$[\zeta, v] = 0, \quad [\pi_1, v] = 0,$$

unless it is a constant; and if $\frac{\rho_1}{\sigma_1}$ is a constant, say equal to a, then

$$[\zeta, \rho - a\sigma] = 0, \quad [\pi_1, \rho - a\sigma] = 0,$$
$$[\pi_2, \rho - a\sigma] = \rho_1 - a\sigma_1 = 0,$$

so that $\rho - a\sigma$ would be a common integral of the three equations determining v.

Writing $\tau = \frac{\rho_1}{\sigma_1}$, and

$$[\pi_2, \tau] = \tau_1,$$

then if τ_1 vanishes, a common integral of the three equations is $v = \tau$; while if τ_1 does not vanish, we have

$$[\zeta, \tau_1] = \tau_1, \quad [\pi_1, \tau_1] = 0,$$

and therefore

$$\left[\zeta, \frac{\tau_1}{\sigma_1}\right] = 0, \quad \left[\pi_1, \frac{\tau_1}{\sigma_1}\right] = 0,$$

shewing that $\frac{\tau_1}{\sigma_1}$ is an integral common to the two equations $[\zeta, v] = 0$, $[\pi_1, v] = 0$.

We proceed as in the former stage: sooner or later, an integral of the two equations $[\zeta, v] = 0$ and $[\pi_1, v] = 0$ is obtained which is expressible in terms of the earlier integrals, or an integral is obtained which also satisfies $[\pi_2, v] = 0$. In the former alternative, we construct (as in the earlier stage) a combination of all these independent integrals of $[\zeta, v] = 0$ and $[\pi_1, v] = 0$ which shall also satisfy $[\pi_2, v] = 0$. Let it be

$$v = v(x_1, \ldots, x_n, p_3, \ldots, p_m);$$

then the equation

$$v(x_1, \ldots, x_n, p_3, \ldots, p_m) = a_3$$

coexists with the equations

$$\zeta = 0, \quad \pi_1 = 0, \quad \pi_2 = 0.$$

Let it be resolved for one of the variables p, say p_3, in the form

$$p_3 - \theta_3 = 0,$$

where θ_3 involves $a_3, x_1, \ldots, x_n, p_4, \ldots, p_m$; when this value is substituted in χ, χ_1, χ_2, let them become $\theta, \theta_1, \theta_2$; then our equations are

$$z - \theta = 0, \quad p_1 - \theta_1 = 0, \quad p_2 = \theta_2, \quad p_3 = \theta_3.$$

So we proceed from stage to stage. In each stage the construction of the new equation requires, in the least favourable combination of circumstances, the assignment of two integrals of the subsidiary system associated with the initial equation

$$[f, u] = 0.$$

This subsidiary system contains $2n$ differential equations: its integral equivalent must therefore contain $2n$ integral equations, that is, it possesses $2n$ integrals. Hence there are sufficient integrals for the achievement of n stages; at the end of the last, we shall have

$$z = \text{function of } x_1, \ldots, x_n, a_1, \ldots, a_n,$$

(where $a_1, \ldots, a_n$ are arbitrary constants) as the integral of the original equation. Or at the completion of the $(n-1)$th stage, we can resolve the n equations then coexisting, and express $p_1, \ldots, p_n$ in terms of $z, x_1, \ldots, x_n, a_1, \ldots, a_{n-1}$; substitution in the relation

$$dz = p_1 dx_1 + \ldots + p_n dx_n,$$

and quadrature, lead to the integral required.

Ex. Let Z denote $z - p_1x_1 - \ldots - p_nx_n$; and suppose that a set of equations

$$F_\mu = F_\mu (p_1, \ldots, p_n, Z) = 0, \qquad (\mu = 1, \ldots, m),$$

where $m < n$, is propounded for solution.

We have

$$\frac{dF_\mu}{dx_i} = \frac{\partial F_\mu}{\partial x_i} + p_i \frac{\partial F_\mu}{\partial z} = 0,$$

for all values of μ and of i: consequently

$$[F_r, F_s] = 0,$$

for all values of r and s, so that the system is in involution.

To obtain other equations consistent with the system, we need simultaneous integrals of

$$[F_1,\ u]=0,\ \ldots,\ [F_m,\ u]=0.$$

The equations subsidiary to the solution of $[F_1,\ u]=0$ are

$$\ldots=\frac{dp_i}{\dfrac{\partial F_1}{\partial x_i}+p_i\dfrac{\partial F_1}{\partial z}}=\ldots=\frac{dz}{-p_1\dfrac{\partial F_1}{\partial p_1}-\ldots-p_n\dfrac{\partial F_1}{\partial p_n}}=\ldots;$$

but $\dfrac{\partial F_1}{\partial x_i}+p_i\dfrac{\partial F_1}{\partial z}=0$, and so an integral of these equations is given by

$$p_1=\text{constant}.$$

Also

$$[p_1,\ F_r]=0,$$

for $r=2,\ \ldots,\ m$; so that $u=p_1$ is an integral common to all the equations $[F_\mu,\ u]=0$. We therefore associate the equation

$$p_1=a_1$$

with the given set; the new system is

$$F_1=0,\ \ldots,\ F_\mu=0,\ \ p_1=a_1,$$

and it is easily seen to be in involution.

Similarly, we may associate the equations

$$p_2=a_2,\ \ldots,\ p_{n-m+1}=a_{n-m+1},$$

where $a_2,\ \ldots,\ a_{n-m+1}$ are arbitrary constants, with the amplified system and with one another: and the whole system thus extended, viz.

$$F_1=0,\ \ldots,\ F_m=0,\ \ p_1=a_1,\ \ldots,\ p_{n-m+1}=a_{n-m+1},$$

is in involution. If therefore the quantities $p_1,\ \ldots,\ p_n$ can be eliminated from the system, the eliminant will give an integral of the original set.

Now the $n+1$ equations thus obtained are independent of one another, and they involve the $n+1$ quantities $p_1,\ \ldots,\ p_n,\ Z$; when resolved with regard to these quantities, they give

$$Z=c,\ \ p_i=a_i,$$

that is,

$$z-a_1x_1-a_2x_2-\ldots-a_nx_n=c,$$

where the constants $a_1,\ \ldots,\ a_{n-m+1}$ are arbitrary, and the remaining constants $a_{n-m+2},\ \ldots,\ a_n,\ c$ satisfy the m relations

$$F_\mu(a_1,\ \ldots,\ a_n,\ c)=0,$$

for the values $\mu=1,\ \ldots,\ m$. The equation

$$z=a_1x_1+a_2x_2+\ldots+a_nx_n+c,$$

with the limitations upon the constants, provides an integral of the propounded system.

CHARPIT'S METHOD: INTEGRALS WHEN THE CONDITIONS IN CAUCHY'S THEOREM ARE NOT SATISFIED.

68. Naturally, the simplest case of the preceding method arises when the number of independent variables is two. With the usual notation in this case, the equation may be written

$$f(x, y, z, p, q) = 0;$$

and the condition $[f, u] = 0$, which must be satisfied by u if $u =$ constant is to coexist with $f = 0$, is

$$\left(\frac{\partial f}{\partial x} + p\frac{\partial f}{\partial z}\right)\frac{\partial u}{\partial p} + \left(\frac{\partial f}{\partial y} + q\frac{\partial f}{\partial z}\right)\frac{\partial u}{\partial q} - \frac{\partial f}{\partial p}\frac{\partial u}{\partial x} - \frac{\partial f}{\partial q}\frac{\partial u}{\partial y}$$
$$- \left(p\frac{\partial f}{\partial p} + q\frac{\partial f}{\partial q}\right)\frac{\partial u}{\partial z} = 0.$$

To obtain an integral of this homogeneous linear equation which shall involve p or q or both, the system of ordinary equations

$$\frac{dp}{\dfrac{\partial f}{\partial x} + p\dfrac{\partial f}{\partial z}} = \frac{dq}{\dfrac{\partial f}{\partial y} + q\dfrac{\partial f}{\partial z}} = \frac{dx}{-\dfrac{\partial f}{\partial p}} = \frac{dy}{-\dfrac{\partial f}{\partial q}} = \frac{dz}{-p\dfrac{\partial f}{\partial p} - q\dfrac{\partial f}{\partial q}}$$

is formed: if

$$u(x, y, z, p, q) = \text{constant}$$

be any integral, distinct from $f = 0$, involving p or q or both, then the equations

$$f(x, y, z, p, q) = 0, \quad u(x, y, z, p, q) = a,$$

where a is an arbitrary constant, are resolved with respect* to p and q. These values make the equation

$$dz - p\,dx - q\,dy = 0$$

exact. For from the equations $f = 0$, $u = a$, we find

$$\frac{\partial(f, u)}{\partial(x, p)} + p\frac{\partial(f, u)}{\partial(z, p)} + \frac{\partial(f, u)}{\partial(y, q)} + q\frac{\partial(f, u)}{\partial(z, q)} = \left(\frac{dq}{dx} - \frac{dp}{dy}\right)\frac{\partial(f, u)}{\partial(p, q)};$$

and because $u(x, y, z, p, q) =$ constant is an integral of the system of ordinary equations, the left-hand side of this equation vanishes, so that

$$\left(\frac{dq}{dx} - \frac{dp}{dy}\right)\frac{\partial(f, u)}{\partial(p, q)} = 0;$$

* Or with respect to other variables, with a modification in the rest of the process, similar to that in §§ 58, 59.

and therefore, as the Jacobian of f and u with regard to p and q does not vanish *, we have

$$\frac{dq}{dx} - \frac{dp}{dy} = 0,$$

the necessary and sufficient condition. Effecting the necessary quadrature of the equation

$$dz - pdx - qdy = 0,$$

we have an equation giving z in terms of x, y, and two arbitrary constants.

This mode of obtaining the integral of the original equation by means of a single integral of the subsidiary system was first devised by Charpit †.

The method of Jacobi, whether in its original form as developed by himself or in the amplified form as developed by Lie and Mayer, and (for the case of two independent variables) the method of Charpit, aim at the construction of an integral containing a number of arbitrary constants; and the results do not indicate any particular suggestion of Cauchy's existence-theorem. The association will be made later, partly by a modified use of the equations of the characteristics; and it will be necessary to indicate the kinds of integrals which can be deduced from those provided by the methods of Jacobi and of Charpit.

69. All the examples, that follow, have been chosen, so as to give some initial indications of one investigation hitherto practically omitted by mathematicians. When an equation

$$f(x, y, z, p, q) = 0$$

is resolved with regard to p, or is given in a resolved form, so that it may be written

$$p = g(x, y, z, q),$$

Cauchy's existence-theorem can be applied only if the function $g(x, y, z, q)$ is a regular function of its arguments within the

* It would vanish if u involved neither p nor q.

† In a memoir, presented 30 June, 1784, to the Académie des Sciences, Paris; he died soon afterwards, and the memoir was never printed: see Lacroix, *Traité du calcul différentiel et du calcul intégral*, 2e éd., 1814, t. II, p. 548. Lacroix indicates (*ib.*, p. 567) that Charpit tried to extend his method to partial differential equations of the first order and degree higher than the first, involving more than two dependent variables.

domains of the initial values adopted: it ceases to apply if initial values are selected in the domains of which the function $g(x, y, z, q)$ is regular.

In all these examples, it is possible to choose initial values which make p infinite or indeterminate: the known method of constructing an integral has been used so as to give indications of the kind of integral (if any) which exists in association with such initial conditions.

What is required for the full discussion of an equation

$$f(x, y, z, p, q)=0,$$

(and, *à fortiori*, of an equation in more than two independent variables), is a classification of all the non-regular forms arising out of the resolution of the equation with regard to p or, what is the same thing, a classification of all the non-regular forms of $g(x, y, z, q)$ in an equation

$$p=g(x, y, z, q).$$

Each of these would need to be considered in turn, as was done* for the non-regular forms of an equation

$$\frac{dw}{dz}=f(w, z);$$

the following set of examples give a few of the simplest types.

Meanwhile, some indications of results can be given: the methods of Charpit and of Jacobi are entirely independent even of the results given by Cauchy's theorem.

Ex. 1. Consider the equation

$$p(ax+by+cz)=1.$$

It is clear that Cauchy's general theorem will not apply to this equation if, when $x=0$, we require z to acquire the value of a function of y regular in the vicinity of $y=0$ and vanishing there: the initial value of p is infinite and the proof no longer is valid.

But an integral can be obtained by Charpit's method. One of the subsidiary equations is

$$\frac{dp}{ap+cp^2}=\frac{dq}{bp+cpq},$$

so that

$$\frac{dp}{a+cp}=\frac{dq}{b+cq},$$

* In Chapters III and IV in Part II of this work.

an integral of which is

$$\frac{b+cq}{a+cp}=\alpha,$$

where α is an arbitrary constant. Accordingly, we combine this equation with the original equation, and we resolve them for p and q: substituting these found values in $dz-pdx-qdy=0$, we have

$$dz=\frac{dx}{ax+by+cz}+\left\{-\frac{b}{c}+\frac{a}{c}\alpha+\left(\frac{\alpha}{ax+by+cz}\right)\right\}dy,$$

and therefore

$$a\,dx+b\,dy+c\,dz=\left(\frac{c}{ax+by+cz}+a\right)(dx+\alpha dy).$$

Writing

$$u=ax+by+cz,$$

a simple quadrature leads to the equation

$$u-\frac{c}{a}\log(c+au)=\beta+a\,(x+\alpha y).$$

The value of z thus provided is an integral which contains the two arbitrary constants α and β.

In order to see whether any integral z exists, which vanishes when $x=0$ and $y=0$, these being values which make p infinite initially, we note that the foregoing equation is satisfied by $z=0$, $x=0$, $y=0$, provided

$$\beta=-\frac{c}{a}\log c.$$

Assuming this value of β, we have

$$e^u\left(1+\frac{a}{c}u\right)^{-\frac{c}{a}}=e^{a\,(x+\alpha y)};$$

and therefore, in the vicinity of the initial values assigned, we have

$$\left(1+\frac{a}{c}u\right)e^{-\frac{a}{c}u}=e^{-\frac{a^2}{c}(x+\alpha y)},$$

that is,

$$u^2+\ldots=2c\,(x+\alpha y)+\ldots,$$

so that, unless $c=0$ (and this will be excluded), we have

$$ax+by+cz=u=(x+\alpha y)^{\frac{1}{2}}\,R\,\{(x+\alpha y)^{\frac{1}{2}}\},$$

where R is a regular function of its argument and does not vanish when $x=0$ and $y=0$.

Ex. 2. In the same way it may be proved that an integral of the equation

$$p\,(ax+by+cz)^m=1,$$

where m is a positive integer, is given by

$$\int\frac{u^m}{c+au^m}\,du=\beta+x+\alpha y,$$

where α and β are arbitrary constants, and

$$u=ax+by+cz;$$

and that an integral, which vanishes when $x=0$ and $y=0$, is given by

$$ax+by+cz=(x+ay)^{\frac{1}{1+m}} R\{(x+ay)^{\frac{1}{1+m}}\},$$

where R is a regular function of its argument and does not vanish with x and y.

Ex. 3. It is easy to see that the integral of the equation

$$(p+a'q)(ax+by+cz)^m=1,$$

where a' is a constant and m is a positive integer, is of the same type as in the preceding example: obtain the integral.

Ex. 4. Consider the equation

$$p(ax+by+cz+kq)=1,$$

where a, b, c, k are constants.

Proceeding from subsidiary equations as in Ex. 1, we find that they have an integral

$$\frac{b+cq}{a+cp}=\alpha,$$

where α is an arbitrary constant.

There are two ways of continuing. We may either resolve the original equation and the new equation for p and y, and introduce a new dependent variable ζ, where

$$\zeta=z-qy,$$

and then we have

$$d\zeta=pdx-ydq:$$

we substitute for p and y; and, effecting the necessary quadrature, we eliminate q by the relation

$$\frac{\partial\zeta}{\partial q}=-y.$$

Or we may resolve the two equations for p and q, substitute in $dz=pdx+qdy$, and effect the quadrature. The result is

$$-\frac{1}{c}\left(1+\frac{c^2}{a^2k\alpha}\right)\log\{\Delta^{\frac{1}{2}}-cu+k(b-a\alpha)\}+\frac{c}{a^2k\alpha}\log\{\Delta^{\frac{1}{2}}-cu+k(b+a\alpha)\}$$

$$=\frac{-2\frac{c}{a}}{\Delta^{\frac{1}{2}}-cu+k(b-a\alpha)}+\frac{x+\alpha y}{ak}+\beta,$$

where α and β are arbitrary constants, $u=ax+by+cz$, and

$$\Delta=(cu+bk-ak\alpha)^2-4ck\{(b-a\alpha)u-c\alpha\}.$$

It is possible (but the analysis is somewhat laborious) to deduce, from this result when $k=0$, the integral of the equation in Ex. 1.

We can make one more inference. If it were possible that the equation could possess an integral such that when $x=0$, the dependent variable acquires the value of a function of y such that z and q vanish when $y=0$, then p would become infinite for the initial values $x=0$ and $y=0$: Cauchy's theorem no longer applies. Now we are to have

$$\frac{b+cq}{a+cp}=\alpha;$$

therefore for such integral (if any) we have $\alpha=0$ because initially p is infinite, and then $b+cq=0$. But q is to vanish initially, so that $b=0$; and thus $q=0$ always; or z is merely a function of x, vanishing with x and given by

$$x=\frac{c}{\alpha^2}e^{\alpha z}-\frac{c}{\alpha^2}(1+\alpha z).$$

Excluding this trivial case, it follows that the given equation has no integral of the kind indicated, provided c is different from zero.

Ex. 5. Integrate the equation

$$p(ax+by+cz+kyq)=1;$$

and discuss the question whether it possesses an integral which, when $x=0$, acquires the value of a regular function of y that vanishes when $y=0$.

[An integral is given by eliminating q between the two equations

$$\left(z-qy+\frac{ax}{c}+\frac{1}{a}\right)(b+c'q)^{-\frac{c}{c'}}$$
$$=\beta+\frac{x}{ac}-\frac{1}{a\alpha^2}\log\left\{1-a\alpha(b+c'q)^{-\frac{c}{c'}}\right\},$$

$$\beta+\frac{x}{ac}+\frac{y}{c}(b+c'q)^{\frac{k}{c'}}-\frac{1}{\alpha}\,\frac{1}{(b+c'q)^{\frac{c}{c'}}-a\alpha}=\frac{1}{a\alpha^2}\log\left\{1-a\alpha(b+c'q)^{-\frac{c}{c'}}\right\},$$

where α and β are arbitrary constants, and $c'=c+k$.]

Ex. 6. Obtain an integral of the equation

$$p=\frac{a}{yq-\frac{1}{2}z}$$

in the form

$$\tfrac{1}{4}z^2=(x-\beta)(ay-\alpha),$$

where α and β are arbitrary constants; and discuss the integrals of the equation (if any) which are such that yq and z vanish when $x=0$.

Ex. 7. As another example, consider the equation

$$p=\frac{cq}{yq-\frac{1}{2}z},$$

with a view to inquiring whether it possesses an integral which, when $x=0$, can be a function of y that vanishes, when $y=0$, in an order higher than the first, so that then q may vanish when $x=0$, $y=0$.

Forming the subsidiary equations in Charpit's method, we find one integral of them in a form

$$pq = a^2,$$

where a is an arbitrary constant. Resolving this equation and the original equation with regard to p and q, substituting in

$$dz = p\,dx + q\,dy,$$

and effecting the quadrature, we find

$$3ac\,(2cx - yz) - \beta = (a^2y^2 - 2cz)^{\frac{3}{2}} - a^3y^3,$$

where β is an arbitrary constant. This equation gives an integral involving two arbitrary constants.

If the equation is to provide an integral of the kind indicated, it is clear that $\beta = 0$. To discuss the consequent value of z when $x = 0$, we proceed from the equation

$$(a^2y^2 - 2cz)^{\frac{3}{2}} = a^3y^3 - 3acyz.$$

This equation certainly gives a value of z which vanishes when $y = 0$; two roots are zero, and the third is

$$z = \frac{3a^2}{8c}\,y^2,$$

which is of the required type.

Accordingly, the equation possesses an infinitude of integrals (because of the parameter a) which, when $x = 0$, give z and q as functions of y that vanish when $y = 0$; these integrals are provided by the equation

$$(a^2y^2 - 2cz)^3 = (a^3y^3 - 3acyz + 6ac^2x)^2,$$

where a is an arbitrary constant, that is, by the equation

$$8c^3z^3 - 3a^2c^2y^2z^2 - x\,(36a^2c^3yz - 12a^4c^2y^3) + 36a^2c^4x^2 = 0.$$

It is easy to see that, though, when $x = 0$, the integral becomes the simple regular function for the vicinity of $y = 0$, the integral itself is not a regular function of x and y in the specified domains.

Ex. 8. Prove that an integral of the equation

$$pz = aq + x,$$

where a is a constant, can be obtained by eliminating p and q between the equation itself and the equations

$$z\,(p^2 - 1)^{\frac{1}{2}} + ay + \beta = aa\left\{p + \tfrac{1}{2}\log\left(\frac{p-1}{p+1}\right)\right\}, \quad q = a\,(p^2 - 1)^{\frac{1}{2}},$$

where a and β are arbitrary constants. Discuss the integrals in the vicinity of $x = 0$.

Ex. 9. Consider the equation

$$p\,(ax + by + cz) + a'x + b'y + c'z = 0.$$

Changing the dependent variable so that

$$z' = z - a''x - b''y,$$

where a'' and b'' are constants, we can choose a'' and b'' so that the new equation has the form

$$p'(ax+\beta y+\gamma z')+\gamma' z'=0.$$

Accordingly, we consider the equation in the form

$$p(ax+by+cz)+c'z=0;$$

as it is homogeneous in the constants a, b, c, c', we can imagine it multiplied by such a constant factor as to make $a+c'=1$ unless $a+c'=0$.

Firstly, if $a+c'=1$, prove that an integral is given by the elimination of p between the equations

$$\left.\begin{array}{c} p(ax+by+cz)+c'z=0 \\ \dfrac{z-px}{p^{c'}\left(p+\dfrac{1}{c}\right)^{a}}=Ay+B+\dfrac{by}{c}\displaystyle\int p^{-c'}\left(p+\frac{1}{c}\right)^{-a-1}dp \end{array}\right\},$$

where A and B are arbitrary constants.

Secondly, if $a+c'=0$, prove that an integral is given by the elimination of p between the equations

$$\left.\begin{array}{c} p(ax+by+cz)+c'z=0 \\ \dfrac{z}{x}-p+\dfrac{by}{c'}-(Ay+B)e^{-\frac{c'}{cp}}=0 \end{array}\right\},$$

where A and B are arbitrary constants.

Discuss these integrals in the vicinity of $x=0$.

CHAPTER V.

Classes of Integrals possessed by Equations of the first order: Generalisation of Integrals.

The customary classification of integrals of a partial differential equation of the first order into three kinds was first made by Lagrange: see his *Œuvres Complètes*, t. III, p. 572, t. IV, pp. 65, 74. A full exposition is given in Imschenetsky's memoir, quoted on p. 100: it will be found in chapter I of the memoir. Other expositions are given by Goursat, *Leçons sur l'intégration...premier ordre*, by Mansion, *Théorie des équations...premier ordre*, and by Jordan, *Cours d'Analyse*, t. III.

That the theory is not complete even for the simplest case is pointed out by Goursat, in the book just quoted, § 18. Some further exceptions are indicated in the present chapter.

70. Before proceeding to the exposition of further methods of integration, and partly in order to facilitate the discussion of characteristics in particular, it is convenient to develop the relations, to one another, of the different integrals that have been obtained or have been proved to exist.

We have seen that, in the case of a homogeneous linear equation of the first order, it is possible to construct an integral which, on appropriate determination of its arbitrary elements, comprehends any integral of the equation: also that, in the case of a linear non-homogeneous equation of the first order, it is possible to construct an integral which similarly comprehends any integral that is not of the type called special. Consequently, no further discussion is necessary in those cases.

But in the case of equations that are not linear, it has been seen that there certainly are two kinds of integrals. On the one hand, there is Cauchy's existence-theorem according to which an arbitrary functional element occurs in the expression of the integral proved to exist. On the other hand, Jacobi's method of integration, either in its original form or in any of its modified forms, has led to integrals which contain arbitrary constants in their expression. It is natural to enquire what is the relation, if any, between integrals of such widely distinct types and, further, whether integrals of other types exist.

Variation of Parameters.

71. Accordingly, beginning with a single equation which (after the preceding explanations) may be taken as not linear, we shall suppose it given in the form

$$f(x_1, \ldots, x_n, z, p_1, \ldots, p_n) = 0;$$

and we may imagine that it has been integrated by the Jacobian method, with a result that z is given as a function of the variables and of n arbitrary constants $a_1, \ldots, a_n$ by means of an equation

$$\phi(z, x_1, \ldots, x_n, a_1, \ldots, a_n) = 0.$$

The values of the derivatives are given by equations

$$\phi_m = \frac{\partial \phi}{\partial z} p_m + \frac{\partial \phi}{\partial x_m} = 0,$$

for $m = 1, \ldots, n$; these values of p_m, together with the value of z deduced from $\phi = 0$, will, when substituted in the differential equation, make it satisfied identically. Moreover, the elimination of the n arbitrary constants between the $n+1$ equations

$$\phi = 0, \quad \phi_1 = 0, \ \ldots, \ \phi_n = 0$$

leads to the differential equation, and to that differential equation alone, provided that not all the Jacobians

$$J\left(\left(\frac{\phi, \phi_1, \ldots, \phi_n}{a_1, \ldots, a_n}\right)\right)$$

vanish; and conversely, when there is only a single differential equation, the Jacobians do not all vanish.

In the process of returning from the $n+1$ equations

$$\phi = 0, \quad \phi_1 = 0, \ \ldots, \ \phi_n = 0$$

to the differential equation, the quantities $a_1, \ldots, a_n$ are to be eliminated: but no regard is paid, during the operation, to their constant values; and the resulting differential equation will be the same, provided the $n+1$ equations have the same form, when these quantities are made variable. We therefore make $a_1, \ldots, a_n$ functions of $x_1, \ldots, x_n$, subject to this proviso. This change leaves the equation $\phi = 0$ unaltered in form: in order that $\phi_m = 0$ (for $m = 1, \ldots, n$) may remain unaltered in form, it is necessary that the equation

$$\frac{\partial \phi}{\partial a_1}\frac{\partial a_1}{\partial x_m} + \ldots + \frac{\partial \phi}{\partial a_n}\frac{\partial a_n}{\partial x_m} = 0$$

should be satisfied, for each of the n values of m: and if these equations are satisfied, then $\phi = 0$ (with the changed values of $a_1, \ldots, a_n$) will still give an integral of the differential equation.

Multiplying the n equations by $dx_1, \ldots, dx_n$ respectively and adding, we find

$$\frac{\partial \phi}{\partial a_1} da_1 + \ldots + \frac{\partial \phi}{\partial a_n} da_n = 0,$$

where $da_1, \ldots, da_n$ are the complete variations of the quantities $a_1, \ldots, a_n$; and conversely this equation, when satisfied, yields the n conditions. The coefficients of the differential elements are functions of $z, x_1, \ldots, x_n, a_1, \ldots, a_n$ in general: but z is given by $\phi = 0$ in terms of the other quantities; and, as $a_1, \ldots, a_n$ are (unknown) functions of $x_1, \ldots, x_n$, so the latter may be regarded in the most general case as functions of $a_1, \ldots, a_n$: that is, the coefficients may, in the most general case, be regarded as functions of $a_1, \ldots, a_n$. Thus we have a Pfaffian equation: by the general theory of Pfaffian equations*, the integral equivalent consists of one equation or of several equations connecting the quantities $a_1, \ldots, a_n$.

In the argument, one exceptional case has been omitted: it may be that the Pfaffian equation is evanescent, on account of vanishing coefficients: we then have

$$\frac{\partial \phi}{\partial a_1} = 0, \ \ldots, \ \frac{\partial \phi}{\partial a_n} = 0,$$

concurrently with $\phi = 0$.

After noting this exceptional case, we return to the integral equivalent of the Pfaffian equation. Let it consist of μ equations

$$g_1(a_1, \ldots, a_n) = 0, \ \ldots, \ g_\mu(a_1, \ldots, a_n) = 0,$$

* See Part I of this work, *passim*.

and of these solely: then the only relations among the differential elements are

$$dg_1 = 0, \ \ldots, \ dg_\mu = 0,$$

and the Pfaffian equation must be satisfied in virtue of these. Thus μ quantities $\lambda_1, \ldots, \lambda_\mu$ must exist such that

$$\frac{\partial \phi}{\partial a_1} da_1 + \ldots + \frac{\partial \phi}{\partial a_n} da_n = \lambda_1 dg_1 + \ldots + \lambda_\mu dg_\mu;$$

and therefore

$$\frac{\partial \phi}{\partial a_m} = \lambda_1 \frac{\partial g_1}{\partial a_m} + \ldots + \lambda_\mu \frac{\partial g_\mu}{\partial a_m},$$

for the n values of m. These n equations, together with

$$\phi = 0, \quad g_1 = 0, \ \ldots, \ g_\mu = 0,$$

make up $n + \mu + 1$ equations, involving $a_1, \ldots, a_n, \lambda_1, \ldots, \lambda_\mu$: eliminating the quantities a and λ, we have a single equation as the result, and it expresses z in terms of $x_1, \ldots, x_n$. The value of z determined by this final equation is an integral of the original differential equation: the functional forms $g_1, \ldots, g_\mu$ are involved in its expression.

72. It might appear as if there were integrals of a character intermediate between those of the two kinds considered. Thus we might have $a_{m+1}, \ldots, a_n$ as constants, so that the differential relation would then be

$$\frac{\partial \phi}{\partial a_1} da_1 + \ldots + \frac{\partial \phi}{\partial a_m} da_m = 0.$$

If the integral equivalent of this relation consists of σ equations in the form

$$g_1(a_1, \ldots, a_m) = 0, \ \ldots, \ g_\sigma(a_1, \ldots, a_m) = 0,$$

and of these only, then the same argument as before leads to equations

$$\frac{\partial \phi}{\partial a_i} = \rho_1 \frac{\partial g_1}{\partial a_i} + \ldots + \rho_\sigma \frac{\partial g_\sigma}{\partial a_i},$$

for $i = 1, \ldots, m$. These m equations, together with

$$\phi = 0, \quad g_1 = 0, \ \ldots, \ g_\sigma = 0,$$

are $m + \sigma + 1$ equations involving $z, x_1, \ldots, x_n, a_1, \ldots, a_m, \rho_1, \ldots, \rho_\sigma$: eliminating these m quantities $a_1, \ldots, a_m$ and the σ quantities ρ, we have a single equation between $z, x_1, \ldots, x_n$. The value of z

thus given is an integral of the original equation. The functional forms $g_1, \ldots, g_\sigma$ are involved in its expression; and the arbitrary constants $a_{m+1}, \ldots, a_n$ also occur. The latter can be regarded as given by $n - m$ relations

$$h_1(a_{m+1}, \ldots, a_n) = 0, \ldots, h_{n-m}(a_{m+1}, \ldots, a_n) = 0,$$

involving the $n - m$ constants: they are such that the equations

$$dh_1 = 0, \ldots, dh_{n-m} = 0$$

are satisfied identically. Now it is known from the theory of Pfaffian equations that

$$\sigma + n - m \geqslant \mu,$$

so that the total number of equations among the quantities $a_1, \ldots, a_n$ is greater than before: their range of value is therefore more restricted than in the preceding case. Accordingly, we can regard the present mode of satisfying the differential relation as a specialisation of the preceding mode or as a special instance of the preceding mode involving a greater number of relations some of which are of restricted forms.

In this argument, as in the preceding argument in § 71, one exceptional case is omitted: it may be that the reduced Pfaffian equation is evanescent, on account of vanishing coefficients: we then have

$$\frac{\partial \phi}{\partial a_1} = 0, \ldots, \frac{\partial \phi}{\partial a_m} = 0,$$

concurrently with $\phi = 0$.

It thus appears that, while the completed process leads in every case to a single equation providing an integral, there are intrinsic differences according to the circumstances of the cases. It is clear that distinctions will arise according to the number of relations postulated among the quantities $a_1, \ldots, a_n$; it is customary to regard a class of integrals as being defined according to the number of relations so postulated. When μ relations of the indicated character occur, the corresponding class of integrals is frequently called the μth *class*: and if

$$0 < \mu < n,$$

the integrals of all the classes may be regarded as falling within the category of what will presently be called general integrals. Thus there will be $n - 1$ classes of general integrals.

The extreme cases must also be taken into consideration. It is possible that $\mu = n$: there are then n functional relations connecting the n quantities $a_1, \ldots, a_n$, independent of one another; all these quantities are constants and, when the relations are quite arbitrary, the constants are arbitrary: the integral then provided is what will be called the complete integral. It is possible that $\mu = 0$: if the equations can be satisfied, and an integral is provided, we have what will be called the singular integral.

Of the general integrals, the most comprehensive is that in which only a single functional form occurs, say

$$a_1 = \psi (a_2, \ldots, a_n),$$

and ψ can be taken as the most general and arbitrary function of its arguments. The equations which determine the integral are

$$\phi = 0, \quad a_1 = \psi (a_2, \ldots, a_n),$$

$$\frac{\partial \phi}{\partial a_m} + \frac{\partial \phi}{\partial a_1} \frac{\partial \psi}{\partial a_m} = 0,$$

for $m = 2, \ldots, n$; and the integral itself is given by the elimination of $a_1, \ldots, a_n$ among these $n + 1$ equations.

That it is the most extensive class of general integral can easily be seen by the following argument, whereby it is proved to include all the other classes. When μ relations are postulated among the n quantities $a_1, \ldots, a_n$ in the form

$$g_r (a_1, \ldots, a_n) = 0,$$

for $r = 1, \ldots, \mu$, the integral is given by these equations, together with

$$\phi = 0,$$

$$\frac{\partial \phi}{\partial a_m} = \lambda_1 \frac{\partial g_1}{\partial a_m} + \ldots + \lambda_\mu \frac{\partial g_\mu}{\partial a_m},$$

for $m = 1, \ldots, n$. Let

$$\theta (a_1, \ldots, a_n) = \lambda_1 g_1 + \ldots + \lambda_\mu g_\mu,$$

so that the relation $\theta = 0$ is certainly satisfied for the integral in question; moreover, the equations

$$\frac{\partial \phi}{\partial a_m} = \frac{\partial \theta}{\partial a_m}$$

are certainly satisfied for this integral. Now let $\theta=0$ be resolved for a_1 so as to express it in terms of $a_2, \ldots, a_n$ in a form

$$a_1=\chi(a_2, \ldots, a_n):$$

we have

$$\frac{\partial\theta}{\partial a_m}+\frac{\partial\theta}{\partial a_1}\frac{\partial\chi}{\partial a_m}=0.$$

Hence, for the integral in question, the equations

$$\frac{\partial\phi}{\partial a_m}+\frac{\partial\phi}{\partial a_1}\frac{\partial\chi}{\partial a_m}=0$$

are satisfied: and conversely, when these are satisfied, the original set of equations also is satisfied. Now in the case when there is only a single relation

$$a_1=\psi(a_2, \ldots, a_n),$$

ψ is the most general function possible: so that the relation

$$a_1=\chi(a_2, \ldots, a_n)$$

is included as a special case, and consequently the equations

$$\frac{\partial\phi}{\partial a_m}+\frac{\partial\phi}{\partial a_1}\frac{\partial\chi}{\partial a_m}=0$$

are a special case of the equations

$$\frac{\partial\phi}{\partial a_m}+\frac{\partial\phi}{\partial a_1}\frac{\partial\psi}{\partial a_m}=0;$$

that is, the general integral in question is a special case of the general integral, which arises when there is only a single relation between the quantities $a_1, \ldots, a_n$. The latter general integral is accordingly the most comprehensive.

In passing, we may note that the general integral includes the exceptional case noted, in which $a_{m+1}, \ldots, a_n$ are arbitrary constants and the equations

$$\frac{\partial\phi}{\partial a_1}=0, \ldots, \frac{\partial\phi}{\partial a_m}=0$$

are satisfied. We can represent it by relations

$$a_\mu=\psi_\mu(a_1, \ldots, a_m),$$

for $\mu=m+1, \ldots, n$, and by restricting the functions ψ_μ to be constants; for then

$$\frac{\partial\psi_\mu}{\partial a_i}=0,$$

for $i = 1, \ldots, m$, and the relation

$$\frac{\partial \phi}{\partial a_i} + \sum_{\mu} \frac{\partial \phi}{\partial a_\mu} \frac{\partial \psi_\mu}{\partial a_i} = 0$$

simply becomes

$$\frac{\partial \phi}{\partial a_i} = 0,$$

which (for $i = 1, \ldots, m$) are the equations for the exceptional case.

Classes of Integrals.

73. Three kinds of integrals may thus arise. One of them is given by an equation containing n arbitrary constants; it is called the *complete integral*. Another of them is given by equations that involve a functional form or several functional forms, and in the most general type these forms are arbitrary; these integrals are called *general integrals* and often, when there is only a single functional form so that the widest range of variation is provided, the integral is called *the general integral*. And, lastly, the equations

$$\phi = 0, \quad \frac{\partial \phi}{\partial a_1} = 0, \ldots, \frac{\partial \phi}{\partial a_n} = 0,$$

may be possible and be consistent with one another; if the result of eliminating $a_1, \ldots, a_n$ among them provides a single equation involving no arbitrary element, and if the equation determines an integral*, the integral thus furnished is called the *singular integral*.

It must however be noticed that an integral, containing the appropriate number of arbitrary constants, is not necessarily the complete integral, any more than one which contains no arbitrary element is necessarily a singular integral. On the one hand, since an arbitrary function can be regarded as containing any number of arbitrary constants, a general integral may be simply specialised so as to contain the appropriate number of arbitrary constants: it will not thereby necessarily become a complete integral, for it may

* The reason for this limitation will appear subsequently: meanwhile, it may be sufficient to point out that, while the equations $\frac{\partial \phi}{\partial a_1} = 0, \ldots, \frac{\partial \phi}{\partial a_n} = 0$ are consistent with the existence of an integral, it has not been proved (and, indeed, cannot be proved) that their significance is only co-extensive with that existence. Even in the case of ordinary equations of the first order, the corresponding process frequently gives rise to relations that do not provide integrals of the equations in question: and the same holds, to a wider extent, in partial equations.

be only a special case of the general integral. On the other hand, by assigning particular values to the arbitrary constants in a complete integral, the latter becomes free from all arbitrary elements: it will not thereby become a singular integral (even if such an integral is possessed by the equation), for it is only a special case of the complete integral. It is therefore important to devise tests which shall shew to what category any given integral should, if possible, be assigned: and this necessity raises a further question as to how comprehensive is the retained aggregate of integrals.

Special Integrals.

74. Suppose, then, that we have an integral of the differential equation

$$f(x_1, \ldots, x_n, z, p_1, \ldots, p_n) = 0$$

given by the equation

$$\theta(x_1, \ldots, x_n, z) = 0;$$

and let the values of z thus determined be denoted by ζ. Also, let a complete integral be given in the form

$$g(x_1, \ldots, x_n, z, a_1, \ldots, a_n) = 0;$$

and let the value of z thus determined be denoted by Z. We have to consider whether it is possible to associate with $g=0$ equations or relations which will change Z into ζ; if this should be possible, then the character of the added equations or relations will indicate the character of the integral ζ.

In order to obtain the tests that may be both sufficient and necessary, assume that $a_1, \ldots, a_n$ are changed into functions of $x_1, \ldots, x_n$, such that Z is still an integral of the differential equation and such that, if possible, it becomes the integral ζ. As the two integrals are now hypothetically the same functions of $x_1, \ldots, x_n$, the derivatives of these functions with regard to the variables are respectively the same. For the integral ζ, they are given by

$$\frac{\partial \theta}{\partial x_m} + p_m \frac{\partial \theta}{\partial z} = 0,$$

for $m = 1, \ldots, n$, when z is replaced by ζ in these equations; and for the integral Z, they are given by

$$\frac{\partial g}{\partial x_m} + p_m \frac{\partial g}{\partial z} = 0, \quad \frac{\partial g}{\partial a_1}\frac{\partial a_1}{\partial x_m} + \ldots + \frac{\partial g}{\partial a_n}\frac{\partial a_n}{\partial x_m} = 0,$$

for $m=1, \ldots, n$, when z is replaced by Z in these equations. Consequently, we must have

$$\frac{\partial g}{\partial x_m}\frac{\partial \theta}{\partial z}-\frac{\partial g}{\partial z}\frac{\partial \theta}{\partial x_m}=0,$$

$$\frac{\partial g}{\partial a_1}\frac{\partial a_1}{\partial x_m}+\ldots+\frac{\partial g}{\partial a_n}\frac{\partial a_n}{\partial x_m}=0,$$

for $m=1, \ldots, n$, when z is replaced by the supposed common value of ζ and Z.

Now when this common value is substituted, the n equations

$$\frac{\partial g}{\partial x_m}\frac{\partial \theta}{\partial z}-\frac{\partial g}{\partial z}\frac{\partial \theta}{\partial x_m}=0$$

are a set of equations involving the quantities $a_1, \ldots, a_n$. If they determine values for these quantities, we can proceed to the identification of the integral; but they do not necessarily determine such values, and then we cannot proceed.

Suppose that such values are determined. If they are constants, then ζ is a more or less particular form of the complete integral: all the equations

$$\frac{\partial g}{\partial a_1}\frac{\partial a_1}{\partial x_m}+\ldots+\frac{\partial g}{\partial a_n}\frac{\partial a_n}{\partial x_m}=0$$

are satisfied. If values are found, so that some at least have the form of functions of $x_1, \ldots, x_n$, there may be some functional relation or several functional relations among them: let these be denoted by

$$g_1(a_1, \ldots, a_n)=0, \ldots, g_\mu(a_1, \ldots, a_n)=0.$$

Then the other n equations are satisfied by means of the equations

$$\frac{\partial g}{\partial a_m}=\lambda_1\frac{\partial g_1}{\partial a_m}+\ldots+\lambda_\mu\frac{\partial g_\mu}{\partial a_m},$$

for $m=1, \ldots, n$, with appropriately determinate values of $\lambda_1, \ldots, \lambda_\mu$. All the conditions then are satisfied; and ζ then is a more or less particular form of the general integral. If on the other hand the variable values found (say m in number) are such that no functional relation subsists among these m quantities, the n remaining equations can only be satisfied by having

$$\frac{\partial g}{\partial a_i}=0,$$

for each of the m quantities a_i found to be variable; the integral ζ would then be a degenerate form of the general integral of the differential equation. Lastly, if all the quantities a are variable and if there is no functional relation among them, the n remaining equations can only be satisfied by having

$$\frac{\partial g}{\partial a_1} = 0, \ldots, \frac{\partial g}{\partial a_n} = 0;$$

the integral ζ would then be a singular integral of the differential equation.

It thus appears that, subject to the determination of the quantities $a_1, \ldots, a_n$ from the equations

$$\frac{\partial g}{\partial x_m}\frac{\partial \theta}{\partial z} - \frac{\partial g}{\partial z}\frac{\partial \theta}{\partial x_m} = 0,$$

the integral ζ is comprehended within the aggregate of the complete integral, the general integral, and the singular integral. This aggregate is widely comprehensive: it cannot be declared to be completely comprehensive, because occasions arise in which the equations refuse to provide a consistent set of values of $a_1, \ldots, a_n$ needed to secure inclusion. The whole of this theory is formal: it does not take account of the peculiarities of equations: and examples will be indicated to which it fails to apply.

Such integrals, as do occur but are not included in any of the three classes, will be called *special*.

Ex. 1. It is easy to see that the equation

$$p_1x_1 + \ldots + p_nx_n = z$$

has an integral

$$z = a_1x_1 + \ldots + a_nx_n,$$

which is a complete integral. To obtain a general integral, the most general possible, we take only a single relation among the quantities $a_1, \ldots, a_n$ in the form

$$a_1 = f(a_2, \ldots, a_n),$$

where f is an arbitrary function of its arguments. The associated equations are

$$x_s + x_1\frac{\partial f}{\partial a_s} = 0,$$

for $s = 2, \ldots, n$; these give

$$a_r = g_r\left(\frac{x_2}{x_1}, \ldots, \frac{x_n}{x_1}\right),$$

where the character of g_r is dominated by the arbitrary form of f. Inserting these values of a, we have

$$\frac{z}{x_1}=F\left(\frac{x_2}{x_1}, \dots, \frac{x_n}{x_1}\right),$$

where F is an arbitrary function.

This is the integral which would be obtained by the process of § 30; accordingly, the most comprehensive integral given by that process is the general integral.

The equations, which would give the singular integral if it existed, are

$$z=a_1x_1+\dots+a_nx_n,$$
$$x_1=0, \dots, x_n=0:$$

clearly there is no singular integral of the equation, though $z=0$ is a particular case of the complete integral.

Ex. 2. The equation

$$xp+2yq=2\left(z-\frac{x^2}{y}\right)^2$$

has been discussed (§ 34, Ex. 3); in particular, it was shewn that the integral

$$z=\frac{x^2}{y}$$

was not derivable from the general integral there obtained. The equation does not possess a singular integral.

Is the integral $z=\dfrac{x^2}{y}$ comprehended in the complete integral?

Ex. 3. At the end of § 59, it was shewn that the equation

$$3px+qy+q^3x^2=0$$

possesses two complete integrals

$$z=a-\tfrac{1}{3}b^3x+bx^{-\frac{1}{3}}y,$$
$$z=A+\tfrac{2}{3}y^{\frac{3}{2}}(x^2+2Bx)^{-\frac{1}{2}}.$$

The general integral deduced from the first of these complete integrals is obtained by associating with it the equations

$$a=\phi(b), \quad \phi'(b)-xb^2+x^{-\frac{1}{3}}y=0,$$

where ϕ is arbitrary: the general integral deduced from the second of them is obtained by associating with it the equations

$$A=\psi(B), \quad \psi'(B)-\tfrac{2}{3}y^{\frac{3}{2}}x(x^2+2Bx)^{-\frac{3}{2}},$$

where ψ is arbitrary. Clearly there is no singular integral.

To obtain the relations to one another of the two complete integrals, we adopt the method in the text. When we equate the different respective derivatives, we have the relations

$$-\tfrac{1}{3}b^3-\tfrac{1}{3}bx^{-\frac{4}{3}}y=-\tfrac{2}{3}y^{\frac{3}{2}}(x^2+2Bx)^{-\frac{3}{2}}(x+B),$$
$$bx^{-\frac{1}{3}}=y^{\frac{1}{2}}(x^2+2Bx)^{-\frac{1}{2}};$$

these relations are consistent with one another, in virtue of the single relation

$$b=y^{\frac{1}{2}}x^{\frac{1}{3}}(x^2+2Bx)^{-\frac{1}{2}}.$$

When we equate the two integrals themselves, we find

$$a=A-\tfrac{2}{3}Bxy^{\frac{3}{2}}(x^2+2Bx)^{-\frac{3}{2}}.$$

The values of a and b are thus variable quantities; and it is easy to see that they are connected by the relation

$$a-A=-\tfrac{2}{3}Bb^3.$$

In virtue of this relation, and of the values of a and b, the other necessary relations

$$\frac{\partial z}{\partial a}\frac{\partial a}{\partial x}+\frac{\partial z}{\partial b}\frac{\partial b}{\partial x}=0,$$

$$\frac{\partial z}{\partial a}\frac{\partial a}{\partial y}+\frac{\partial z}{\partial b}\frac{\partial b}{\partial y}=0,$$

are satisfied.

Hence each of the two complete integrals is a particular case of the general integral deduced from the other: the generalising relation is

$$a-A+\tfrac{2}{3}Bb^3=0.$$

Ex. 4. The equation

$$pq=4xy$$

has

$$z=\frac{x^2}{a}+ay^2+b$$

for a complete integral; it has no singular integral: and its general integral is given by

$$z=\frac{x^2}{a}+ay^2+b,\quad 0=-\frac{x^2}{a^2}+y^2+f'(a),\quad b=f(a).$$

Another integral is given by

$$z=2xy+b.$$

To investigate its relation to the complete integral, we proceed as before. Equating the derivatives, we find

$$\frac{2x}{a}=2y,\quad 2ay=2x,$$

giving

$$a=\frac{x}{y};$$

with this value, the two quantities z are the same.

The other equations

$$\frac{\partial g}{\partial a}\frac{\partial a}{\partial x}+\frac{\partial g}{\partial b}\frac{\partial b}{\partial x}=0,\quad \frac{\partial g}{\partial a}\frac{\partial a}{\partial y}+\frac{\partial g}{\partial b}\frac{\partial b}{\partial y}=0,$$

are satisfied by

$$\frac{\partial g}{\partial a}=0,\quad b=\text{arbitrary constant}.$$

The new integral is a special case of the general integral; for we have

$$z=\frac{x^2}{a}+ay^2+b, \quad b=f(a), \quad -\frac{x^2}{a^2}+y^2+f'(a)=0,$$

as the equations of the general integral; and they lead to the new integral, when $f(a)$ is regarded as a pure constant.

Ex. 5. Classify the integral $z=3x_1^{\frac{1}{3}}x_2^{\frac{1}{3}}x_3^{\frac{1}{3}}+b$ of the equation

$$p_1p_2p_3=1.$$

Ex. 6. Consider the equation

$$\{1+(z-x-y)^{\frac{1}{2}}\}p+q=2,$$

which has already (§ 34, Ex. 4) been discussed from the point of view of the general integral. The equation is clearly satisfied by

$$z=x+y:$$

the question is, does this integral fall within the three classes of integrals considered?

Proceeding to integrate by Charpit's method, we find

$$\frac{q-1}{p-1}=a,$$

as one integral of the subsidiary equations. When this relation is combined with the original equation, we have values of p and q: these are substituted in

$$dz=p\,dx+q\,dy,$$

and the quadrature is effected: the result is

$$z+(a-1)y+2(a+1)(z-x-y)^{\frac{1}{2}}=b,$$

where a and b are arbitrary. Writing

$$u=z+(a-1)y+2(a+1)(z-x-y)^{\frac{1}{2}},$$

the singular integral (if any) is given by

$$\frac{\partial(u-b)}{\partial a}=0, \quad \frac{\partial(u-b)}{\partial b}=0;$$

the latter equation shews that the singular integral does not exist: consequently

$$z=x+y$$

is not a singular integral.

The general integral is given by the elimination of a between

$$u=\phi(a),$$

$$\frac{\partial u}{\partial a}=\phi'(a).$$

If ϕ can be determined, so that the result of the elimination is to give

$$z=x+y,$$

the second of the equations for the elimination must become

$$y=\phi'(a),$$

and the first of them must become

$$x+ay=\phi(a).$$

The former of these, for any function ϕ, makes a a function of y only; the latter, instead of being identically satisfied (as it should be) if the integral $z=x+y$ could thus arise, leads to a relation between x and y. Any such relation is excluded. Hence $z=x+y$ is not a particular case of the general integral.

It is clear that constant values of a and b cannot be chosen such that the equation $u=b$ leads to the equation $z=x+y$: hence the integral is not a particular case of the complete integral.

It follows therefore that, while $z=x+y$ is an integral of the equation

$$\{1+(z-x-y)^{\frac{1}{2}}\}\,p+q=2,$$

it does not belong to any of the three usual classes of integrals: an instance is thus provided in which the general theorem due to Lagrange does not hold.

If the differential equation is rationalised, so that it takes the form

$$z-x-y=\left(\frac{2-q}{p}-1\right)^2,$$

the complete integral is

$$\{z+(a-1)\,y-b\}^2=4\,(a+1)^2\,(z-x-y);$$

and $z=x+y$ is easily seen to be a singular integral. The explanation of the difference is left to the student as an exercise.

Ex. 7. Given the equation

$$Ap^2+Bpq+Cq^2=D,$$

where A, B, C, D are functions of x and y only, investigate the conditions necessary and sufficient to secure that it possesses a complete integral of the form

$$z=au^2+\frac{1}{a}\,v^2+b,$$

where u and v are functions of x and y, and a, b are constants.

Verify that, if the conditions are satisfied, it also possesses an integral

$$z=uv+b.$$

What is the character of this integral?

Tests for a Complete Integral.

75. In the preceding investigation, it has been assumed that a complete integral of the differential equation is known, so that it is possible to proceed from that integral to the differential equation, and to that equation alone: and it has been pointed out that an integral, containing the proper number of arbitrary constants, is not necessarily complete. The important limitation is that elimination among the equations, denoted in § 71 by

$$\phi=0,\quad \phi_1=0,\ \ldots,\ \phi_n=0,$$

should lead to one, and to only one, equation.

For this purpose, it is necessary that not all the Jacobians should vanish: if they do vanish, then the elimination of the n quantities $a_1, \ldots, a_n$ will lead to at least two equations.

Again, if all the Jacobians but one, say

$$J\left(\frac{\phi, \phi_2, \ldots, \phi_n}{a_1, \ldots\ldots, a_n}\right),$$

are known to vanish, then either that Jacobian vanishes or else

$$\frac{\partial \phi_1}{\partial a_1} = 0, \ldots, \frac{\partial \phi_1}{\partial a_n} = 0,$$

that is, either that Jacobian vanishes, or ϕ_1 involves none of the constants. The first of these two alternatives is the preceding case. As regards the second alternative, we at once have

$$\phi_1 = 0$$

as an equation. The constants $a_1, \ldots, a_n$ may or may not be eliminable between $\phi = 0$, $\phi_2 = 0, \ldots, \phi_n = 0$; so that there would be only one equation if they cannot be eliminated, and there would be at least two equations if they can be eliminated. If there is only one equation, the integral is complete; if there is more than one, the integral is not complete.

If a Jacobian, say

$$J\left(\frac{\phi, \phi_2, \ldots, \phi_n}{a_1, \ldots\ldots, a_n}\right),$$

is known not to vanish, then the equations

$$\phi = 0, \quad \phi_2 = 0, \ \ldots, \ \phi_n = 0$$

can be resolved for $a_1, \ldots, a_n$; their values, substituted in $\phi_1 = 0$, if it involves any of them, lead to a single equation; while, if $\phi_1 = 0$ does not involve any of the constants $a_1, \ldots, a_n$, it is itself one equation involving derivatives. We have only a single equation: the integral is complete.

Ex. 1. Consider an integral equation

$$z = (x_1 - a_1)^2 + (x_2 - a_2)^2 + (x_3 - a_3)^2;$$

it is easily seen to be a complete integral of the differential equation

$$4z = p_1^2 + p_2^2 + p_3^2,$$

the elimination being immediate.

An integral equation

$$z = (x_1 - a_1)^2 + (x_2 - a_2)^2 + x_3 - a_3$$

leads to a single equation

$$p_3=1,$$

and no other elimination is possible: the integral is complete.

An integral equation*

$$z=\{(x_1-a_1)^2+(x_2-a_2)^2\}^{\frac{1}{2}}+x_3-a_3$$

leads to two equations

$$p_1^2+p_2^2=1, \quad p_3=1;$$

it is not a complete integral of either equation, nor of an equation such as

$$p_1^2+p_2^2+(p_3-1)^2=1.$$

Ex. 2. The equation

$$z=ax_1+bx_2+cx_3$$

is a complete integral of the equation

$$x_1p_1+x_2p_2+x_3p_3=z.$$

Another integral, containing three arbitrary constants, is

$$z=\alpha x_1+\beta x_2+\gamma\frac{x_2^2}{x_1}.$$

To determine its significance, we equate the values of p_1, p_2, p_3 derived from the two values; and we have

$$a=\alpha-\gamma\frac{x_2^2}{x_1^2},$$

$$b=\beta+2\gamma\frac{x_2}{x_1},$$

$$c=0,$$

giving variable values for a and b. These variable values are subject to the two equations

$$4\gamma(a-\alpha)+(b-\beta)^2=0, \quad c=0;$$

and these, as two equations connecting the assumed variable magnitudes, shew that

$$z=\alpha x_1+\beta x_2+\gamma\frac{x_2^2}{x_1}$$

is not a complete integral of the equation, but is a special case of the general integral derived from the complete integral

$$z=ax_1+bx_2+cx_3.$$

In point of fact, the equation

$$z=\alpha x_1+\beta x_2+\gamma\frac{x_2^2}{x_1}$$

leads to two equations

$$x_1p_1+x_2p_2=z, \quad p_3=0,$$

thus verifying the conclusion that it is not a complete integral of the original equation.

Ex. 3. To illustrate a different aspect of the relations of integrals, consider the equation

$$4z=p_1^2+p_2^2+p_3^2,$$

* This example is given by Goursat, *Leçons*, p. 98.

which occurred in Ex. 1. It possesses an integral

$$z=(x_1-a_1)^2+(x_2-a_2)^2+(x_3-a_3)^2,$$

which is easily seen to be complete: it possesses an integral

$$z=\frac{(x_1+\mathrm{a}_1x_2+\mathrm{a}_2x_3+\mathrm{a}_3)^2}{1+\mathrm{a}_1{}^2+\mathrm{a}_2{}^2},$$

which also is easily seen to be complete. What is the relation, if any, between the two integrals?

To determine it, we first equate the values of p_1, p_2, p_3 for the two values of z, and resolve the three equations for (say) a_1, a_2, a_3; and we find the three variable values

$$\mathrm{a}_1=\frac{x_2-a_2}{x_1-a_1},\qquad \mathrm{a}_2=\frac{x_3-a_3}{x_1-a_1},$$

$$\mathrm{a}_3=a_1-a_2\frac{x_2-a_2}{x_1-a_1}-a_3\frac{x_3-a_3}{x_1-a_1}.$$

These are connected by a functional relation

$$a_1+a_2\mathrm{a}_1+a_3\mathrm{a}_2+\mathrm{a}_3=0.$$

If then we construct the general integral to be associated with the complete integral

$$z=(x_1-a_1)^2+(x_2-a_2)^2+(x_3-a_3)^2,$$

the integral

$$z=\frac{(x_1+\mathrm{a}_1x_2+\mathrm{a}_2x_3+\mathrm{a}_3)^2}{1+\mathrm{a}_1{}^2+\mathrm{a}_2{}^2}$$

is a particular case of that general integral given by the particular equations

$$\left.\begin{aligned} a_1&=-\mathrm{a}_1a_2-\mathrm{a}_2a_3-\mathrm{a}_3\\ z&=(x_1-a_1)^2+(x_2-a_2)^2+(x_3-a_3)^2\\ 0&=(x_1-a_1)\,\mathrm{a}_1-(x_2-a_2)\\ 0&=(x_1-a_1)\,\mathrm{a}_2-(x_3-a_3)\end{aligned}\right\},$$

when a_1, a_2, a_3 are eliminated among them.

On the other hand, if we construct the general integral to be associated with the complete integral

$$z=\frac{(x_1+\mathrm{a}_1x_2+\mathrm{a}_2x_3+\mathrm{a}_3)^2}{1+\mathrm{a}_1{}^2+\mathrm{a}_2{}^2},$$

the integral

$$z=(x_1-a_1)^2+(x_2-a_2)^2+(x_3-a_3)^2$$

is a particular case of that general integral given by the particular equations

$$\left.\begin{aligned} \mathrm{a}_3&=-a_2\mathrm{a}_1-a_3\mathrm{a}_2-a_1\\ z&=\frac{(x_1+\mathrm{a}_1x_2+\mathrm{a}_2x_3+\mathrm{a}_3)^2}{1+\mathrm{a}_1{}^2+\mathrm{a}_2{}^2}\\ 0&=x_2-a_2-\mathrm{a}_1\,\frac{x_1+\mathrm{a}_1x_2+\mathrm{a}_2x_3+\mathrm{a}_3}{1+\mathrm{a}_1{}^2+\mathrm{a}_2{}^2}\\ 0&=x_3-a_3-\mathrm{a}_2\,\frac{x_1+\mathrm{a}_1x_2+\mathrm{a}_2x_3+\mathrm{a}_3}{1+\mathrm{a}_1{}^2+\mathrm{a}_2{}^2}\end{aligned}\right\},$$

when a_1, a_2, a_3 are eliminated among them.

It thus appears that a single equation, of degree higher than the first, may have quite distinct complete integrals; and that a complete integral may be a particular case of a general integral derived from another complete integral, and, *à fortiori*, may be a particular case of the general integral derived from itself. (See also Ex. 3, § 74.)

Ex. 4. Discuss the character of the integral of

$$(p_1-1)^2+p_2^2+\dots+p_n^2=1,$$

as given by the equation

$$z-a_1=x_1+\{(x_2-a_2)^2+\dots+(x_n-a_n)^2\}^{\frac{1}{2}}.$$

Singular Integrals.

76. We have seen that, when a singular integral of the equation

$$f(x_1, \dots, x_n, z, p_1, \dots, p_n)=0$$

exists, it can be obtained from the complete integral

$$\phi(z, x_1, \dots, x_n, a_1, \dots, a_n)=0,$$

by eliminating $a_1, \dots, a_n$ between the equations

$$\phi=0, \quad \frac{\partial\phi}{\partial a_1}=0, \quad \dots, \quad \frac{\partial\phi}{\partial a_n}=0.$$

But when it exists, it may also be obtained from the differential equation itself: the formal argument is as follows.

The values of $p_1, \dots, p_n$ belonging to any integral given by $\phi=0$ are

$$\frac{\partial\phi}{\partial z}p_r+\frac{\partial\phi}{\partial x_r}=0,$$

for $r=1, \dots, n$; when these are substituted in the differential equation $f=0$, the latter becomes a relation between $z, x_1, \dots, x_n$ and the quantities $a_1, \dots, a_n$ introduced by the derivatives of ϕ. When the value of z given by $\phi=0$ is substituted in this relation, it becomes an identity: for it is thus that the original differential equation is satisfied in connection with $\phi=0$. Hence some value of z given by the changed form of $f=0$ is the same as a value of z given by $\phi=0$; for all such values, the two equations

$$f=0, \quad \phi=0$$

are equivalent to one another, f being transformed by the introduction of the values of $p_1, \dots, p_n$.

Now suppose that the integral is the singular integral, assumed to exist; we know that the equations

$$\frac{\partial \phi}{\partial a_1} = 0, \ \ldots, \ \frac{\partial \phi}{\partial a_n} = 0$$

are satisfied. As the transformed expression for f is equivalent to ϕ for this integral, we must therefore have

$$\frac{\partial f}{\partial a_1} = 0, \ \ldots, \ \frac{\partial f}{\partial a_n} = 0:$$

hence, as the quantities $a_1, \ldots, a_n$ have been introduced into this transformed expression solely through $p_1, \ldots, p_n$, we have

$$\frac{\partial f}{\partial p_1}\frac{\partial p_1}{\partial a_r} + \frac{\partial f}{\partial p_2}\frac{\partial p_2}{\partial a_r} + \ldots + \frac{\partial f}{\partial p_n}\frac{\partial p_n}{\partial a_r} = 0,$$

for $r = 1, \ldots, n$.

These are an aggregate of n equations, linear and homogeneous in the n derivatives of f with regard to p.

They could be satisfied by non-zero values for these derivatives, if

$$J\begin{pmatrix} p_1, \ldots, p_n \\ a_1, \ldots, a_n \end{pmatrix} = 0;$$

when this is the case, there exist m relations, where $m \leqslant n-1$, connecting these derivatives of f linearly and homogeneously. As our purpose is the derivation (if possible) of an integral from the differential equation itself without assuming knowledge of the actual form of the complete integral, we shall omit any further discussion of this alternative.

The aggregate of n equations could also be satisfied (and if the preceding alternative were inadmissible, the aggregate could only be satisfied) by

$$\frac{\partial f}{\partial p_1} = 0, \ \ldots, \ \frac{\partial f}{\partial p_n} = 0:$$

and these must coexist with $f = 0$. It may be possible that these $n+1$ equations determine $z, p_1, \ldots, p_n$ as functions of $x_1, \ldots, x_n$; but the value of z so obtained cannot be an integral of the original equation, unless the values of $p_1, \ldots, p_n$ are the same as the values of $\frac{\partial z}{\partial x_1}, \ldots, \frac{\partial z}{\partial x_n}$ derived from that value of z. To test this possibility, suppose that the n equations

$$\frac{\partial f}{\partial p_1} = 0, \ \ldots, \ \frac{\partial f}{\partial p_n} = 0$$

can be resolved* with regard to $p_1, \ldots, p_n$, and that the values of $p_1, \ldots, p_n$ thence deduced are substituted in $f=0$; the latter then becomes a relation between $z, x_1, \ldots, x_n$. If the relation provides an integral of the original equation, then

$$p_r = \frac{\partial z}{\partial x_r},$$

for $r=1, \ldots, n$, the values of $p_1, \ldots, p_n$ being the above values, and the values of $\frac{\partial z}{\partial x_1}, \ldots, \frac{\partial z}{\partial x_n}$ being deduced from the integral relation. The latter are given by

$$\frac{\partial f}{\partial x_r} + \frac{\partial f}{\partial z}\frac{\partial z}{\partial x_r} + \sum_{i=1}^{n} \left\{ \frac{\partial f}{\partial p_i} \left(\frac{\partial p_i}{\partial x_r} + \frac{\partial p_i}{\partial z}\frac{\partial z}{\partial x_r} \right) \right\} = 0,$$

for $r=1, \ldots, n$; hence we must have

$$\frac{\partial f}{\partial x_r} + \frac{\partial f}{\partial z} p_r = 0.$$

Conversely, if this equation is satisfied, and if the initial assumption that the $n+1$ equations determine $z, p_1, \ldots, p_n$ as functions of $x_1, \ldots, x_n$ is justified, we have

$$\frac{\partial z}{\partial x_r} = p_r,$$

for $r=1, \ldots, n$, provided $\frac{\partial f}{\partial z}$ is not zero. In that case, we have an integral of the differential equation: it is the singular integral. But if the values of $z, p_1, \ldots, p_n$ make $\frac{\partial f}{\partial z}$ vanish, the inference cannot be made: separate investigation is then required and will come later. We thus have the following theorem:

If the equations

$$f(x_1, \ldots, x_n, z, p_1, \ldots, p_n) = 0,$$

$$\frac{\partial f}{\partial p_1} = 0, \ldots, \frac{\partial f}{\partial p_n} = 0, \quad \frac{\partial f}{\partial x_1} + p_1 \frac{\partial f}{\partial z} = 0, \ldots, \frac{\partial f}{\partial x_n} + p_n \frac{\partial f}{\partial z} = 0,$$

are consistent with one another, and if they determine $z, p_1, \ldots, p_n$ as functions of $x_1, \ldots, x_n$, such that $\frac{\partial f}{\partial z}$ does not vanish identically,

* This supposition requires that the Hessian of f does not vanish simultaneously with the $n+1$ quantities $f, \frac{\partial f}{\partial p_1}, \ldots, \frac{\partial f}{\partial p_n}$.

the value of z thus given is an integral of the equation, being the singular integral.

Of course, if the $2n+1$ equations are not consistent with one another, no integral of the differential equation can be found by this avenue.

And it must not be assumed that the locus, given by the elimination of $p_1, \ldots, p_n$ among the equations

$$f=0, \quad \frac{\partial f}{\partial p_1}=0, \ldots, \frac{\partial f}{\partial p_n}=0,$$

is the singular integral: if it exists, it will be included in the locus, but the locus may include other equations which do not provide integrals.

Ex. Discuss in the preceding manner, so as to obtain singular integrals (if any), the equations

(i) $z=p_1x_1+\ldots+p_nx_n+ap_1\ldots p_n$;

(ii) $(ap_1-z)(ap_2-z)(ap_3-z)=a^3p_1p_2p_3$;

(iii) $z=f(p_1, \ldots, p_n)$,

where, in the last equation, f is a polynomial in its arguments.

Exceptional Integrals.

77. Now it may happen that the $2n+1$ equations

$$f=0, \quad \frac{\partial f}{\partial p_1}=0, \ldots, \frac{\partial f}{\partial p_n}=0, \quad \frac{\partial f}{\partial x_1}+p_1\frac{\partial f}{\partial z}=0, \ldots, \frac{\partial f}{\partial x_n}+p_n\frac{\partial f}{\partial z}=0$$

are consistent with one another, but that (contrary to the hypothesis in the preceding theorem) they do not determine all the quantities $z, p_1, \ldots, p_n$ in terms of $x_1, \ldots, x_n$; they may determine only a number of these quantities in terms of the remainder, say

$$p_r=g_r(x_1, \ldots, x_n, z, p_{m+1}, \ldots, p_n),$$

for $r=1, \ldots, m$. When these values are substituted in the above equations, each of them becomes an identity, $z, p_{m+1}, \ldots, p_n$ being regarded as functions of $x_1, \ldots, x_n$. In particular, $f=0$ is an identity; and therefore

$$\frac{\partial f}{\partial x_s}+\frac{\partial f}{\partial z}\frac{\partial z}{\partial x_s}+\sum_{r=1}^{m}\left[\frac{\partial f}{\partial p_r}\left\{\frac{\partial g_r}{\partial x_s}+\frac{\partial g_r}{\partial z}\frac{\partial z}{\partial x_s}+\sum_{\mu}\left(\frac{\partial g_r}{\partial p_{m+\mu}}\frac{dp_{m+\mu}}{dx_s}\right)\right\}\right]$$
$$+\sum_{\mu}\frac{\partial f}{\partial p_{m+\mu}}\frac{dp_{m+\mu}}{dx_s}=0.$$

But the equations

$$\frac{\partial f}{\partial p_r} = 0, \quad \frac{\partial f}{\partial p_{m+\mu}} = 0,$$

are satisfied identically, for $r = 1, \ldots, m$, and $\mu = 1, \ldots, n - m$, by the values in question; also

$$\frac{\partial f}{\partial x_s} + \frac{\partial f}{\partial z}\frac{\partial z}{\partial x_s} = 0,$$

for $s = 1, \ldots, n$: hence, unless $\dfrac{\partial f}{\partial z}$ vanishes for the values in question, we have

$$p_s = \frac{\partial z}{\partial x_s},$$

for all the values of s.

Thus the set of $2n + 1$ equations may be replaced by a set

$$p_r = g_r(x_1, \ldots, x_n, z, p_{m+1}, \ldots, p_n),$$

for $r = 1, \ldots, m$: and, from their source, we have seen that

$$p_s = \frac{\partial z}{\partial x_s}, \qquad (s = 1, \ldots, n),$$

that is, the quantities p are the derivatives of z. Thus the set of m equations is a complete system: it possesses an integral containing $n - m + 1$ arbitrary constants.

Although such an integral has affinities with the complete integral, it can hardly be claimed as a specialised case of the complete integral: and although it has affinities with the singular integral, it can hardly be claimed as a generalised case of the singular integral. It may be regarded as belonging to the, as yet, unclassified aggregate of special integrals.

Examples will be given later.

Integrals of Equations of first order in two independent variables.

78. After the general discussion for equations in n independent variables, it is unnecessary to enter upon the similar discussion for equations in two independent variables: but the results are so important for the latter set of equations, particularly in connection with the geometry of ordinary space, that they are worthy of separate statement.

Accordingly, an equation of the first order in two independent variables, represented as usual by

$$f(x, y, z, p, q) = 0,$$

possesses a *complete integral* involving two arbitrary constants, which may be represented by

$$\phi(x, y, z, a, b) = 0.$$

In addition, it possesses a *general integral*, obtained by the elimination of a and b between the equations

$$\left.\begin{aligned} \phi(x, y, z, a, b) &= 0 \\ b - \theta(a) &= 0 \\ \frac{\partial \phi}{\partial a} + \frac{\partial \phi}{\partial b}\theta'(a) &= 0 \end{aligned}\right\},$$

where θ is an arbitrary function: frequently, the elimination cannot explicitly be performed, and then the three equations give the general integral.

The differential equation may, but does not necessarily, possess another integral derivable from the complete integral. If the equations

$$\phi = 0, \quad \frac{\partial \phi}{\partial a} = 0, \quad \frac{\partial \phi}{\partial b} = 0,$$

furnish values of z, a, b in terms of x and y, such that z is an integral of the differential equation, then if b can be expressed in terms of a alone, the integral so furnished is a particular case of the general integral: but, if b cannot be expressed in terms of a alone, the integral is a *singular integral*.

Moreover, a differential equation may possess integrals of the unclassified aggregate called *special*; they are not derivable from the complete integral.

Further, if the equation possesses a singular integral, it is given by the equations

$$f = 0, \quad \frac{\partial f}{\partial p} = 0, \quad \frac{\partial f}{\partial q} = 0, \quad \frac{\partial f}{\partial x} + p\frac{\partial f}{\partial z} = 0, \quad \frac{\partial f}{\partial y} + q\frac{\partial f}{\partial z} = 0,$$

provided these equations are consistent with one another and determine p, q, z as functions of x and y, such as to leave $\frac{\partial f}{\partial z}$ different from zero: the value of z so determined is the singular integral. If the five equations are inconsistent with one another, there is no singular integral. If the five equations are consistent with one another but are equivalent to two equations only, which may be regarded as determining p and q in terms of x, y, z, then the equation

$$dz = p dx + q dy$$

is exact after the values of p and q have been substituted: a quadrature leads to an equation, involving one arbitrary constant and providing an integral of the equation. Such an integral will be called *special*.

Ex. Examples of the ordinary integrals that occur most frequently in connection with the simplest forms of equations are found freely in text-books.

As an illustration of the integrals here called special, when they arise through the process that, if otherwise favourable, allows the deduction of the singular integral from the equation itself, consider the equation

$$f = (px + qy - z)^2 + \frac{z^2}{x^2 + y^2 - 1} - p^2 - q^2 = 0.$$

The five equations

$$f = 0, \quad \frac{\partial f}{\partial p} = 0, \quad \frac{\partial f}{\partial q} = 0, \quad \frac{\partial f}{\partial x} + p\frac{\partial f}{\partial z} = 0, \quad \frac{\partial f}{\partial y} + q\frac{\partial f}{\partial z} = 0,$$

are satisfied by $z = 0$, which is a singular integral. They also are satisfied by

$$p = \frac{xz}{x^2 + y^2 - 1}, \quad q = \frac{yz}{x^2 + y^2 - 1};$$

and they then do not determine z in terms of x and y. When these values of p and q are substituted in

$$dz = p\, dx + q\, dy,$$

and the quadrature is effected, we have

$$z^2 = a^2 (x^2 + y^2 - 1),$$

where a is an arbitrary constant. It is easy to verify that this value of z satisfies the differential equation, and therefore is an integral.

In order to consider the relation of the integral thus deduced to other integrals of the equation, we use Charpit's method (§ 68) for the solution of the equation. Writing

$$\zeta = px + qy - z, \qquad u = x^2 + y^2 - 1,$$

and equating to $\frac{1}{2}dt$ each of the fractions in Charpit's subsidiary equations these become

$$\frac{dx}{dt} = p - x\zeta, \quad \frac{dy}{dt} = q - y\zeta,$$

$$\frac{dz}{dt} = \frac{z^2}{u} - z\zeta,$$

$$\frac{dp}{dt} = \frac{z}{u}\left(p - \frac{z}{u}x\right), \quad \frac{dq}{dt} = \frac{z}{u}\left(q - \frac{z}{u}y\right),$$

after slight reduction and using the equation $f=0$. Hence

$$\frac{d\zeta}{dt}=\left(\zeta-\frac{z}{u}\right)\frac{z}{u},\quad \frac{du}{dt}=2z-2\zeta u,$$

and therefore

$$\frac{d}{dt}\left(\zeta-\frac{z}{u}\right)=0,$$

that is, one integral of Charpit's equations, involving the derivatives p and q, is given by

$$\zeta-\frac{z}{u}=a,$$

where a is an arbitrary constant. When this is combined with the original equation, and the two equations are resolved for p and q, we find

$$p\,(x^2+y^2)=xv-yau^{\frac{1}{2}},$$
$$q\,(x^2+y^2)=yv+xau^{\frac{1}{2}},$$

where

$$v=a+z+\frac{z}{u};$$

and these are to be substituted in

$$dz=p\,dx+q\,dy,$$

which then becomes an exact equation. We have

$$dz=\tfrac{1}{2}\frac{du}{u+1}\left(a+z\frac{u+1}{u}\right)+\frac{xdy-ydx}{u+1}au^{\frac{1}{2}},$$

that is,

$$d\left(\frac{z}{u^{\frac{1}{2}}}\right)=\tfrac{1}{2}a\frac{du}{(u+1)\,u^{\frac{1}{2}}}+a\frac{xdy-ydx}{x^2+y^2};$$

after a quadrature, we have

$$\frac{z}{u^{\frac{1}{2}}}-\beta=a\left\{\tan^{-1}u^{\frac{1}{2}}+\tan^{-1}\frac{y}{x}\right\},$$

which may be regarded as the complete integral, expressed in a form that is both transcendental and irrational.

Writing

$$x=r\cos\theta,\quad y=r\sin\theta,$$

this complete integral becomes

$$\frac{z}{(r^2-1)^{\frac{1}{2}}}=\beta+a\,[\theta+\tan^{-1}\{(r^2-1)^{\frac{1}{2}}\}].$$

The general integral is expressible in the form

$$\frac{z^2}{r^2-1}=F\,[\theta+\tan^{-1}\{(r^2-1)^{\frac{1}{2}}\}].$$

The special integral, which was obtained in the form

$$z^2=a^2\,(x^2+y^2-1),$$

can be deduced from the complete integral by assuming $a=0$, $\beta=a$, and rationalising the result: it can also be obtained from the general integral by assuming $F(\xi)=a^2$.

The singular integral $z=0$ can be derived from the complete integral, taken in the form

$$z^2=(r^2-1)[\beta+a\theta+a\tan^{-1}\{(r^2-1)^{\frac{1}{2}}\}]^2,$$

by the customary process: it can also be deduced as a particular case of the special integral, by the assumption $a=0$.

79. A whole class of equations possessing special integrals of the indicated type can be constructed as follows*. Let

$$f(x, y, z, p, q)=0$$

be an equation which has a singular integral according to the formal Lagrangian theory: the values of z, p, q given by this integral must satisfy the equations

$$\frac{\partial f}{\partial p}=0, \quad \frac{\partial f}{\partial q}=0, \quad \frac{\partial f}{\partial x}+p\frac{\partial f}{\partial z}=0, \quad \frac{\partial f}{\partial y}+q\frac{\partial f}{\partial z}=0.$$

Let the first two (or any two) of the last four consistent equations be resolved so as to express p and q in terms of the rest of the variables; and let the result of substituting these expressions for p and q in $f(x, y, z, p, q)$ be $g(x, y, z)$: then

$$g(x, y, z)=0$$

provides the singular integral of the equation

$$f(x, y, z, p, q)=0,$$

on the Lagrangian theory.

The equations

$$\frac{\partial g}{\partial x}+p\frac{\partial g}{\partial z}=0, \quad \frac{\partial g}{\partial y}+q\frac{\partial g}{\partial z}=0,$$

are consistent with the preceding five equations. Moreover, as $g(x, y, z)$ is the value of $f(x, y, z, p, q)$ when the values of p and q given by $\frac{\partial f}{\partial p}=0$ and $\frac{\partial f}{\partial q}=0$ are substituted in $f(x, y, z, p, q)$, we have

$$\begin{aligned}\frac{\partial g}{\partial x}+p\frac{\partial g}{\partial z}&=\frac{\partial f}{\partial x}+p\frac{\partial f}{\partial z}+\frac{\partial p}{\partial x}\frac{\partial f}{\partial p}+\frac{\partial q}{\partial x}\frac{\partial f}{\partial q}\\&=\frac{\partial f}{\partial x}+p\frac{\partial f}{\partial z},\end{aligned}$$

and, similarly,

$$\frac{\partial g}{\partial y}+q\frac{\partial g}{\partial z}=\frac{\partial f}{\partial y}+q\frac{\partial f}{\partial z},$$

* The process was suggested to me by a remark in a letter from Prof. Chrystal, dated 18 May, 1896.

both identically; that is, these equations are satisfied identically by values of p and q given by

$$\frac{\partial f}{\partial p} = 0, \quad \frac{\partial f}{\partial q} = 0,$$

when in addition to these two relations, we take

$$f(x, y, z, p, q) = g(x, y, z),$$

where $g(x, y, z)$ is the result of substituting the values of p and q in $f(x, y, z, p, q)$.

Now consider the equation

$$F(x, y, z, p, q) = f(x, y, z, p, q) - g(x, y, z) = 0.$$

If it possesses a singular integral according to Lagrange's formal theory, this integral must be such as to satisfy, not merely $F = 0$, but also

$$\frac{\partial F}{\partial p} = 0, \quad \frac{\partial F}{\partial q} = 0, \quad \frac{\partial F}{\partial x} + p\frac{\partial F}{\partial z} = 0, \quad \frac{\partial F}{\partial y} + q\frac{\partial F}{\partial z} = 0.$$

The first two of the last four are

$$\frac{\partial f}{\partial p} = 0, \quad \frac{\partial f}{\partial q} = 0;$$

when these are resolved for p and q in terms of x, y, z, the values of p and q are such that

$$f(x, y, z, p, q) = g(x, y, z),$$

that is, the three equations

$$F = 0, \quad \frac{\partial F}{\partial p} = 0, \quad \frac{\partial F}{\partial q} = 0,$$

are equivalent to two equations only, expressing p and q in terms of x, y, z.

Moreover, it appears that when the specified values of p and q are substituted in $f(x, y, z, p, q)$, the latter becomes $g(x, y, z)$: hence, after the preceding explanations and taking account of the source of $g(x, y, z)$, the two equations

$$\frac{\partial f}{\partial x} + p\frac{\partial f}{\partial z} = \frac{\partial g}{\partial x} + p\frac{\partial g}{\partial z},$$

$$\frac{\partial f}{\partial y} + q\frac{\partial f}{\partial z} = \frac{\partial g}{\partial y} + q\frac{\partial g}{\partial z},$$

are satisfied identically, that is, the equations

$$\frac{\partial F}{\partial x} + p\frac{\partial F}{\partial z} = 0, \quad \frac{\partial F}{\partial y} + q\frac{\partial F}{\partial z} = 0,$$

are satisfied identically. Hence the five equations

$$F=0,\quad \frac{\partial F}{\partial p}=0,\quad \frac{\partial F}{\partial q}=0,\quad \frac{\partial F}{\partial x}+p\frac{\partial F}{\partial z}=0,\quad \frac{\partial F}{\partial y}+q\frac{\partial F}{\partial z}=0,$$

are equivalent to two equations only, expressing p and q as functions of x, y, z: the equation

$$F=0$$

has a special integral.

Note. In what precedes, there is a tacit assumption that $F=0$ is irreducible; if, however, $F=0$ can be resolved into distinct equations, the argument is no longer valid.

Ex. 1. Apply the preceding process to construct from the equation

$$z^2(1+p^2+q^2)=\lambda^2\{(x+pz)^2+(y+qz)^2\},$$

which has

$$z^2(1-\lambda^2)=\lambda^2(x^2+y^2)$$

for a singular integral, another equation which has a special integral.

Ex. 2. Can the method be applied to the equation

$$z=px+qy+p^nq^n,$$

for any value of n?

80. The preceding discussion has been concerned with the integrals that are derivable from the complete integral of a partial differential equation; a distinctive property of the complete integral is that the number of parameters which it involves is the same as the number of independent variables. But integral equations, not distinguished by this property, may be propounded for consideration: thus the number of parameters may exceed the number of independent variables; and we have seen how an equation can arise as an integral of a set of simultaneous differential equations, and then the number of parameters involved is less than the number of independent variables.

A very brief discussion is sufficient to deal with an equation

$$\phi(z, x_1, \ldots, x_n, a_1, \ldots, a_m)=0,$$

when $m>n$. It may, of course, be assumed that the m parameters are essential*, that is, are not reducible to a smaller number: the necessary and sufficient test is that the equation

$$\alpha_1\frac{\partial\phi}{\partial a_1}+\ldots+\alpha_m\frac{\partial\phi}{\partial a_m}=0$$

* In the sense adopted in Lie's theory of groups.

is not identically satisfied for any non-zero values of $\alpha_1, \ldots, \alpha_m$ as functions of $a_1, \ldots, a_m$. Forming the equations

$$\phi_r = \frac{\partial \phi}{\partial x_r} + p_r \frac{\partial \phi}{\partial z} = 0,$$

for $r = 1, \ldots, n$, it usually is not possible to eliminate a number of constants greater than n among the $n + 1$ equations

$$\phi = 0, \quad \phi_1 = 0, \ldots, \phi_n = 0;$$

so that usually, on proceeding to the eliminant equation, we should find that it contained $m - n$ parameters. If, however, the appropriate Jacobian conditions

$$J\left(\left(\frac{\phi, \phi_1, \ldots, \phi_n}{a_1, a_2, \ldots, a_m}\right)\right) = 0$$

are satisfied for each selection of $n + 1$ parameters from the set of m, then the single equation resulting from elimination contains no parameters. The integral equation is then a special case of the general integral of the partial equation.

Ex. The equation

$$z = ax + by + c\frac{y^2}{x} + g\frac{x^2}{y}$$

leads to

$$xp + yq = z:$$

it is a special form of

$$z = xf\left(\frac{y}{x}\right),$$

which is the general integral.

CLASSES OF INTEGRALS OF A COMPLETE SYSTEM.

81. Coming next to an equation

$$\phi(z, x_1, \ldots, x_n, a_1, \ldots, a_m) = 0,$$

for which $m < n$, we shall assume, as before in § 80, that the parameters $a_1, \ldots, a_m$ are essential. We shall also assume that the m constants can be eliminated between $\phi = 0$ and the n derived equations

$$\phi_r = \frac{\partial \phi}{\partial x_r} + p_r \frac{\partial \phi}{\partial z} = 0,$$

for $r = 1, \ldots, n$, so as to give $n - m + 1$ equations involving $z, x_1, \ldots, x_n, p_1, \ldots, p_n$. And we assume that $\frac{\partial \phi}{\partial z}$ does not vanish in association with $\phi = 0$. We then have the case, which has already been considered, of a number of simultaneous differential equations; these equations form a complete set, because of their source; and the Jacobian method of integration has shewn how to construct an integral $\phi = 0$ containing m constants that are arbitrary. Having regard to the investigation in the case when $m = n$, which made the derivation of other integrals from $\phi = 0$ possible, we proceed to a similar quest and seek to derive other integrals from $\phi = 0$ in the present case when $m < n$.

We proceed as before. The $n - m + 1$ differential equations are the result of eliminating $a_1, \ldots, a_m$ among the equations

$$\phi = 0, \quad \phi_1 = 0, \ldots, \phi_n = 0.$$

The course of the elimination takes no account of the quality of $a_1, \ldots, a_m$: it will lead to the same result if these quantities be changed in such a way that each of the $n + 1$ equations is unaltered in form. Accordingly, subject to this limitation, we make $a_1, \ldots, a_m$ functions of the independent variables $x_1, \ldots, x_n$; and the limitation requires that the n relations

$$\sum_{i=1}^{m} \frac{\partial \phi}{\partial a_i} \frac{\partial a_i}{\partial x_r} = 0,$$

for $r = 1, \ldots, n$, shall be satisfied, conditions that clearly are sufficient as well as necessary to secure the invariability of form of the $n + 1$ equations. Multiplying the n relations by $dx_1, \ldots, dx_n$ respectively, and adding, we obtain a single relation

$$\sum_{i=1}^{m} \frac{\partial \phi}{\partial a_i} da_i = 0$$

in the differential elements: it is equivalent to the n relations and therefore, when satisfied, it suffices for the present purpose.

This differential relation can be satisfied in various ways.

In the first place, all the quantities $da_1, \ldots, da_m$ may vanish, so that all the quantities $a_1, \ldots, a_m$ are constant. We then resume the original integral: on the analogy of the corresponding integral for a single equation, we call it the *complete integral*.

In the second place, an integral equivalent of the differential relation may consist of μ equations

$$g_1(a_1, \ldots, a_m) = 0, \ldots, g_\mu(a_1, \ldots, a_m) = 0,$$

and of these only. Obviously μ cannot be greater than m. If μ be equal to m, then there are m equations involving m quantities a: each of these quantities is a constant, and so we fall back upon the preceding case. Hence we need consider only values of μ that are less than m. As there are μ integral equations in the complete equivalent of the differential relation, the only relations among the differential elements are

$$dg_1 = 0, \ldots, dg_\mu = 0;$$

and the differential relation

$$\sum_{i=1}^{m} \frac{\partial \phi}{\partial a_i} da_i = 0$$

must be satisfied in virtue of them. Consequently, μ quantities $\lambda_1, \ldots, \lambda_\mu$ must exist such that

$$\sum_{i=1}^{m} \frac{\partial \phi}{\partial a_i} da_i = \lambda_1 dg_1 + \ldots + \lambda_\mu dg_\mu;$$

and therefore

$$\frac{\partial \phi}{\partial a_i} = \lambda_1 \frac{\partial g_1}{\partial a_i} + \ldots + \lambda_\mu \frac{\partial g_\mu}{\partial a_i},$$

for $i = 1, \ldots, m$. These m equations, together with

$$\phi = 0, \quad g_1 = 0, \ldots, g_\mu = 0,$$

make up $m + \mu + 1$ equations: eliminating the m parameters $a_1, \ldots, a_m$ and the μ multipliers $\lambda_1, \ldots, \lambda_\mu$ among them, we obtain a single equation among $z, x_1, \ldots, x_n$. The value of z thus determined is an integral of the original differential equation: as before, we call it a *general integral*.

In the expression of a general integral, the functional forms $g_1, \ldots, g_\mu$ occur; and so there are various classes of general integrals, which arise according to the number of postulated relations. It is clear that

$$0 < \mu < m;$$

and it is customary to describe a general integral, associated with μ forms, as of *class* μ.

The extreme case, when $\mu = m$, has already been mentioned: the integral is then complete. The other extreme case, when $\mu = 0$, will be discussed immediately.

As in the case of a single differential equation, it might be supposed that a class of integral, intermediate between the complete integral and the general integrals, would be obtained by taking

$$a_{i+1} = \text{constant}, \ldots, a_m = \text{constant},$$

and then postulating a number of relations

$$g_1(a_1, \ldots, a_i) = 0, \ldots, g_\tau(a_1, \ldots, a_i) = 0,$$

where $\tau < i$, among the remaining parameters $a_1, \ldots, a_i$. Effectively, these equations amount to $m + \tau - i$ relations among the parameters $a_1, \ldots, a_m$, of which $m - i$ are special in form: the corresponding integral would be a specialised general integral of class $m + \tau - i$.

Moreover, it can be proved, as in the case of a single differential equation, that the most comprehensive general integral is of the first class when the single relation is quite arbitrary: on this account, it is sometimes called *the general integral*.

There is one other mode of securing that the differential relation is satisfied. It is possible that the equations

$$\frac{\partial \phi}{\partial a_1} = 0, \ldots, \frac{\partial \phi}{\partial a_m} = 0$$

could hold; the differential relation then becomes evanescent, and so it ceases to have any necessary influence upon the organic variations under consideration. The equations

$$\phi = 0, \quad \frac{\partial \phi}{\partial a_1} = 0, \ldots, \frac{\partial \phi}{\partial a_m} = 0$$

may coexist and may be consistent with one another: if the result of the elimination of $a_1, \ldots, a_m$ among them provides a single equation involving no arbitrary element, and if that single equation determines an integral*, the integral thus furnished is called the *singular integral*.

Ex. 1. The simplest cases arise when there are only two independent variables. Thus let

$$\phi = z - a^2 y + ax = 0;$$

the value of z thus provided satisfies the two equations

$$\left.\begin{array}{l} z = px + p^2 y \\ z = px + qy \end{array}\right\}.$$

* The reason for this limitation is similar to the reason in the former case (§ 73). Even when the process is possible, the locus provided by the eliminant frequently is composite: some of its components, even all the components, may not be integrals of the differential equation but may be loci of singularities on the complete integral.

The elimination of a, between $\phi=0$ and

$$\frac{\partial\phi}{\partial a}=x-2ay=0,$$

leads to an equation

$$x^2+4yz=0.$$

It is easy to verify that the value of z given by this last equation satisfies the two differential equations: accordingly, it is a singular integral.

Geometrically interpreted, the complete integral is a family of planes through the origin touching a cone which is the singular integral.

Ex. 2. Obtain the simplest differential equations satisfied by the value of z given by

$$2z=a^2y^2-ax^2:$$

and prove that they possess a singular integral represented by

$$8y^2z+x^4=0.$$

Ex. 3. Discuss similarly the equation

$$z=ax+a^2y+a^3,$$

obtaining the two simplest differential equations which it satisfies, in the form

$$\left.\begin{aligned} z&=px+qy+pq \\ z&=px+p^2y+p^3 \end{aligned}\right\}.$$

Prove that the equation

$$27z^2+z(18xy-4y^3)-x^2y^2+4x^3=0$$

provides a singular integral.

Shew also that the value of z given by the original equation satisfies the two partial equations

$$\left.\begin{aligned} z&=px+p^2y+p^3 \\ (z-qy)^2&=q(x+q)^2 \end{aligned}\right\}.$$

Is the original equation the complete integral of these two equations?

Ex. 4. Integrate the equations

$$\left.\begin{aligned} z&=p_1x_1+p_2x_2+p_3x_3 \\ p_3&=p_1p_2 \end{aligned}\right\};$$

and shew that they possess a singular integral

$$zx_3+x_1x_2=0.$$

Ex. 5. It has been seen (§ 57) that the simultaneous equations

$$p_1p_2-x_3x_4=0, \qquad p_3p_4-x_1x_2=0$$

possess two complete integrals

$$z=\frac{x_1x_3}{a}+ax_2x_4+b, \qquad z=\frac{x_2x_3}{A}+Ax_1x_4+B.$$

The respective general integrals are evidently given by

$$\left.\begin{aligned} z&=\frac{x_1x_3}{a}+ax_2x_4+\phi(a)\\ 0&=-\frac{x_1x_3}{a^2}+x_2x_4+\phi'(a)\end{aligned}\right\},\qquad \left.\begin{aligned} z&=\frac{x_2x_3}{A}+Ax_1x_4+\psi(A)\\ 0&=-\frac{x_2x_3}{A^2}+x_1x_4+\psi'(A)\end{aligned}\right\};$$

and there is no singular integral. There is a question as to the relation of the two complete integrals to one another.

If they can be brought into such relation that one of them can be changed into the other, then (after the preceding discussion) we must have

$$p_1=\frac{x_3}{a}=Ax_4,\quad p_3=\frac{x_1}{a}=\frac{x_2}{A},$$

$$p_2=ax_4=\frac{x_3}{A},\quad p_4=ax_2=Ax_1,$$

all of which are satisfied by the values

$$a=\left(\frac{x_1x_3}{x_2x_4}\right)^{\frac{1}{2}},\quad A=\left(\frac{x_2x_3}{x_1x_4}\right)^{\frac{1}{2}}.$$

In order that the two values of z may be equal, we must (with these values of a and A) have

$$b=B.$$

The other equations require that the relation

$$\frac{\partial z}{\partial a}da+\frac{\partial z}{\partial b}db=0$$

be satisfied: that this may be the case, we must have

$$db=0,$$

that is, b must be a pure constant.

We thus see that the conditions, necessary to secure that each of the complete integrals can be transformed into the other, are not satisfied: the values of a and b, which have been obtained, do not lead after substitution to the other integral. The complete integrals are distinct from one another: it will be seen, on reference to the construction of the integrals, that they belong to different resolutions of the original system.

But, on the other hand, by the substitution of the values of a and A, both complete integrals lead to a new integral

$$z=2(x_1x_2x_3x_4)^{\frac{1}{2}}+b,$$

which is a particular form of each of the general integrals.

82. The aggregate of integrals, composed of the complete integral, the general integrals, and (when it exists) the singular integral, is widely comprehensive: but for a complete set of differential equations, as for a single equation in the earlier discussion, the aggregate cannot be declared wholly comprehensive. The argument is similar to the argument in the case of a single equation and therefore hardly needs to be repeated in the present connection.

If any given integral is included in the above aggregate, the determination of its character is easily effected. Let

$$\phi(z, x_1, \ldots, x_n, a_1, \ldots, a_m) = 0$$

be the complete integral of a given complete set of $n-m+1$ partial differential equations; and let

$$\psi(z, x_1, \ldots, x_n) = 0$$

be any other integral of that set. If $\psi = 0$ is included in the aggregate, it must be possible to assign values (constant or variable) so that the equations are satisfied, and so also that the values of z given by the two equations are the same. If the latter condition be satisfied, the values of $p_1, \ldots, p_n$ must be the same. For $\psi = 0$, they are given by

$$\frac{\partial \psi}{\partial x_r} + \frac{\partial \psi}{\partial z} p_r = 0,$$

for $r = 1, \ldots, n$: and for $\phi = 0$, with values assigned to $a_1, \ldots, a_m$ such that the differential equations still are satisfied, the quantities $p_1, \ldots, p_n$ are given by

$$\frac{\partial \phi}{\partial x_r} + \frac{\partial \phi}{\partial z} p_r = 0,$$

for $r = 1, \ldots, n$. If they are the same for the two integrals, we have

$$\frac{\partial \phi}{\partial x_r} \frac{\partial \psi}{\partial z} - \frac{\partial \psi}{\partial x_r} \frac{\partial \phi}{\partial z} = 0,$$

for $r = 1, \ldots, n$. Hence the quantities $a_1, \ldots, a_m$ must be such as to satisfy these n equations and also the condition that the value of z given by $\phi = 0$ is the same as the value of z given by $\psi = 0$.

If these $n+1$ conditions give constant values for $a_1, \ldots, a_m$, the integral furnished by $\psi = 0$ is a particular case of the complete integral.

If the $n+1$ conditions express $a_1, \ldots, a_m$ as functions of the variables, such that these functions are connected by a number of relations of the type

$$g(a_1, \ldots, a_m) = 0,$$

the number of these relations being less than m, the integral furnished by $\psi = 0$ is a general integral.

If the $n+1$ conditions express $a_1, \ldots, a_m$ in terms of the variables, the values being unconnected by any functional relation or relations, the integral furnished by $\psi = 0$ is a singular integral.

But as $m < n$, we cannot affirm that the $n+1$ equations must certainly determine the quantities $a_1, \ldots, a_m$ without the introduction of relations among the independent variables. In all instances, when $a_1, \ldots, a_m$ are not determined in one or other of the foregoing forms, the integral furnished by $\psi = 0$ is not included in the aggregate of integrals associated with $\phi = 0$; it belongs to the unclassified set of integrals previously called *special*. In such an event, the retained aggregate of integrals associated with $\phi = 0$ is not wholly comprehensive.

Ex. 1. The two equations

$$f_1 = p_2 - p_3 = 0$$
$$f_2 = x_1 p_1 + 2x_2 p_2 + 2x_3 p_3 - 2\left(z - \frac{x_1^2}{x_2 + x_3}\right)^2 = 0,$$

are a complete set: for

$$[f_1, f_2] = 0,$$

in virtue of $f_1 = 0$.

A complete integral is furnished by

$$\frac{x_2 + x_3}{z(x_2 + x_3) - x_1^2} = a + b\frac{x_1^2}{x_2 + x_3} - \log(x_2 + x_3);$$

a general integral is furnished by

$$y e^{\frac{x_2 + x_3}{z(x_2 + x_3) - x_1^2}} = g\left(\frac{x_1^2}{x_2 + x_3}\right),$$

where g is an arbitrary function; and there is no singular integral.

It is easy to verify that the two equations are satisfied by

$$z = \frac{x_1^2}{x_2 + x_3};$$

no definite values can be assigned to a and b, and no definite form can be assigned to g, so that this integral can be included in the foregoing aggregate.

Ex. 2. Discuss the character of the integral of

$$p_2 - p_3 = 0, \quad \{1 + (z - x_1 - x_2 - x_3)^{\frac{1}{2}}\} p_1 + p_2 + p_3 = 3,$$

which is given by

$$z = x_1 + x_2 + x_3.$$

SINGULAR INTEGRAL OF A COMPLETE SYSTEM.

83. When a complete set of simultaneous equations possesses a singular integral, the knowledge of the property of a single equation in similar circumstances makes it natural to enquire whether the singular integral can be derived from the complete set itself, without the intervention of the complete integral. It is possible to do so in cases when the equations possess (or when, without extension of their significance, they can be transformed so as to possess) a particular form, as will now be proved. Partly owing to the elaboration of the conditions even when the particular form is possessed, and partly owing to the fact that a set of simultaneous equations in one dependent variable has nothing like uniqueness of form, an investigation into the general case will not be pursued.

Suppose that, by appropriate combinations of the members of a complete set of $n - m + 1$ equations, it is possible to deduce one equation (or more than one equation) in an equivalent set which involves m, and not more than m, of the derivatives. As the complete set is assumed to possess a singular integral, we shall further suppose that the equation in question is not resolved with regard to any of those derivatives*; and so we may take the equation in the form

$$f(z, x_1, \ldots, x_n, p_1, \ldots, p_m) = 0,$$

the remaining $n - m$ equations involving $p_{m+1}, \ldots, p_n$, or some of them in each equation, as well as possibly $p_1, \ldots, p_m$. Let the complete integral be denoted by

$$\phi(z, x_1, \ldots, x_n, a_1, \ldots, a_m) = 0:$$

the values of the derivatives are given by

$$\phi_r = \frac{\partial \phi}{\partial x_r} + p_r \frac{\partial \phi}{\partial z} = 0,$$

for $r = 1, \ldots, n$; and the complete set of $n - m + 1$ equations results from the elimination of $a_1, \ldots, a_m$ among the $n + 1$ equations

$$\phi = 0, \quad \phi_1 = 0, \ \ldots, \ \phi_n = 0.$$

* It will appear from the analysis that a resolved equation of the indicated type would exclude the existence of a singular integral, because it would lead to impossible conditions.

In particular, the equation $f=0$ selected from the set must result from the elimination: as it involves only $p_1, \ldots, p_m$ but not $p_{m+1}, \ldots, p_n$, it must result from the elimination of $a_1, \ldots, a_m$ among the $m+1$ equations

$$\phi=0, \quad \phi_1=0, \ldots, \phi_m=0.$$

If we assume that all the quantities $\phi, \phi_1, \ldots, \phi_m$ are rational and integral, quantities $\lambda, \lambda_1, \ldots, \lambda_m$ will exist such that

$$f=\lambda\phi+\lambda_1\phi_1+\ldots+\lambda_m\phi_m.$$

Now f does not involve any of the parameters $a_1, \ldots, a_m$: hence, in connection with our integral, we have

$$0=\lambda\frac{\partial\phi}{\partial a_i}+\lambda_1\frac{\partial\phi_1}{\partial a_i}+\ldots+\lambda_m\frac{\partial\phi_m}{\partial a_i},$$

for $i=1, \ldots, m$. The singular integral is given, in connection with $\phi=0$, by

$$\frac{\partial\phi}{\partial a_1}=0, \ldots, \frac{\partial\phi}{\partial a_m}=0:$$

and therefore the singular integral, in connection with $\phi=0$, requires that the equations

$$0=\lambda_1\frac{\partial\phi_1}{\partial a_i}+\ldots+\lambda_m\frac{\partial\phi_m}{\partial a_i},$$

for $i=1, \ldots, m$, be satisfied. These m equations are linear and homogeneous in $\lambda_1, \ldots, \lambda_m$, and the determinant of the coefficients is

$$\frac{\partial(\phi_1, \ldots, \phi_m)}{\partial(a_1, \ldots, a_m)},$$

which does not vanish (the elimination would not be possible if this quantity were to vanish); hence the singular integral requires that

$$\lambda_1=0, \ldots, \lambda_m=0.$$

Again, f involves $p_1, \ldots, p_m$, which do not occur in ϕ and occur only individually in $\phi_1, \ldots, \phi_m$ respectively: hence, in connection with the integral $\phi=0$,

$$\frac{\partial f}{\partial p_r}=\lambda_r\frac{\partial\phi}{\partial z},$$

for $r=1, \ldots, m$. Consequently, the singular integral requires that

$$\frac{\partial f}{\partial p_1}=0, \ldots, \frac{\partial f}{\partial p_m}=0,$$

and the singular integral satisfies

$$f = 0:$$

hence it may be possible to obtain the singular integral by the elimination of $p_1, \ldots, p_m$ among these $m+1$ equations.

The equation that results from the elimination is not necessarily an integral of $f=0$: if it is, the relations

$$\frac{\partial f}{\partial x_1} + p_1 \frac{\partial f}{\partial z} = 0, \ \ldots, \ \frac{\partial f}{\partial x_n} + p_n \frac{\partial f}{\partial z} = 0,$$

must be satisfied, when the derived values of $p_1, \ldots, p_n$ are substituted. And in order that it may be an integral of the remaining $n-m$ equations of the system, each of those equations must be satisfied when the values of $p_1, \ldots, p_n, z$ are substituted.

It therefore appears that *when a complete set of $n-m+1$ equations possesses a singular integral, and when an equation can be selected or compounded from the set involving only m of the derivatives in a form*

$$f(z, x_1, \ldots, x_n, p_1, \ldots, p_m) = 0,$$

the singular integral may be given by the elimination of $p_1, \ldots, p_m$ between the equations

$$f = 0, \quad \frac{\partial f}{\partial p_1} = 0, \ \ldots, \ \frac{\partial f}{\partial p_m} = 0;$$

the conditions indicated must be satisfied in order that the eliminant may provide an integral; and, when they are satisfied, the eliminant proves the singular integral.

Ex. 1. Consider the system

$$\left.\begin{aligned} z &= px + p^2 y \\ z &= px + qy \end{aligned}\right\},$$

in Ex. 1, § 81.

An equation of the required type is furnished by

$$f = p^2 y + px - z = 0;$$

the associated equation is

$$\frac{\partial f}{\partial p} = 2py + x = 0.$$

Eliminating p between these equations, we have

$$4yz + x^2 = 0;$$

the value of z thus given satisfies both equations of the system, and therefore it is a singular integral.

Another equation of the required type is furnished by

$$g=(z-qy)^2-x^2q=0;$$

the associated equation is

$$\frac{\partial g}{\partial q}=-2y(z-qy)-x^2=0.$$

Eliminating q between these equations, we have

$$x^2(x^2+4yz)=0.$$

The equation $x=0$ does not provide an integral. The equation $x^2+4yz=0$, as before, does provide an integral which accordingly is the singular integral.

Ex. 2. Obtain all the integrals of the system

$$\left.\begin{array}{l} 2z=x_1p_1+x_2p_2+x_3p_3 \\ 0=p_1p_2x_3-2x_1x_2p_3 \end{array}\right\};$$

and shew that the singular integral can be deduced from the differential equations.

CHAPTER VI.

The Method of Characteristics for Equations in two independent variables: Geometrical Relations of the various Integrals.

For the material of this chapter and the next, which are limited to equations in two independent variables, reference should be made in the first place to Cauchy's discussion as given in the section of his *Exercices d'analyse et de physique mathématique* (quoted in § 84), and to the exposition of Cauchy's method given by Mansion, in his treatise already quoted (p. 100).

A considerable portion of the chapter is devoted to the geometrical interpretation of the analysis, particularly to the interpretation of results by the geometry of ordinary space. For this portion, reference should be made to Darboux's memoir, *Mém. des Sav. Étrang.*, t. XXVII (1880), dealing with the singular solutions of partial equations of the first order; ample use has been made of the memoir. Reference may also be made to Monge's treatise, quoted in § 97; to Goursat's treatise, (quoted on p. 55), particularly chapter IX which is based upon Darboux's memoir; and especially to a memoir by Lie, *Math. Ann.*, t. V (1872), pp. 145—256.

84. It has been seen that, in the method of Charpit as applied to any equation of the first order in two independent variables, and in the method of Jacobi for any equation in n independent variables, the first step towards the solution of the equation consists in the construction of an integral of a simultaneous system of ordinary equations.

As introduced by these methods, the system of ordinary equations is subsidiary to the integration of a homogeneous linear partial equation of the first order; there are, however, other ways in which they can arise. Two of these will now be expounded; in presentation, and in significance, they seem distinct from one another, but they will be found to be fundamentally the same. Partly for the sake of simplicity, and partly because of the associated geometry, we shall begin with equations that involve only two independent variables; the discussion of equations involving n independent variables can be made briefer, after the explanations in the simplest case.

The method adopted* by Cauchy for the construction of the equations was originally propounded† by Ampère; it is based upon a change of independent variables, chosen so as to simplify relations. Let these be changed from x and y to x and u; then y, z, p, q can be regarded as functions of x and u, and, whatever be the differential equation, we have

$$\frac{\partial z}{\partial x} = p + q\frac{\partial y}{\partial x},$$

$$\frac{\partial z}{\partial u} = \quad q\frac{\partial y}{\partial u}.$$

As u and x are independent variables, it follows that

$$\frac{\partial}{\partial u}\left(\frac{\partial z}{\partial x}\right) = \frac{\partial}{\partial x}\left(\frac{\partial z}{\partial u}\right),$$

$$\frac{\partial}{\partial u}\left(\frac{\partial y}{\partial x}\right) = \frac{\partial}{\partial x}\left(\frac{\partial y}{\partial u}\right);$$

substituting in the former relation and using the latter, we have

$$\frac{\partial p}{\partial u} = \frac{\partial q}{\partial x}\frac{\partial y}{\partial u} - \frac{\partial q}{\partial u}\frac{\partial y}{\partial x}.$$

When the proper values of y, z, p, q as functions of x and u are substituted in a differential equation

$$f(x, y, z, p, q) = 0,$$

it must become an identity; hence, writing

$$\frac{\partial f}{\partial x}, \frac{\partial f}{\partial y}, \frac{\partial f}{\partial z}, \frac{\partial f}{\partial p}, \frac{\partial f}{\partial q} = X, Y, Z, P, Q,$$

respectively, we have

$$X + Y\frac{\partial y}{\partial x} + Z\frac{\partial z}{\partial x} + P\frac{\partial p}{\partial x} + Q\frac{\partial q}{\partial x} = 0,$$

$$Y\frac{\partial y}{\partial u} + Z\frac{\partial z}{\partial u} + P\frac{\partial p}{\partial u} + Q\frac{\partial q}{\partial u} = 0.$$

* *Exercices d'analyse et de physique mathématique*, t. II, pp. 238—272. This is dated 1841; but the memoir, which contained a first exposition of Cauchy's theory, was published in 1819.

† In his memoir of 1814, to which references will be given subsequently.

Inserting the above values of $\frac{\partial z}{\partial u}$ and $\frac{\partial p}{\partial u}$ in the second of these equations, it becomes

$$\frac{\partial y}{\partial u}\left(Y + qZ + P\frac{\partial q}{\partial x}\right) + \frac{\partial q}{\partial u}\left(Q - P\frac{\partial y}{\partial x}\right) = 0.$$

As the variable u is thus far at our disposal let it be chosen, if possible, so that

$$Q - P\frac{\partial y}{\partial x} = 0;$$

then, as y cannot be a function of x alone so that $\frac{\partial y}{\partial u}$ cannot be zero, and as $\frac{\partial q}{\partial u}$ cannot be a permanent infinity, we have

$$Y + qZ + P\frac{\partial q}{\partial x} = 0.$$

Inserting in the first of the two derived equations the above value of $\frac{\partial z}{\partial x}$, and the values of $\frac{\partial y}{\partial x}$ and $\frac{\partial q}{\partial x}$ given by the equations just obtained, we find

$$X + pZ + P\frac{\partial p}{\partial x} = 0.$$

Hence there are four equations involving derivatives of y, z, p, q with regard to x alone: they can be taken in the form

$$P\frac{\partial y}{\partial x} = Q, \qquad P\frac{\partial p}{\partial x} = -(X + pZ),$$

$$P\frac{\partial z}{\partial x} = pP + qQ, \qquad P\frac{\partial q}{\partial x} = -(Y + qZ).$$

These do not contain derivatives with regard to u nor do they contain u itself; hence, if we take these equations in the form

$$\frac{dx}{P} = \frac{dy}{Q} = \frac{dz}{pP + qQ} = \frac{dp}{-(X + pZ)} = \frac{dq}{-(Y + qZ)},$$

and obtain their integrals, the arbitrary elements that arise in those integrals will be functions of u in the most general case. So far as these equations are concerned, the arbitrary elements may be made arbitrary functions of u: but other equations must be satisfied, viz.

$$f(x, y, z, p, q) = 0, \quad \frac{\partial z}{\partial u} = q \frac{\partial y}{\partial u},$$

and these may impose limitations*.

It will be noted that the above set of ordinary equations is the same as the set in Charpit's method (§ 68).

85. To satisfy the requirements, we proceed from the known theory of ordinary equations. If P is not an identical zero for our problem, that is, if it does not vanish for all values of five arguments x, y, z, p, q, tied by a condition

$$f(x, y, z, p, q) = 0,$$

then values x_0, y_0, z_0, p_0, q_0 can be assigned to the arguments such that P does not vanish, provided

$$f(x_0, y_0, z_0, p_0, q_0) = 0.$$

Further, suppose that P, Q, X, Y, Z are regular functions of x, y, z, p, q in the vicinity of these values. Then it is known, by Cauchy's theorem† on the integrals of ordinary equations, that a unique system of integrals exists; they give y, z, p, q as regular functions of x and these acquire values y_0, z_0, p_0, q_0 when $x = x_0$.

If P is an identical zero in the sense explained, then we consider Q. If Q is not similarly an identical zero, we proceed as above making y the independent variable for the ordinary system: and with the same hypotheses, we obtain a set of integrals.

If Q is an identical zero in the sense explained, then we should proceed to consider $X + pZ$, making p the independent variable; and if that were to fail, we should consider $Y + qZ$, making q the independent variable. We should obtain a set of integrals save only in the case, when all the equations

$$P = 0, \quad Q = 0, \quad X + pZ = 0, \quad Y + qZ = 0$$

are satisfied for values of x, y, z, p, q, which obey the relation

$$f(x, y, z, p, q) = 0.$$

* The equation

$$\frac{\partial p}{\partial u} = \frac{\partial q}{\partial x}\frac{\partial y}{\partial u} - \frac{\partial y}{\partial x}\frac{\partial q}{\partial u}$$

imposes no additional limitation: it is satisfied in virtue of the equations retained, being a mere deduction from them.

† See vol. II of this work, ch. II.

This case is obviously exceptional: it provides what has already been recognised as the singular integral, and consequently it will be set aside from the present investigation. It may or may not occur: when it does occur, its relation to other integrals will be considered subsequently.

If then, setting this case aside, we take y_0, z_0, p_0, q_0 as functions of u satisfying the relation

$$f(x_0, y_0, z_0, p_0, q_0) = 0,$$

we have a set of integrals of the system of ordinary equations; and these are further to satisfy the equations

$$f(x, y, z, p, q) = 0, \qquad \frac{\partial z}{\partial u} = q\frac{\partial y}{\partial u}.$$

Let

$$E = \frac{\partial z}{\partial u} - q\frac{\partial y}{\partial u},$$

so that E is to be zero: thus

$$\frac{\partial E}{\partial x} = \frac{\partial^2 z}{\partial x \partial u} - q\frac{\partial^2 y}{\partial x \partial u} - \frac{\partial q}{\partial x}\frac{\partial y}{\partial u}.$$

But the quantities already obtained satisfy the equation

$$\frac{\partial z}{\partial x} = p + q\frac{\partial y}{\partial x},$$

and therefore

$$\frac{\partial^2 z}{\partial x \partial u} = \frac{\partial p}{\partial u} + q\frac{\partial^2 y}{\partial x \partial u} + \frac{\partial q}{\partial u}\frac{\partial y}{\partial x}.$$

Hence

$$\frac{\partial E}{\partial x} = \frac{\partial p}{\partial u} + \frac{\partial q}{\partial u}\frac{\partial y}{\partial x} - \frac{\partial y}{\partial u}\frac{\partial q}{\partial x}.$$

Now

$$\frac{\partial y}{\partial x} = \frac{Q}{P}, \qquad \frac{\partial q}{\partial x} = -\frac{Y + qZ}{P},$$

so that

$$\frac{\partial E}{\partial x} = \frac{1}{P}\left(Y\frac{\partial y}{\partial u} + qZ\frac{\partial y}{\partial u} + P\frac{\partial p}{\partial u} + Q\frac{\partial q}{\partial u}\right).$$

Our quantities are required to satisfy $f(x, y, z, p, q) = 0$, and therefore

$$Y\frac{\partial y}{\partial u} + Z\frac{\partial z}{\partial u} + P\frac{\partial p}{\partial u} + Q\frac{\partial q}{\partial u} = 0;$$

hence

$$\frac{\partial E}{\partial x} = \frac{1}{P}Z\left(q\frac{\partial y}{\partial u} - \frac{\partial z}{\partial u}\right)$$

$$= -E\frac{Z}{P}.$$

Now after the earlier explanations, we may assume that P is not identically zero, that it does not vanish for the values x_0, y_0, z_0, p_0, q_0 of the variables, and that Z and P are regular functions of the variables in the vicinity of those values: hence, if

$$I = \int_{x_0}^{x} \frac{Z}{P}\, dx,$$

the quantity I is regular in the vicinity of the arguments, and it vanishes when $x = x_0$. If we denote by E_0 the value of E for the values x_0, y_0, z_0, p_0, q_0 of the variables, then

$$E = E_0 e^{-I}.$$

Now E is to vanish, and e^{-I} does not vanish: hence we must have

$$E_0 = 0,$$

that is, we must have

$$\frac{\partial z_0}{\partial u} - q_0 \frac{\partial y_0}{\partial u} = 0.$$

86. This equation can be satisfied in two ways. First, suppose that y_0 is not independent of u: as u has not hitherto been made precise, we take

$$y_0 = u.$$

Let $z_0 = \phi(u)$, where ϕ is an arbitrary function of u; the equation will be satisfied if

$$q_0 = \phi'(u);$$

and the value of p_0 is then determined by

$$f(x_0, y_0, z_0, p_0, q_0) = 0.$$

With these values, the equation $E = 0$ is satisfied.

In the second place, suppose that y_0 is independent of u: as it is a value of y when $x = x_0$, we take it to be an arbitrary constant. The equation is then satisfied if $\dfrac{\partial z_0}{\partial u}$ vanishes, that is, if z_0 is similarly an arbitrary constant. Then q_0 can be any function of u; and p_0 is given as a function of u satisfying the relation

$$f(x_0, y_0, z_0, p_0, q_0) = 0.$$

With these values, the equation $E = 0$ is satisfied.

Now the integrals of the system of ordinary equations are

$$\begin{aligned} y &= y(x, x_0, z_0, p_0, q_0), \\ z &= z(x, x_0, z_0, p_0, q_0), \\ p &= p(x, x_0, z_0, p_0, q_0), \\ q &= q(x, x_0, z_0, p_0, q_0). \end{aligned}$$

In the former of the above cases, when u is eliminated between the first two equations, a relation is obtained involving x, y, z and an arbitrary function; if it gives an integral of the original equation, the integral so given is *general.* In the latter of the cases, when p_0 and q_0 are eliminated between the first two equations and $f(x_0, y_0, z_0, p_0, q_0) = 0$, a relation is obtained involving x, y, z and two arbitrary constants: if it gives an integral of the original equation, the integral so given is *complete.*

It therefore remains to prove that the relations thus obtained actually satisfy the equation

$$f(x, y, z, p, q) = 0.$$

Now y and z, obtained above as functions of x and u, satisfy the equations

$$\frac{\partial z}{\partial x} = p + q\frac{\partial y}{\partial x}, \quad \frac{\partial z}{\partial u} = q\frac{\partial y}{\partial u};$$

hence, when u is eliminated so as to give a single relation between x, y, z, the quantities p and q are the derivatives of z with regard to x and y respectively. Further, the values of x, y, z, p, q are such that the equation

$$X + Y\frac{\partial y}{\partial x} + Z\frac{\partial z}{\partial x} + P\frac{\partial p}{\partial x} + Q\frac{\partial q}{\partial x} = 0$$

is satisfied identically, that is,

$$\frac{\partial f}{\partial x} = 0.$$

Also, as $E = 0$, we have

$$q\frac{\partial y}{\partial u} - \frac{\partial z}{\partial u} = 0:$$

and as $\dfrac{\partial E}{\partial x} = 0$, we have seen that

$$Y\frac{\partial y}{\partial u} + qZ\frac{\partial y}{\partial u} + P\frac{\partial p}{\partial u} + Q\frac{\partial q}{\partial u} = 0,$$

that is,

$$Y\frac{\partial y}{\partial u} + Z\frac{\partial z}{\partial u} + P\frac{\partial p}{\partial u} + Q\frac{\partial q}{\partial u} = 0,$$

or

$$\frac{\partial f}{\partial u} = 0.$$

Hence f is constant: its value when $x = x_0$ is zero: hence its value is zero always, that is, the quantities so obtained satisfy the equation

$$f(x, y, z, p, q) = 0.$$

We thus obtain the *general integral* and the *complete integral* in the respective cases. In the course of the proof, it was seen how the *singular integral* could arise exceptionally: when it arose, it was recognised: and the analysis proceeded with the alternatives. Moreover, it was assumed that X, Y, Z, P, Q are regular functions; if this is not true, the results are not necessarily applicable to the equation and then integrals may arise which are *special integrals*, though these are not the only special integrals that may arise*.

DARBOUX'S MODIFICATION OF CAUCHY'S METHOD.

87. The preceding exposition is substantially due to Cauchy: a modification in the treatment of the ordinary equations has been introduced† by Darboux, which has the further advantage of permitting an easier discussion of singularities. The equations are taken in the form

$$\frac{dx}{P}=\frac{dy}{Q}=\frac{dz}{pP+qQ}=\frac{dp}{-(X+pZ)}=\frac{dq}{-(Y+qZ)}=dt,$$

and they are regarded as determining the five variables x, y, z, p, q in terms of t, the arbitrary quantities that occur being made functions of a variable u, as before. The establishment of the results is simpler than in the Cauchy treatment.

We assume that, when‡ $t=0$, the five variables acquire values x_0, y_0, z_0, p_0, q_0, subject to the relation

$$f(x_0, y_0, z_0, p_0, q_0)=0,$$

and that, in the vicinity of these initial values, the functions P, Q, $pP+qQ$, $X+pZ$, $Y+qZ$ are regular. Then, by the theory of systems of ordinary equations, a unique set of integrals exists, being regular functions of t and acquiring the assigned initial values when $t=0$: let them be

$$\begin{aligned} x &= x\,(t, x_0, y_0, z_0, p_0, q_0),\\ y &= y\,(t, x_0, y_0, z_0, p_0, q_0),\\ z &= z\,(t, x_0, y_0, z_0, p_0, q_0),\\ p &= p\,(t, x_0, y_0, z_0, p_0, q_0),\\ q &= q\,(t, x_0, y_0, z_0, p_0, q_0). \end{aligned}$$

* See, for instance, § 34.

† "Mémoire sur les solutions singulières...premier ordre," *Mém. de l'Inst. de France*, t. XXVII (1880), § 24.

‡ No generality is gained by taking t_0 as an initial value for t: the only difference is that $t-t_0$ comes in place of t.

Moreover, we have

$$\begin{aligned}\frac{df}{dt} &= X\frac{dx}{dt} + Y\frac{dy}{dt} + Z\frac{dz}{dt} + P\frac{dp}{dt} + Q\frac{dq}{dt}\\ &= 0,\end{aligned}$$

on substitution: thus

$$\begin{aligned}f(x, y, z, p, q) &= \text{a quantity independent of } t\\ &= \text{its value when } t=0\\ &= f(x_0, y_0, z_0, p_0, q_0)\\ &= 0,\end{aligned}$$

so that the above values are connected by this relation.

If this is to be equivalent to the original differential equation, then p and q must be equal to the derivatives of z with regard to x and y: the requirement is met as follows. Let the quantities x_0, y_0, z_0, p_0, q_0 be functions of u, a new parametric variable, so that x, y, z, p, q are now functions of t and u. So far as variations with regard to t are concerned, the system of equations gives

$$\frac{\partial z}{\partial t} = p\frac{\partial x}{\partial t} + q\frac{\partial y}{\partial t}.$$

Let

$$E = \frac{\partial z}{\partial u} - p\frac{\partial x}{\partial u} - q\frac{\partial y}{\partial u};$$

then

$$\frac{\partial^2 z}{\partial t\partial u} = p\frac{\partial^2 x}{\partial t\partial u} + q\frac{\partial^2 y}{\partial t\partial u} + \frac{\partial p}{\partial u}\frac{\partial x}{\partial t} + \frac{\partial q}{\partial u}\frac{\partial y}{\partial t},$$

and

$$\begin{aligned}\frac{\partial E}{\partial t} &= \frac{\partial^2 z}{\partial t\partial u} - p\frac{\partial^2 x}{\partial t\partial u} - q\frac{\partial^2 y}{\partial t\partial u} - \frac{\partial p}{\partial t}\frac{\partial x}{\partial u} - \frac{\partial q}{\partial t}\frac{\partial y}{\partial u}\\ &= \frac{\partial p}{\partial u}\frac{\partial x}{\partial t} + \frac{\partial q}{\partial u}\frac{\partial y}{\partial t} - \frac{\partial p}{\partial t}\frac{\partial x}{\partial u} - \frac{\partial q}{\partial t}\frac{\partial y}{\partial u}\\ &= P\frac{\partial p}{\partial u} + Q\frac{\partial q}{\partial u} + (X+pZ)\frac{\partial x}{\partial u} + (Y+qZ)\frac{\partial y}{\partial u},\end{aligned}$$

on substituting from the equations. But

$$X\frac{\partial x}{\partial u} + Y\frac{\partial y}{\partial u} + Z\frac{\partial z}{\partial u} + P\frac{\partial p}{\partial u} + Q\frac{\partial q}{\partial u} = 0,$$

because the values of x, y, z, p, q satisfy the equation

$$f(x, y, z, p, q) = 0;$$

hence

$$\frac{\partial E}{\partial t} = -Z\left(\frac{\partial z}{\partial u} - p\frac{\partial x}{\partial u} - q\frac{\partial y}{\partial u}\right)$$
$$= -ZE,$$

so that

$$E = E_0 e^{-\int_0^t Z dt},$$

where E_0 denotes the value of E when $t=0$. Now Z is a regular function of t in the vicinity of $t=0$, and therefore $\int_0^t Z dt$ is finite. Hence it is necessary and sufficient that the equation

$$E_0 = 0$$

should be satisfied in order that E may vanish, that is,

$$\frac{\partial z_0}{\partial u} = p_0 \frac{\partial x_0}{\partial u} + q_0 \frac{\partial y_0}{\partial u}.$$

88. This equation may be satisfied in three distinct ways. In the first of these ways, both x_0 and y_0 involve u; in that case, let

$$x_0 = f(u), \quad y_0 = g(u), \quad z_0 = h(u),$$

so that the relation

$$h'(u) = p_0 f'(u) + q_0 g'(u)$$

becomes an equation which, with $f(x_0, y_0, z_0, p_0, q_0) = 0$, determines p_0 and q_0. In the second of the ways, only one of the variables x_0 and y_0 involves u, say y_0: in that case, let $x_0 = \alpha$, $y_0 = u$, $z_0 = g(u)$; then

$$g'(u) = q_0,$$

and the equation $f(x_0, y_0, z_0, p_0, q_0) = 0$ determines p_0. In the third of the ways, neither of the variables x_0 and y_0 involves u: in that case, let $x_0 = \alpha$, $y_0 = \beta$: then z_0 does not involve u, and we may write $z_0 = \gamma$, where α, β, γ are constants, and $f(\alpha, \beta, \gamma, p_0, q_0) = 0$.

In all these cases, we have $E_0 = 0$ and therefore $E = 0$, that is,

$$\frac{\partial z}{\partial u} = p\frac{\partial x}{\partial u} + q\frac{\partial y}{\partial u};$$

and we had

$$\frac{\partial z}{\partial t} = p\frac{\partial x}{\partial t} + q\frac{\partial y}{\partial t}.$$

These relations shew that, when z is expressed in terms of x and y consistently with the relations obtained, its derivatives with regard to x and y are p and q. Moreover, we can prove, exactly as in § 86, that the equation

$$f(x, y, z, p, q) = 0$$

is satisfied: thus we have an integral of the original equation.

Consider the three cases in turn: for all of them, we have to make the respective substitutions in

$$x = x(t, x_0, y_0, z_0, p_0, q_0),$$
$$y = y(t, x_0, y_0, z_0, p_0, q_0),$$
$$z = z(t, x_0, y_0, z_0, p_0, q_0).$$

In the first case, we eliminate t and u between the three equations: there results a single relation between x, y, z. The assigned initial conditions are such that $x = x_0$, $y = y_0$, $z = z_0$, functions of a variable parameter u: or getting rid of u by elimination, we can say that, when some relation

$$\phi(x, y) = 0$$

is satisfied, then z becomes some function of x and y. No limitation except regularity has been imposed upon the functions: thus z becomes an assigned arbitrary function of x and y, when these are connected by any assigned relation $\phi(x, y) = 0$. The integral is *general*.

In the second case, for the initial conditions, we have $x = x_0 = \alpha$, $y = y_0 = u$, $z = z_0 = g(u)$, that is, when $x = \alpha$, $z = g(y)$. Here g can be an arbitrary function. We have a special form of the first case, obtained by writing $\phi(x, y) = x - \alpha$. The integral is *general*.

In the third case, we have to eliminate three quantities t, p_0, q_0 between the equations

$$x = x(t, \alpha, \beta, \gamma, p_0, q_0),$$
$$y = y(t, \alpha, \beta, \gamma, p_0, q_0),$$
$$z = z(t, \alpha, \beta, \gamma, p_0, q_0),$$
$$0 = f(\alpha, \beta, \gamma, p_0, q_0).$$

The resulting equation involves α, β, γ arising as values of x, y, z. One of these may be taken as an initial value: the other two may be arbitrary. When they are quite arbitrary, the integral is *complete*.

As regards the derivation of the result, it is to be noted that the values of p_0 and q_0 are given by

$$f(x_0, y_0, z_0, p_0, q_0) = 0,$$

$$p_0 \frac{\partial x_0}{\partial u} + q_0 \frac{\partial y_0}{\partial u} = \frac{\partial z_0}{\partial u};$$

if f be a regular function of its arguments, the two equations can be resolved for p_0 and q_0, provided the magnitude

$$\frac{\partial f}{\partial p_0} \frac{\partial y_0}{\partial u} - \frac{\partial f}{\partial q_0} \frac{\partial x_0}{\partial u}$$

does not vanish. We may assume this to be the case for the first two of the three alternatives, because we have excluded the hypothesis that P_0 and Q_0 are zero: it does not arise for the third alternative.

Again, the equations for x and y may be resolved for t and u in the vicinity of $t = 0$ and $u = u_0$, provided

$$\frac{\partial x}{\partial t} \frac{\partial y}{\partial u} - \frac{\partial y}{\partial t} \frac{\partial x}{\partial u}$$

does not vanish there, that is, provided

$$P_0 \frac{\partial y_0}{\partial u} - Q_0 \frac{\partial x_0}{\partial u}$$

does not vanish: this is the above magnitude, assumed not to vanish. The values of t and u so obtained are substituted in the expression for z, which becomes the integral.

It might of course happen that $P_0 = 0$, $Q_0 = 0$, without $X_0 + p_0 Z_0$ and $Y_0 + q_0 Z_0$ vanishing: the possibility is discussed later (Chapter VII, § 109) and will be seen to provide a singularity.

89. The preceding analysis and argument are valid in establishing this result, save in one set of circumstances controlling the hypotheses adopted. It has been assumed that P, Q, $X + pZ$, $Y + qZ$ are regular functions of their arguments in the vicinity of the values x_0, y_0, z_0, p_0, q_0; but if, for all sets of values satisfying the necessary relation

$$f(x_0, y_0, z_0, p_0, q_0) = 0,$$

it should happen that

$$P = 0, \quad Q = 0, \quad X + pZ = 0, \quad Y + qZ = 0,$$

then the only regular integrals of the system of ordinary equations are

$$x = x_0, \quad y = y_0, \quad z = z_0, \quad p = p_0, \quad q = q_0,$$

and the results can no longer be established.

The case, as before, is usually that of the *singular integral*: it is put on one side, for it admits of other treatment as follows. The equations

$$f = 0, \quad P = 0, \quad Q = 0, \quad X + pZ = 0, \quad Y + qZ = 0$$

coexist. If it is possible to eliminate p and q between them, there will result an equation between x, y, z, which is the singular integral. If it is not possible to eliminate p and q, then, f being supposed irreducible, they will serve to determine p and q: the values of p and q are substituted in

$$dz = p\,dx + q\,dy,$$

and quadrature leads to an integral involving one arbitrary constant: it may (§ 78) be regarded as a specialised or particular form of the complete integral.

Further, if P, Q, $X + pZ$, $Y + qZ$ are not regular functions of their arguments, the inference as regards the ordinary equations cannot be made and so the result cannot be established. Special integrals can thus arise: but it is not the only source of such integrals.

90. One other exceptional case requires special mention, viz. that in which

$$pP + qQ = 0,$$

though neither P nor Q vanishes: it is a case which occurs when f is homogeneous in p and q of any degree. An integral of the ordinary equations is then

$$z = \text{quantity independent of } t$$
$$= z_0;$$

and if it should happen that we are dealing with conditions leading to a complete integral, it is clear that the equation $z = z_0$ cannot be used for purposes of elimination.

As in corresponding difficulties (§§ 58, 59), we use a Legendrian transformation, introducing a new variable z' by the relation

$$z' = z - px,$$

and assigning other associated variables in the form

$$y' = y, \quad q' = q, \quad x' = -p, \quad p' = x.$$

The quantity $p'P' + q'Q'$ which occurs in the modified system is obtained from

$$x(X + pZ) + qQ,$$

by making the above substitutions; this does not vanish in general, and so the process can be applied to the modified equation: the integral of the original equation can be deduced as before.

Note 1. Both in Cauchy's method and in Darboux's modification, an integral has been obtained which has been declared a general integral. Its expression certainly contains an arbitrary function in the least restricted case; but this property is not, of itself, sufficient to secure the character in the customary form. A general integral is given by the elimination of a between the equations

$$g\{x, y, z, a, f(a)\} = 0,$$

$$\frac{\partial g}{\partial a} + \frac{\partial g}{\partial f} f'(a) = 0,$$

that is, one of the equations is the derivative of the other with regard to the parameter which is to be eliminated. Of course, when the elimination can be actually achieved, this relation between the two equations disappears; usually, however, the expression of the general integral must be left in this form.

In particular instances, the result can be verified by bringing the last two equations into an equivalent form which exhibits the relation characteristic of the general integral.

Note 2. The general integral that has thus been obtained is the integral specified in Cauchy's existence-theorem.

Taking the simpler form, we have an integral such that $y = y_0$ and $z = z_0 = \phi(y_0)$, when $x = x_0$, where ϕ can be any function subject to the conditions involved in the existence-theorem for ordinary differential equations: in other words, the integral is such that z acquires a value $\phi(y)$, when $x = x_0$. The conditions have relation to the regularity of the coefficients in the ordinary equations and of the integrals of those equations; they are the same as those set out in Cauchy's existence-theorem, and so need not be repeated here.

Of course, it must not be assumed that, if all the conditions required in the proof of the theorem are not satisfied, the integral does not exist: in such a case, we merely are not in a position to affirm its existence.

Ex. 1. An example is given by Cauchy*: he discusses the equation

$$pq - xy = 0.$$

The initial values must be such that

$$p_0 q_0 - x_0 y_0 = 0.$$

The subsidiary system of ordinary equations is

$$\frac{dx}{q} = \frac{dy}{p} = \frac{dz}{2pq} = \frac{dp}{y} = \frac{dq}{x};$$

and a unique system of regular integrals satisfying all the conditions is given by

$$y = y_0 \left\{1 + \tfrac{1}{2}\frac{x^2 - x_0^2}{q_0^2} - \tfrac{1}{8}\frac{(x^2 - x_0^2)^2}{q_0^2} - \dots\right\}$$

$$= y_0 \left\{1 + \frac{x^2 - x_0^2}{q_0^2}\right\}^{\frac{1}{2}},$$

$$z = z_0 + \frac{y_0}{q_0}(x^2 - x_0^2),$$

$$p = x\frac{y_0}{q_0},$$

$$q = q_0 \left\{1 + \frac{x^2 - x_0^2}{q_0^2}\right\}^{\frac{1}{2}},$$

the same branch of the irrational quantity being taken in q as in y. We have

$$(z - z_0) q_0 = y_0 (x^2 - x_0^2),$$

$$(z - z_0)^2 = (y^2 - y_0^2)(x^2 - x_0^2);$$

when we take, in accordance with the preceding theory,

$$y_0 = u, \quad z_0 = \phi(u), \quad q_0 = \phi'(u_0),$$

we have a general integral given by the two equations: and when, also in accordance with the preceding theory, we take

$$y_0 = a, \quad z_0 = \beta,$$

where a and β are arbitrary constants, the second equation becomes

$$(z - \beta)^2 = (y - a)^2 (x - x_0)^2,$$

giving rise to a complete integral. And there is no singular integral.

If the subsidiary equations are treated by Darboux's method, they can be taken in the form†

$$\frac{dx}{q} = \frac{dy}{p} = \frac{dz}{2pq} = \frac{dp}{y} = \frac{dq}{x} = \frac{dt}{2pq\,(=2xy)}.$$

* *L.c.*, t. II, p. 249.

† The new variable t is at our disposal, and any modification tending to simplify the equations may be adopted; accordingly, $\dfrac{dt}{2pq}$ is chosen instead of dt, as the common value of the fractions in the subsidiary equations.

Integrals of these, which give x_0, y_0, z_0, p_0, q_0 as the values of x, y, z, p, q, when $t=0$, are

$$x^2 = x_0^2 + t\frac{q_0}{y_0},$$

$$y^2 = y_0^2 + t\frac{y_0}{q_0},$$

$$z = z_0 + t,$$

$$p^2 = p_0^2 + t\frac{y_0}{q_0},$$

$$q^2 = q_0^2 + t\frac{q_0}{y_0}.$$

The first three equations give

$$(z-z_0)\,q_0 = y_0\,(x^2 - x_0^2),$$

$$(z-z_0)^2 = (y^2 - y_0^2)\,(x^2 - x_0^2):$$

the assumptions

$$y_0 = u, \quad z_0 = \phi(u), \quad q_0 = \phi'(u),$$

lead to the general integral as before, on the elimination of u: the assumptions

$$y_0 = a, \quad z_0 = \beta,$$

lead to the complete integral as contained in the second of the two equations.

Ex. 2. Discuss similarly the equation

$$2xz - px^2 + qxy + q^2x = 0,$$

obtaining its complete integral and its general integral. Are there any limitations upon the initial conditions caused by the form of the equation? (Mansion.)

Ex. 3. Integrate the equation

$$pqy - pz + aq = 0,$$

where a is a constant. (Mansion.)

Ex. 4. As a slight variation in the details of working in any particular question when a Cauchy integral is required, we obtain the integral of

$$xzp + yzq = xy,$$

which is such that z becomes $\phi(y)$, when $x = x_0$.

The ordinary equations are

$$\frac{dx}{xz} = \frac{dy}{yz} = \frac{dz}{pxz + qyz} = \ldots;$$

both y and z can be expressed in terms of x, on using the original equation, by integrals

$$\frac{y}{x} = \text{constant}, \quad z^2 - xy = \text{constant}.$$

We take $y = u$ and $z = \phi(u)$, when $x = x_0$: thus

$$\frac{y}{x} = \frac{u}{x_0}, \quad z^2 - xy = \{\phi(u)\}^2 - x_0 u.$$

Eliminating u, the relation is

$$z^2 - xy = \left\{\phi\left(\frac{x_0}{x}y\right)\right\}^2 - \frac{{x_0}^2}{x}y,$$

which gives the integral in question.

In this case, there is no limitation upon the form of the function ϕ.

Geometry of the Integrals.

91. It is found convenient, particularly for equations involving two independent variables, to associate geometrical considerations with the analysis. A slight use of this association has already been made (§ 21), by way of interpreting an integral of an equation: we shall now proceed to greater detail, particularly in order to indicate the significance of the various equations, to illustrate the relations of different integrals to one another, and to discuss some at least of the singularities which have received no more than passing mention in the preceding sections. The differential equation, when there are two independent variables, is taken to be

$$f(x, y, z, p, q) = 0;$$

its complete integral is taken to be

$$\phi(x, y, z, a, b) = 0,$$

where a and b are arbitrary constants; and other integrals can be deduced by methods already explained.

The equation $\phi = 0$ is the equation of a double family of surfaces. All these surfaces are such that $f = 0$ is satisfied; consequently, they are said to satisfy the differential equation. Also, $f = 0$ is given by the elimination of a and b between the equations

$$\phi = 0, \quad \frac{\partial \phi}{\partial x} + p\frac{\partial \phi}{\partial z} = 0, \quad \frac{\partial \phi}{\partial y} + q\frac{\partial \phi}{\partial z} = 0.$$

Through any point of space x, y, z, there passes a simple infinitude of the surfaces. The tangent planes at the point can be represented by the equation

$$z' - z = p(x' - x) + q(y' - y),$$

where x', y', z' are current coordinates; and the normals to the surfaces are given by the equations

$$\frac{x' - x}{p} = \frac{y' - y}{q} = \frac{z' - z}{-1}.$$

All these normals lie on the cone

$$f\left(x, y, z, -\frac{x'-x}{z'-z}, -\frac{y'-y}{z'-z}\right) = 0;$$

this cone of normals will be denoted by N. The reciprocal of the cone of normals is the envelope of all the tangent planes at the point to the surfaces satisfying the differential equation: this cone, the envelope of the tangent planes, will be denoted by T.

The equation of the cone T can be simply constructed as follows. The tangent plane is

$$z' - z = p(x' - x) + q(y' - y);$$

we require the envelope of this plane, subject to the condition $f(x, y, z, p, q) = 0$, and therefore we have

$$0 = (x' - x)\,dp + (y' - y)\,dq,$$

$$0 = \frac{\partial f}{\partial p}dp + \frac{\partial f}{\partial q}dq,$$

so that the equation of the cone T is given by the elimination of p and q between the equations

$$\left.\begin{array}{c} z' - z = p(x' - x) + q(y' - y) \\ \dfrac{x' - x}{\dfrac{\partial f}{\partial p}} = \dfrac{y' - y}{\dfrac{\partial f}{\partial q}} \\ f(x, y, z, p, q) = 0 \end{array}\right\}.$$

The result will obviously be of the form

$$F\left(x, y, z, \frac{z'-z}{x'-x}, \frac{y'-y}{x'-x}\right) = 0.$$

Moreover, the equations of the generator of T, which lies in the tangent plane in question, are

$$\frac{x' - x}{\dfrac{\partial f}{\partial p}} = \frac{y' - y}{\dfrac{\partial f}{\partial q}} = \frac{z' - z}{p\dfrac{\partial f}{\partial p} + q\dfrac{\partial f}{\partial q}};$$

this generator is the line along which the tangent plane touches the cone T, and it obviously is perpendicular to the corresponding normal on the reciprocal cone N.

Again, take any plane

$$z' = \alpha x' + \beta y' + \gamma$$

in space: if this plane touches an integral surface

$$\phi(x, y, z, a, b) = 0,$$

the point of contact is given by

$$\phi = 0, \quad \frac{\partial \phi}{\partial x} + \alpha \frac{\partial \phi}{\partial z} = 0, \quad \frac{\partial \phi}{\partial y} + \beta \frac{\partial \phi}{\partial z} = 0.$$

The coordinates of the point of contact satisfy the equation

$$f(x, y, z, \alpha, \beta) = 0,$$

whatever a and b may be: and, of course, as the point lies in the plane, we have

$$z = \alpha x + \beta y + \gamma.$$

But the two equations

$$z = \alpha x + \beta y + \gamma, \quad f(x, y, z, \alpha, \beta) = 0,$$

determine a plane curve: owing to its source, it is the locus of points of contact of integral surfaces with the assumed plane. This curve lying in the plane will be denoted by C.

Moreover, as this plane touches a surface at the point x, y, z, it touches the cone T which is associated with the point; we may therefore regard the curve C as the locus of those points in the plane at which the plane touches the associated cones T. And, conversely, a cone T associated with a point is the envelope of those planes whose curves C pass through the point.

Consequently, the integral surfaces which satisfy the differential equation can be regarded in two ways. On the one hand, all those which pass through a given point have their tangent planes enveloped by a cone T: on the other hand, all those which touch a given plane have their points of contact with the plane lying upon a curve C in the plane. Each of these is obviously deducible from the other by reciprocal polars.

Ex. 1. Shew that if the curve C is a degenerate curve, composed of a number of straight lines, and if the (Legendre) transformation

$$\zeta = px + qy - z$$

is applied to the partial differential equation, the transformed differential equation is linear.

(Goursat.)

Ex. 2. Shew that, if integral surfaces be given by an equation

$$x^2 + y^2 + z^2 = 2ax + 2by + 2cz,$$

where

$$(a\alpha + b\beta + c\gamma - 1)^2 = (a^2 + b^2 + c^2)(\alpha^2 + \beta^2 + \gamma^2),$$

α, β, γ being given constants, and a, b, c arbitrary constants subject to this condition, the curves C are circles.

(Goursat.)

CHARACTERISTICS: THEIR PROPERTIES.

92. Now consider a general integral as deduced from a complete integral; it is given by the equations

$$\phi \equiv \phi\{x, y, z, a, g(a)\} = 0,$$

$$\frac{d\phi}{da} \equiv \frac{\partial\phi}{\partial a} + \frac{\partial\phi}{\partial g} g'(a) = 0,$$

where $g(a)$ is an arbitrary function: the surface, represented by the general integral, is obtained by eliminating a between the two equations. The equation $\phi = 0$ is one of the surfaces included in the complete integral; the equation $\frac{d\phi}{da} = 0$ then determines a curve on the surface $\phi = 0$, being in fact the intersection of $\phi = 0$ with the surface that arises from a consecutive value of a; the curve thus given, for any value of a, is called a *characteristic* of the surface $\phi = 0$. The general integral, arising from the elimination of a between $\phi = 0$ and $\frac{d\phi}{da} = 0$, is thus the locus of the characteristics of the surfaces $\phi = 0$, which arise for any one function g and for all values of a.

On any surface represented by a complete integral, there is an infinitude of characteristics: they arise because $g(a)$ is an arbitrary function which can be assigned in an infinitude of ways. Through any ordinary point on such a surface, there passes certainly one characteristic: for, at that point, there are two independent equations to determine a and $g(a)$. Moreover, through any ordinary point there passes only one characteristic in general: because the two equations, $\phi = 0$ and $\frac{d\phi}{da} = 0$, in general give unique values for the ratios $dx : dy : dz$.

At any point on a characteristic of $\phi = 0$, the curve is touched by the tangent plane there: it is also touched there by the tangent plane at that point to the consecutive surface, because it lies on that surface. Hence the tangent line to the characteristic is the intersection of the tangent planes to two consecutive surfaces through the point, that is, it coincides with the generator of the cone T along which the cone touches the tangent plane to $\phi = 0$.

The general integral touches the surface $\phi=0$ along the characteristic. On the surface $\phi=0$, the tangent to the surface is determined by the values of p and q which are given by

$$\frac{\partial\phi}{\partial x}+p\frac{\partial\phi}{\partial z}=0,\quad \frac{\partial\phi}{\partial y}+q\frac{\partial\phi}{\partial z}=0.$$

At that point on the general integral, the value of a is given, in terms of x, y, z, by $\frac{d\phi}{da}=0$: when this value of a is substituted in $\phi=0$, the values of p and q determining the tangent plane to the general integral at the point are given by

$$\frac{\partial\phi}{\partial x}+p\frac{\partial\phi}{\partial z}+\frac{d\phi}{da}\frac{da}{dx}=0,\quad \frac{\partial\phi}{\partial y}+q\frac{\partial\phi}{\partial z}+\frac{d\phi}{da}\frac{da}{dy}=0,$$

that is, they are the same as for the tangent plane to $\phi=0$, because $\frac{d\phi}{da}=0$. As the tangent planes are the same at every point, the general integral touches the surface $\phi=0$ everywhere along the characteristic.

Further, if two integral surfaces touch at a point, they touch everywhere along the characteristic through the point: for each of them touches the general integral everywhere along the characteristic.

Again, general integrals arising from the assumption of different forms of arbitrary function represent different surfaces, being the loci of the characteristics: it is natural to enquire whether two different surfaces representing general integrals have any characteristics in common. It is obvious that the equations

$$\phi(x, y, z, a, b)=0,\quad b=g(a),\quad \frac{\partial\phi}{\partial a}+\frac{\partial\phi}{\partial b}g'(a)=0,$$

$$\phi(x, y, z, a, c)=0,\quad c=h(a),\quad \frac{\partial\phi}{\partial a}+\frac{\partial\phi}{\partial c}h'(a)=0,$$

will represent the same curve on the two different surfaces for each value of a, which satisfies the relations

$$g(a)=h(a),\quad g'(a)=h'(a);$$

and that, if there are m such values of a, the two surfaces will have m characteristics in common.

Now pass to the limit when these m characteristics coincide:

then the common value of a is a root of multiple order $m+1$ of the equation

$$g(a) = h(a),$$

so that we have

$$g(a) = h(a), \quad g'(a) = h'(a), \quad \ldots, \quad g^{(m)}(a) = h^{(m)}(a).$$

Then along this common characteristic the two surfaces representing the general integral have contact of order m. The derivatives of z, up to and including those of order m, belonging to the surface

$$\phi(x, y, z, a, b) = 0, \quad b = g(a), \quad \frac{\partial \phi}{\partial a} + \frac{\partial \phi}{\partial b} g'(a) = 0,$$

involve derivatives of $g(a)$ of order not higher than m; and similarly for the derivatives of z, given by

$$\phi(x, y, z, a, c) = 0, \quad c = h(a), \quad \frac{\partial \phi}{\partial a} + \frac{\partial \phi}{\partial c} h'(a) = 0,$$

derivatives of $h(a)$ of order not higher than m occur. Owing to the relations between $g(a)$ and $h(a)$, the derivatives of z at a common point are the same for the two surfaces up to order m; and therefore the two surfaces have contact of order m along the characteristic.

It is an immediate corollary that, if two integral surfaces have contact of order m at any point, they have contact of that order along the whole characteristic through the point.

Ex. 1. These properties are used by Darboux to determine the integral surface or surfaces which pass through any given curve K.

We may assume that K does not lie on the singular integral, if any: otherwise, a single surface is at once given; and other surfaces will occur among those which represent complete integrals or general integrals.

We may also assume that K does not lie on a complete integral: otherwise, a single surface is at once given; and, if K be a characteristic, an infinitude of surfaces satisfies the required condition.

Accordingly, take any point P on the curve K and draw the tangent PQ to the curve at the point P: through the line PQ draw a plane to touch the cone T associated with the point. Then, for the point, we have values of x, y, z, p, q, the two latter being given by the tangent plane: by means of the integrals of the equations of the characteristics, we construct the complete integral having these for initial values. This complete integral touches the curve K at the point P. Through this point P draw the characteristic on the complete integral; it touches the curve K at the point.

Now let P travel along the curve: for each successive position, we have a characteristic; the locus of these characteristics is the integral surface required, and (being such a locus) it is a general integral. It can also be obtained as the envelope of the complete integrals touching K.

If only a single tangent plane can be drawn through PQ to the cone T, the general integral thus obtained constitutes the whole of the surface required. If several tangent planes can be drawn, the general integral consists of a corresponding number of sheets.

Ex. 2. Shew that, if the curve K lies on a surface representing a complete integral though it is not a characteristic on that surface, no other integral surface can be drawn through K.

(Darboux.)

93. It is convenient to consider the developable surface circumscribed to an integral surface along a characteristic: it is called the *characteristic developable.*

To obtain the simplest properties, consider less particularly any two surfaces S and S': let a plane touch them in A and A' respectively; then the developable surface circumscribed to S and S' is the envelope of such planes. A generator of the developable is the intersection of two consecutive planes: hence AA' is a generator, because a plane, consecutive to the supposed plane, touches S in A and S' in A'.

Take a plane P and, in that plane, the curve C which is the locus of its points of contact with integral surfaces. Suppose that the two preceding surfaces S and S' touch the plane, and let their points of contact be Q and Q', being points on C; then QQ' is a generator of the developable circumscribed to S and S'. Now let S and S' be consecutive surfaces: the circumscribed developable becomes the characteristic developable circumscribed along the characteristic which is the ultimate intersection of S and S': the points Q and Q' coincide, and the line QQ' becomes the tangent to C. Hence the generator of the characteristic developable through a point on the characteristic is the tangent to C at that point.

Moreover, the cone T, being the envelope of the tangent planes at the point to all the integral surfaces through the point, touches P at the point: the generator along which it touches P is, as seen above, the tangent to the characteristic line.

Take the characteristic, being the intersection of two consecutive surfaces, and draw the tangent planes along it: the envelope of these is the characteristic developable, and the generator of this developable at any point is the tangent there to the curve C. Accordingly, at any point on the characteristic line, draw the tangent: construct the plane which touches the cone T along this tangent: in this plane, obtain the curve C: the tangent to C at the point is the generator of the characteristic developable through the point, being the intersection of two consecutive tangent planes to the surface along the characteristic.

94. On the basis of these geometrical properties we can obtain the differential equations of the characteristics: these will, of course, be a set of ordinary equations, because each characteristic is a curve and therefore admits of only one independent variable.

The tangent line to the characteristic at any point x, y, z, is the generator of the cone T lying in the tangent plane; the equations of this generator were obtained (§ 91) in the form

$$\frac{x'-x}{\frac{\partial f}{\partial p}} = \frac{y'-y}{\frac{\partial f}{\partial q}} = \frac{z'-z}{p\frac{\partial f}{\partial p} + q\frac{\partial f}{\partial q}},$$

and therefore the direction of the tangent to the characteristic satisfies the equations

$$\frac{dx}{\frac{\partial f}{\partial p}} = \frac{dy}{\frac{\partial f}{\partial q}} = \frac{dz}{p\frac{\partial f}{\partial p} + q\frac{\partial f}{\partial q}} = du,$$

say. Next, the curve C in the plane

$$z' - z = p(x' - x) + q(y' - y)$$

satisfies the equation

$$f(x', y', z', p, q) = 0,$$

where p and q define the plane: hence the tangent to C at the point is given by

$$\delta z = p\,\delta x + q\,\delta y,$$

$$\frac{\partial f}{\partial x}\delta x + \frac{\partial f}{\partial y}\delta y + \frac{\partial f}{\partial z}\delta z = 0;$$

so that

$$\left(\frac{\partial f}{\partial x} + p\frac{\partial f}{\partial z}\right)\delta x + \left(\frac{\partial f}{\partial y} + q\frac{\partial f}{\partial z}\right)\delta z = 0.$$

This direction is to be the same as the intersection of the tangent plane at the point with the tangent plane at a consecutive point on the characteristic: at this consecutive point, the tangent plane is

$$z' - (z + dz) = (p + dp)\{x' - (x + dx)\} + (q + dq)\{y' - (y + dy)\},$$

and therefore, along the intersection, we have

$$-dz = (x' - x)\,dp - p\,dx + (y' - y)\,dq - q\,dy.$$

But, along the characteristic,

$$dz = p\,dx + q\,dy;$$

and therefore

$$(x' - x)\,dp + (y' - y)\,dq = 0,$$

that is,

$$\delta x\,dp + \delta y\,dq = 0.$$

Hence

$$\frac{dp}{\dfrac{\partial f}{\partial x} + p\dfrac{\partial f}{\partial z}} = \frac{dq}{\dfrac{\partial f}{\partial y} + q\dfrac{\partial f}{\partial z}} = dv,$$

say. Also, the variations along the characteristic must be subject to the equation

$$f(x, y, z, p, q) = 0,$$

and therefore

$$\frac{\partial f}{\partial x}dx + \frac{\partial f}{\partial y}dy + \frac{\partial f}{\partial z}dz + \frac{\partial f}{\partial p}dp + \frac{\partial f}{\partial q}dq = 0.$$

Substituting, we find

$$du + dv = 0;$$

consequently, the equations of the characteristic are

$$\frac{dx}{\dfrac{\partial f}{\partial p}} = \frac{dy}{\dfrac{\partial f}{\partial q}} = \frac{dz}{p\dfrac{\partial f}{\partial p} + q\dfrac{\partial f}{\partial q}} = \frac{dp}{-\left(\dfrac{\partial f}{\partial x} + p\dfrac{\partial f}{\partial z}\right)} = \frac{dq}{-\left(\dfrac{\partial f}{\partial y} + q\dfrac{\partial f}{\partial z}\right)}.$$

But these are the equations which occur in Cauchy's method of integration, whether in its original form or in Darboux's presentation: accordingly, it is usually described as the *method of characteristics.*

The integration of these equations gives an integrated form of the equations of the characteristic, coupled with the assignment of initial values satisfying

$$f(x, y, z, p, q) = 0, \quad dz = p\,dx + q\,dy;$$

these initial values are regarded as functions of a variable u: when this variable is eliminated between the equations

$$\left.\begin{aligned} y &= y(x, x_0, y_0, z_0, p_0, q_0) \\ z &= z(x, x_0, y_0, z_0, p_0, q_0) \end{aligned}\right\},$$

the result is a single equation representing a surface which, whether the integral be general or be complete, is a locus of characteristics.

Ex. Denoting any integral of the partial equation

$$f(x, y, z, p, q)=0$$

by z, the derivatives of x, y, z, with respect to a variable t by x', y', z', and any arbitrary function of t by T, verify that the conditions, necessary and sufficient to secure a stationary (zero) value for the integral

$$\int_{t_0}^{t} T(px'+qy'-z')\,dt,$$

are the equations of the characteristics of the partial differential equation.

Apply this method to deduce the condition that two equations

$$f(x, y, z, p, q)=0, \quad g(x, y, z, p, q)=0,$$

may possess a common integral. (Hilbert.)

95. In the preceding investigation of the equations of the characteristics, they have been associated analytically with the original partial differential equation: it is desirable to associate them also with the integral of the differential equation.

The differential equation $f(x, y, z, p, q)=0$ is the result of eliminating a and b between the equations

$$\left.\begin{aligned} \phi &= \phi(x, y, z, a, b) = 0 \\ \phi_1 &= \frac{\partial\phi}{\partial x} + p\frac{\partial\phi}{\partial z} = 0 \\ \phi_2 &= \frac{\partial\phi}{\partial y} + q\frac{\partial\phi}{\partial z} = 0 \end{aligned}\right\};$$

and $f=0$ is the only equation which results from that elimination. The only independent relations, that connect differential elements dx, dy, dz, dp, dq, are

$$d\phi=0, \quad d\phi_1=0, \quad d\phi_2=0;$$

and $df=0$ is a relation connecting these differential elements: hence quantities λ, μ, ρ, free from differential elements, must exist such that the relation

$$df=\lambda d\phi + \mu d\phi_1 + \rho d\phi_2$$

is satisfied*, and therefore we have

$$f = \lambda\phi + \mu\phi_1 + \rho\phi_2,$$

because f, ϕ, ϕ_1, ϕ_2 vanish together.

As a and b do not occur in f and do occur in ϕ, ϕ_1, ϕ_2, we have

$$\lambda\frac{\partial\phi}{\partial a} + \mu\frac{\partial\phi_1}{\partial a} + \rho\frac{\partial\phi_2}{\partial a} = 0,$$

$$\lambda\frac{\partial\phi}{\partial b} + \mu\frac{\partial\phi_1}{\partial b} + \rho\frac{\partial\phi_2}{\partial b} = 0,$$

parts such as $\phi\frac{\partial\lambda}{\partial a} + \phi_1\frac{\partial\mu}{\partial a} + \phi_2\frac{\partial\rho}{\partial a}$ vanishing, because of the integral equations under consideration. Moreover, p occurs in ϕ_1, but not in ϕ or ϕ_2: and q occurs in ϕ_2, but not in ϕ or ϕ_1; hence

$$\frac{\partial f}{\partial p} = \mu\frac{\partial\phi}{\partial z}, \qquad \frac{\partial f}{\partial q} = \rho\frac{\partial\phi}{\partial z}.$$

Again,

$$\frac{\partial f}{\partial x} = \lambda\frac{\partial\phi}{\partial x} + \mu\left(\frac{\partial^2\phi}{\partial x^2} + p\frac{\partial^2\phi}{\partial x\partial z}\right) + \rho\left(\frac{\partial^2\phi}{\partial x\partial y} + q\frac{\partial^2\phi}{\partial x\partial z}\right),$$

$$\frac{\partial f}{\partial y} = \lambda\frac{\partial\phi}{\partial y} + \mu\left(\frac{\partial^2\phi}{\partial x\partial y} + p\frac{\partial^2\phi}{\partial y\partial z}\right) + \rho\left(\frac{\partial^2\phi}{\partial y^2} + q\frac{\partial^2\phi}{\partial y\partial z}\right),$$

$$\frac{\partial f}{\partial z} = \lambda\frac{\partial\phi}{\partial z} + \mu\left(\frac{\partial^2\phi}{\partial x\partial z} + p\frac{\partial^2\phi}{\partial z^2}\right) + \rho\left(\frac{\partial^2\phi}{\partial y\partial z} + q\frac{\partial^2\phi}{\partial z^2}\right);$$

hence

$$\frac{\partial f}{\partial x} + p\frac{\partial f}{\partial z} = \mu\left(\frac{\partial^2\phi}{\partial x^2} + 2p\frac{\partial^2\phi}{\partial x\partial z} + p^2\frac{\partial^2\phi}{\partial z^2}\right)$$
$$+ \rho\left(\frac{\partial^2\phi}{\partial x\partial y} + q\frac{\partial^2\phi}{\partial x\partial z} + p\frac{\partial^2\phi}{\partial y\partial z} + pq\frac{\partial^2\phi}{\partial z^2}\right),$$

$$\frac{\partial f}{\partial y} + q\frac{\partial f}{\partial z} = \mu\left(\frac{\partial^2\phi}{\partial x\partial y} + q\frac{\partial^2\phi}{\partial x\partial z} + p\frac{\partial^2\phi}{\partial y\partial z} + pq\frac{\partial^2\phi}{\partial z^2}\right)$$
$$+ \rho\left(\frac{\partial^2\phi}{\partial y^2} + 2q\frac{\partial^2\phi}{\partial y\partial z} + q^2\frac{\partial^2\phi}{\partial z^2}\right).$$

Now the integral equations of the characteristics are

$$\phi = \phi(x, y, z, a, b) = 0, \quad b = g(a),$$

$$\psi = \frac{\partial\phi}{\partial a} + b'\frac{\partial\phi}{\partial b} \qquad = 0, \quad b' = g'(a).$$

* The equation

$$f = \lambda\phi + \mu\phi_1 + \rho\phi_2$$

can also be obtained by the ordinary theory of elimination.

We have, from foregoing equations,

$$\lambda\left(\frac{\partial\phi}{\partial a}+b'\frac{\partial\phi}{\partial b}\right)+\mu\left(\frac{\partial\phi_1}{\partial a}+b'\frac{\partial\phi_1}{\partial b}\right)+\rho\left(\frac{\partial\phi_2}{\partial a}+b'\frac{\partial\phi_2}{\partial b}\right)=0,$$

and therefore, along any characteristic,

$$\mu\left(\frac{\partial\phi_1}{\partial a}+b'\frac{\partial\phi_1}{\partial b}\right)+\rho\left(\frac{\partial\phi_2}{\partial a}+b'\frac{\partial\phi_2}{\partial b}\right)=0,$$

that is,

$$\mu\left(\frac{\partial\psi}{\partial x}+p\frac{\partial\psi}{\partial z}\right)+\rho\left(\frac{\partial\psi}{\partial y}+q\frac{\partial\psi}{\partial z}\right)=0.$$

But as $\psi=0$ permanently along the characteristics, we have

$$\frac{\partial\psi}{\partial x}dx+\frac{\partial\psi}{\partial y}dy+\frac{\partial\psi}{\partial z}dz=0,$$

that is,

$$\left(\frac{\partial\psi}{\partial x}+p\frac{\partial\psi}{\partial z}\right)dx+\left(\frac{\partial\psi}{\partial y}+q\frac{\partial\psi}{\partial z}\right)dy=0,$$

and therefore

$$\frac{dx}{\mu}=\frac{dy}{\rho}=u,$$

say, where dx and dy are elements of a characteristic. Consequently,

$$\begin{aligned}u\left(\frac{\partial f}{\partial x}+p\frac{\partial f}{\partial z}\right)&=\left(\frac{\partial^2\phi}{\partial x^2}+2p\frac{\partial^2\phi}{\partial x\partial z}+p^2\frac{\partial^2\phi}{\partial z^2}\right)dx\\&\quad+\left(\frac{\partial^2\phi}{\partial x\partial y}+q\frac{\partial^2\phi}{\partial x\partial z}+p\frac{\partial^2\phi}{\partial y\partial z}+pq\frac{\partial^2\phi}{\partial z^2}\right)dy\\&=d\left(\frac{\partial\phi}{\partial x}+p\frac{\partial\phi}{\partial z}\right)-\frac{\partial\phi}{\partial z}dp\\&=-\frac{\partial\phi}{\partial z}dp,\end{aligned}$$

because the relation $\phi_1=\frac{\partial\phi}{\partial x}+p\frac{\partial\phi}{\partial z}=0$ is satisfied in connection with our equations; and similarly

$$u\left(\frac{\partial f}{\partial y}+q\frac{\partial f}{\partial z}\right)=-\frac{\partial\phi}{\partial z}dq.$$

Also

$$u\frac{\partial f}{\partial p}=\frac{\partial\phi}{\partial z}dx,\qquad u\frac{\partial f}{\partial q}=\frac{\partial\phi}{\partial z}dy.$$

Hence, gathering together the various results, we have

$$\frac{dx}{\frac{\partial f}{\partial p}}=\frac{dy}{\frac{\partial f}{\partial q}}=\frac{dz}{p\frac{\partial f}{\partial p}+q\frac{\partial f}{\partial q}}=\frac{dp}{-\left(\frac{\partial f}{\partial x}+p\frac{\partial f}{\partial z}\right)}=\frac{dq}{-\left(\frac{\partial f}{\partial y}+q\frac{\partial f}{\partial z}\right)}.$$

These equations are satisfied along the characteristics; and they have been derived from the integral equations.

Ex. As an illustration of the elimination process, consider the equation

$$\phi=z-ax^2-by^2-ab=0:$$

then

$$\phi_1=p-2ax=0, \quad \phi_2=q-2by=0,$$

and so

$$f=z-\tfrac{1}{2}px-\tfrac{1}{2}qy-\frac{pq}{4xy}=0.$$

Now it is easy to verify that

$$\begin{aligned} z-\tfrac{1}{2}px-\tfrac{1}{2}qy-\frac{pq}{4xy} &\\ =z-ax^2-by^2-ab-\tfrac{1}{2}\left(\frac{b}{x}+x\right)(p-2ax)&\\ -\tfrac{1}{2}\left(\frac{a}{y}+y\right)(q-2by)-\frac{1}{4xy}(p-2ax)(q-2by),& \end{aligned}$$

identically: and therefore

$$f=\phi-\tfrac{1}{2}\left(\frac{b}{x}+x\right)\phi_1-\tfrac{1}{2}\left(\frac{a}{y}+y\right)\phi_2-\frac{1}{4xy}\phi_1\phi_2.$$

Thus

$$-\tfrac{1}{2}x-\frac{q}{4xy}=\frac{\partial f}{\partial p}=-\tfrac{1}{2}\left(\frac{b}{x}+x\right)-\frac{1}{4xy}\phi_2,$$

$$-\tfrac{1}{2}y-\frac{p}{4xy}=\frac{\partial f}{\partial q}=-\tfrac{1}{2}\left(\frac{a}{y}+y\right)-\frac{1}{4xy}\phi_1,$$

which are satisfied in virtue of the equations $\phi_1=0$, $\phi_2=0$: these, together with $\phi=0$, lead to $f=0$.

96. In these discussions the surface, which is connected with a complete integral, has been given by a single equation

$$\phi(x, y, z, a, b)=0.$$

In the general theory of surfaces, it often is convenient to have a parametric representation, whereby x, y, z are expressed in terms of two independent parameters: when this mode is adopted for the representation of the complete integral, we should have equations of the type

$$x=h(u, v, a, b), \quad y=k(u, v, a, b), \quad z=l(u, v, a, b).$$

The single equation of the complete surface would be obtained by the elimination of u and v between these three equations; hence, taking

$$\phi(x, y, z, a, b) = \alpha(x-h) + \beta(y-k) + \gamma(z-l),$$

we have

$$\frac{\partial\phi}{\partial x} = \alpha, \quad \frac{\partial\phi}{\partial y} = \beta, \quad \frac{\partial\phi}{\partial z} = \gamma,$$

$$0 = \alpha\frac{\partial h}{\partial u} + \beta\frac{\partial k}{\partial u} + \gamma\frac{\partial l}{\partial u},$$

$$0 = \alpha\frac{\partial h}{\partial v} + \beta\frac{\partial k}{\partial v} + \gamma\frac{\partial l}{\partial v},$$

$$-\frac{\partial\phi}{\partial a} = \alpha\frac{\partial h}{\partial a} + \beta\frac{\partial k}{\partial a} + \gamma\frac{\partial l}{\partial a},$$

$$-\frac{\partial\phi}{\partial b} = \alpha\frac{\partial h}{\partial b} + \beta\frac{\partial k}{\partial b} + \gamma\frac{\partial l}{\partial b}.$$

The characteristic is given by

$$\frac{\partial\phi}{\partial a} + \frac{\partial\phi}{\partial b}f'(a) = 0, \quad b = f(a);$$

hence

$$\alpha\left(\frac{\partial h}{\partial a} + \frac{\partial h}{\partial b}b'\right) + \beta\left(\frac{\partial k}{\partial a} + \frac{\partial k}{\partial b}b'\right) + \gamma\left(\frac{\partial l}{\partial a} + \frac{\partial l}{\partial b}b'\right) = 0.$$

Accordingly, the equations of the characteristic are

$$\left.\begin{array}{c} x = h, \quad y = k, \quad z = l \\ b = f(a) \\ \dfrac{\partial(h, k, l)}{\partial(u, v, a)} + \dfrac{\partial(h, k, l)}{\partial(u, v, b)}f'(a) = 0 \end{array}\right\};$$

the form earlier considered would be given by the elimination of u, v, b among these five equations.

As regards p and q, we have

$$dz = p\,dx + q\,dy$$

for all variations; hence p and q are given by the equations

$$\left.\begin{array}{l} \dfrac{\partial l}{\partial u} = p\dfrac{\partial h}{\partial u} + q\dfrac{\partial k}{\partial u} \\ \dfrac{\partial l}{\partial v} = p\dfrac{\partial h}{\partial v} + q\dfrac{\partial k}{\partial v} \end{array}\right\}.$$

The differential equation

$$f(x, y, z, p, q) = 0$$

results from the elimination of u, v, a, b between these two equations and

$$x = h, \quad y = k, \quad z = l.$$

Taking

$$f(x, y, z, p, q) = A(x-h) + B(y-k) + C(z-l) + D\left(p\frac{\partial h}{\partial u} + q\frac{\partial k}{\partial u} - \frac{\partial l}{\partial u}\right) + E\left(p\frac{\partial h}{\partial v} + q\frac{\partial k}{\partial v} - \frac{\partial l}{\partial v}\right),$$

and noting the explicit occurrences of variables, as well as the disappearances of parameters and constants, we have

$$\frac{\partial f}{\partial x} = A, \quad \frac{\partial f}{\partial y} = B, \quad \frac{\partial f}{\partial z} = C,$$

$$\frac{\partial f}{\partial p} = D\frac{\partial h}{\partial u} + E\frac{\partial h}{\partial v},$$

$$\frac{\partial f}{\partial q} = D\frac{\partial k}{\partial u} + E\frac{\partial k}{\partial v};$$

$$0 = -A\frac{\partial h}{\partial u} - B\frac{\partial k}{\partial u} - C\frac{\partial l}{\partial u} + D\left(p\frac{\partial^2 h}{\partial u^2} + q\frac{\partial^2 k}{\partial u^2} - \frac{\partial^2 l}{\partial u^2}\right) + E\left(p\frac{\partial^2 h}{\partial u\partial v} + q\frac{\partial^2 k}{\partial u\partial v} - \frac{\partial^2 l}{\partial u\partial v}\right),$$

$$0 = -A\frac{\partial h}{\partial v} - B\frac{\partial k}{\partial v} - C\frac{\partial l}{\partial v} + D\left(p\frac{\partial^2 h}{\partial u\partial v} + q\frac{\partial^2 k}{\partial u\partial v} - \frac{\partial^2 l}{\partial u\partial v}\right) + E\left(p\frac{\partial^2 h}{\partial v^2} + q\frac{\partial^2 k}{\partial v^2} - \frac{\partial^2 l}{\partial v^2}\right),$$

$$0 = -A\frac{\partial h}{\partial a} - B\frac{\partial k}{\partial a} - C\frac{\partial l}{\partial a} + D\left(p\frac{\partial^2 h}{\partial u\partial a} + q\frac{\partial^2 k}{\partial u\partial a} - \frac{\partial^2 l}{\partial u\partial a}\right) + E\left(p\frac{\partial^2 h}{\partial v\partial a} + q\frac{\partial^2 k}{\partial v\partial a} - \frac{\partial^2 l}{\partial v\partial a}\right),$$

$$0 = -A\frac{\partial h}{\partial b} - B\frac{\partial k}{\partial b} - C\frac{\partial l}{\partial b} + D\left(p\frac{\partial^2 h}{\partial u\partial b} + q\frac{\partial^2 k}{\partial u\partial b} - \frac{\partial^2 l}{\partial u\partial b}\right) + E\left(p\frac{\partial^2 h}{\partial v\partial b} + q\frac{\partial^2 k}{\partial v\partial b} - \frac{\partial^2 l}{\partial v\partial b}\right).$$

The last four equations determine the ratios $A : B : C : D : E$. The equations of the characteristic, being

$$\frac{dx}{D\frac{\partial h}{\partial u} + E\frac{\partial h}{\partial v}} = \frac{dy}{D\frac{\partial k}{\partial u} + E\frac{\partial k}{\partial v}} = \frac{dp}{-(A+pC)} = \frac{dq}{-(B+qC)} = \frac{dz}{D\frac{\partial l}{\partial u} + E\frac{\partial l}{\partial v}},$$

on substituting for $\frac{\partial f}{\partial x}, \frac{\partial f}{\partial y}, \frac{\partial f}{\partial z}, \frac{\partial f}{\partial p}, \frac{\partial f}{\partial q}$, take the simple form

$$\frac{du}{D} = \frac{dv}{E} = \frac{dp}{-(A + pC)} = \frac{dq}{-(B + qC)},$$

when the values of x, y, z are inserted. While these are the general equations of the characteristic, it is clear that the direction of the characteristic at any point on the surface is given by

$$\frac{du}{D} = \frac{dv}{E},$$

where the quantity $D:E$ is given by the preceding equations and where the magnitudes p and q, which occur in that quantity, are

$$\frac{p}{\dfrac{\partial(l, k)}{\partial(u, v)}} = \frac{q}{\dfrac{\partial(h, l)}{\partial(u, v)}} = \frac{1}{\dfrac{\partial(h, k)}{\partial(u, v)}}.$$

Edge of Regression: Integral Curves.

97. Returning now to a complete integral surface

$$\phi(x, y, z, a, b) = 0,$$

we have the characteristics given by

$$\phi = 0, \quad b = g(a), \quad \frac{d\phi}{da} = \frac{\partial\phi}{\partial a} + \frac{\partial\phi}{\partial b} g'(a) = 0.$$

As has been seen, these equations represent the general integral surface when a and b are eliminated, this surface being the locus of the characteristics. When a and b are not eliminated, the equations represent the characteristics which lie on this general integral surface.

Now take a characteristic on the general integral determined by a value a, and a neighbouring characteristic determined by a value $a + \delta a$, where δa is infinitesimal: the equations of the former are

$$\phi = 0, \quad \frac{d\phi}{da} = 0,$$

and those of the latter can be taken

$$\phi + \frac{d\phi}{da}\delta a = 0, \quad \frac{d\phi}{da} + \frac{d^2\phi}{da^2}\delta a = 0,$$

keeping δa sufficiently small; hence the ultimate intersection of two consecutive characteristics is given by

$$\phi = 0, \quad \frac{d\phi}{da} = 0, \quad \frac{d^2\phi}{da^2} = 0.$$

The three equations determine the point of intersection, giving its coordinates as functions of a. When a is eliminated between the three equations, the two resulting equations give a curve, which is the locus of the ultimate intersections of the characteristics. This curve lies on the surface

$$\phi = 0, \quad \frac{d\phi}{da} = 0,$$

which represents the general integral; on the analogy of developable surfaces, Monge called* it the *edge of regression* of the general integral surface.

This locus may be regarded as the envelope of the characteristics on the general integral: for it touches a characteristic at its point of ultimate intersection with its neighbour. To verify this statement, we note that the curve passes through the point: for it is the locus of such points. Further, to obtain its tangent at the point, we assume that, for it, a is determined as a function of x, y, z by means of the equation

$$\frac{d^2\phi}{da^2} = 0,$$

and that the value of a is substituted in the other two equations: the values of $dx : dy : dz$, derived from them, are then given by

$$d\phi + \frac{d\phi}{da}\,da = 0, \quad d\frac{d\phi}{da} + \frac{d^2\phi}{da^2}\,da = 0,$$

where

$$d\phi = \frac{\partial\phi}{\partial x}\,dx + \frac{\partial\phi}{\partial y}\,dy + \frac{\partial\phi}{\partial z}\,dz,$$

and so for the others: that is, the values are given by

$$d\phi = 0, \quad d\frac{d\phi}{da} = 0,$$

which are the equations determining the ratios $dx : dy : dz$ for the characteristic at the point: hence the two curves touch at the point. Thus all the characteristics on the general integral touch the edge of regression.

* *Application de l'Analyse à la Géométrie*, § VI: the last edition was edited by Liouville in 1850.

Ex. The simplest example of all arises when the complete integral is a double family of planes

$$z = ax + by + f(a, b).$$

The general integral is of the form

$$\left.\begin{aligned} z &= x\theta(a) + y\phi(a) + \psi(a) \\ 0 &= x\theta'(a) + y\phi'(a) + \psi'(a) \end{aligned}\right\},$$

being a developable surface, for it is the envelope of a plane whose equation contains one parameter. The characteristics are the generators: they are the intersections of consecutive planes, and each is the curve of contact of a plane with the general integral. The envelope of the generators is the edge of regression of the developable surface.

98. We have seen that the tangent to the characteristic at any point P is a generator of the cone T associated with P; hence, as the equation of this cone is

$$F\left(x, y, z, \frac{z' - z}{x' - x}, \frac{y' - y}{x' - x}\right) = 0,$$

P being the point x, y, z, the tangents to the characteristic, and therefore the characteristic itself at the point, satisfy the equation

$$F\left(x, y, z, \frac{dz}{dx}, \frac{dy}{dx}\right) = 0.$$

Curves satisfying this equation are called *integral curves.*

It is clear that characteristics are included among these integral curves: it is equally clear that they are not the most general integral curves, because $F = 0$ is only a single equation involving two unknown quantities and one of these can be assumed arbitrarily, the equation then determining the other. Properties sufficient to distinguish characteristics among integral curves have already (§§ 92, 93) been given.

The edge of regression of the general integral is easily seen to be an integral curve: for, at any point on it, the tangent is the same as that of the characteristic which touches it there. The latter at the point satisfies $F = 0$: hence, also, the equations of the edge of regression satisfy $F = 0$.

Conversely, every integral curve can be obtained as an edge of regression. Owing to the equation, the tangents to the integral curves through the point are the generators of the cone associated with the point. Taking any integral curve, its tangent is a generator of the cone, and the direction of the generator determines a characteristic through the point, which characteristic accordingly touches the integral curve at the point: moreover, it is the only characteristic which can be drawn through the point. Now it has been proved (Ex. 1, § 92) that a general integral passing through a curve is generated as the locus of the characteristics drawn through the points of the curve: and the result is not affected by the angle of intersection between the curve and the characteristic. Hence in the present case, taking the aggregate of the characteristics tangent to the integral curve, we have a general integral surface: on that surface, the integral curve is the envelope of the characteristics and therefore is an edge of regression.

Hence we may take

$$\phi = 0, \quad \frac{d\phi}{da} = 0, \quad \frac{d^2\phi}{da^2} = 0,$$

as the comprehensive integral curve satisfying the equation $F = 0$.

The contact relations of the various integral surfaces with one another will be discussed later: it is worth noting the contact relations of these integral curves with the surface $\phi = 0$.

At a point along the integral curve determined by the relations

$$\phi = 0, \quad \frac{d\phi}{da} = 0, \quad \frac{d^2\phi}{da^2} = 0,$$

we have

$$d\phi + \frac{d\phi}{da}\, da = 0,$$

$$d\,\frac{d\phi}{da} + \frac{d^2\phi}{da^2}\, da = 0,$$

that is,

$$d\phi = 0, \quad d\,\frac{d\phi}{da} = 0.$$

Again, because $d\phi = 0$, we have

$$d\,(d\phi) = 0,$$

that is,

$$d^2\phi + \left(d\,\frac{d\phi}{da}\right) da = 0,$$

and therefore

$$d^2\phi = 0.$$

Consequently, along the integral curve, we have

$$\phi = 0, \quad d\phi = 0, \quad d^2\phi = 0:$$

shewing that the integral surface $\phi = 0$ is cut in three consecutive points by the integral curve. Thus the edge of regression of the general integral has contact of the second order with the complete integral surface from which it originates.

The edge of regression of a general integral is given by the equations

$$\phi = 0, \quad \frac{d\phi}{da} = 0, \quad \frac{d^2\phi}{da^2} = 0;$$

the quantity b is any function of a, and the equations involve b', b'', the first and the second derivatives of b. As there thus arises a curve associated with any assumed function, there thus will arise an infinitude of such curves associated with all forms of the function.

Let those curves among this infinitude be selected which are such that

$$\frac{d^3\phi}{da^3} = 0;$$

that such curves do, in general, exist can be seen as follows. For variations along any edge of regression, the ratios $dx : dy : dz$ satisfy the equations

$$d\phi + \frac{d\phi}{da} da = 0,$$

$$d\frac{d\phi}{da} + \frac{d^2\phi}{da^2} da = 0,$$

$$d\frac{d^2\phi}{da^2} + \frac{d^3\phi}{da^3} da = 0;$$

hence, if the edge of regression be such that

$$\frac{d^3\phi}{da^3} = 0,$$

its direction at the point is given by

$$d\phi = 0, \quad d\frac{d\phi}{da} = 0, \quad d\frac{d^2\phi}{da^2} = 0.$$

Between these three equations and the equations

$$\phi = 0, \quad \frac{d\phi}{da} = 0, \quad \frac{d^2\phi}{da^2} = 0,$$

we eliminate a, b, b', b'', and we have two resulting equations involving $dx : dy : dz$. The two differential equations (which will be algebraical in form when ϕ is algebraical) define curves: so that curves of the specified type do exist in general. Further, along such a curve x, y, z, $dx : dy : dz$ are expressible in terms of a parametric variable, say t: hence at the point, the foregoing equations, when resolved, will express both a and b in terms of t and so, on the elimination of t between the expressions, will give b in terms of a.

To find the order of contact of this curve with the integral surface from which it originates, we proceed as before. Along the curve, we have

$$d\phi = 0, \quad d\frac{d\phi}{da} = 0, \quad d\frac{d^2\phi}{da^2} = 0.$$

Also, since $d\phi = 0$, we have

$$d\,(d\phi) = 0,$$

that is,

$$d^2\phi + \left(d\frac{d\phi}{da}\right) da = 0,$$

and therefore

$$d^2\phi = 0.$$

And, since $d\dfrac{d\phi}{da} = 0$, we have

$$d\left(d\frac{d\phi}{da}\right) = 0,$$

that is,

$$d^2\frac{d\phi}{da} + \left(d\frac{d^2\phi}{da^2}\right) da = 0,$$

so that

$$d^2\frac{d\phi}{da} = 0.$$

And, since $d^2\phi = 0$, we have

$$d\,(d^2\phi) = 0,$$

that is,

$$d^3\phi + \left(d^2\frac{d\phi}{da}\right) da = 0,$$

so that

$$d^3\phi = 0.$$

Hence along the curve in question, we have

$$\phi = 0, \quad d\phi = 0, \quad d^2\phi = 0, \quad d^3\phi = 0,$$

at the point: consequently, the curve in question meets the integral surface in four consecutive points: that is, the curve has contact of the third order with the surface.

Ex. 1. The equation of a sphere, that is absolutely unrestricted, contains four arbitrary independent constants: hence, when it is made subject to two independent conditions, the equation will contain two arbitrary constants and may be regarded as giving the complete integral of some partial differential equation.

To obtain a general integral, we select a family of these spheres and construct their envelope. The characteristics, being the intersections of consecutive spheres, are circles: each sphere touches the envelope surface along a circle. The edge of regression on the envelope surface, which is the general integral, is itself the envelope of the characteristic circles; and the earlier investigation shewed that, where the sphere meets this edge of regression, it meets the edge in three consecutive points.

But there is one general integral for which the contact is closer. By an appropriate choice of a functional relation between the two constants in the complete integral, we obtain the envelope of one selected family of spheres: each sphere, where it meets the edge of regression on this envelope surface, meets it in four consecutive points, and therefore is its osculating sphere. The characteristics are the circles which are the intersections of consecutive spheres: hence, for this general integral, they are the osculating circles of the edge of regression.

Ex. 2. As a particular instance of the last example, construct the various equations of the complete integral, the general integral, and the selected general integral, when the edge of regression is a regular helix; and obtain the partial differential equation satisfied by the integrals.

Ex. 3. In the explanations in the general theory, all the surfaces and curves concerned are unrestricted in properties; it has been assumed that the various contacts are possible.

Consider, in particular, a plane. Its equation involves three independent constants when completely unrestricted: if the plane be subject to one condition, two independent constants will remain in the equation, which then can be regarded as giving the complete integral of a partial differential equation. To obtain a general integral, we make one of the constants an arbitrary function of the other and proceed to obtain the envelope of the planes so selected. As their equation involves one arbitrary parameter, this envelope is a developable surface: the characteristics are the generators, being the intersections of consecutive planes; and each plane osculates the edge of regression of the surface.

But it is not possible to select a general integral so as to have closer contact between the plane and its edge of regression: because not more than three consecutive points of a curve can lie in a plane, unless at a singular point, or unless the curve be a plane curve.

In this case, the selected general integral of § 98 does not occur.

99. In discussing the selected general integral, and its edge of regression with which a complete integral has triple contact, the complete integral is supposed known; and the equations of the edge of regression in question are deduced from that of the complete integral. It is, however, possible to deduce equations of the particular edge of regression from the original differential equation.

For the purpose it is sufficient to note that, at a point on the particular edge of regression, there is triple contact between the curve and the complete integral surface; and therefore, when the equations

$$f(x, y, z, p, q) = 0, \quad dz = p\,dx + q\,dy,$$

are regarded as determining p and q in terms of x, y, z, dx, dy, dz, (supposed known, as belonging to the required curve), they must provide a triple root. Hence, on writing

$$z' = \frac{dz}{dx}, \quad y' = \frac{dy}{dx},$$

the equation

$$f(x, y, z, z' - qy', q) = 0$$

must provide a triple root; so that

$$\frac{\partial f}{\partial q} - y' \frac{\partial f}{\partial p} = 0,$$

$$\frac{\partial^2 f}{\partial q^2} - 2y' \frac{\partial^2 f}{\partial p \partial q} + y'^2 \frac{\partial^2 f}{\partial p^2} = 0.$$

The elimination of q, between the last two equations and

$$f(x, y, z, z' - qy', q) = 0,$$

leads to two equations which are the (ordinary differential) equations of the curve in question.

It is clear that, if the equation is rational and integral in p and q, it must be of at least the third degree if the curve in question is to arise.

Ex. 1. Prove that, if such a curve exists, it touches the edges of regression of the cones T and that its tangents are perpendicular to the planes of inflexion of the cones N.

(Darboux.)

Ex. 2. Discuss the various loci, indicated in the preceding sections, to be associated with the partial differential equation, the complete integral of which is

$$(1 - a^2)\,x + (1 + a^2)\,kz + 2ay + b = 0,$$

where a and b are arbitrary constants, and k is a pure constant.

(Goursat.)

Discuss also the integrals of the equation which has

$$\left.\begin{aligned} ax^2+by^2+cz^2=1 \\ a^3+b^3+c^3=1 \end{aligned}\right\}$$

for its complete primitive.

Ex. 3. It has been proved that every integral curve (§ 98) can be represented as an edge of regression of a general integral: it has also been proved that an integral surface touching an edge of regression has contact of the second order: hence every integral curve, touching an integral surface, has contact of the second order with that surface.

Verify this proposition directly from the equations.

(Lie.)

Lie's Classification of Equations.

100. In the discussion of the characteristics, regard has been paid chiefly to their association with the surface represented by the general integral: but they can be considered also in their association with the surface represented by the complete integral. It is in this connection that Lie has considered them*, in particular, classifying partial differential equations according to the nature of the characteristics as curves upon the complete integral surface. Some of his results can be obtained very simply as follows.

Among the various curves that can be drawn upon a surface, three of the most important classes are asymptotic lines (being the lines touched by the principal tangents of the surface at successive points), lines of curvature, and geodesics; and, accordingly, Lie investigates those partial differential equations of the first order, the surface integrals of which have characteristics belonging to one of these three classes of curves upon the surface.

(i) The directions of the asymptotic lines upon a surface at any point, being the directions of the principal tangents at the point, are given by

$$pdx+qdy=dz,$$

$$rdx^2+2sdxdy+tdy^2= 0,$$

in the usual notation: as

$$dp=rdx+sdy, \quad dq=sdx+tdy,$$

the latter equation is

$$dpdx+dqdy=0.$$

* *Math. Ann.* t. v (1872), pp. 188—200.

If these are the characteristics of the surface integral of an equation

$$f(x, y, z, p, q) = 0,$$

they must accord with the ordinary differential equations of the characteristics (§ 94). Hence

$$\left(\frac{\partial f}{\partial x} + p\frac{\partial f}{\partial z}\right)\frac{\partial f}{\partial p} + \left(\frac{\partial f}{\partial y} + q\frac{\partial f}{\partial z}\right)\frac{\partial f}{\partial q} = 0;$$

and this condition is sufficient, as well as necessary, to secure the property. Accordingly, any function f, which satisfies this condition, will lead to a differential equation possessing the property that *the characteristics are asymptotic lines upon the integral surface.*

Ex. Verify that a complete integral of the foregoing equation, which must be satisfied by f, is

$$f = ax + by - \frac{ac+b}{cp+q}z + g(cp+q) + h,$$

where a, b, c, g, h are arbitrary constants: obtain other integrals of that equation: and discuss the surface integrals of the equation $f=0$ for the respective forms of f.

(ii) The directions of the lines of curvature at any point of a surface are given by

$$pdx + qdy = dz,$$

$$(dx + pdz)\,dq = (dy + q\,dz)\,dp.$$

If these are the characteristics of the surface integral of an equation

$$f(x, y, z, p, q) = 0,$$

they must accord with the ordinary differential equations of the characteristics; hence (§ 94)

$$\left(\frac{\partial f}{\partial y} + q\frac{\partial f}{\partial z}\right)\left\{\frac{\partial f}{\partial p} + p\left(p\frac{\partial f}{\partial p} + q\frac{\partial f}{\partial q}\right)\right\}$$
$$= \left(\frac{\partial f}{\partial x} + p\frac{\partial f}{\partial z}\right)\left\{\frac{\partial f}{\partial q} + q\left(p\frac{\partial f}{\partial p} + q\frac{\partial f}{\partial q}\right)\right\},$$

that is,

$$\frac{\partial f}{\partial x}\left\{\frac{\partial f}{\partial q} + q\left(p\frac{\partial f}{\partial p} + q\frac{\partial f}{\partial q}\right)\right\} - \frac{\partial f}{\partial y}\left\{\frac{\partial f}{\partial p} + p\left(p\frac{\partial f}{\partial p} + q\frac{\partial f}{\partial q}\right)\right\}$$
$$+ \frac{\partial f}{\partial z}\left(p\frac{\partial f}{\partial q} - q\frac{\partial f}{\partial p}\right) = 0;$$

and this condition is sufficient, as well as necessary, to secure the property. Accordingly, any function f, which satisfies this condition, will lead to a differential equation possessing the property that *the characteristics are lines of curvature upon the integral surface.*

(iii) The determination of equations such that the integral surfaces have geodesics for their characteristics is, at first sight, a problem leading to equations of the second order. Geodesics are given by the equations

$$\frac{1}{p}\frac{d^2x}{ds^2} = \frac{1}{q}\frac{d^2y}{ds^2} = -\frac{d^2z}{ds^2};$$

and these can be replaced by the equations

$$\frac{dz}{dt} = p\frac{dx}{dt} + q\frac{dy}{dt},$$

$$(1 + p^2 + q^2)\left(\frac{d^2x}{dt^2}\frac{dy}{dt} - \frac{d^2y}{dt^2}\frac{dx}{dt}\right)$$
$$= \left(q\frac{dx}{dt} - p\frac{dy}{dt}\right)\left(\frac{dp}{dt}\frac{dx}{dt} + \frac{dq}{dt}\frac{dy}{dt}\right),$$

where t is any variable. When the partial differential equation is

$$f(x, y, z, p, q) = 0,$$

the characteristics are given by

$$\frac{dx}{dt} = \frac{\partial f}{\partial p}, \quad \frac{dy}{dt} = \frac{\partial f}{\partial q}, \quad \frac{dz}{dt} = p\frac{\partial f}{\partial p} + q\frac{\partial f}{\partial q},$$

$$\frac{dp}{dt} = -\frac{\partial f}{\partial x} - p\frac{\partial f}{\partial z}, \quad \frac{dq}{dt} = -\frac{\partial f}{\partial y} - q\frac{\partial f}{\partial z}.$$

The first of the two equations for the geodesics is satisfied identically; when substitution is effected in the other equation, it becomes a partial differential equation of the second order—a result to be expected, when the curvature property of the geodesic is used.

But an equation of the first order can be obtained, by using the known properties of geodesic parallels and their orthogonal geodesics*. The element of arc upon the surface is given by

$$ds^2 = (1 + p^2)\,dx^2 + 2pq\,dx\,dy + (1 + q^2)\,dy^2;$$

* Darboux, *Théorie générale des surfaces*, t. II, pp. 424 et seq.

if θ be an integral of the equation

$$(1+q^2)\left(\frac{\partial\theta}{\partial x}\right)^2 - 2pq\,\frac{\partial\theta}{\partial x}\,\frac{\partial\theta}{\partial y} + (1+p^2)\left(\frac{\partial\theta}{\partial y}\right)^2 = 1 + p^2 + q^2,$$

containing a non-additive constant a, the geodesics are given by

$$\frac{\partial\theta}{\partial a} = \text{constant}.$$

On the basis of this property, the equation of the required surfaces can be constructed.

Let $\phi(x, y, z)$ denote any function of x, y, z: and let the (unknown) value of z of the surface be supposed substituted in ϕ, so that it becomes the foregoing function θ; then

$$\frac{\partial\theta}{\partial x} = \frac{\partial\phi}{\partial x} + p\,\frac{\partial\phi}{\partial z}, \quad \frac{\partial\theta}{\partial y} = \frac{\partial\phi}{\partial y} + q\,\frac{\partial\phi}{\partial z}.$$

When these values are substituted in the above equation, it becomes

$$(1+q^2)\left(\frac{\partial\phi}{\partial x}\right)^2 + (1+p^2)\left(\frac{\partial\phi}{\partial y}\right)^2 + (p^2+q^2)\left(\frac{\partial\phi}{\partial z}\right)^2$$
$$+ 2p\,\frac{\partial\phi}{\partial x}\,\frac{\partial\phi}{\partial z} + 2q\,\frac{\partial\phi}{\partial y}\,\frac{\partial\phi}{\partial z} - 2pq\,\frac{\partial\phi}{\partial x}\,\frac{\partial\phi}{\partial y} = 1 + p^2 + q^2,$$

or, what is an equivalent form,

$$\left(p\,\frac{\partial\phi}{\partial x} + q\,\frac{\partial\phi}{\partial y} - \frac{\partial\phi}{\partial z}\right)^2 = (1+p^2+q^2)\left\{\left(\frac{\partial\phi}{\partial x}\right)^2 + \left(\frac{\partial\phi}{\partial y}\right)^2 + \left(\frac{\partial\phi}{\partial z}\right)^2 - 1\right\}.$$

When a value of z is obtained so as to satisfy this equation, and when it is substituted in the assumed function $\phi(x, y, z)$, the latter becomes a quantity θ, such that $\theta = \text{constant}$ gives a family of parallel curves. The orthogonal geodesics on the surface are given by

$$\frac{\partial\theta}{\partial a} = \text{constant}$$
$$= 0,$$

say, for one geodesic, where a is a non-additive constant in θ. Now

$$\phi(x, y, z) = \theta(x, y, a),$$

and the quantity a has entered only through z, as given by the integral of the foregoing equation: hence

$$\frac{\partial\phi}{\partial z}\,\frac{dz}{da} = \frac{\partial\theta}{\partial a} = 0,$$

so that, as $\frac{\partial\phi}{\partial z}$ is not zero, we have

$$\frac{dz}{da}=0;$$

and this gives a geodesic on the surface. But

$$z=\text{function }(x, y, a),$$
$$\frac{dz}{da}=0,$$

are the equations of the characteristic: hence *the surfaces, obtained by integrating the equation*

$$\left(p\frac{\partial\phi}{\partial x}+q\frac{\partial\phi}{\partial y}-\frac{\partial\phi}{\partial z}\right)^2=(1+p^2+q^2)\left\{\left(\frac{\partial\phi}{\partial x}\right)^2+\left(\frac{\partial\phi}{\partial y}\right)^2+\left(\frac{\partial\phi}{\partial z}\right)^2-1\right\},$$

where $\phi(x, y, z)$ *is any function of* x, y, z, *have geodesics for their characteristics**.

Ex. 1. Shew that, if the normals to a surface touch the sphere $x^2+y^2+z^2=1$, its equation satisfies the partial differential equation

$$(x^2+y^2+z^2-1)(p^2+q^2+1)=(z-px-qy)^2;$$

integrate this equation, discussing the characteristics and the edge of regression. (Monge.)

Ex. 2. If the characteristics of a non-linear equation are straight lines, the equation is of the form

$$z=px+qy+f(p, q).$$

(Goursat.)

Ex. 3. Shew that the characteristics of the equation

$$p+a=(q+b)f(x, y, z)$$

are curves in parallel planes; and indicate how to form the equation of surfaces whose characteristics are plane curves.

* These results are due to Lie, who gives other properties in his memoir quoted on p. 244. The method of establishment differs from Lie's, which is based upon properties of complexes of lines and curves.

CHAPTER VII.

Singular Integrals and their Geometrical Properties: Singularities of the Characteristics.

The authorities used in the construction of this chapter have been quoted at the beginning of the preceding chapter.

101. We now proceed to consider the relations of the Singular Integral (which will be assumed to exist) with the various integral surfaces and curves that have been discussed. That Singular Integral is given by the equation, which results from the elimination* of a and b between the equations

$$\phi(x, y, z, a, b) = 0, \quad \frac{\partial \phi}{\partial a} = 0, \quad \frac{\partial \phi}{\partial b} = 0,$$

or by part of that resulting equation; the value of z must be such that the differential equation is satisfied. In that case, the resulting equation (or the part of the resulting equation) represents the envelope which then is possessed by the family of complete integral surfaces: as the values of p and q are the same

* It is to be borne in mind that the singular integral is assumed to exist, so that it will arise as indicated in the text. The result of the elimination in general is not to give an integral of the equation; the eliminant contains the locus of conical points (if any), the locus of double lines (if any), and other loci, which are not integrals of the differential equation. For a discussion of such matters, which are not our concern at this stage, reference may be made to a memoir by M. J. M. Hill, *Phil. Trans.* A (1892), pp. 141—278.

Moreover, it is assumed that the elimination is possible, so that the three equations are independent of one another. This condition, however, is not always satisfied; and it is easy to construct exceptions of the type

$$\phi = \lambda \frac{\partial \phi}{\partial a} + \mu \frac{\partial \phi}{\partial b},$$

which would make the three equations equivalent to two only. Again, such matters are not our present concern: we assume that the elimination is possible, and that it leads to the singular integral.

at any point, common to the envelope and the complete integral surface, the latter surface is touched by the envelope at a common point. Let the envelope be denoted by E.

The three equations are equivalent to the equation of E and to values (or to sets of values) of a and b: by each such set of values of a and b, there is given a point on E where it is touched by $\phi = 0$.

Take any point P on E, and let the values of a and b at P be denoted by a_0 and b_0. A characteristic through P is given by

$$\phi(x, y, z, a_0, b_0) = 0, \quad b_0 = f(a_0),$$

$$\frac{\partial \phi}{\partial a_0} + f'(a_0)\frac{\partial \phi}{\partial b_0} = 0;$$

but these equations are satisfied, whatever be the form $f(a_0)$, because the equations

$$\frac{\partial \phi}{\partial a_0} = 0, \quad \frac{\partial \phi}{\partial b_0} = 0$$

are satisfied at the point; hence an infinitude of characteristics passes through any point on the envelope. Moreover, all these characteristics through P touch E there: for they pass through P as a point on the complete surface which touches E at the point.

Take any two curves PT and PT' on E passing through P; let T and T' denote points consecutive to P on those curves respectively; also let $a_0 + da_0$, $b_0 + db_0$ be the values of a and b for T, and $a_0 + \delta a_0$, $b_0 + \delta b_0$ be their values for T'. Then along PT, the values at P of the differential elements dx, dy, dz are given by

$$0 = \frac{\partial \phi}{\partial x} dx + \frac{\partial \phi}{\partial y} dy + \frac{\partial \phi}{\partial z} dz + \frac{\partial \phi}{\partial a_0} da_0 + \frac{\partial \phi}{\partial b_0} db_0,$$

$$0 = \frac{\partial^2 \phi}{\partial a_0 \partial x} dx + \frac{\partial^2 \phi}{\partial a_0 \partial y} dy + \frac{\partial^2 \phi}{\partial a_0 \partial z} dz + \frac{\partial^2 \phi}{\partial a_0^2} da_0 + \frac{\partial^2 \phi}{\partial a_0 \partial b_0} db_0,$$

$$0 = \frac{\partial^2 \phi}{\partial b_0 \partial x} dx + \frac{\partial^2 \phi}{\partial b_0 \partial y} dy + \frac{\partial^2 \phi}{\partial b_0 \partial z} dz + \frac{\partial^2 \phi}{\partial a_0 \partial b_0} da_0 + \frac{\partial^2 \phi}{\partial b_0^2} db_0;$$

and the last two terms in the first equation vanish. Now the complete integral touching E at T' cuts the complete integral touching E at P in the characteristic whose equations are

$$\phi = 0, \quad \frac{\partial \phi}{\partial a_0} + \frac{\delta b_0}{\delta a_0}\frac{\partial \phi}{\partial b_0} = 0;$$

and the direction of the tangent to this characteristic is given by the two equations

$$0 = \frac{\partial \phi}{\partial x} dx + \frac{\partial \phi}{\partial y} dy + \frac{\partial \phi}{\partial z} dz,$$

$$0 = \frac{\partial^2 \phi}{\partial a_0 \partial x} dx + \frac{\partial^2 \phi}{\partial a_0 \partial y} dy + \frac{\partial^2 \phi}{\partial a_0 \partial z} dz$$
$$+ \frac{\delta b_0}{\delta a_0} \left(\frac{\partial^2 \phi}{\partial b_0 \partial x} dx + \frac{\partial^2 \phi}{\partial b_0 \partial y} dy + \frac{\partial^2 \phi}{\partial b_0 \partial z} dz \right).$$

If then this characteristic touches PT at P, the last two equations giving the ratios $dx : dy : dz$ must be satisfied by the values of dx, dy, dz belonging to PT: hence

$$\frac{\partial^2 \phi}{\partial a_0^2} da_0 + \frac{\partial^2 \phi}{\partial a_0 \partial b_0} db_0 + \frac{\delta b_0}{\delta a_0} \left(\frac{\partial^2 \phi}{\partial a_0 \partial b_0} da_0 + \frac{\partial^2 \phi}{\partial b_0^2} db_0 \right) = 0,$$

that is,

$$\frac{\partial^2 \phi}{\partial a_0^2} da_0 \delta a_0 + \frac{\partial^2 \phi}{\partial a_0 \partial b_0} (db_0 \delta a_0 + da_0 \delta b_0) + \frac{\partial^2 \phi}{\partial b_0^2} db_0 \delta b_0 = 0.$$

Moreover, this relation is symmetrical between the two sets of differential elements that are associated with T and T' respectively. Hence we have Darboux's theorem*:

If we take any direction PT' through a point P on E which represents the singular integral, and if PT be the direction of the characteristic which is the intersection of the complete integrals touching E at consecutive points P and T', then PT' is the direction of the characteristic which is the intersection of the complete integrals touching E at consecutive points P and T.

Characteristics, in directions such as PT and PT' at P, may be called *conjugate*: obviously, any characteristic has a conjugate.

When a characteristic coincides with its conjugate, so that it may be called *self-conjugate*, its direction at the point on the envelope is given by

$$\frac{\partial^2 \phi}{\partial a_0^2} da_0^2 + 2 \frac{\partial^2 \phi}{\partial a_0 \partial b_0} da_0 db_0 + \frac{\partial^2 \phi}{\partial b_0^2} db_0^2 = 0.$$

Hence, in general, there are two sets of curves upon the envelope such that, at every point upon each of them, the tangent characteristic is self-conjugate; such curves may be called *asym-*

* *L.c.*, p. 60.

ptotic curves. The two curves through any point touch one another, if the equation

$$\frac{\partial^2\phi}{\partial a_0^2}\frac{\partial^2\phi}{\partial b_0^2}=\left(\frac{\partial^2\phi}{\partial a_0\partial b_0}\right)^2$$

is satisfied at the point. If the equation is satisfied everywhere upon the envelope, then there is only one asymptotic curve through a point, and there is only a single set of such curves upon the surface.

These results are the analogue of the results in the ordinary theory of surfaces: the singular integral corresponds to a surface, the complete integrals correspond to the tangent planes, the characteristics to the tangent lines, conjugate directions to conjugate directions, and asymptotic curves to asymptotic curves.

102. These properties of the singular integral have been derived from the complete integral with its associated curves and surfaces: they can be used to bring the singular integral into relation with the original partial equation

$$f(x, y, z, p, q)=0.$$

We have seen that, through any point on the singular integral surface, there passes an infinitude of tangent characteristics: the direction therefore of a characteristic through such a point, as given by

$$\frac{dx}{\frac{\partial f}{\partial p}}=\frac{dy}{\frac{\partial f}{\partial q}}=\frac{dz}{p\frac{\partial f}{\partial p}+q\frac{\partial f}{\partial q}},$$

must be indeterminate, and so

$$\frac{\partial f}{\partial p}=0,\quad \frac{\partial f}{\partial q}=0.$$

Moreover, the equation must be identically satisfied when the values of z, p, q belonging to the singular integral at the point are substituted; hence

$$\frac{\partial f}{\partial x}+p\frac{\partial f}{\partial z}+r\frac{\partial f}{\partial p}+s\frac{\partial f}{\partial q}=0,$$

$$\frac{\partial f}{\partial y}+q\frac{\partial f}{\partial z}+s\frac{\partial f}{\partial p}+t\frac{\partial f}{\partial q}=0,$$

that is, in connection with the former equations, we must have

$$\frac{\partial f}{\partial x}+p\frac{\partial f}{\partial z}=0,\quad \frac{\partial f}{\partial y}+q\frac{\partial f}{\partial z}=0.$$

We thus have the former aggregate of equations: every singular integral must satisfy the equations

$$f=0,\quad \frac{\partial f}{\partial p}=0,\quad \frac{\partial f}{\partial q}=0,\quad \frac{\partial f}{\partial x}+p\frac{\partial f}{\partial z}=0,\quad \frac{\partial f}{\partial y}+q\frac{\partial f}{\partial z}=0.$$

It does not follow that these equations definitely determine a singular integral. We have already indicated one class of exceptions, when the five equations are satisfied in virtue of two only, so that p and q cannot be eliminated among them. But, as pointed* out by Darboux, there may be other exceptions. Thus, suppose the five equations do determine z, p, q as functions of x and y; then, along the surface expressing z in terms of x and y, the equation

$$\frac{\partial f}{\partial x}dx+\frac{\partial f}{\partial y}dy+\frac{\partial f}{\partial z}dz+\frac{\partial f}{\partial p}dp+\frac{\partial f}{\partial q}dq=0$$

is satisfied, that is, in virtue of the five equations, we must have

$$\frac{\partial f}{\partial z}(dz-p\,dx-q\,dy)=0.$$

If $\frac{\partial f}{\partial z}$ does not vanish, by virtue of the values of z, p, q in terms of x and y, then

$$dz-p\,dx-q\,dy=0;$$

the relation between z, x, y is then a singular integral. But if $\frac{\partial f}{\partial z}$ does vanish by virtue of these values, we are not justified in making the inference; the general case then requires separate consideration, though in any particular instance the test of satisfying the equation can be applied immediately.

Ex. 1. Prove that the normal to the singular integral surface is a double line on the cone N of normals at the point. (Darboux.)

Ex. 2. Obtain the partial differential equation of the first order which has

$$(z+ax)(z+by)=x^2+y^2$$

for a complete integral. Does it possess a singular integral?

Discuss the characteristics: and, in particular, prove that there is a rectilinear edge of regression of the type indicated in § 97.

[The edge of regression in question is given by

$$\alpha x=\beta y=\gamma z,$$

* *L.c.*, p. 67; see also § 76, *ante*.

where

$$\gamma^2(a^2+\beta^2)+4\beta\gamma=0,$$

and the relation between a and b is

$$(a\gamma+a)(b\gamma+\beta)+4=0.]$$

Ex. 3. Discuss the integrals of the equation

$$(z-\tfrac{1}{2}p^2)^2-\tfrac{1}{9}q^3=0;$$

and indicate the character of the integral

$$z=\tfrac{1}{2}(x-a)^2. \qquad \text{(Darboux.)}$$

Ex. 4. If an equation of the first order possesses singular integrals and can be expressed in a form

$$au^m+bv^n+cw^p=0,$$

where m, n, p are integers, the singular integrals satisfy the equations

$$u=0, \quad v=0, \quad w=0. \qquad \text{(Darboux.)}$$

103. The relation between the surfaces represented by the general integral and the singular integral respectively can be indicated simply. In the equations of the singular integral, which are

$$\phi(x, y, z, a, b)=0, \quad \frac{\partial\phi}{\partial a}=0, \quad \frac{\partial\phi}{\partial b}=0,$$

the quantities a, b are functions of x, y, z: if, therefore, we take

$$b=f(a),$$

where f is any function, we are selecting a curve through a point on the singular integral. At every point on this curve, the equations

$$\phi=0, \quad b=f(a), \quad \frac{\partial\phi}{\partial a}+f'(a)\frac{\partial\phi}{\partial b}=0,$$

are satisfied: as these are the equations of the general integral, the curve lies upon the surface represented by that integral. At any point on this curve common to the two surfaces, the values of p and q are the same, being given by

$$\frac{\partial\phi}{\partial x}+p\frac{\partial\phi}{\partial z}=0, \quad \frac{\partial\phi}{\partial y}+q\frac{\partial\phi}{\partial z}=0,$$

together with the other equations: hence the two surfaces touch along a curve.

It may happen that the equation

$$b=f(a)$$

determines not a single curve alone but a number of curves on the singular integral: in that case, each of the curves is common to the singular integral and the appropriate general integral determined in connection with the above equation, and the two surfaces touch one another at every point on each of the curves. Hence the general integral and the singular integral touch one another along a curve or curves: or, what is the same property, through any curve on the surface, represented by the singular integral, there passes the surface, represented by the general integral which is the envelope of the complete integrals that touch the singular integral along the curve.

Order of Contact of the Singular Integral with the General Integral and the Complete Integral.

104. We have seen that the singular integral, when it exists, is the envelope of the complete integrals, each of which touches it at one or more points: it also touches the general integrals along a curve or curves. We have to consider the order of the contact, a matter already (§§ 92, 98) discussed for characteristics and complete integrals.

The assumed singular integral is given by the single equation which results from the elimination of a and b between

$$\phi(x, y, z, a, b) = 0, \quad \frac{\partial \phi}{\partial a} = 0, \quad \frac{\partial \phi}{\partial b} = 0;$$

let this single equation be supposed resolved with regard to z, so that it has the form

$$z = \psi(x, y).$$

We introduce a new dependent variable ζ, defined by the equation

$$\zeta = z - \psi(x, y);$$

the complete integral now is

$$\phi(x, y, \zeta + \psi, a, b) = 0.$$

Derivatives of the first order (and of all orders) with regard to a and b are the same as before: hence, when the elimination of a and b is performed between the equations

$$\phi = 0, \quad \frac{\partial \phi}{\partial a} = 0, \quad \frac{\partial \phi}{\partial b} = 0,$$

and the eliminant is resolved with regard to ζ, the resolved equation of the singular integral is

$$\zeta = 0.$$

Now it may be that, in both forms, there are several branches of the singular integral leading to resolved equations; in that case, one of them is certainly $\zeta = 0$ in the second form; and though this will not simultaneously represent all branches of the singular integral, it will suffice for the discussion of the order of contact at the point. Accordingly, without loss of generality for the immediate purpose, we may take the singular integral in the form

$$z = 0.$$

The values of p and q, in general, are given by

$$\frac{\partial \phi}{\partial x} + p \frac{\partial \phi}{\partial z} = 0, \quad \frac{\partial \phi}{\partial y} + q \frac{\partial \phi}{\partial z} = 0:$$

for our present purpose, $z = 0$, so that $p = 0$, $q = 0$: and therefore

$$\frac{\partial \phi}{\partial x} = 0, \quad \frac{\partial \phi}{\partial y} = 0.$$

Thus we have, at all points of the singular integral taken in the form $z = 0$, the five equations

$$\phi = 0, \quad \frac{\partial \phi}{\partial a} = 0, \quad \frac{\partial \phi}{\partial b} = 0, \quad \frac{\partial \phi}{\partial x} = 0, \quad \frac{\partial \phi}{\partial y} = 0.$$

But we do not have $\dfrac{\partial \phi}{\partial z} = 0$, in addition: for the point would then be a singularity, conical or otherwise, on the complete integral: and circumstances would be exceedingly special if such a singularity of the complete integrals were to lie on the envelope of those integrals.

Take, first, a general integral in the form

$$\phi(x, y, z, a, b) = 0, \quad b = f(a),$$

$$\frac{d\phi}{da} = \frac{\partial \phi}{\partial a} + f'(a) \frac{\partial \phi}{\partial b} = 0:$$

it touches the singular integral $z = 0$ at one point and along a curve through the point. Consider variations in the vicinity of any such point. On the singular integral $z = 0$, they will be represented by dx and dy; the variations da and db, as determined for the singular integral by dx and dy, are not at once required for the present purpose. On the general integral, they will be represented by dz, da, and by the same values of dx and dy as for the singular integral: and the order of contact of the two surfaces at the point will be measured by the order of dz, expressed in terms of the small quantities dx and dy. Thus, for the general integral, we have

$$\frac{\partial\phi}{\partial x}dx + \frac{\partial\phi}{\partial y}dy + \frac{\partial\phi}{\partial z}dz + \frac{d\phi}{da}da$$
$$+ \tfrac{1}{2}\left\{\frac{\partial^2\phi}{\partial x^2}dx^2 + \ldots\right\} = 0,$$

where the unexpressed terms are the aggregate of terms bilinear in dx, dy, dz, da. But at the point of contact of the singular integral and the general integral, we have

$$\frac{\partial\phi}{\partial x} = 0, \quad \frac{\partial\phi}{\partial y} = 0, \quad \frac{d\phi}{da} = 0,$$

so that dz is at least of the second order of small quantities; hence dz is given, accurately to the second order of small quantities inclusive, by the equation

$$\frac{\partial\phi}{\partial z}dz + \tfrac{1}{2}\left\{\frac{\partial^2\phi}{\partial x^2}dx^2 + 2\frac{\partial^2\phi}{\partial x\partial y}dx\,dy + \frac{\partial^2\phi}{\partial y^2}dy^2\right\}$$
$$+ \tfrac{1}{2}\left\{2\frac{\partial}{\partial x}\left(\frac{d\phi}{da}\right)da\,dx + 2\frac{\partial}{\partial y}\left(\frac{d\phi}{da}\right)da\,dy + \frac{d^2\phi}{da^2}da^2\right\} = 0.$$

But, because $\dfrac{d\phi}{da} = 0$ for the general integral, we have

$$\frac{\partial}{\partial x}\left(\frac{d\phi}{da}\right)dx + \frac{\partial}{\partial y}\left(\frac{d\phi}{da}\right)dy + \frac{d^2\phi}{da^2}da = 0,$$

accurately to the first order of small quantities: thus

$$-2\frac{\partial\phi}{\partial z}dz = \frac{\partial^2\phi}{\partial x^2}dx^2 + 2\frac{\partial^2\phi}{\partial x\partial y}dx\,dy + \frac{\partial^2\phi}{\partial y^2}dy^2 - \frac{d^2\phi}{da^2}da^2$$
$$= \frac{\partial^2\phi}{\partial x^2}dx^2 + 2\frac{\partial^2\phi}{\partial x\partial y}dx\,dy + \frac{\partial^2\phi}{\partial y^2}dy^2$$
$$- \frac{1}{\dfrac{d^2\phi}{da^2}}\left\{\frac{\partial}{\partial x}\left(\frac{d\phi}{da}\right)dx + \frac{\partial}{\partial y}\left(\frac{d\phi}{da}\right)dy\right\}^2,$$

accurately to the second order inclusive.

When the contact between the surfaces is of the first order, dz is of the second order in the small quantities dx and dy. The right-hand side of the above expression for dz does not vanish for all values of dx and dy, except under special conditions; and therefore *the contact between the general integral and the singular integral at any point is usually of the first order.* There are apparently two directions, given by

$$t = \frac{dy}{dx},$$

where t satisfies the quadratic

$$\frac{d^2\phi}{da^2}\left(\frac{\partial^2\phi}{\partial x^2} + 2t\frac{\partial^2\phi}{\partial x\partial y} + t^2\frac{\partial^2\phi}{\partial y^2}\right) = \left\{\frac{\partial}{\partial x}\left(\frac{d\phi}{da}\right) + t\frac{\partial}{\partial y}\left(\frac{d\phi}{da}\right)\right\}^2,$$

along which dz is of the third order; but these two directions will later* be proved to be the same, a property that can be verified by means of the analysis that follows.

105. When the contact between the surfaces is of the second order, dz is of the third order in the small quantities dx and dy; and therefore the right-hand side of the expression obtained for dz, being generally accurate up to the second order, must vanish up to that order for all values of dx and dy. That this may be the case, we must have

$$\frac{d^2\phi}{da^2}\frac{\partial^2\phi}{\partial x^2} = \left(\frac{\partial^2\phi}{\partial a\partial x} + b'\frac{\partial^2\phi}{\partial b\partial x}\right)^2,$$

$$\frac{d^2\phi}{da^2}\frac{\partial^2\phi}{\partial x\partial y} = \left(\frac{\partial^2\phi}{\partial a\partial x} + b'\frac{\partial^2\phi}{\partial b\partial x}\right)\left(\frac{\partial^2\phi}{\partial a\partial y} + b'\frac{\partial^2\phi}{\partial b\partial y}\right),$$

$$\frac{d^2\phi}{da^2}\frac{\partial^2\phi}{\partial y^2} = \left(\frac{\partial^2\phi}{\partial a\partial y} + b'\frac{\partial^2\phi}{\partial b\partial y}\right)^2;$$

so that, if we take

$$\frac{\partial^2\phi}{\partial x^2} = \mu\frac{\partial^2\phi}{\partial x\partial y},$$

we must have

$$\frac{\partial^2\phi}{\partial x\partial y} = \mu\frac{\partial^2\phi}{\partial y^2},$$

and also

$$\frac{\partial^2\phi}{\partial a\partial x} + b'\frac{\partial^2\phi}{\partial b\partial x} = \mu\frac{\partial^2\phi}{\partial a\partial y} + b'\mu\frac{\partial^2\phi}{\partial b\partial y}.$$

We can prove that the last equation is satisfied in virtue of the other two, as follows.

* See § 125.

Consider the variations along the singular integral in the immediate vicinity of the point of contact; the quantities dx and dy determine variations da and db along the singular integral, and conversely: and because $\frac{\partial\phi}{\partial x}=0$ and $\frac{\partial\phi}{\partial y}=0$, these are subject to the relations

$$\frac{\partial^2\phi}{\partial x^2}dx+\frac{\partial^2\phi}{\partial x\partial y}dy+\frac{\partial^2\phi}{\partial a\partial x}da+\frac{\partial^2\phi}{\partial b\partial x}db=0,$$

$$\frac{\partial^2\phi}{\partial x\partial y}dx+\frac{\partial^2\phi}{\partial y^2}dy+\frac{\partial^2\phi}{\partial a\partial y}da+\frac{\partial^2\phi}{\partial b\partial y}db=0.$$

Multiplying the second equation by μ and subtracting from the first, we have

$$\left(\frac{\partial^2\phi}{\partial a\partial x}-\mu\frac{\partial^2\phi}{\partial a\partial y}\right)da+\left(\frac{\partial^2\phi}{\partial b\partial x}-\mu\frac{\partial^2\phi}{\partial b\partial y}\right)db=0.$$

In order to take all directions through the point, we must keep dx and dy independent of one another; and therefore da and db are independent of one another, so that

$$\frac{\partial^2\phi}{\partial a\partial x}-\mu\frac{\partial^2\phi}{\partial a\partial y}=0,$$

$$\frac{\partial^2\phi}{\partial b\partial x}-\mu\frac{\partial^2\phi}{\partial b\partial y}=0.$$

Hence the equation

$$\frac{\partial^2\phi}{\partial a\partial x}+b'\frac{\partial^2\phi}{\partial b\partial x}=\mu\frac{\partial^2\phi}{\partial a\partial y}+b'\mu\frac{\partial^2\phi}{\partial b\partial y}$$

is satisfied for all values of b': and this holds in virtue of the equations

$$\frac{\partial^2\phi}{\partial x^2}=\mu\frac{\partial^2\phi}{\partial x\partial y},\quad \frac{\partial^2\phi}{\partial x\partial y}=\mu\frac{\partial^2\phi}{\partial y^2}.$$

Consequently, *if one general integral has contact of the second order with the singular integral, all general integrals have contact of that order with the singular integral.*

Ex. 1. Prove that, if the equation

$$\frac{\partial^2\phi}{\partial x^2}\frac{\partial^2\phi}{\partial y^2}=\left(\frac{\partial^2\phi}{\partial x\partial y}\right)^2$$

is satisfied at any point on the singular integral, then the equation

$$\frac{\partial^2\phi}{\partial a^2}\frac{\partial^2\phi}{\partial b^2}=\left(\frac{\partial^2\phi}{\partial a\partial b}\right)^2$$

is also satisfied: and conversely. (Darboux.)

Ex. 2. Discuss the preceding proposition in the text when $\frac{\partial^2\psi}{\partial x\partial y}$ vanishes.

106. Now consider the order of contact between the complete integral and the singular integral at a point. In the vicinity of the point, variations along the complete integral are given by dx, dy, dz alone; for a and b are constants along the complete integral. As its equation is

$$\phi(x, y, z, a, b) = 0,$$

those variations are given by

$$\frac{\partial\phi}{\partial x}dx + \frac{\partial\phi}{\partial y}dy + \frac{\partial\phi}{\partial z}dz + \tfrac{1}{2}\left\{\frac{\partial^2\phi}{\partial x^2}dx^2 + \ldots\right\} = 0.$$

But $\frac{\partial\phi}{\partial x} = 0$ and $\frac{\partial\phi}{\partial y} = 0$ at the point in question: hence, accurately to the second order inclusive,

$$-2\frac{\partial\phi}{\partial z}dz = \frac{\partial^2\phi}{\partial x^2}dx^2 + 2\frac{\partial^2\phi}{\partial x\partial y}dx\,dy + \frac{\partial^2\phi}{\partial y^2}dy^2.$$

The expression on the right-hand side does not in general vanish for all values of dx and dy, except under very special conditions: hence *the contact between the complete integral and the singular integral is usually of the first order.* There are however two directions given by

$$\frac{dy}{dx} = t,$$

where t satisfies the quadratic

$$\frac{\partial^2\phi}{\partial x^2} + 2t\frac{\partial^2\phi}{\partial x\partial y} + t^2\frac{\partial^2\phi}{\partial y^2} = 0,$$

along which dz is of the third order; the point is usually a double point on the curve of intersection of the two surfaces, and these directions are the tangents to this curve at the point.

Special interest attaches to the case when these two directions coincide: the two surfaces are then said to *osculate.* In that case, we have

$$\frac{\partial^2\phi}{\partial x^2} = \mu\frac{\partial^2\phi}{\partial x\partial y}, \quad \frac{\partial^2\phi}{\partial x\partial y} = \mu\frac{\partial^2\phi}{\partial y^2};$$

and conversely, if these conditions hold, the two surfaces osculate. Taking account of the earlier result relating to general integrals, we have the theorem: *if the complete integrals do not osculate the singular integral, then no integral has contact of the second order with the singular integral: but if the complete integrals do osculate the singular integral, then all the general integrals have contact of the second order with the singular integral.*

Ex. Prove that, if the complete integral osculates the singular integral, all the characteristics passing through the point of contact have the same tangent.

Illustrate this property by reference to the spheres that osculate a surface. (Darboux.)

Singularities on the Characteristics.

107. In the preceding discussion of the surfaces and curves associated with the integrals of the differential equation

$$f(x, y, z, p, q)=0,$$

we have been concerned mainly with regions that are devoid of singularities for those surfaces and curves; the effect of possible singularities must now be considered. Also, it will be convenient to consider at the same time the exceptional cases of the preceding investigations that were noted (§§ 85, 89) but not discussed. We shall proceed, as before, from the characteristics.

It may be assumed that the equation $f=0$ has been transformed so as to be free from irrationalities; and we shall discount the loss of generality in assuming, as will be done, that f is a regular function of its arguments. Take the tangent plane to any integral surface as the plane $z=0$, and the point of contact for origin: then at that point $p=0$, $q=0$; in the vicinity of the point, f is a regular function of x, y, z, p, q, which vanishes at the point. The equations of the characteristic are

$$\frac{dx}{P}=\frac{dy}{Q}=\frac{dz}{pP+qQ}=\frac{dp}{-(X+pZ)}=\frac{dq}{-(Y+qZ)}=dt;$$

we propose to discuss the form of the characteristic in the vicinity of the origin, according to the varieties of form of $f=0$.

If P vanishes but not Q, the axes of x and y can be changed to another set such that neither the new P nor the new Q shall vanish at the point: similarly, if either $X+pZ$ or $Y+qZ$ should vanish but not both, a change can be made such that neither of the quantities in the new position shall vanish. Also, taking account of earlier exceptions, we thus have the following cases:

I. All the denominators are different from zero at the point:

II. The quantities P and Q vanish at the point, but not the others:

III. The quantities $X+pZ$ and $Y+qZ$ vanish at the point, but not the others:

IV. All the denominators vanish at the point, but Z is not zero:

V. All the quantities X, Y, Z, P, Q vanish.

Of these, the first is included in order to make the set complete: it is the assumption that was made in the earlier investigations and, as there, it will be found to constitute the origin an ordinary point. The second has been left over from § 88, and the fifth from § 78; and the fourth has given a singular integral, if such an integral exists. Let

$$f = ax + by + cz + gp + hq + \ldots.$$

108. *Case I.* As $X+pZ$, $Y+qZ$, P, Q do not vanish at the origin, a, b, g, h are not zero. Hence, assuming that t vanishes at the origin, we have, in the immediate vicinity,

$$\frac{dx}{g+\ldots} = \frac{dy}{h+\ldots} = \frac{-dp}{a+\ldots} = \frac{-dq}{b+\ldots} = dt,$$

$$dz = p\,dx + q\,dy,$$

and therefore

$$x = gt + \ldots, \qquad y = ht + \ldots,$$

$$p = -at + \ldots, \quad q = -bt + \ldots,$$

$$dz = -(ag+bh)\,t\,dt + \ldots,$$

so that

$$z = -\tfrac{1}{2}(ag+bh)\,t^2 + \ldots.$$

Hence, in the vicinity of the origin, the characteristic is given by

$$y = \frac{h}{g}x + \ldots,$$

$$z = -\frac{ag+bh}{2g^2}x^2 + \ldots;$$

the origin is an ordinary point on the characteristic, which touches the plane $z=0$ there. This, as already indicated, is the former result.

109. *Case II.* As P and Q vanish at the origin, we have

$$g=0, \quad h=0.$$

Let

$$f=ax+by+cz+(a'x+b'y+c'z)p+(a''x+b''y+c''z)q$$
$$+\tfrac{1}{2}(\alpha p^2+2\beta pq+\gamma q^2)+\phi(x, y, z, p, q),$$

where $\phi(x, y, z, p, q)$ contains terms of the second degree in x, y, z alone, and all other terms in all the quantities of higher degree in the aggregate. Two of the equations of the characteristic now are

$$\frac{-dp}{a+\ldots}=\frac{-dq}{b+\ldots}=dt,$$

so that

$$p=-at+\ldots, \quad q=-bt+\ldots,$$

in the vicinity of the origin; and the other equations are

$$\frac{dx}{dt}=a'x+b'y+c'z+\alpha p+\beta q+\ldots,$$

$$\frac{dy}{dt}=a''x+b''y+c''z+\beta p+\gamma q+\ldots,$$

$$\frac{dz}{dt}=p\frac{dx}{dt}+q\frac{dy}{dt}.$$

(A) First, suppose that neither of the quantities $\alpha a+\beta b$ and $\beta a+\gamma b$ vanishes; this is the most general case as it involves no restricting relation between constants. Then we have

$$x=-\tfrac{1}{2}(\alpha a+\beta b)t^2+\ldots, \quad y=-\tfrac{1}{2}(\beta a+\gamma b)t^2+\ldots,$$

$$\frac{dz}{dt}=(\alpha a^2+2\beta ab+\gamma b^2)t^2+\ldots,$$

so that

$$z=\tfrac{1}{3}(\alpha a^2+2\beta ab+\gamma b^2)t^3+\ldots;$$

and therefore, in the immediate vicinity of the origin, the characteristic is given by

$$y=\mu x+\ldots, \quad z=\mu' x^{\frac{3}{2}}+\ldots,$$

where μ and μ' are determinate constants. The origin is a cusp on the characteristic; the tangent at the cusp lies in the plane $z=0$, along the line $y=\mu x$.

It thus appears that, when $f=0$, $P=0$, $Q=0$, while $X+pZ$ and $Y+qZ$ are not zero, and when they allow the elimination of p and q between the three equations (which is the most general case, as not involving relations among the equations), the surface given by the eliminant is a locus of cusps of the characteristics. Hence, *in general, the surface obtained by eliminating p and q between the equations*

$$f=0,\quad P=0,\quad Q=0,$$

is a locus of cusps of the characteristics; through every point of the surface there passes a characteristic having a cusp at that point.

If it should happen that $\alpha a^2 + 2\beta ab + \gamma b^2$ vanishes though $\alpha a + \beta b$ and $\beta a + \gamma b$ do not vanish, the only difference is that, in the vicinity of the origin,

$$y = \mu x + \dots,\quad z = \lambda x^2 + \lambda_1 x^{\frac{5}{2}} + \dots;$$

the origin is still a singularity on the curve in general, of an order higher than a cusp.

Ex. Prove that the tangent plane of the characteristic developable is given by

$$z + atx + bty + \tfrac{1}{6}(\alpha a^2 + 2\beta ab + \gamma b^2)\, t^3 = 0,$$

keeping the most important terms, and that therefore the point is an ordinary point for the developable. (Darboux.)

(B) Next, suppose that one (but not both) of the two quantities $\alpha a + \beta b$, $\beta a + \gamma b$ vanishes: let

$$\beta a + \gamma b = 0.$$

Proceeding as before, we have

$$x = -\tfrac{1}{2}(\alpha a + \beta b)\, t^2 + \dots,$$
$$y = \lambda t^3 + \dots,$$
$$z = \tfrac{1}{3}(\alpha a + \beta b)\, a t^3 + \dots;$$

and therefore, in the vicinity of the origin, we have

$$y = \mu z + \dots,\quad z = \mu x^{\frac{3}{2}} + \dots.$$

The origin is a cusp on the characteristic; the tangent to the cusp is the axis of y.

If $\lambda = 0$, or if $a = 0$, or if both λ and a vanish, the origin is still a singularity on the curve of an order higher than a cusp: the curve still touches the plane of z at the origin.

(C) Lastly, suppose that

$$\alpha a + \beta b = 0,\quad \beta a + \gamma b = 0:$$

then, in the most general case, that is subject to these conditions, we find

$$x=\lambda_1 t^3+\ldots,\quad y=\lambda_2 t^3+\ldots,\quad z=\lambda_3 t^4+\ldots,$$

so that, in the vicinity of the origin,

$$y=\mu x+\ldots,\quad z=\mu x^{\frac{4}{3}}+\ldots.$$

The origin is a singularity on the characteristic: and the tangent to the characteristic lies in the plane of $z=0$ along the line $y=\mu x$.

Similarly, for other special relations among the constants, we obtain a corresponding result: in every case, the equation obtained by eliminating p and q between $f=0$, $P=0$, $Q=0$, when $X+pZ$ and $Y+qZ$ do not vanish, is a locus of singularities on the characteristics.

110. *Case III.* As $X+pZ$ and $Y+qZ$ vanish at the origin, we have

$$a=0,\quad b=0;$$

but P and Q do not vanish there, so that g and h are different from zero. Hence

$$\begin{aligned}f=cz+gp+hq+(a'x+b'y+c'z)p+(a''x+b''y+c''z)q\\ +\tfrac{1}{2}(\alpha p^2+2\beta pq+\gamma q^2)+\tfrac{1}{2}(Ax^2+2Bxy+Cy^2)+\psi(x,y,z,p,q),\end{aligned}$$

where ψ contains terms of the third and higher orders. The equations are

$$\frac{dx}{dt}=g+\ldots,\quad \frac{dy}{dt}=h+\ldots,$$

so that

$$x=gt+\ldots,\quad y=ht+\ldots;$$

also

$$-\frac{dp}{dt}=(c+a')p+a''q+Ax+By+\ldots,$$

$$-\frac{dq}{dt}=b'p+(b''+c)q+Bx+Cy+\ldots,$$

so that

$$p=-\tfrac{1}{2}(Ag+Bh)t^2+\ldots,$$
$$q=-\tfrac{1}{2}(Bg+Ch)t^2+\ldots.$$

Hence

$$\begin{aligned}\frac{dz}{dt}&=p\frac{dx}{dt}+q\frac{dy}{dt}\\ &=-\tfrac{1}{2}(Ag^2+2Bgh+Ch^2)t^2+\ldots,\end{aligned}$$

and so

$$z = -\tfrac{1}{6}(Ag^2 + 2Bgh + Ch^2)\,t^3 + \dots .$$

Consequently the characteristic, in the vicinity of the origin, is given by

$$y = \mu x + \dots, \quad z = \rho x^3 + \dots;$$

the point is an inflexion on the characteristic, and the inflexional tangent to the characteristic lies in the plane of z along the line $y = \mu x$.

Ex. Prove, by means of reciprocal polars applied to Case II or otherwise, that the surface, obtained by eliminating p and q between the equations

$$f=0, \quad X+pZ=0, \quad Y+qZ=0,$$

is a locus such that, at every point on it, the characteristic developable of the equation $f=0$ has a singular plane. Sketch the characteristic developable in the vicinity of the point. (Darboux.)

111. *Case IV.* Here we have

$$f=0, \quad P=0, \quad Q=0, \quad X+pZ=0, \quad Y+qZ=0,$$

while Z is not zero. These are five equations, which involve five quantities x, y, z, p, q: hence if f, already supposed regular, be a polynomial function of its arguments, there would in general be only a limited number of sets of values of the five variables, and therefore only a limited number of such points.

Taking the origin to be such a point, we assume (as before) that x, y, z, p, q all vanish there: hence

$$a=0, \quad b=0, \quad g=0, \quad h=0,$$

and the equations of the characteristic, in the immediate vicinity of the origin, are

$$\frac{dx}{dt} = a'x + b'y + c'z + \alpha p + \beta q + \dots,$$

$$\frac{dy}{dt} = a''x + b''y + c''z + \beta p + \gamma q + \dots,$$

$$\frac{dp}{dt} = Ax + By \qquad + (c + a')\,p + a''q + \dots,$$

$$\frac{dq}{dt} = Bx + Cy \qquad + b'p + (b'' + c)\,q + \dots,$$

$$\frac{dz}{dt} = p\frac{dx}{dt} + q\frac{dy}{dt}.$$

As the only regular integrals of these equations, which vanish when $t=0$, are

$$x=0, \quad y=0, \quad z=0, \quad p=0, \quad q=0,$$

it is simpler to make x the independent variable of the characteristic and omit t completely. It is clear that z is of a higher order of small quantities than x or y in the vicinity of the origin; hence, retaining only the most important terms within this vicinity, the equations may be taken in the form

$$\frac{dy}{dx}=\frac{a''x+b''y+\beta p+\gamma q}{a'x+b'y+\alpha p+\beta q},$$

$$\frac{dp}{dx}=\frac{Ax+By+(c+a')p+a''q}{a'x+b'y+\alpha p+\beta q},$$

$$\frac{dq}{dx}=\frac{Bx+Cy+b'p+(b''+c)q}{a'x+b'y+\alpha p+\beta q}.$$

Then* there are integrals of these equations of the form

$$y=x(\rho+\eta),$$
$$p=x(\sigma+\pi),$$
$$q=x(\tau+\kappa),$$

where ρ, σ, τ are constants, and η, π, κ are functions of x that vanish with x. As regards the constants, they are given by the equations

$$\rho=\frac{a''+b''\rho+\beta\sigma+\gamma\tau}{a'+b'\rho+\alpha\sigma+\beta\tau},$$

$$\sigma=\frac{A+B\rho+(c+a')\sigma+a''\tau}{a'+b'\rho+\alpha\sigma+\beta\tau},$$

$$\tau=\frac{B+C\rho+b'\sigma+(b''+c)\tau}{a'+b'\rho+\alpha\sigma+\beta\tau}.$$

Writing

$$\theta=a'+b'\rho+\alpha\sigma+\beta\tau,$$

we have

$$\rho\theta=a''+b''\rho+\beta\sigma+\gamma\tau,$$
$$\sigma\theta=A+B\rho+(c+a')\sigma+a''\tau,$$
$$\tau\theta=B+C\rho+b'\sigma+(b''+c)\tau.$$

Hence θ is a root of the equation

$$\begin{vmatrix} a'-\theta, & b', & \alpha, & \beta \\ a'', & b''-\theta, & \beta, & \gamma \\ A, & B, & c+a'-\theta, & a'' \\ B, & C, & b', & b''+c-\theta \end{vmatrix}=0;$$

* See Part III of this treatise, chapters XI, XII.

when θ is known, any three of the earlier equations determine ρ, σ, τ.

For the quantities η, π, κ, we have

$$\rho + \eta + x\frac{d\eta}{dx} = \frac{\rho\theta + b''\eta + \beta\pi + \gamma\kappa}{\theta + b'\eta + \alpha\pi + \beta\kappa},$$

so that

$$x\frac{d\eta}{dx} = -\eta + \frac{(b'' - b'\rho)\eta + (\beta - \alpha\rho)\pi + (\gamma - \beta\rho)\kappa}{\theta + b'\eta + \alpha\pi + \beta\kappa}$$

$$= \frac{1}{\theta}\{(b'' - b'\rho - \theta)\eta + (\beta - \alpha\rho)\pi + (\gamma - \beta\rho)\kappa\},$$

to the order of quantities retained; and similarly

$$x\frac{\partial\pi}{\partial x} = \frac{1}{\theta}\{(B - b'\sigma)\eta + (c + a' - \alpha\sigma - \theta)\pi + (a'' - \beta\sigma)\kappa\},$$

$$x\frac{\partial\kappa}{\partial x} = \frac{1}{\theta}\{(C - b'\tau)\eta + (b' - \alpha\tau)\pi + (b'' + c - \beta\tau - \theta)\kappa\};$$

and the quantities η, π, κ are to vanish with x. The characters of these functions depend upon the roots of the critical cubic

$$\begin{vmatrix} b'' - b'\rho - \theta - \mu, & \beta - \alpha\rho, & \gamma - \beta\rho \\ B - b'\sigma, & c + a' - \alpha\sigma - \theta - \mu, & a'' - \beta\sigma \\ C - b'\tau, & b' - \alpha\tau, & b'' + c - \beta\tau - \theta - \mu \end{vmatrix} = 0$$

in μ: and these can be expressed in terms of the roots of the preceding quartic.

Let θ_1, θ_2, θ_3, θ_4 be the roots of that quartic: of these, suppose that θ_1 is the root of the quartic chosen, and that the corresponding values of ρ, σ, τ have been obtained. Multiply the columns of the quartic determinant by 1, ρ, σ, τ and subtract the sum of the last three from the first: the determinant is

$$\begin{vmatrix} \theta_1 - \theta, & b', & \alpha, & \beta \\ \rho(\theta_1 - \theta), & b'' - \theta, & \beta, & \gamma \\ \sigma(\theta_1 - \theta), & B, & c + a' - \theta, & a'' \\ \tau(\theta_1 - \theta), & C, & b', & b'' + c - \theta \end{vmatrix}.$$

Multiply the first row by ρ, σ, τ in turn and subtract the products from the second row, the third row, and the fourth row in respective succession: the determinant is

$$\begin{vmatrix} \theta_1 - \theta, & b', & \alpha, & \beta \\ 0, & b'' - b'\rho - \theta, & \beta - \alpha\rho, & \gamma - \beta\rho \\ 0, & B - b'\sigma, & c + a' - \alpha\sigma - \theta, & a'' - \beta\sigma \\ 0, & C - b'\tau, & b' - \alpha\tau, & b'' + c - \beta\tau - \theta \end{vmatrix},$$

and it is equal to the determinant in the quartic equation. Hence the roots of

$$\begin{vmatrix} b''-b'\rho-\theta, & \beta-\alpha\rho \quad , & \gamma-\beta\rho \\ B-b'\sigma \quad , & c+a'-\alpha\sigma-\theta, & a''-\beta\sigma \\ C-b'\tau \quad , & b'-\alpha\tau \quad , & b''+c-\beta\tau-\theta \end{vmatrix}=0$$

are θ_2, θ_3, θ_4. But the cubic in μ is

$$\begin{vmatrix} b''-b'\rho-\theta_1-\mu, & \beta-\alpha\rho \quad , & \gamma-\beta\rho \\ B-b'\sigma \quad , & c+a'-\alpha\sigma-\theta_1-\mu, & a''-\beta\sigma \\ C-b'\tau \quad , & b'-\alpha\tau \quad , & b''+c-\beta\tau-\theta_1-\mu \end{vmatrix}=0;$$

hence the roots are

$$\mu+\theta_1=\theta_2,\ \theta_3,\ \theta_4,$$

that is, the roots of the cubic in μ are

$$\theta_2-\theta_1, \quad \theta_3-\theta_1, \quad \theta_4-\theta_1,$$

where θ_1 is the root of the quartic selected for the case under consideration.

We shall denote these roots by μ_1, μ_2, μ_3.

112. Of the various sub-cases, it will be sufficient to mention some of the more important.

(i) Let the three roots of the critical cubic be unequal to one another, no one of them being a positive integer; then the preceding equations possess integrals, expressing η, π, κ as unique regular functions of x which vanish with x. For this characteristic, we have

$$y=(\rho+\eta)\,x,$$
$$z=\tfrac{1}{2}(\sigma+\rho\tau)\,x^2+\dots;$$

the point in question is an ordinary point on the curve, which has its tangent lying in the plane $z=0$ along the line $y=\rho x$.

Further integrals may be possessed by the equations, on the preceding assumption as to the roots: but the integrals are not regular functions of x.

If the real parts of μ_1, μ_2, μ_3 are positive and are such that no one of the quantities

$$(m_1-1)\,\mu_1+m_2\mu_2+m_3\mu_3+m_4,$$
$$m_1\mu_1+(m_2-1)\,\mu_2+m_3\mu_3+m_4,$$
$$m_1\mu_1+m_2\mu_2+(m_3-1)\,\mu_3+m_4,$$

vanishes for positive integer values m_1, m_2, m_3, m_4, such that $m_1+m_2+m_3+m_4 \geqslant 2$, the equations possess a triple infinitude of non-regular integrals that vanish with x: these integrals are regular functions of x, x^{μ_1}, x^{μ_2}, x^{μ_3}. There is thus a triple infinitude of curves through the point: it is easily seen to be a singularity on each of them.

If the real parts of μ_1 and μ_2 be positive, if that of μ_3 be negative, and if no one of the quantities

$$(m_1-1)\,\mu_1+m_2\mu_2+m_3, \quad m_1\mu_1+(m_2-1)\,\mu_2+m_3$$

vanishes for positive integer values m_1, m_2, m_3, such that

$$m_1+m_2+m_3 \geqslant 2,$$

the equations possess a double infinitude of non-regular integrals that vanish with x: these integrals are regular functions of x, x^{μ_1}, x^{μ_2}. There is then a double infinitude of curves through the origin: it is easily seen to be a singularity on each of them.

If the real part of μ_1 be positive, and if the real parts of μ_2 and μ_3 be negative, the equations possess a single infinitude of non-regular integrals that vanish with x; these integrals are regular functions of x and x^{μ_1}. There is then a single infinitude of curves through the origin; the point is easily seen to be a singularity on each of them.

If the real parts of μ_1, μ_2, μ_3 be each negative, the regular integrals first indicated are the only integrals of the equation which vanish with x. As already proved, the origin is then an ordinary point upon the sole characteristic, which passes through it touching the plane $z=0$ at the point.

(ii) Let the three roots of the critical cubic be unequal to one another, and let one (but only one) of them, say μ_1, be a positive integer.

Unless a particular condition among the coefficients in the equations be satisfied, the equations possess no regular integrals that vanish with x. When that condition is not satisfied, they possess a simple infinitude of non-regular integrals that vanish with x, being regular functions of x and $x \log x$, if the real parts of μ_2 and μ_3 are negative: they possess a double infinitude of non-regular integrals that vanish with x, being regular functions of x, $x \log x$, and x^{μ_2}, if the real part of μ_2 is positive and the real part of μ_3 is negative; they possess a triple infinitude of non-regular integrals that vanish with x, being regular functions of x, $x \log x$, x^{μ_2}, x^{μ_3}, if the real parts of μ_2 and μ_3 are positive. Every curve passing through the point has the origin for a singularity.

When the particular condition among the coefficients in the equation is satisfied, the equations possess a simple infinitude of regular integrals that vanish with x: each of these gives a characteristic touching the plane $z=0$ along the same tangent $y=\rho x$ in that plane, and the point of contact is an ordinary point for each curve in the infinitude. Further, when the condition is satisfied, the equations possess either a double infinitude or a single infinitude of non-regular integrals according as the real parts of μ_2 and μ_3, or of only one of them, are positive, these integrals being regular functions of x, x^{μ_2} and x^{μ_3}, or of x and either x^{μ_2} or x^{μ_3}, according to the respective cases; for each of these curves, the origin is a singularity. But if the real parts of μ_2 and μ_3 are negative, the equations are devoid of non-regular integrals.

And so for other cases: the results depend upon the characters of the integrals of the equations for η, π, κ: and these characters are known by the critical conditions* as regards the roots μ_1, μ_2, μ_3. Regular integrals make the origin an ordinary point on the corresponding characteristics: non-regular integrals make the origin a singularity on the corresponding characteristics.

113. At the beginning of the discussion, it was pointed out that the five equations

$$f=0, \quad P=0, \quad Q=0, \quad X+pZ=0, \quad Y+qZ=0$$

will, in general, give a finite number of determinations of sets of values of the variables involved: so that, in general, there will be a finite number of points in space at which the characteristics must be considered for the present purpose. The preceding discussion is typical of the discussion to be effected at each such point for any particular equation of the assumed character.

But the five equations may be of such a form, or they may be so related, that they do not determine a set of values or a limited

* These are set out for equations of the form in question in n variables, in § 187, chapter XII, vol. III, of this treatise. They were proved in full detail for all the cases, that arise in a system of two equations, in my memoir, *On the integrals of systems of differential equations*, published in the Stokes 1899 Commemoration volume (vol. XVIII, 1900) of the Transactions of the Cambridge Philosophical Society.

number of sets of values for the five variables involved. In that event, there are three possible alternatives, always on the adopted hypothesis that the equation $f=0$ is irreducible:

(a) they may determine four of the variables in terms of the remaining one, say, y, z, p, q, in terms of x; there then is a curve-locus in space, and values of p and q are associated with each point on the curve:

(b) they may determine three of the variables in terms of the remaining two, say z, p, q in terms of x, y; there then is a surface-locus in space, and values of p and q are associated with each point on the surface:

(c) they may determine two of the variables in terms of the remaining three, say p, q in terms of x, y, z; values of p and q are then associated with each point of space.

The third of these alternatives has already been discussed (§§ 78, 79): substitution of p and q in

$$dz = p\,dx + q\,dy,$$

followed by a quadrature, leads to a surface integral of the equation. A characteristic through any point on such a surface lies in the surface. The second of these alternatives leads to a singular integral of the equation, unless $Z=0$; we shall come to the consideration of the characteristics through a point on the surface after the discussion of the first alternative, which can be discussed briefly after the earlier analysis.

114. Suppose, then, that the five equations

$$f=0, \quad P=0, \quad Q=0, \quad X+pZ=0, \quad Y+qZ=0$$

determine a curve-locus

$$y = u(x), \quad z = v(x),$$

together with values of p and q as functions of x in the form

$$p = \pi(x), \quad q = \kappa(x).$$

Take any point on this locus, say $x = x_0$; transfer the origin to that point, and let

$$\pi(x_0) = \pi_0, \quad \kappa(x_0) = \kappa_0.$$

Then the form of f must be such that the five equations give

$$y=0, \quad z=0, \quad p=\pi_0, \quad q=\kappa_0,$$

when $x=0$. As may easily be verified, we have

$$f=c(z-x\pi_0-y\kappa_0)+\text{higher powers of } x,\ y,\ z$$
$$+(p-\pi_0)R(x,\ y,\ z,\ p-\pi_0,\ q-\kappa_0)$$
$$+(q-\kappa_0)S(x,\ y,\ z,\ p-\pi_0,\ q-\kappa_0),$$

where R and S are regular functions of their arguments: these functions must vanish when $x=0$, $y=0$, $z=0$, $p=\pi_0$, $q=\kappa_0$: and there must be limitations on the form of f sufficient to make the five equations equivalent to four only, though the precise form of the limitation is not necessary for the present purpose. What is required is the nature of this point as a point on the characteristic.

The point is the origin: as usual, we take the tangent plane to the integral surface to be $z=0$, so that $p=0$, $q=0$ at the point on the characteristic. Let

$$\lambda_0=R(0,\ 0,\ 0,\ -\pi_0,\ -\kappa_0)-\pi_0 R_0'-\kappa_0 S_0',$$
$$\mu_0=S(0,\ 0,\ 0,\ -\pi_0,\ -\kappa_0)-\pi_0 R_0''-\kappa_0 S_0'',$$

where R_0', R_0'', S_0', S_0'' are the values of $\frac{\partial R}{\partial p}$, $\frac{\partial R}{\partial q}$, $\frac{\partial S}{\partial p}$, $\frac{\partial S}{\partial q}$, when x, y, z, p, q all are made zero: then, in the vicinity of the origin along the characteristic, we have

$$\frac{dx}{dt}=\lambda_0+\ldots,\qquad \frac{dy}{dt}=\mu_0+\ldots,$$

as two of the equations. Also, let R_1, R_2, S_1, S_2 denote the values of $\frac{\partial R}{\partial x}$, $\frac{\partial R}{\partial y}$, $\frac{\partial S}{\partial x}$, $\frac{\partial S}{\partial y}$, when x, y, z, p, q all vanish: then two other equations of the characteristic are

$$-\frac{dp}{dt}=-c\pi_0-\pi_0 R_1-\kappa_0 S_1=\rho_0,\text{ say},$$
$$-\frac{dq}{dt}=-c\kappa_0-\pi_0 R_2-\kappa_0 S_2=\sigma_0,\text{ say},$$

in the immediate vicinity of the origin. Hence in that vicinity,

$$x=\lambda_0 t+\ldots,\qquad y=\mu_0 t+\ldots,$$
$$z=-\tfrac{1}{2}(\rho_0\lambda_0+\mu_0\sigma_0)t^2+\ldots$$

are the most important terms in the equations of the characteristic. It is clear that, unless λ_0 and μ_0 both vanish, the origin is an ordinary point on the characteristic, and that its tangent lies in the plane $z=0$ along the line $\lambda_0 y-\mu_0 x=0$.

The inference cannot be made if $\lambda_0 = 0$ and $\mu_0 = 0$. The further consideration, in this event, will not be undertaken; in the complicated analysis that would be necessary, particularly as regards the forms of the integrals of the differential equations, the details would be substantially similar to those which have been given for the case when the five equations determine sets of values for the five variables.

115. Coming to the remaining alternative, in which the five equations determine z, p, q, as functions of x and y, and assuming (as has been assumed throughout) that Z does not vanish for such relations, we know (§ 78) that the equations define a singular integral; and our quest is the examination of the characteristics at and near any point on this integral surface.

Let the singular integral be given by $z = \theta(x, y)$: then, when a new variable ζ is defined by the relation $z - \theta(x, y) = \zeta$, the singular integral will be given by $\zeta = 0$, as in § 104. Hence we may take the plane $z = 0$ as the singular integral. Moreover, Z does not vanish on account of $z = 0$ and it does not vanish identically: consequently, we may take the differential equation in a resolved form

$$f = z - \phi(x, y, p, q) = 0.$$

Take any point on the singular integral

$$z = 0,$$

and make it the origin; moreover, as z is steadily zero along the singular integral, the associated values of p and q are

$$p = 0, \quad q = 0.$$

The singular integral, thus defined by

$$z = 0, \quad p = 0, \quad q = 0,$$

is to satisfy the equations

$$f = 0, \quad \frac{\partial \phi}{\partial p} = 0, \quad \frac{\partial \phi}{\partial q} = 0, \quad \frac{\partial \phi}{\partial x} - p = 0, \quad \frac{\partial \phi}{\partial y} - q = 0,$$

that is, $\phi, \dfrac{\partial \phi}{\partial p}, \dfrac{\partial \phi}{\partial q}, \dfrac{\partial \phi}{\partial x}, \dfrac{\partial \phi}{\partial y}$ must vanish when $p = 0$, $q = 0$; and therefore ϕ is of the form

$$\begin{aligned} \phi = p^2 (\alpha + ax + a'y + Ap + A'q + \dots) \\ + 2pq (\beta + bx + b'y + \dots) \\ + q^2 (\gamma + cx + c'y + Cp + C'q + \dots). \end{aligned}$$

We require the equations of the characteristics in the immediate vicinity of the point $x=0$, $y=0$, $z=0$ on the singular integral: in differential form, they are

$$\frac{dx}{2\alpha p + 2\beta q + \dots} = \frac{dy}{2\beta p + 2\gamma q + \dots} = \frac{dp}{p + \dots} = \frac{dq}{q + \dots},$$

integrals of which for the immediate vicinity of the origin are

$$\begin{aligned} p &= \kappa q + \dots, \\ x &= 2(\alpha\kappa + \beta) q + \dots, \\ y &= 2(\beta\kappa + \gamma) q + \dots, \end{aligned}$$

κ being an arbitrary constant. Also, as

$$\begin{aligned} dz &= p\,dx + q\,dy \\ &= 2(\alpha\kappa^2 + 2\beta\kappa + \gamma)\, q\,dq + \dots, \end{aligned}$$

we have

$$z = (\alpha\kappa^2 + 2\beta\kappa + \gamma)\, q^2 + \dots.$$

Hence the equations of the characteristic in the immediate vicinity of the origin are

$$\left.\begin{aligned} (\alpha\kappa + \beta)\, y &= (\beta\kappa + \gamma)\, x \\ 4(\alpha\gamma - \beta^2)\, z &= \gamma x^2 - 2\beta xy + \alpha y^2 \end{aligned}\right\},$$

where κ is an arbitrary quantity: and the solution is satisfactory unless $\alpha\gamma - \beta^2$ vanishes. Now the curve touches $z=0$ at the origin, and the tangent to the curve lies in this plane along the line

$$y = \frac{\beta\kappa + \gamma}{\alpha\kappa + \beta}\, x;$$

when κ is arbitrary, the quantity $\dfrac{\beta\kappa + \gamma}{\alpha\kappa + \beta}$ also is arbitrary unless $\alpha\gamma - \beta^2$ vanishes: and therefore a characteristic can be drawn touching every line in the plane. We thus have the former result:—

When an equation has a singular integral, then through any point on it there passes an infinitude of characteristics unless

$$\frac{\partial^2 f}{\partial p^2}\frac{\partial^2 f}{\partial q^2} - \left(\frac{\partial^2 f}{\partial p \partial q}\right)^2$$

vanishes at the point; each characteristic touches the singular integral there and has the point for an ordinary point, and no two characteristics have the same tangent. Moreover, all these characteristics lie on a surface which touches the singular integral and has the point of contact for an ordinary point. For, in order to obtain this integral surface from the equations of the characteristics, all that needs (§ 84) to be done is to make the initial values satisfy the equations

$$f=0, \quad dz=p\,dx+q\,dy.$$

In the present case, both of these equations are satisfied at the point by taking

$$\alpha\kappa^2+2\beta\kappa+\gamma=0.$$

Hence, in the immediate vicinity of the point of intersection of the surface with the singular integral, we have

$$x=2(\alpha\kappa+\beta)q+\dots,$$
$$y=2(\beta\kappa+\gamma)q+\dots,$$
$$z=\lambda q^3+\dots,$$

on the surface; and therefore

$$\kappa x+y=0, \quad z=\mu x^3+\dots;$$

the point of intersection is an ordinary point on the surface, and the two surfaces touch there.

If however $\alpha\gamma-\beta^2=0$, without α, β, γ separately vanishing, we may take

$$\beta=\alpha\theta, \quad \gamma=\alpha\theta^2.$$

We still have an infinitude of characteristics through the point, given by the equations

$$p=\kappa q+\dots,$$
$$x=2(\kappa+\theta)\alpha q+\dots,$$
$$y=2\theta(\kappa+\theta)\alpha q+\dots,$$
$$z=\alpha(\kappa+\theta)^2 q^2+\dots;$$

they touch the plane $z=0$ at the origin, and the tangents are given by

$$y=\theta x,$$

that is, the infinitude of characteristics have a common tangent at the origin: and as

$$4\alpha z=x^2+\dots,$$

the origin is an ordinary point for each of the characteristics.

This infinitude of characteristics still lies on a surface (which of course gives an integral of the differential equation). To obtain its equation from the equations of the characteristics, we make the initial values satisfy the equations

$$f=0, \quad dz = p\,dx + q\,dy.$$

In the present case, both of these equations are satisfied at the point by taking

$$\kappa + \theta = 0.$$

With these initial values, we have, for the characteristics,

$$\begin{aligned} p &= \kappa q + \ldots, \\ \frac{dx}{dq} &= \{3A\kappa^2 + (2A' + 4B)\kappa + 2B' + C\}\,q + \ldots \\ &= \lambda q + \ldots, \\ \frac{dy}{dq} &= \{(A' + 2B)\kappa^2 + (4B' + 2C)\kappa + 3C'\}\,q + \ldots \\ &= \mu q + \ldots, \end{aligned}$$

and therefore

$$\begin{aligned} x &= \tfrac{1}{2}\lambda q^2 + \ldots, \\ y &= \tfrac{1}{2}\mu q^2 + \ldots, \\ z &= \tfrac{1}{3}(\kappa\lambda + \mu)\,q^3 + \ldots. \end{aligned}$$

Hence, at the point of intersection of the singular integral and the surface that is the locus of the characteristics, we have

$$\begin{aligned} y &= \rho x + \ldots, \\ z &= \sigma x^{\frac{3}{2}} + \ldots, \end{aligned}$$

along the surface: and therefore the point is a singularity on the surface that is the locus of the characteristics. The two surfaces touch at the point which, on the assumptions implicitly made that neither λ nor μ nor $\kappa\lambda + \mu$ vanishes, is a cusp on the locus containing the characteristics through the point.

Even if α, β, γ be zero, the same kind of result holds: for it is sufficient in that event to make $\kappa = 0$: we merely have different values of λ and μ. Hence we have the result:

When an equation has a singular integral, and when the relation

$$\frac{\partial^2 f}{\partial p^2}\frac{\partial^2 f}{\partial q^2} = \left(\frac{\partial^2 f}{\partial p\,\partial q}\right)^2$$

is satisfied at any point of it, an infinitude of characteristics passes through the point having a common tangent that also touches (or lies in) the singular integral. All the characteristics lie on a surface which touches the singular integral at the point and has the point for a singularity.

Ex. 1. Prove that, when an equation $f(x, y, z, p, q)=0$ possesses a singular integral in the form

$$z=\phi(x, y),$$

such that $\frac{\partial f}{\partial z}$ does not vanish at all points on the surface represented by that integral, the general integral which touches the singular integral along a curve can be represented in the form

$$z=\phi(x, y)+\zeta^2,$$

where ζ is a regular function of x and y, and the complete integral can be represented in the form

$$z=\phi(x, y)+u^2+v^2,$$

where u and v are regular functions of x and y. (Darboux.)

Ex. 2. Shew that the characteristics of the equation

$$(pz-x)^2=q^2(x^2+z^2-1)$$

are plane curves, and that the locus of their cusps is

$$x^2+z^2=1.$$ (Goursat.)

Ex. 3. Discuss the various integrals of the equation

$$\{p(x^2+z^2-1)+qxy\}^2=q^2(1-z^2)(x^2+y^2+z^2-1).$$

(Goursat.)

116. *Case V.* Here we have

$$f=0,\quad P=0,\quad Q=0,\quad X=0,\quad Y=0,\quad Z=0.$$

These are six equations involving five variables: hence they can coexist only if connected by certain relations. If so connected, they may be equivalent to five equations, or to four, or to three, or to two: on the assumption that $f=0$ is irreducible, they cannot all be satisfied in virtue of one equation alone. We shall assume that f is a polynomial function of its arguments.

When the six equations are equivalent to five, the equations determine a limited number of sets of values for the five variables. Taking any one of these, we have to consider the form of the characteristic at the point. As before, we take the point for origin, and the tangent plane of the complete integral is made the plane $z=0$: so that we have

$$x=0,\quad y=0,\quad z=0,\quad p=0,\quad q=0$$

at the point. It is clear that there can be no terms of the first order in f: so that the constant c, of § 110, is zero. In other respects the analysis of §§ 111, 112 applies in the present sub-case: and the nature of the point on the characteristic is the same as for the corresponding alternatives in that discussion.

When the six equations are equivalent to four, determining y, z, p, q as functions of x, the discussion of § 111 will suffice for the present sub-case. When they are equivalent to two only, their significance is similar to that of the corresponding equations already (§ 109) discussed.

It remains therefore to discuss them when they are equivalent to three equations only, expressing z, p, q as functions of x and y: we have seen that it was not possible* definitely to declare that the relation between x, y, z is a singular integral, because of the vanishing of Z. The method of § 104 is not applicable: for the equation $f=0$ cannot be resolved with regard to z because $Z=0$: indeed, it cannot be resolved with regard to any of the variables so as to give a regular equation because X, Y, Z, P, Q all vanish. Suppose that

$$f=0, \quad P=0, \quad Q=0,$$

are three independent equations giving z, p, q as functions of x and y: and that these values make X, Y, Z all vanish. When the values are substituted, they make $f=0$, $P=0$, $Q=0$ satisfied identically: hence†

$$\frac{\partial P}{\partial x}+\frac{\partial P}{\partial p}\frac{\partial p}{\partial x}+\frac{\partial P}{\partial q}\frac{\partial q}{\partial x}+\frac{\partial P}{\partial z}\frac{\partial z}{\partial x}=0,$$

$$\frac{\partial P}{\partial y}+\frac{\partial P}{\partial p}\frac{\partial p}{\partial y}+\frac{\partial P}{\partial q}\frac{\partial q}{\partial y}+\frac{\partial P}{\partial z}\frac{\partial z}{\partial y}=0.$$

If then $\dfrac{\partial P}{\partial z}$ does not vanish when the values of z, p, q are substituted in it, it will be sufficient that the equations

$$\frac{\partial P}{\partial x}+\frac{\partial P}{\partial p}\frac{\partial p}{\partial x}+\frac{\partial P}{\partial q}\frac{\partial q}{\partial x}+\frac{\partial P}{\partial z}p=0,$$

$$\frac{\partial P}{\partial y}+\frac{\partial P}{\partial p}\frac{\partial p}{\partial y}+\frac{\partial P}{\partial q}\frac{\partial q}{\partial y}+\frac{\partial P}{\partial z}q=0,$$

shall be satisfied in order to secure that the relation between x, y, z is an integral of the partial equation. The equation $Q=0$

* Except, of course, by direct substitution in the partial differential equation.

† The equation $f=0$ gives no further information when thus treated, because P, Q, X, Y, Z all vanish for the values in question.

can be treated similarly, and will give a similar result if $\frac{\partial Q}{\partial z}$ does not vanish for the values in question.

But the obvious tests, which really are the equivalent of substituting in the differential equation, are that values of p, q, z should satisfy

$$p=\frac{\partial}{\partial x}(z), \qquad q=\frac{\partial}{\partial y}(z):$$

if these relations are satisfied, the relation between x, y, z is an integral.

The subject, in the case when $Z=0$, admits of considerable expansion: but it will not be pursued further in this place. Darboux has given some discussion* of a limited class of equations; there seems plenty of opening for further investigation.

Ex. 1. Consider the equation

$$f=(px+qy-z)^2-p^2-q^2+\frac{z^2}{x^2+y^2-1}=0,$$

which already (§ 78) has been discussed. The equations

$$f=0, \quad X=0, \quad Y=0, \quad Z=0, \quad P=0, \quad Q=0$$

are satisfied in virtue of the two relations

$$\frac{p}{x}=\frac{q}{y}=\frac{z}{x^2+y^2-1};$$

and these lead to an integral

$$z^2=a^2(x^2+y^2-1),$$

which is not a singular integral as it certainly is not the envelope of the complete integrals of the equation. But, though it satisfies

$$Z=0,$$

it is an integral of the original equation.

Ex. 2. Consider the equation

$$64(p^2+q^2)^3=27z^2.$$

The equations

$$f=0, \quad X=0, \quad Y=0, \quad Z=0, \quad P=0, \quad Q=0$$

are satisfied by a relation

$$z=0$$

(with $p=0$, $q=0$), which is an integral: and the complete integral is

$$\{(x-a)^2+(y-b)^2\}^3=27z^4.$$

What is the relation between the two integrals?

* In § 35 of his memoir.

Ex. 3. Consider the equation

$$a^2\{(p-1)^2+(q-1)^2\}=z^2.$$

The six equations

$$f=0,\quad X=0,\quad Y=0,\quad Z=0,\quad P=0,\quad Q=0$$

are satisfied by

$$z=0,\quad p-1=0,\quad q-1=0:$$

but $z=0$ is clearly not an integral of the partial differential equation.

The same holds of any equation

$$f(p+a,\ q+b)=z^m,$$

where the constant m is greater than unity, a and b are constants which do not vanish, and f is a regular function of its arguments containing no terms of order lower than two in $p+a$ and $q+b$ combined. All the six equations are satisfied by

$$z=0,\quad p+a=0,\quad q+b=0:$$

but $z=0$ is not an integral of the equation.

Ex. 4. Discuss the relation of the locus $z=0$ to the complete integral and to the general integral of

$$a^2\{(p-1)^2+(q-1)^2\}=z^2.$$

Ex. 5. Shew that all the integrals of

$$pq=z^2,$$

which touch the integral $z=0$ along the axis of y, are given by

$$z=Ax^n\frac{e^{(n^2+4xy)^{\frac{1}{2}}}}{\{n+(n^2+4xy)^{\frac{1}{2}}\}^n},$$

where A and n are arbitrary. (Darboux.)

Ex. 6. Obtain the complete integral and the general integral of the equation in the preceding example. Is $z=0$ a singular integral?

Ex. 7. Integrate the equations:

(i) $p^2+q^2+2zpq=z^2$:

(ii) $p^3+q^3+z(p+q)=z^2$:

(iii) $pq(p+q)+z(p^2+q^2)=z^2$,

discussing, for each of them, the relations between the integral $z=0$ and the other integrals.

CHAPTER VIII.

THE METHOD OF CHARACTERISTICS IN ANY NUMBER OF INDEPENDENT VARIABLES.

THE present chapter gives an account of Cauchy's method of characteristics as applied to a single equation in n independent variables: and the account is made brief, because the process and the results are a generalisation of the process and the results for two independent variables, as expounded at considerable length in the two preceding chapters. Moreover, as the geometry of ordinary space has been amply used for illustration of the simpler case, it is not deemed necessary to enter at any length into illustrations of the more general case derived from the hypergeometry of $n+1$ dimensions.

Reference may be made to the works of Cauchy and of Darboux, quoted at the beginning of chapter VI. Many of the results towards the end of this chapter are believed to be new: and the subject admits of considerable development.

117. The method of characteristics can be employed when there are n independent variables $x_1, \ldots, x_n$. Adopting Cauchy's use of Ampère's practice, we change the independent variables so that they become $x_1, u_2, \ldots, u_n$: the new variables are functions of $x_2, \ldots, x_n$ (and, it may be, of x_1 also) which are independent of one another, and they will be chosen so as to simplify relations. Conversely, $x_2, \ldots, x_n, z, p_1, \ldots, p_n$ can be regarded as functions of $x_1, u_2, \ldots, u_n$; and, whatever the differential equation may be, we have

$$\frac{\partial z}{\partial x_1} = p_1 + \sum_{r=2}^{n} p_r \frac{\partial x_r}{\partial x_1},$$

$$\frac{\partial z}{\partial u_i} = \sum_{r=2}^{n} p_r \frac{\partial x_r}{\partial u_i}, \qquad (i = 2, \ldots, n).$$

Differentiating the former with regard to u_i and the latter with regard to x_1, and subtracting, we find

$$\frac{\partial p_1}{\partial u_i} = \sum_{r=2}^{n} \left(\frac{\partial p_r}{\partial x_1} \frac{\partial x_r}{\partial u_i} - \frac{\partial p_r}{\partial u_i} \frac{\partial x_r}{\partial x_1} \right),$$

holding for $i = 2, \ldots, n$.

When proper values of $x_2, \ldots, x_n, z, p_1, \ldots, p_n$, in terms of $x_1, u_2, \ldots, u_n$, are substituted in the differential equation, which may be taken in the form

$$f(x_1, \ldots, x_n, z, p_1, \ldots, p_n) = 0,$$

it becomes an identity: hence, if

$$\frac{\partial f}{\partial z} = Z, \quad \frac{\partial f}{\partial x_i} = X_i, \quad \frac{\partial f}{\partial p_j} = P_j,$$

for i and $j = 1, \ldots, n$, we have

$$X_1 + Z\frac{\partial z}{\partial x_1} + \sum_{s=2}^{n} X_s \frac{\partial x_s}{\partial x_1} + \sum_{i=1}^{n} P_i \frac{\partial p_i}{\partial x_1} = 0,$$

$$Z\frac{\partial z}{\partial u_\mu} + \sum_{s=2}^{n} X_s \frac{\partial x_s}{\partial u_\mu} + \sum_{i=1}^{n} P_i \frac{\partial p_i}{\partial u_\mu} = 0,$$

the latter holding for $\mu = 2, \ldots, n$. Substituting in the latter for $\frac{\partial z}{\partial u_\mu}$ and $\frac{\partial p_1}{\partial u_\mu}$, and rearranging the terms, we find

$$\sum_{r=2}^{n} \left\{\left(X_r + Zp_r + P_1 \frac{\partial p_r}{\partial x_1}\right) \frac{\partial x_r}{\partial u_\mu}\right\} + \sum_{r=2}^{n} \left\{\left(P_r - P_1 \frac{\partial x_r}{\partial x_1}\right) \frac{\partial p_r}{\partial u_\mu}\right\} = 0.$$

Thus far, the new variables $u_2, \ldots, u_n$ are at our disposal: let them, if possible, be chosen so that

$$P_r - P_1 \frac{\partial x_r}{\partial x_1} = 0,$$

for $r = 2, \ldots, n$, these $n-1$ equations being formally independent of one another. On the choice thus made, the foregoing equation becomes

$$\sum_{r=2}^{n} \left\{\left(X_r + Zp_r + P_1 \frac{\partial p_r}{\partial x_1}\right) \frac{\partial x_r}{\partial u_\mu}\right\} = 0;$$

and it holds for $\mu = 2, \ldots, n$. There is thus a set of $n-1$ equations, homogeneous and linear in $n-1$ quantities; the determinant of their coefficients, being

$$J\left(\frac{x_2, \ldots, x_n}{u_2, \ldots, u_n}\right),$$

does not vanish, and therefore the quantities themselves vanish, that is,

$$X_r + Zp_r + P_1 \frac{\partial p_r}{\partial x_1} = 0,$$

for $r=2, \ldots, n$. Substituting the values of $X_2, \ldots, X_n$ thus given, and also the value of $\dfrac{\partial z}{\partial x_1}$, in the equation

$$X_1 + Z\frac{\partial z}{\partial x_1} + \sum_{s=2}^{n} X_s \frac{\partial x_s}{\partial x_1} + \sum_{i=1}^{n} P_i \frac{\partial p_i}{\partial x_1} = 0,$$

and reducing, we have

$$X_1 + Zp_1 + P_1\frac{\partial p_1}{\partial x_1} = 0.$$

Consequently, the equations

$$P_r - P_1\frac{\partial x_r}{\partial x_1} = 0, \quad X_i + Zp_i + P_1\frac{\partial p_i}{\partial x_1} = 0,$$

for $r = 2, \ldots, n$, and $i = 1, \ldots, n$, are satisfied: it will be noticed that they involve no derivatives with regard to $u_2, \ldots, u_n$.

Now this aggregate of $2n-1$ equations can be taken in the form

$$\frac{dx_1}{P_1} = \frac{dx_2}{P_2} = \ldots = \frac{dx_n}{P_n} = \frac{dp_1}{-(X_1+Zp_1)} = \ldots = \frac{dp_n}{-(X_n+Zp_n)},$$

which are a set of ordinary equations; so far as they are concerned, the arbitrary quantities that arise in the integration can be made functions of the variables $u_2, \ldots, u_n$, which do not occur explicitly in the set. If we equate each member of the above aggregate to

$$\frac{dz}{p_1P_1 + \ldots + p_nP_n},$$

the equation

$$\frac{\partial z}{\partial x_1} = p_1 + \sum_{r=2}^{n} p_r \frac{\partial x_r}{\partial x_1}$$

will be satisfied by the integrals of the set; no limitation will thereby be imposed upon the arbitrary functions of $u_2, \ldots, u_n$ that occur in the integrals. But the equations

$$\frac{\partial z}{\partial u_\mu} = \sum_{r=2}^{n} p_r \frac{\partial x_r}{\partial u_\mu}$$

have also to be satisfied; these will obviously impose limitations upon the arbitrary functions of $u_2, \ldots, u_n$.

118. Accordingly, we take the equations in the form

$$\frac{dx_1}{P_1} = \frac{dx_2}{P_2} = \ldots = \frac{dx_n}{P_n} = \frac{dz}{p_1P_1 + \ldots + p_nP_n}$$

$$= \frac{dp_1}{-(X_1+Zp_1)} = \ldots = \frac{dp_n}{-(X_n+Zp_n)} = dt,$$

introducing a new variable t so as to secure the general symmetry in Darboux's presentation of Cauchy's method: t is the independent variable for the system of ordinary equations. Let

$$z=\zeta, \quad x_1=\xi_1, \ \ldots, \ x_n=\xi_n, \quad p_1=\pi_1, \ \ldots, \ p_n=\pi_n,$$

be a set of initial values assumed by the variables in the aggregate of ordinary equations, subject at this stage to the sole condition

$$f(\xi_1, \ldots, \xi_n, \zeta, \pi_1, \ldots, \pi_n)=0.$$

We shall suppose that not all the quantities $P_1, \ldots, P_n, X_1+Zp_1, \ldots, X_n+Zp_n$ vanish for these initial values and that each of these quantities is a regular function of its arguments in the vicinity of the initial values. From the theory of ordinary equations, it is known that the equations possess a unique set of integrals which are regular functions of t and are such that, when $t=0$, they acquire the assigned initial values respectively: let these integrals be

$$\begin{aligned}
x_1 &= x_1(t, \xi_1, \ldots, \xi_n, \zeta, \pi_1, \ldots, \pi_n), \\
&\ldots\ldots\ldots\ldots\ldots\ldots\ldots\ldots\ldots\ldots \\
x_n &= x_n(t, \xi_1, \ldots, \xi_n, \zeta, \pi_1, \ldots, \pi_n), \\
z &= z(t, \xi_1, \ldots, \xi_n, \zeta, \pi_1, \ldots, \pi_n), \\
p_1 &= p_1(t, \xi_1, \ldots, \xi_n, \zeta, \pi_1, \ldots, \pi_n), \\
&\ldots\ldots\ldots\ldots\ldots\ldots\ldots\ldots\ldots\ldots \\
p_n &= p_n(t, \xi_1, \ldots, \xi_n, \zeta, \pi_1, \ldots, \pi_n).
\end{aligned}$$

When in the expressions thus obtained we make the quantities $\xi_1, \ldots, \xi_n, \zeta, \pi_1, \ldots, \pi_n$ functions of $u_2, \ldots, u_n$, at present arbitrary subject solely to the condition

$$f(\xi_1, \ldots, \xi_n, \zeta, \pi_1, \ldots, \pi_n)=0,$$

it is necessary that they should satisfy the relation

$$\frac{\partial z}{\partial t}=p_1\frac{\partial x_1}{\partial t}+\ldots+p_n\frac{\partial x_n}{\partial t},$$

(which the differential equations in the ordinary system shew to be satisfied), and the relations

$$\frac{\partial z}{\partial u_\mu}=p_1\frac{\partial x_1}{\partial u_\mu}+\ldots+p_n\frac{\partial x_n}{\partial u_\mu},$$

for $\mu=2, \ldots, n$, if the above integrals thus modified are to provide integrals of the original partial equation. Let

$$L_\mu=\frac{\partial z}{\partial u_\mu}-p_1\frac{\partial x_1}{\partial u_\mu}-\ldots-p_n\frac{\partial x_n}{\partial u_\mu};$$

then, as

$$\frac{\partial z}{\partial t} = p_1 \frac{\partial x_1}{\partial t} + \dots + p_n \frac{\partial x_n}{\partial t},$$

so that

$$\frac{\partial^2 z}{\partial t \partial u_\mu} = \sum_{r=1}^{n} \left(p_r \frac{\partial^2 x_r}{\partial t \partial u_\mu} + \frac{\partial p_r}{\partial u_\mu} \frac{\partial x_r}{\partial t} \right),$$

we have

$$\begin{aligned} \frac{\partial L_\mu}{\partial t} &= \frac{\partial^2 z}{\partial t \partial u_\mu} - \sum_{r=1}^{n} \left(p_r \frac{\partial^2 x_r}{\partial t \partial u_\mu} + \frac{\partial p_r}{\partial t} \frac{\partial x_r}{\partial u_\mu} \right) \\ &= \sum_{r=1}^{n} \left(\frac{\partial p_r}{\partial u_\mu} \frac{\partial x_r}{\partial t} - \frac{\partial x_r}{\partial u_\mu} \frac{\partial p_r}{\partial t} \right) \\ &= \sum_{r=1}^{n} \left\{ P_r \frac{\partial p_r}{\partial u_\mu} + (X_r + p_r Z) \frac{\partial x_r}{\partial u_\mu} \right\}. \end{aligned}$$

The quantities must satisfy the equation

$$f(x_1, \dots, x_n, z, p_1, \dots, p_n) = 0,$$

when their values are substituted: hence

$$Z \frac{\partial z}{\partial u_\mu} + \sum_{r=1}^{n} \left(X_r \frac{\partial x_r}{\partial u_\mu} + P_r \frac{\partial p_r}{\partial u_\mu} \right) = 0,$$

and therefore

$$\begin{aligned} \frac{\partial L_\mu}{\partial t} &= -Z \frac{\partial z}{\partial u_\mu} + \sum_{r=1}^{n} p_r Z \frac{\partial x_r}{\partial u_\mu} \\ &= -Z L_\mu, \end{aligned}$$

so that

$$L_\mu = \Lambda_\mu e^{-\int_0^t Z\,dt},$$

where Λ_μ is the value of L_μ, when $t = 0$. Now Z is a regular function of t in the vicinity of $t = 0$, so that $\int_0^t Z dt$ is finite. Consequently, in order to satisfy the relation

$$L_\mu = 0,$$

it is necessary and sufficient that the relation

$$\Lambda_\mu = 0$$

should be satisfied, that is,

$$\frac{\partial \zeta}{\partial u_\mu} = \pi_1 \frac{\partial \xi_1}{\partial u_\mu} + \dots + \pi_n \frac{\partial \xi_n}{\partial u_\mu};$$

and this must hold for $\mu = 2, \dots, n$. We thus have $n - 1$ further conditions imposed upon the quantities $\xi_1, \dots, \xi_n, \zeta, \pi_1, \dots, \pi_n$, regarded as functions of $u_2, \dots, u_n$.

Further, when these conditions are satisfied, and when the quantities are substituted in $f(x_1, \ldots, x_n, z, p_1, \ldots, p_n) = 0$, the equation is satisfied identically. For we have

$$\begin{aligned}\frac{\partial f}{\partial t} &= Z\frac{\partial z}{\partial t} + \sum_{r=1}^{n} X_r \frac{\partial x_r}{\partial t} + \sum_{r=1}^{n} P_r \frac{\partial p_r}{\partial t} \\ &= 0,\end{aligned}$$

from the differential equations that led to the construction of the variables as functions of $t, u_2, \ldots, u_n$. Also, as $L_\mu = 0$, we have

$$\frac{\partial z}{\partial u_\mu} = p_1 \frac{\partial x_1}{\partial u_\mu} + \ldots + p_n \frac{\partial x_n}{\partial u_\mu};$$

and as now

$$\frac{\partial L_\mu}{\partial t} = 0,$$

and as always

$$\frac{\partial L_\mu}{\partial t} = \sum_{r=1}^{n} \left\{P_r \frac{\partial p_r}{\partial u_\mu} + (X_r + p_r Z)\frac{\partial x_r}{\partial u_\mu}\right\},$$

we now have

$$\sum_{r=1}^{n} \left(P_r \frac{\partial p_r}{\partial u_\mu} + X_r \frac{\partial x_r}{\partial u_\mu}\right) + Z \sum_{r=1}^{n} p_r \frac{\partial x_r}{\partial u_\mu} = 0,$$

that is,

$$Z\frac{\partial z}{\partial u_\mu} + \sum_{r=1}^{n} \left(P_r \frac{\partial p_r}{\partial u_\mu} + X_r \frac{\partial x_r}{\partial u_\mu}\right) = 0,$$

and therefore

$$\frac{\partial f}{\partial u_\mu} = 0.$$

Thus

$$\frac{\partial f}{\partial t} = 0, \quad \frac{\partial f}{\partial u_2} = 0, \quad \ldots, \quad \frac{\partial f}{\partial u_n} = 0,$$

and therefore

$$\begin{aligned}f(x_1, \ldots, x_n, z, p_1, \ldots, p_n) &= \text{constant} \\ &= f(\xi_1, \ldots, \xi_n, \zeta, \pi_1, \ldots, \pi_n) \\ &= 0.\end{aligned}$$

Consequently, the expressions obtained for $x_1, \ldots, x_n, z, p_1, \ldots, p_n$ satisfy the equation

$$f(x_1, \ldots, x_n, z, p_1, \ldots, p_n) = 0$$

identically when their values are substituted, provided only that the relations

$$\Lambda_2 = 0, \quad \ldots, \quad \Lambda_n = 0$$

are satisfied. Moreover, they are such that

$$\frac{\partial z}{\partial t} = \sum_{r=1}^{n} p_r \frac{\partial x_r}{\partial t},$$

$$\frac{\partial z}{\partial u_\mu} = \sum_{r=1}^{n} p_r \frac{\partial x_r}{\partial u_\mu};$$

and therefore, if $z, p_1, \ldots, p_n$ can be expressed in terms of $x_1, \ldots, x_n$ alone, on the elimination of $t, u_2, \ldots, u_n$, the n quantities p are the derivatives of z with regard to the n variables x, that is, the relation between $z, x_1, \ldots, x_n$ provides an integral of the original partial equation. It therefore is sufficient that the initial quantities $\xi_1, \ldots, \xi_n, \zeta, \pi_1, \ldots, \pi_n$ should satisfy the $n-1$ relations

$$\frac{\partial \zeta}{\partial u_\mu} = \sum_{r=1}^{n} \pi_r \frac{\partial \xi_r}{\partial u_\mu},$$

together with the condition

$$f(\xi_1, \ldots, \xi_n, \zeta, \pi_1, \ldots, \pi_n) = 0;$$

and then the integral will be given by the elimination of the parametric quantities among the $n+1$ equations

$$x_1 = x_1(t, \xi_1, \ldots, \xi_n, \zeta, \pi_1, \ldots, \pi_n),$$

$$\ldots\ldots\ldots\ldots\ldots\ldots\ldots\ldots$$

$$x_n = x_n(t, \xi_1, \ldots, \xi_n, \zeta, \pi_1, \ldots, \pi_n),$$

$$z = z(t, \xi_1, \ldots, \xi_n, \zeta, \pi_1, \ldots, \pi_n).$$

119. The conditions as regards the initial values may be satisfied in various ways.

In the first of these ways, all the quantities ζ and ξ involve $u_2, \ldots, u_n$; the $n-1$ relations, together with

$$f(\xi_1, \ldots, \xi_n, \zeta, \pi_1, \ldots, \pi_n) = 0,$$

become n equations for the determination of $\pi_1, \ldots, \pi_n$. These n equations can be resolved for these magnitudes, unless

$$\begin{vmatrix} \Pi_1, & \ldots, & \Pi_n \\ \frac{\partial \xi_1}{\partial u_2}, & \ldots, & \frac{\partial \xi_n}{\partial u_2} \\ \ldots & \ldots & \ldots \\ \frac{\partial \xi_1}{\partial u_n}, & \ldots, & \frac{\partial \xi_n}{\partial u_n} \end{vmatrix}$$

should vanish, where

$$\Pi_r = \frac{\partial f(\xi_1, \ldots, \xi_n, \zeta, \pi_1, \ldots, \pi_n)}{\partial \pi_r},$$

so that Π_r is the initial value of P_r. We shall assume that this quantity does not vanish, recognising that an exceptional case occurs when $\Pi_1, \ldots, \Pi_n$ all vanish.

As the quantities $\zeta, \xi_1, \ldots, \xi_n$ involve the $n-1$ variables $u_2, \ldots, u_n$, then on the elimination of $u_2, \ldots, u_n$, we have two relations which may be represented in the form

$$\phi(\xi_1, \ldots, \xi_n) = 0,$$
$$g(\xi_1, \ldots, \xi_n) = \zeta:$$

in other words, the initial conditions are such that, when a relation

$$\phi(x_1, \ldots, x_n) = 0$$

is satisfied, z is to acquire a value $g(x_1, \ldots, x_n)$. All the requirements are now satisfied, without any further restrictions: so that g can be taken quite arbitrarily, as also can ϕ.

The integral thus obtained is the *general integral*.

In the second of the ways, some (but not all) of the quantities $\zeta, \xi_1, \ldots, \xi_n$ involve the $n-1$ variables $u_2, \ldots, u_n$. Suppose that $\xi_{i+1}, \ldots, \xi_n$ do not involve any of the variables: then we clearly have

$$\xi_{i+1} = a_{i+1}, \ldots, \xi_n = a_n,$$

where the quantities $a_{i+1}, \ldots, a_n$ are constants. The relations are

$$\frac{\partial \zeta}{\partial u_\mu} = \sum_{r=1}^{i} \pi_r \frac{\partial \xi_r}{\partial u_\mu},$$

for $\mu = 2, \ldots, n$; these shew that some functional relation exists in a form

$$\zeta = g(\xi_1, \ldots, \xi_i),$$

and they determine the values of $\pi_1, \ldots, \pi_i$: and then $\pi_{i+1}, \ldots, \pi_n$ can be taken as arbitrary functions of $u_2, \ldots, u_n$, subject to the equation

$$f(\xi_1, \ldots, \xi_i, a_{i+1}, \ldots, a_n, \zeta, \pi_1, \ldots, \pi_n) = 0.$$

All the requirements now are satisfied, without further restrictions; so that the function g can be taken quite arbitrarily. The initial conditions are such that, when

$$x_{i+1} = a_{i+1}, \ldots, x_n = a_n,$$

z is to acquire a value $g(x_1, \ldots, x_i)$.

The integral is of the general type: clearly it is a specialised case of the general integral.

In the third of the ways, all the quantities ζ, $\xi_1, \ldots, \xi_n$ may be independent of the variables: they then are constants. Among the $n+2$ equations

$$
\begin{aligned}
x_1 &= x_1(t, \xi_1, \ldots, \xi_n, \zeta, \pi_1, \ldots, \pi_n), \\
&\cdots\cdots\cdots\cdots\cdots\cdots \\
x_n &= x_n(t, \xi_1, \ldots, \xi_n, \zeta, \pi_1, \ldots, \pi_n), \\
z &= z\ (t, \xi_1, \ldots, \xi_n, \zeta, \pi_1, \ldots, \pi_n), \\
0 &= f\ (\ \xi_1, \ldots, \xi_n, \zeta, \pi_1, \ldots, \pi_n),
\end{aligned}
$$

we eliminate the $n+1$ quantities $t, \pi_1, \ldots, \pi_n$. The eliminant is an integral.

As there are $n+1$ constants, which are values of the variables, one of them may be looked upon as an initial value, and the rest of them may be regarded as arbitrary. The integral so obtained is a *complete integral.*

It is clear that the integral, provided by the second of the ways, is intermediate in character between the complete integral and the general integral. If in its expression, $i = n-1$, it effectively is the general integral; if, on the other hand, $i=0$, it effectively is the complete integral.

As regards the possibility of giving (or even expecting) an explicit form, by the elimination of $t, u_2, \ldots, u_n$ among the $n+1$ equations

$$
\begin{aligned}
x_1 &= x_1\ (t, \xi_1, \ldots, \xi_n, \zeta, \pi_1, \ldots, \pi_n), \\
&\cdots\cdots\cdots\cdots\cdots\cdots \\
x_n &= x_n(t, \xi_1, \ldots, \xi_n, \zeta, \pi_1, \ldots, \pi_n), \\
z &= z\ \ (t, \xi_1, \ldots, \xi_n, \zeta, \pi_1, \ldots, \pi_n),
\end{aligned}
$$

we may resolve the first n of the equations with regard to $t, u_2, \ldots, u_n$, and substitute the values so obtained in the last: the resulting form will be the integral. This process is theoretically possible, unless the quantity

$$
\begin{vmatrix}
\dfrac{\partial x_1}{\partial t}, & \ldots, & \dfrac{\partial x_n}{\partial t} \\
\dfrac{\partial x_1}{\partial u_2}, & \ldots, & \dfrac{\partial x_n}{\partial u_2} \\
\cdots & \cdots & \cdots \\
\dfrac{\partial x_1}{\partial u_n}, & \ldots, & \dfrac{\partial x_n}{\partial u_n}
\end{vmatrix}
$$

vanishes identically, that is, unless the quantity

$$\begin{vmatrix} P_1, & \dots, & P_n \\ \frac{\partial x_1}{\partial u_2}, & \dots, & \frac{\partial x_n}{\partial u_2} \\ \dots & \dots & \dots \\ \frac{\partial x_1}{\partial u_n}, & \dots, & \frac{\partial x_n}{\partial u_n} \end{vmatrix}$$

vanishes identically. In assuming this, we make no new assumption: for if it vanishes identically, its value, when $t = 0$, is zero: and this has already been assumed not to be the fact.

120. Various assumptions have been made which, as in the case of only two independent variables, restrict the application of the theorem and the process.

Thus, it has been assumed that the quantities $P_1, \dots, P_n$, $X_1 + p_1 Z, \dots, X_n + p_n Z$ are regular functions of their arguments within the vicinity of the assigned initial values. If, therefore, any, or some, or all, of these quantities are characterised by deviations from regularity, whether by singularities or by algebraic irrationalities or by places of indeterminate values (to mention only the more familiar examples), then the theorems relating to a set of ordinary equations no longer apply of necessity: and the further inferences are then not necessarily valid.

Again, it has been recognised that, if $P_1, \dots, P_n$ vanish simultaneously for the initial values, the argument is not completely effective: as in the case of two independent variables, deviation from regularity can be caused thereby: and the equations require further consideration.

Again, it has been assumed that $P_1, \dots, P_n, X_1 + p_1 Z, \dots, X_n + p_n Z$ do not simultaneously vanish for values of the variables connected by the relation $f = 0$; but instances are known in which this assumption is not justified, the equations

$$f = 0, \quad P_1 = 0, \dots, P_n = 0, \quad X_1 + p_1 Z = 0, \dots, X_n + p_n Z = 0,$$

being consistent with one another. In such an event, there are two alternatives. Either the quantities $p_1, \dots, p_n$ can be eliminated, and the eliminant is a relation between $z, x_1, \dots, x_n$: this relation then provides the *singular integral* of the equation. Or, though the elimination is not possible, the equations are satisfied ($f = 0$

being irreducible) by proper values for at least two of the quantities p in terms of the remainder: let there be m such equations, where $m \leqslant n$; then these equations form a complete system, and they have an integral involving $n-m+1$ constants, which integral is a specialised case of the complete integral.

Moreover, there is no guarantee at any stage that every possible integral of the equation can be derived by the processes adopted: and it has, in fact, been found to be the case that a partial equation can be satisfied by an integral of the type called *special*, not falling within any of the indicated classes.

121. As in the case of two independent variables, so in the case of n independent variables, one exceptional instance requires consideration. It may happen that, though no one of the quantities $P_1, \ldots, P_n$ vanishes, still the relation

$$p_1 P_1 + \ldots + p_n P_n = 0$$

might be satisfied: it would, for example, be satisfied if f were homogeneous in the derivatives p. One of the integrals of the ordinary equations then would be

$$\begin{aligned} z &= \text{quantity independent of } t \\ &= z_0, \end{aligned}$$

a relation which would be useless for purposes of elimination if the complete integral were being sought.

In such an event, we adopt (as in the case of two variables in the corresponding event) a Legendrian transformation of the type

$$z' = z - p_1 x_1,$$

or of some similar type. For the particular transformation, the associated variables are

$$\begin{aligned} x_2' &= x_2, \ldots, x_n' = x_n, \quad x_1' = -p_1, \\ p_2' &= p_2, \ldots, p_n' = p_n, \quad p_1' = x_1: \end{aligned}$$

the quantity $p_1'P_1' + \ldots + p_n'P_n'$ in the transformed system is then obtained from

$$x_1(X_1 + p_1 Z) + p_2 P_2 + \ldots + p_n P_n,$$

that is, from

$$x_1(X_1 + p_1 Z) - p_1 P_1,$$

by making the above substitutions. This quantity does not vanish, and so the process can be applied to the modified system; the integral of the original equation can be deduced as before.

This general method of integration has been illustrated for the case of two independent variables: it is unnecessary to illustrate it in detail for the general case.

Ex. Integrate, by Cauchy's method, the equations

(i) $p_1 p_2 \ldots p_n = x_1 x_2 \ldots x_n$;

(ii) $(p_1 - z)(p_2 - z)\ldots(p_n - z) = p_1 p_2 \ldots p_n$;

(iii) $p_1 x_1 + \ldots + p_n x_n = p_1 p_2 \ldots p_n$;

obtaining in each case an integral z which acquires an assigned value $\phi(x_2, \ldots, x_n)$, when $x_1 = a_1$.

122. The ordinary equations subsidiary to the integration of the partial equation can also be obtained as follows. In space of $n+1$ dimensions, the integral represents a hypersurface, which can be regarded as the envelope of its tangent planes. The equation of any tangent plane is

$$\zeta - z = p_1(\xi_1 - x_1) + \ldots + p_n(\xi_n - x_n);$$

when the envelope of this plane is formed, subject to the law

$$f(x_1, \ldots, x_n, z, p_1, \ldots, p_n) = 0,$$

we have

$$0 = (\xi_1 - x_1)\,\delta p_1 + \ldots + (\xi_n - x_n)\,\delta p_n,$$

$$0 = P_1 \delta p_1 + \ldots + P_n \delta p_n,$$

and therefore

$$\frac{\xi_1 - x_1}{P_1} = \ldots = \frac{\xi_n - x_n}{P_n},$$

that is, in the vicinity of the point

$$\frac{dx_1}{P_1} = \ldots = \frac{dx_n}{P_n},$$

$$= dt,$$

say, giving equations for a direction through the point. Moreover, as this direction belongs to an integral which satisfies the equation, the equation $f = 0$ will be satisfied identically when the proper values of $z, p_1, \ldots, p_n$ are inserted: so that

$$X_r + Zp_r + P_1 \frac{\partial p_1}{\partial x_r} + \ldots + P_n \frac{\partial p_n}{\partial x_r} = 0,$$

for $r = 1, \ldots, n$. Thus

$$(X_r + Zp_r)\,dt + \frac{\partial p_1}{\partial x_r} dx_1 + \ldots + \frac{\partial p_n}{\partial x_r} dx_n = 0.$$

Now

$$\frac{\partial p_n}{\partial x_r} = \frac{\partial^2 z}{\partial x_r \partial x_n} = \frac{\partial p_r}{\partial x_n};$$

and therefore

$$\frac{\partial p_1}{\partial x_r} dx_1 + \ldots + \frac{\partial p_n}{\partial x_r} dx_n = \frac{\partial p_r}{\partial x_1} dx_1 + \ldots + \frac{\partial p_r}{\partial x_n} dx_n$$
$$= dp_r,$$

so that

$$(X_r + Zp_r)\, dt + dp_r = 0,$$

for $r = 1, \ldots, n$. Also

$$dz = p_1 dx_1 + \ldots + p_n dx_n;$$

hence, gathering together the various equations, we have

$$\frac{dx_1}{P_1} = \ldots = \frac{dx_n}{P_n} = \frac{dz}{p_1 P_1 + \ldots + p_n P_n} = \frac{-dp_1}{X_1 + p_1 Z} = \ldots = \frac{-dp_n}{X_n + p_n Z}$$
$$= dt,$$

which are the equations in question.

Next, consider the various integrals. There is a complete integral, which may be taken in the form

$$\phi(x_1, \ldots, x_n, z, a_1, \ldots, a_n) = 0.$$

The least restricted general integral is obtained by eliminating the n constants among the $n + 1$ equations

$$\phi = 0, \quad a_1 = \theta(a_2, \ldots, a_n),$$
$$\frac{d\phi}{da_r} = \frac{\partial \phi}{\partial a_r} + \frac{\partial \phi}{\partial a_1} \frac{\partial \theta}{\partial a_r} = 0,$$

for $r = 2, \ldots, n$; it will be a single equation, and it represents the envelope of that family of complete surfaces selected by the relation

$$a_1 = \theta(a_2, \ldots, a_n).$$

In the uneliminated form, the equations represent a locus of one dimension, which is the intersection of n consecutive surfaces obtained by varying the $n - 1$ independent parameters in

$$\phi(x_1, \ldots, x_n, z, \theta, a_2, \ldots, a_n) = 0.$$

On the analogy of ordinary space, such a curve is called a *characteristic*: clearly the general integral is a locus of characteristics. As a characteristic is a locus of one dimension, it can be represented by a set of ordinary equations, which are easily found as follows to be the preceding set.

The differential equation is obtained by the elimination of $a_1, \ldots, a_n$ among the $n+1$ equations

$$\left.\begin{aligned} \phi &= \phi(x_1, \ldots, x_n, z, a_1, \ldots, a_n) = 0 \\ \phi_r &= \frac{\partial \phi}{\partial x_r} + p_r \frac{\partial \phi}{\partial z} = 0 \end{aligned}\right\},$$

for $r = 1, \ldots, n$; and it is the sole equation resulting from that elimination. The only independent relations, that connect differential elements $dx_1, \ldots, dx_n, dz, dp_1, \ldots, dp_n$, are

$$d\phi = 0, \quad d\phi_1 = 0, \ldots, d\phi_n = 0;$$

and $df = 0$ is a relation connecting those elements: hence quantities $\mu, \mu_1, \ldots, \mu_n$, free from differential relations, must exist such that the relation

$$df = \mu d\phi + \mu_1 d\phi_1 + \ldots + \mu_n d\phi_n$$

is satisfied: and therefore

$$f = \mu\phi + \mu_1\phi_1 + \ldots + \mu_n\phi_n,$$

because $f, \phi, \phi_1, \ldots, \phi_n$ vanish together.

Because no one of the constants $a_1, \ldots, a_n$ appears in f, we have

$$0 = \mu \frac{\partial \phi}{\partial a_r} + \mu_1 \frac{\partial \phi_1}{\partial a_r} + \ldots + \mu_n \frac{\partial \phi_n}{\partial a_r},$$

for $r = 1, \ldots, n$. Again, p_s occurs in ϕ_s only and in no one of the other quantities ϕ; hence

$$\frac{\partial f}{\partial p_s} = \mu_s \frac{\partial \phi}{\partial z},$$

for $s = 1, \ldots, n$. For the integral in question, we have

$$\frac{\partial f}{\partial x_i} = \mu \frac{\partial \phi}{\partial x_i} + \sum_{j=1}^{n} \mu_j \left(\frac{\partial^2 \phi}{\partial x_i \partial x_j} + p_j \frac{\partial^2 \phi}{\partial x_i \partial z} \right),$$

$$\frac{\partial f}{\partial z} = \mu \frac{\partial \phi}{\partial z} + \sum_{j=1}^{n} \mu_j \left(\frac{\partial^2 \phi}{\partial x_j \partial z} + p_j \frac{\partial^2 \phi}{\partial z^2} \right),$$

and therefore

$$\frac{\partial f}{\partial x_i} + p_i \frac{\partial f}{\partial z} = \sum_{j=1}^{n} \mu_j \left(\frac{\partial^2 \phi}{\partial x_i \partial x_j} + p_j \frac{\partial^2 \phi}{\partial x_i \partial z} + p_i \frac{\partial^2 \phi}{\partial x_j \partial z} + p_i p_j \frac{\partial^2 \phi}{\partial z^2} \right),$$

for $i = 1, \ldots, n$.

The integral equations of the characteristic are

$$\phi = 0, \quad a_1 = \theta(a_2, \ldots, a_n),$$

$$\psi_\kappa = \frac{\partial \phi}{\partial a_\kappa} + \frac{\partial \phi}{\partial a_1}\frac{\partial a_1}{\partial a_\kappa} = 0,$$

for $\kappa = 2, \ldots, n$. From the preceding equations, we have

$$\mu\left(\frac{\partial \phi}{\partial a_r} + \frac{\partial a_1}{\partial a_r}\frac{\partial \phi}{\partial a_1}\right) + \mu_1\left(\frac{\partial \phi_1}{\partial a_r} + \frac{\partial a_1}{\partial a_r}\frac{\partial \phi_1}{\partial a_1}\right) + \ldots + \mu_n\left(\frac{\partial \phi_n}{\partial a_r} + \frac{\partial a_1}{\partial a_r}\frac{\partial \phi_n}{\partial a_1}\right) = 0;$$

therefore, along a characteristic,

$$\mu_1\left(\frac{\partial \phi_1}{\partial a_r} + \frac{\partial a_1}{\partial a_r}\frac{\partial \phi_1}{\partial a_1}\right) + \ldots + \mu_n\left(\frac{\partial \phi_n}{\partial a_r} + \frac{\partial a_1}{\partial a_r}\frac{\partial \phi_n}{\partial a_1}\right) = 0,$$

that is,

$$\mu_1\left(\frac{\partial \psi_r}{\partial x_1} + p_1\frac{\partial \psi_r}{\partial z}\right) + \ldots + \mu_n\left(\frac{\partial \psi_r}{\partial x_n} + p_n\frac{\partial \psi_r}{\partial z}\right) = 0;$$

and this holds for $r = 2, \ldots, n$. Now $\psi_r = 0$ holds permanently along the characteristic, so that

$$\frac{\partial \psi_r}{\partial x_1}dx_1 + \ldots + \frac{\partial \psi_r}{\partial x_n}dx_n + \frac{\partial \psi_r}{\partial z}dz = 0;$$

and

$$dz = p_1 dx_1 + \ldots + p_n dx_n$$

for every curve on $\phi = 0$: thus

$$\left(\frac{\partial \psi_r}{\partial x_1} + p_1\frac{\partial \psi_r}{\partial z}\right)dx_1 + \ldots + \left(\frac{\partial \psi_r}{\partial x_n} + p_n\frac{\partial \psi_r}{\partial z}\right)dx_n = 0;$$

and this holds for $r = 2, \ldots, n$. The $n-1$ equations for the ratios of $\mu_1, \ldots, \mu_n$ are exactly the same as the $n-1$ equations for the ratios of $dx_1, \ldots, dx_n$; hence

$$\frac{dx_1}{\mu_1} = \ldots = \frac{dx_n}{\mu_n} = u,$$

say, where $dx_1, \ldots, dx_n$ are elements of the characteristic. Consequently,

$$\begin{aligned} u\left(\frac{\partial f}{\partial x_i} + p_i\frac{\partial f}{\partial z}\right) &= \sum_{j=1}^{n}\left(\frac{\partial^2 \phi}{\partial x_i \partial x_j} + p_j\frac{\partial^2 \phi}{\partial x_i \partial z} + p_i\frac{\partial^2 \phi}{\partial x_j \partial z} + p_i p_j\frac{\partial^2 \phi}{\partial z^2}\right)dx_j \\ &= d\left(\frac{\partial \phi}{\partial x_i} + p_i\frac{\partial \phi}{\partial z}\right) - \frac{\partial \phi}{\partial z}dp_i \\ &= -\frac{\partial \phi}{\partial z}dp_i, \end{aligned}$$

because the relation

$$\frac{\partial \phi}{\partial x_i} + p_i \frac{\partial \phi}{\partial z} = 0$$

holds permanently in connection with the equation. Also

$$u \frac{\partial f}{\partial p_s} = u\mu_s \frac{\partial \phi}{\partial z}$$
$$= \frac{\partial \phi}{\partial z} dx_s.$$

Hence, gathering together the various equations that are satisfied along the characteristic, we have

$$\frac{dx_1}{\frac{\partial f}{\partial p_1}} = \dots = \frac{dx_n}{\frac{\partial f}{\partial p_n}} = \frac{dz}{p_1 \frac{\partial f}{\partial p_1} + \dots + p_n \frac{\partial f}{\partial p_n}} = \frac{-dp_1}{\frac{\partial f}{\partial x_1} + p_1 \frac{\partial f}{\partial z}} = \dots$$
$$= \frac{-dp_n}{\frac{\partial f}{\partial x_n} + p_n \frac{\partial f}{\partial z}} = \frac{u}{\frac{\partial \phi}{\partial z}}.$$

These are the former subsidiary equations, which accordingly are the differential equations of the characteristics.

Ex. 1. Prove that the envelope of the amplitudes

$$\tfrac{1}{4}z^2 = (a - x_1)(a - x_2)(a - x_3),$$

where a is a variable parameter, is a general integral of the differential equation

$$z = p_1 x_1 + p_2 x_2 + p_3 x_3 + p_1 p_2 p_3;$$

and find the relation among the arbitrary constants in the complete integral which leads to this general integral.

Ex. 2. An amplitude of one dimension is given by the equations

$$x_1 + bc = x_2 + ca = x_3 + ab,$$
$$z + 2abc = 0, \quad a + b + c = 0:$$

find the general form of the partial differential equation of the first order for which this amplitude can be a characteristic, and verify that the equation in the preceding example is a particular case.

Contact of the Integrals.

123. The complete integral is an amplitude of n dimensions, represented by

$$\phi(x_1, \dots, x_n, z, a_1, \dots, a_n) = 0.$$

The general integral is an amplitude also of n dimensions, obtained as the result of eliminating $a_2, \ldots, a_n$ between the equations

$$\phi = 0, \quad \frac{d\phi}{da_2} = 0, \ldots, \frac{d\phi}{da_n} = 0,$$

on taking $a_1 = \theta(a_2, \ldots, a_n)$: the n equations represent an amplitude of one dimension, being the characteristic: and various loci, of dimensions of all orders between unity and n, are given by the elimination of the various sets of constants that can be selected from $a_2, \ldots, a_n$. And there are various classes of general integrals: that general integral, which is represented by means of the foregoing equations, is the most comprehensive of them all.

The singular integral (when it exists) is an amplitude also of n dimensions, obtained as the result of eliminating $a_1, \ldots, a_n$ between the equations

$$\phi = 0, \quad \frac{\partial\phi}{\partial a_1} = 0, \quad \frac{\partial\phi}{\partial a_2} = 0, \ldots, \frac{\partial\phi}{\partial a_n} = 0.$$

The values of $p_1, \ldots, p_n$ at any position on the complete integral are given by

$$\frac{\partial\phi}{\partial x_r} + p_r \frac{\partial\phi}{\partial z} = 0,$$

for $r = 1, \ldots, n$.

The values of $p_1, \ldots, p_n$ at any position on the general integral are given by

$$\frac{\partial\phi}{\partial x_r} + p_r \frac{\partial\phi}{\partial z} + \sum_{s=2}^{n} \frac{d\phi}{da_s}\frac{\partial a_s}{\partial x_r} = 0,$$

that is, by

$$\frac{\partial\phi}{\partial x_r} + p_r \frac{\partial\phi}{\partial z} = 0,$$

for $r = 1, \ldots, n$.

The values of $p_1, \ldots, p_n$ at any position on the singular integral are given by

$$\frac{\partial\phi}{\partial x_r} + p_r \frac{\partial\phi}{\partial z} + \sum_{t=1}^{n} \frac{\partial\phi}{\partial a_t}\frac{\partial a_t}{\partial x_r} = 0,$$

that is, by

$$\frac{\partial\phi}{\partial x_r} + p_r \frac{\partial\phi}{\partial z} = 0,$$

for $r = 1, \ldots, n$.

Hence, at any position common to any two of these three integrals, the values of $p_1, \ldots, p_n$ are the same. Regarding $z, x_1, \ldots,$

$x_n, p_1, \ldots, p_n$ as defining an element of an integral of the differential equation, we can express this last result in the form that, *at any position common to any two of the three amplitudes represented by the complete integral, a general integral, and the singular integral, the two amplitudes have a common element.*

Moreover, the equations

$$\phi = 0, \quad \frac{\partial \phi}{\partial a_1} = 0, \ldots, \frac{\partial \phi}{\partial a_n} = 0$$

usually determine one set of values, or a limited number of sets of values, of $z, x_1, \ldots, x_n$, in terms of $a_1, \ldots, a_n$: that is, the number of positions common to the complete integral and the singular integral is limited and, as has just been proved, the two amplitudes have a common element at each common position. Also, a set of values of $a_1, \ldots, a_n$ determines a position on the singular integral.

Again, a relation

$$a_1 = \theta(a_2, \ldots, a_n)$$

determines an amplitude of one dimension within the singular integral; and at every position on this amplitude within the singular integral, the equations

$$\phi = 0, \quad \frac{d\phi}{da_2} = 0, \ldots, \frac{d\phi}{da_n} = 0$$

are satisfied. These are the equations of the general integral, which accordingly contains the amplitude. Hence the general integral and the singular integral have an amplitude of one dimension in common and, as has been proved, the two integrals have a common element at every position on this amplitude.

The complete integral and a general integral have a characteristic in common, being an amplitude of one dimension. Moreover, the equations of the characteristic determine the relations between z, x, p uniquely from initial values, except when those initial values belong to singularities: hence *if two complete integrals have an element in common at any position, they have in common all the elements along the characteristic through the position.*

Among the special loci to be considered is the amplitude which is the envelope of all the characteristics on any general integral. Taking two consecutive characteristics, we have

$$\phi = 0, \quad \frac{d\phi}{da_2} = 0, \ldots, \frac{d\phi}{da_n} = 0,$$

along one of them; along the other, we have

$$\phi + \frac{d\phi}{da_2} da_2 + \ldots + \frac{d\phi}{da_n} da_n = 0,$$

$$\frac{d^2\phi}{da_r da_2} da_2 + \ldots + \frac{d^2\phi}{da_r da_n} da_n = 0,$$

the latter holding for $r = 2, \ldots, n$. At positions common to both, the first of the second set of equations is satisfied by means of the first set of equations; eliminating the ratios $da_2 : da_3 : \ldots : da_n$ from the rest, we have

$$\begin{vmatrix} \frac{d^2\phi}{da_2^2}, \ldots, \frac{d^2\phi}{da_2 da_n} \\ \cdots\cdots\cdots\cdots \\ \frac{d^2\phi}{da_2 da_n}, \ldots, \frac{d^2\phi}{da_n^2} \end{vmatrix} = 0,$$

say $H(\phi) = 0$. Thus the required envelope is given by the elimination of $a_2, \ldots, a_n$ among the equations

$$\phi = 0, \quad \frac{d\phi}{da_2} = 0, \ldots, \frac{d\phi}{da_n} = 0, \quad H(\phi) = 0;$$

it is an amplitude of $n-1$ dimensions. The $n+1$ equations usually determine a set of values, or a limited number of sets of values, of $z, x_1, \ldots, x_n$ in terms of $a_2, \ldots, a_n$; hence the number of positions common to a complete integral and the envelope of the characteristics on a general integral is limited: and, of course, they are isolated positions on the amplitude of one dimension, along which the complete integral and the general integral have elements in common.

Again, consider a locus intermediate in dimensions between those of a characteristic and a general integral: such an one is obtained, for example, by the elimination of $a_2, \ldots, a_r$ between the equations

$$\phi = 0, \quad \frac{d\phi}{da_2} = 0, \ldots, \frac{d\phi}{da_n} = 0.$$

The result of the elimination will consist of $n-(r-1)$ equations; and therefore it will represent an amplitude of r dimensions in the hyperspace under consideration. Its equations involve the $n-r$ constants $a_{r+1}, \ldots, a_n$. To find its envelope, we take a consecutive amplitude: at any position on the first, we have

$$\phi = 0, \quad \frac{d\phi}{da_2} = 0, \ldots, \frac{d\phi}{da_n} = 0:$$

at any position on the second, we have

$$\phi + \frac{d\phi}{da_{r+1}} da_{r+1} + \ldots + \frac{d\phi}{da_n} da_n = 0,$$

$$\frac{d^2\phi}{da_s da_{r+1}} da_{r+1} + \ldots + \frac{d^2\phi}{da_s da_n} da_n = 0,$$

for $s = 2, \ldots, n$. At a position on the envelope, the first of the latter set is satisfied by the earlier equations: eliminating da_{r+1}, $\ldots$, da_n, we have

$$\left\| \begin{matrix} \dfrac{d^2\phi}{da_2 da_{r+1}}, & \ldots, & \dfrac{d^2\phi}{da_n da_{r+1}} \\ \cdots & \cdots & \cdots \\ \dfrac{d^2\phi}{da_2 da_n}, & \ldots, & \dfrac{d^2\phi}{da_n^2} \end{matrix} \right\| = 0.$$

These equations, which are equivalent to r independent equations in general, together with

$$\phi = 0, \quad \frac{d\phi}{da_2} = 0, \ldots, \frac{d\phi}{da_n} = 0,$$

give the envelope: it is an amplitude of $n - r + 1$ dimensions.

124. It is not proposed to make the complete generalisation of all the properties in ordinary space associated with partial differential equations in two independent variables: only one other property will be generalised here. We shall consider the order of contact in which a common element is possessed by different integrals.

Assuming that the singular integral exists, we know that it is given by the single equation which results from the elimination of $a_1, \ldots, a_n$ among the equations

$$\phi(x_1, \ldots, x_n, z, a_1, \ldots, a_n) = 0,$$

$$\frac{\partial\phi}{\partial a_1} = 0, \ \ldots, \ \frac{\partial\phi}{\partial a_n} = 0.$$

Let this single equation be supposed resolved with regard to z, with the result

$$z = \psi(x_1, \ldots, x_n);$$

and introduce a new dependent variable ζ, defined by the relation

$$\zeta = z - \psi(x_1, \ldots, x_n).$$

The complete integral is now

$$\phi(x_1, \ldots, x_n, \zeta + \psi, a_1, \ldots, a_n) = 0;$$

the derivatives with regard to $a_1, \ldots, a_n$ are unaffected. Hence, when we make the elimination as before and resolve this new eliminant with regard to ζ, the same resolution as before leads to

$$\zeta = 0.$$

We therefore may take the singular integral in the form

$$z = 0.$$

It is true that this equation has arisen out of the resolution of another equation, and that therefore it may not (and generally will not) represent the singular integral in the whole of its extent: but the immediate purpose is the discussion of the closeness of possession of an element, common to the singular integral and to any other integral at a common position, and therefore only the immediate vicinity of any position on $z=0$ need be considered. Let any position on $z=0$ be taken: when chosen, it is made the origin, so that we are considering the immediate vicinity of

$$z=0, \quad x_1=0, \;\ldots,\; x_n=0.$$

Moreover, at that position (and at any other) on the part of the singular integral under consideration, we have

$$p_1=0, \;\ldots,\; p_n=0,$$

because z is steadily zero.

Let

$$\phi(x_1, \ldots, x_n, z, a_1, \ldots, a_n)=0$$

be an integral: the discrimination, as to whether it is general or complete, will depend upon the other equations (if any) that are associated with it. As $z=0$ is the singular integral, then at every position common to $\phi=0$ and the singular integral, we have

$$\frac{\partial \phi}{\partial a_1}=0, \;\ldots,\; \frac{\partial \phi}{\partial a_n}=0;$$

and we know that, at any such position, the two integrals have a common element so that the values of $p_1, \ldots, p_n$ for $\phi=0$, as given by

$$\frac{\partial \phi}{\partial x_r}+p_r\frac{\partial \phi}{\partial z}=0,$$

for $r=1, \ldots, n$, must be the same as those for $z=0$. The latter vanish: hence, at a common element,

$$\frac{\partial \phi}{\partial x_1}=0, \;\ldots,\; \frac{\partial \phi}{\partial x_n}=0.$$

In general, $\frac{\partial\phi}{\partial z}$ will not vanish there: the position then would be a singularity on $\phi=0$, and circumstances would require to be very special in order that a singularity of the amplitude $\phi=0$ should lie upon its envelope.

Assuming the origin to be the common element in question, we thus have

$$\frac{\partial\phi}{\partial a_1}=0,\ \ldots,\ \frac{\partial\phi}{\partial a_n}=0,\quad \frac{\partial\phi}{\partial x_1}=0,\ \ldots,\ \frac{\partial\phi}{\partial x_n}=0,$$

at the position $0, \ldots, 0$ on the integral $\phi=0$.

In order to simplify the consideration of the small variations along two integrals in the immediate vicinity of a common element, Darboux proceeds as follows. As in the case of two independent variables, so, generally, in the case of n independent variables, the discussion centres round an aggregate of terms of the second order of the type

$$\Sigma\left(\frac{\partial^2\phi}{\partial x_i\partial x_j}dx_i dx_j+\frac{\partial^2\phi}{\partial x_i\partial a_\mu}dx_i da_\mu+\frac{\partial^2\phi}{\partial a_\mu\partial a_\nu}da_\mu da_\nu\right),$$

where the values $z=0$, $x_1=0$, $\ldots$, $x_n=0$ and the corresponding values of $a_1, \ldots, a_n$ are to be substituted in the coefficients of the bilinear terms. Let a homogeneous linear change be effected upon the variables x, and another upon the constants a: these do not affect the position of the common element and, among other things, they can be used to render the position of the axes of $x_1, \ldots, x_n$ more precise. The number of constants at our disposal in two such transformations is $2n^2$; let them be chosen so as, if possible, to make all the quantities

$$\frac{\partial^2\phi}{\partial x_i\partial x_j},\quad \frac{\partial^2\phi}{\partial x_i\partial a_j},\quad \frac{\partial^2\phi}{\partial a_i\partial a_j},$$

for all values of i and j from $1, \ldots, n$ that are distinct from one another, vanish at the common element. In general, these conditions amount to $2n(n-1)$ relations among the constants; for these, the $2n^2$ constants more than suffice. Hence, in addition to the equations

$$\frac{\partial\phi}{\partial a_1}=0,\ \ldots,\ \frac{\partial\phi}{\partial a_n}=0,\quad \frac{\partial\phi}{\partial x_1}=0,\ \ldots,\ \frac{\partial\phi}{\partial x_n}=0,$$

satisfied at a common element, we may also suppose that the equations

$$\frac{\partial^2\phi}{\partial a_i \partial a_j} = 0, \quad \frac{\partial^2\phi}{\partial a_i \partial x_j} = 0, \quad \frac{\partial^2\phi}{\partial x_i \partial x_j} = 0,$$

for i and $j = 1, 2, \ldots, n$ but unequal to one another, also are satisfied.

Consider, first, small variations along the singular integral in the vicinity of the common element: these can be represented by $dx_1, \ldots, dx_n$. There is no variation of z, for z is steadily zero: the small variations give rise to variations of $a_1, \ldots, a_n$, which may be represented by $\delta a_1, \ldots, \delta a_n$. All these variations are to be subject to the equations

$$\phi = 0, \quad \frac{\partial\phi}{\partial a_r} = 0, \quad \frac{\partial\phi}{\partial x_r} = 0,$$

for $r = 1, \ldots, n$: hence, taking account of all the quantities that vanish, we have

$$\frac{\partial^2\phi}{\partial a_r^2}\delta a_r + \frac{\partial^2\phi}{\partial a_r \partial x_r} dx_r = 0,$$

$$\frac{\partial^2\phi}{\partial x_r \partial a_r}\delta a_r + \frac{\partial^2\phi}{\partial x_r^2} dx_r = 0,$$

for $r = 1, \ldots, n$. Consequently, other n equations

$$\frac{\partial^2\phi}{\partial a_r^2}\frac{\partial^2\phi}{\partial x_r^2} = \left(\frac{\partial^2\phi}{\partial a_r \partial x_r}\right)^2$$

are satisfied at the common element: these replace n of the equations containing differential elements; and the other n of those equations give the variations $\delta a_1, \ldots, \delta a_n$, belonging to the singular integral and determined by means of $dx_1, \ldots, dx_n$. Hence, at the common element, we may take

$$\frac{\partial^2\phi}{\partial x_r^2} = \mu_r \frac{\partial^2\phi}{\partial x_r \partial a_r} = \mu_r^2 \frac{\partial^2\phi}{\partial a_r^2}.$$

Now consider a general integral possessing the element at the origin: and suppose that it is of class m, that is, such that m relations are postulated among the n quantities $a_1, \ldots, a_n$. We require variations along the general integral, determined by $dx_1, \ldots, dx_n$: these may be denoted by $da_1, \ldots, da_n$: and then the m postulated relations may be taken in the differential form

$$da_1 = c_{1,m+1}\, da_{m+1} + \ldots + c_{1,n}\, da_n,$$

$$\cdots\cdots\cdots\cdots\cdots\cdots\cdots\cdots\cdots\cdots$$

$$da_m = c_{m,m+1}\, da_{m+1} + \ldots + c_{m,n}\, da_n.$$

The equation of the general integral is given by

$$\phi = 0,$$

$$\phi_r = \frac{\partial \phi}{\partial a_1} c_{1,m+r} + \frac{\partial \phi}{\partial a_2} c_{2,m+r} + \ldots + \frac{\partial \phi}{\partial a_m} c_{m,m+r} + \frac{\partial \phi}{\partial a_{m+r}} = 0,$$

for $r = 1, \ldots, n-m$; it results from the elimination of $a_1, \ldots, a_n$ among these $n-m+1$ equations and the m postulated relations.

The order of contact of the common element at the point (being the generalisation of the order of contact in ordinary space) depends upon the magnitude of dz, belonging to the general integral and expressed in terms of $dx_1, \ldots, dx_n$. To find an expression for dz, we expand ϕ along the general integral in the vicinity of the origin, we insert the initial values in the coefficients, and then we retain only the most important terms. The result can be expressed in the form

$$-2\frac{\partial \phi}{\partial z} dz = \sum_{r=1}^{n} \left(\frac{\partial^2 \phi}{\partial x_r^2} dx_r^2 + 2 \frac{\partial^2 \phi}{\partial x_r \partial a_r} dx_r da_r + \frac{\partial^2 \phi}{\partial a_r^2} da_r^2 \right)$$

$$= \sum_{r=1}^{n} \frac{\partial^2 \phi}{\partial a_r^2} (da_r + \mu_r dx_r)^2,$$

on using the former relations. The quantities $da_1, \ldots, da_n$ are given, by means of the m differential relations and by variations of

$$\phi_1 = 0, \ \ldots, \ \phi_{n-m} = 0,$$

in terms of $dx_1, \ldots, dx_n$: these $n-m$ variations are

$$\left(\frac{\partial^2 \phi}{\partial a_1^2} da_1 + \frac{\partial^2 \phi}{\partial a_1 \partial x_1} dx_1 \right) c_{1,m+r} + \ldots + \left(\frac{\partial^2 \phi}{\partial a_m^2} da_m + \frac{\partial^2 \phi}{\partial a_m \partial x_m} dx_m \right) c_{m,m+r}$$

$$+ \frac{\partial^2 \phi}{\partial a^2_{m+r}} da_{m+r} + \frac{\partial^2 \phi}{\partial a_{m+r} \partial x_{m+r}} dx_{m+r} = 0,$$

for $r = 1, \ldots, n-m$, account having been taken of the values of the coefficients of the differential elements at the origin. Using the former relations, we have

$$\frac{\partial^2 \phi}{\partial a_1^2} (da_1 + \mu_1 dx_1) c_{1,m+r} + \ldots + \frac{\partial^2 \phi}{\partial a_m^2} (da_m + \mu_m dx_m) c_{m,m+r}$$

$$+ \frac{\partial^2 \phi}{\partial a^2_{m+r}} (da_{m+r} + \mu_{m+r} dx_{m+r}) = 0,$$

for $r = 1, \ldots, n-m$. Also, the former differential relations can be written in the form

$$da_s + \mu_s dx_s - \sum_{t=1}^{n-m} c_{s,m+t} (da_{m+t} + \mu_{m+t} dx_{m+t})$$

$$= \mu_s dx_s - \sum_{t=1}^{n-m} (c_{s,m+t} \mu_{m+t} dx_{m+t}),$$

for $s=1, \ldots, m$; hence all the quantities $da_r + \mu_r dx_r$ (for $r=1, \ldots, n$) can be expressed linearly in terms of the m quantities

$$\mu_s dx_s - \sum_{t=1}^{n-m} (c_{s,m+t}\,\mu_{m+t}\,dx_{m+t}).$$

Now the quantity dz is given by

$$-2\frac{\partial\phi}{\partial z}dz = \sum_{r=1}^{n} \frac{\partial^2\phi}{\partial a_r^2}(da_r + \mu_r dx_r)^2,$$

which, after substitution, comes to be a bilinear function of the foregoing m quantities. This bilinear function does not vanish for all values of $dx_1, \ldots, dx_n$, except under special conditions; and therefore *the contact of an element, common to a general integral of any class and the singular integral, is usually of the first order.*

While the bilinear function does not vanish for all values of $dx_1, \ldots, dx_n$ except under special conditions, there are certain ratios of the values of the differential elements (which may be called hyperdirections through the origin) for which the function does vanish. Let

$$\theta_\kappa = \frac{\mu_\kappa dx_\kappa - \sum_{t=1}^{n-m} c_{\kappa,m+t}\,\mu_{m+t}\,dx_{m+t}}{\mu_1 dx_1 - \sum_{t=1}^{n-m} c_{1,m+t}\,\mu_{m+t}\,dx_{m+t}},$$

for $\kappa = 2, \ldots, m$; then we have

$$-2\frac{\partial\phi}{\partial z}dz = \left(\mu_1 dx_1 - \sum_{t=1}^{n-m} c_{1,m+t}\,\mu_{m+t}\,dx_{m+t}\right)^2 Q(\theta_2, \ldots, \theta_m),$$

where Q is a quadratic function of its arguments. It is clear that dz will be of the third order whenever

$$Q(\theta_2, \ldots, \theta_m) = 0,$$

an equation which, in general, gives two values of θ_2 in terms of $\theta_3, \ldots, \theta_m$. Hence we may say that, *when variations $dx_2, \ldots, dx_n$ are taken arbitrarily, there are generally two values of dx_1 which can be associated with them so as to make dz belong to a general integral of the third order of small quantities.*

Other results are given in the examples which follow: in particular, exceptions to the last result are indicated.

125. The preceding analysis is not merely a generalisation of that adopted, in § 105, for the case when $n=2$: it is therefore worth while setting it out briefly for that case, in order to allow of comparison with the discussion there given.

We take $z=0$ as the singular solution: any position on it is chosen and made the origin, so that, for an element there, we have

$$z=0, \quad x=0, \quad y=0, \quad p=0, \quad q=0.$$

Hence for any general integral $\phi(x, y, z, a, b)=0$, possessing that element, we have

$$\frac{\partial\phi}{\partial a}=0, \quad \frac{\partial\phi}{\partial b}=0, \quad \frac{\partial\phi}{\partial x}=0, \quad \frac{\partial\phi}{\partial y}=0.$$

Moreover, we may assume that the relations

$$\frac{\partial^2\phi}{\partial a\partial b}=0, \quad \frac{\partial^2\phi}{\partial a\partial y}=0, \quad \frac{\partial^2\phi}{\partial b\partial x}=0, \quad \frac{\partial^2\phi}{\partial x\partial y}=0,$$

are satisfied at the point: if they are not satisfied in the form in which the equation arises, they can be made to be so by making linear transformations of the variables x and y, and linear transformations of the constants a and b.

The critical condition for contact of order closer than the usual contact was found (§ 105) to be

$$\frac{\partial^2\phi}{\partial x^2}\frac{\partial^2\phi}{\partial y^2}=\left(\frac{\partial^2\phi}{\partial x\partial y}\right)^2;$$

and it was indicated (§ 105, Ex. 1) that this equation implies, and is implied by, the equation

$$\frac{\partial^2\phi}{\partial a^2}\frac{\partial^2\phi}{\partial b^2}=\left(\frac{\partial^2\phi}{\partial a\partial b}\right)^2.$$

Thus, for the transformations adopted, these conditions will be

$$\frac{\partial^2\phi}{\partial x^2}\frac{\partial^2\phi}{\partial y^2}=0, \quad \frac{\partial^2\phi}{\partial a^2}\frac{\partial^2\phi}{\partial b^2}=0,$$

respectively.

Let small variations along the singular integral in the immediate vicinity of the element at the origin be denoted by dx, dy, δa, δb: these must satisfy

$$\phi=0, \quad \frac{\partial\phi}{\partial a}=0, \quad \frac{\partial\phi}{\partial b}=0, \quad \frac{\partial\phi}{\partial x}=0, \quad \frac{\partial\phi}{\partial y}=0.$$

Hence, taking account of the various vanishing quantities, we have

$$\left.\begin{aligned}\frac{\partial^2\phi}{\partial a^2}\,\delta a+\frac{\partial^2\phi}{\partial a\partial x}\,dx=0\\ \frac{\partial^2\phi}{\partial a\partial x}\,\delta a+\frac{\partial^2\phi}{\partial x^2}\,dx=0\end{aligned}\right\},\qquad \left.\begin{aligned}\frac{\partial^2\phi}{\partial b^2}\,\delta b+\frac{\partial^2\phi}{\partial b\partial y}\,dy=0\\ \frac{\partial^2\phi}{\partial b\partial y}\,\delta b+\frac{\partial^2\phi}{\partial y^2}\,dy=0\end{aligned}\right\}.$$

Consequently, the two relations

$$\frac{\partial^2\phi}{\partial a^2}\frac{\partial^2\phi}{\partial x^2}=\left(\frac{\partial^2\phi}{\partial a\partial x}\right)^2,\qquad \frac{\partial^2\phi}{\partial b^2}\frac{\partial^2\phi}{\partial y^2}=\left(\frac{\partial^2\phi}{\partial b\partial y}\right)^2,$$

are satisfied at the common element: in virtue of these two, the four equations reduce to two only, which then determine δa and δb in terms of dx and dy. These two relations may be expressed by the equations

$$\frac{\partial^2\phi}{\partial x^2}=\rho\,\frac{\partial^2\phi}{\partial a\partial x}=\rho^2\,\frac{\partial^2\phi}{\partial a^2},$$

$$\frac{\partial^2\phi}{\partial y^2}=\sigma\,\frac{\partial^2\phi}{\partial b\partial y}=\sigma^2\,\frac{\partial^2\phi}{\partial b^2}.$$

Now consider the general integral: let the single relation between a and b, when expressed in differential form, be

$$db=c\,da,$$

where da and db are variations along the general integral; the equations of this integral are

$$\phi=0,\quad \frac{\partial\phi}{\partial a}+c\,\frac{\partial\phi}{\partial b}=0.$$

At the element, $\dfrac{\partial\phi}{\partial a}$ and $\dfrac{\partial\phi}{\partial b}$ vanish: hence da and db are such that

$$\frac{\partial^2\phi}{\partial a^2}\,da+\frac{\partial^2\phi}{\partial a\partial x}\,dx+\left(\frac{\partial^2\phi}{\partial b^2}\,db+\frac{\partial^2\phi}{\partial b\partial y}\,dy\right)c=0,$$

taking account of vanishing quantities. Thus, by the preceding relations, we have

$$\frac{\partial^2\phi}{\partial a^2}(da+\rho\,dx)+c\,\frac{\partial^2\phi}{\partial b^2}(db+\sigma\,dy)=0.$$

Also

$$c\,(da+\rho\,dx)-(db+\sigma\,dy)=c\rho\,dx-\mu\,dy,$$

so that

$$\left.\begin{aligned}\left(\frac{\partial^2\phi}{\partial a^2}+c^2\,\frac{\partial^2\phi}{\partial b^2}\right)(da+\rho\,dx)&=c\,\frac{\partial^2\phi}{\partial b^2}\,(c\rho\,dx-\mu\,dy)\\ \left(\frac{\partial^2\phi}{\partial a^2}+c^2\,\frac{\partial^2\phi}{\partial b^2}\right)(db+\sigma\,dy)&=-\frac{\partial^2\phi}{\partial a^2}\,(c\rho\,dx-\mu\,dy)\end{aligned}\right\},$$

which give the necessary values of da and db. Also, variation of the equation $\phi = 0$ leads, when account is taken of vanishing quantities, to the relation

$$-2\frac{\partial\phi}{\partial z}dz = \frac{\partial^2\phi}{\partial x^2}dx^2 + 2\frac{\partial^2\phi}{\partial a\partial x}da\,dx + \frac{\partial^2\phi}{\partial a^2}da^2$$
$$+\frac{\partial^2\phi}{\partial y^2}dy^2 + 2\frac{\partial^2\phi}{\partial b\partial y}db\,dy + \frac{\partial^2\phi}{\partial b^2}db^2$$
$$=\frac{\partial^2\phi}{\partial a^2}(da+\rho dx)^2 + \frac{\partial^2\phi}{\partial b^2}(db+\sigma dy)^2$$
$$=\frac{\dfrac{\partial^2\phi}{\partial a^2}\dfrac{\partial^2\phi}{\partial b^2}}{\dfrac{\partial^2\phi}{\partial a^2}+c^2\dfrac{\partial^2\phi}{\partial b^2}}(c\rho dx - \mu dy)^2.$$

Hence dz is usually of the second order in dx and dy: and the contact is usually of the first order.

There is a single direction along which dz is of the third order: it is given by

$$c\rho = \mu\frac{dy}{dx},$$

and the point is then a contact-point of the branches of the intersection of the surfaces in question, having this direction for its tangent*.

If, however, the relation

$$\frac{\partial^2\phi}{\partial a^2}\frac{\partial^2\phi}{\partial b^2} = 0$$

is satisfied at the point, dz is of the third order for all variations of x and y through the point: the general integral and the singular integral then have contact of the second order. (This relation agrees with the result to be expected from the earlier case.) Moreover, when this relation is satisfied at the point, it clearly is satisfied independently of any functional relation between a and b: it therefore is satisfied for all such relations and consequently, *if the singular integral has contact of the second order with any general integral, it has contact of the second order with every general integral.* This is the former result (§ 105).

* This establishes the statement made in § 104.

When the contact between the two integrals is only of the first order, the single direction in which the contact is of the second order is given by

$$c\rho = \mu \frac{dy}{dx}.$$

The quantities μ and ρ belong to the singular integral at the point: hence this direction usually changes from one general integral to another.

126. We now proceed to give some examples of the general theory which has just been expounded.

Ex. 1. When we are dealing with what is called the general integral, being the integral for which there is only a single postulated relation among the parameters, the formulæ become simple and lead easily to further results.

Let the postulated relation be

$$da_1 = b_2 da_2 + \ldots + b_n da_n;$$

then the equations of the general integral are

$$\phi = 0, \quad \frac{\partial \phi}{\partial a_2} + b_2 \frac{\partial \phi}{\partial a_1} = 0, \quad \ldots, \quad \frac{\partial \phi}{\partial a_n} + b_n \frac{\partial \phi}{\partial a_1} = 0.$$

Assuming all the properties in the text, we have the expression for dz in the form

$$-2 \frac{\partial \phi}{\partial z} dz = \sum_{r=1}^{n} \frac{\partial^2 \phi}{\partial a_r^2} (da_r + \mu_r dx_r)^2.$$

The quantities $da_1, \ldots, da_n$, determined for the general integral in terms of $dx_1, \ldots, dx_n$, are given by the above relation and by

$$\frac{\partial^2 \phi}{\partial a_r^2} (da_r + \mu_r dx_r) + b_r \frac{\partial^2 \phi}{\partial a_1^2} (da_1 + \mu_1 dx_1) = 0,$$

for $r = 2, \ldots, n$. We take the relation in the form

$$da_1 + \mu_1 dx_1 - \sum_{s=2}^{n} b_s (da_s + \mu_s dx_s) = \mu_1 dx_1 - \sum_{s=2}^{n} b_s \mu_s dx_s:$$

and we have

$$da_1 + \mu_1 dx_1 = \frac{1}{\Delta} \left(\mu_1 dx_1 - \sum_{s=2}^{n} b_s \mu_s dx_s \right),$$

$$da_r + \mu_r dx_r = -\frac{b_r}{\dfrac{\partial^2 \phi}{\partial a_r^2}} \frac{\partial^2 \phi}{\partial a_1^2} \frac{1}{\Delta} \left(\mu_1 dx_1 - \sum_{s=2}^{n} b_s \mu_s dx_s \right),$$

for $r = 2, \ldots, n$, where

$$\Delta = 1 + \frac{\partial^2 \phi}{\partial a_1^2} \left\{ \frac{b_2^2}{\dfrac{\partial^2 \phi}{\partial a_2^2}} + \ldots + \frac{b_n^2}{\dfrac{\partial^2 \phi}{\partial a_n^2}} \right\}.$$

Substituting these values, we have the value of dz in the form

$$-2\frac{\partial\phi}{\partial z}dz=\frac{1}{\Delta}\frac{\partial^2\phi}{\partial a_1^2}\left(\mu_1 dx_1-\sum_{s=2}^{n} b_s\mu_s dx_s\right)^2.$$

Hence, for the general integral (or the integral of the first class, only one relation among the parameters being postulated), an element common with the singular integral has contact of only the first order. This is the former result.

For this general integral, we may take arbitrary variations $dx_2, \ldots, dx_n$: and then there is one variation dx_1, given by

$$\mu_1 dx_1=\sum_{s=2}^{n} b_s\mu_s dx_s,$$

for which dz is of the third order of small quantities; accordingly, for those variations, the element has contact of the second order. The property is special to this general integral: for the general integrals of other classes, where the number of postulated relations is greater than one, there are two variations dx_1 for which the corresponding dz is of the third order.

Ex. 2. The relations defining a general integral of the second class can be taken in the form

$$\left.\begin{aligned}da_1&=b_3da_3+\ldots+b_nda_n\\ da_2&=c_3da_3+\ldots+c_nda_n\end{aligned}\right\}.$$

An element at the origin is possessed in common by this integral and the singular integral: and the various quantities at this position have the same values as in the investigation in the text. Let

$$\phi_{ss}=\frac{\partial^2\phi}{\partial a_s^2},$$

for $s=1, \ldots, n$; also let

$$\Delta=1+\phi_{11}\sum_{r=3}^{n}\frac{b_r^2}{\phi_{rr}}+\phi_{22}\sum_{r=3}^{n}\frac{c_r^2}{\phi_{rr}}+\phi_{11}\phi_{22}\left\{\sum_{r=3}^{n}\frac{b_r^2}{\phi_{rr}}\sum_{r=3}^{n}\frac{c_r^2}{\phi_{rr}}-\left(\sum_{r=3}^{n}\frac{b_rc_r}{\phi_{rr}}\right)^2\right\};$$

and let

$$du=\mu_1dx_1-\mu_3b_3dx_3-\ldots-\mu_nb_ndx_n,$$

$$dv=\mu_2dx_2-\mu_3c_3dx_3-\ldots-\mu_nc_ndx_n.$$

Obtain the value of dz, which measures the order of possession of the element common to this general integral and the singular integral, in the form

$$-2\Delta\frac{\partial\phi}{\partial z}dz$$

$$=\left(\phi_{11}+\phi_{11}\phi_{22}\sum_{r=3}^{n}\frac{c_r^2}{\phi_{rr}}\right)du^2+\left(\phi_{22}+\phi_{11}\phi_{22}\sum_{r=3}^{n}\frac{b_r^2}{\phi_{rr}}\right)dv^2$$

$$-2\phi_{11}\phi_{22}\sum_{r=3}^{n}\frac{b_rc_r}{\phi_{rr}}du\,dv.$$

Hence shew, in general,

(i) that the order of possession of the common element is usually the first:

(ii) that usually there are two distinct sets of variations, having $dx_2, \ldots, dx_n$ arbitrary, for which the element is possessed to the second order:

(iii) that, if either $\frac{\partial^2\phi}{\partial a_1^2}$ or $\frac{\partial^2\phi}{\partial a_2^2}$ should vanish, there is only one set of variations, arbitrary in $dx_2, \ldots, dx_n$, for which the element is possessed to the second order:

(iv) that, if the element is possessed to the second order for all variations, it is sufficient that

$$\frac{\partial^2\phi}{\partial a_1^2}=0, \quad \frac{\partial^2\phi}{\partial a_2^2}=0.$$

Are the two conditions in the last result necessary as well as sufficient?

Shew also that, if

$$\frac{\partial^2\phi}{\partial a_3^2}=0,$$

while none of the other second derivatives vanish, then

$$-2\Delta'\frac{\partial\phi}{\partial z}dz=\phi_{11}\phi_{22}(c_3du-b_3dv)^2,$$

where

$$\Delta'=b_3^2\phi_{11}+c_3^2\phi_{22}+\phi_{11}\phi_{22}\sum_{r=4}^{n}\frac{(b_3c_r-b_rc_3)^2}{\phi_{rr}};$$

and discuss this form.

Ex. 3. Adopting the notation of § 125, and assuming that all the quantities which occur have the values there given, prove that in order to have an element common to a general integral, defined by three relations among the quantities $da_1, \ldots, da_n$, and the singular integral, possessed to the second order, it is necessary and sufficient that three of the quantities

$$\frac{\partial^2\phi}{\partial a_1^2}, \quad \frac{\partial^2\phi}{\partial a_2^2}, \quad \ldots, \quad \frac{\partial^2\phi}{\partial a_n^2}$$

should vanish.

In case fewer than three of these quantities should vanish, what are the sets of variations for which the element is possessed to the second order?

Ex. 4. Shew that, if one general integral of any class has an element common with the singular integral possessed to the second order, then every general integral of that class has the same property.

127. The discussion of the order, in which an element common to a complete integral and the singular integral is possessed, is simpler. Taking the element on the singular integral as before (§ **124**), we require variations along the complete integral: these are given by $dx_1, \ldots, dx_n, dz$ alone, because $a_1, \ldots, a_n$ are constants for this integral. The earlier analysis shews at once that the most important terms in these variations are given by

$$-2\frac{\partial\phi}{\partial z}dz = \frac{\partial^2\phi}{\partial x_1^2}dx_1^2 + \ldots + \frac{\partial^2\phi}{\partial x_n^2}dx_n^2,$$

assuming that the quantities belonging to the singular integral at the point are the same as before. It is clear that, except under very special conditions, *an element common to a complete integral and to the singular integral is usually possessed only to the first order.*

Ex. Discuss the order of possession of an element, common to a complete integral and to a general integral, for which $n-1$ relations are postulated among the parameters.

CHAPTER IX.

Lie's Methods applied to Equations of the First Order.

Many of Lie's investigations are concerned with the integration of partial differential equations of the first order: broadly speaking, he has devised two general methods of proceeding which have considerable features in common. It may be added that they arise as illustrations of processes with wider issues and of analysis having a more extended significance.

One of the methods depends upon the use of tangential transformations (or contact transformations, as they are more frequently called). So far as concerns the properties of these transformations and (as an incident in their application to Pfaff's problem) their application to the integration of a single partial differential equation of the first order, an exposition has already been given* in Part I of the present work; it will be sufficient therefore to give, in this place, merely a statement of the results.

The other of the methods due to Lie depends upon the theory of groups of functions as developed, in part, through the theory of contact transformations. It is applied to a system of simultaneous equations in the first instance and naturally it can be applied to the simplest case when there is only a single equation.

Some references are given in the sections in Part I that have already been mentioned. Of these, the most important are Lie's memoir in the 8th volume (1875) of the *Mathematische Annalen*, as regards the fundamental properties of contact transformations and their simpler applications to the integration of partial differential equations, and the second volume (1890) of his *Theorie der Transformationsgruppen*, which contains many properties, developments, and applications of the transformations in question.

Reference may also be made advantageously to the exposition given by Goursat in chapters XI and XII of his treatise already (p. 55) quoted.

* See chapters IX, X; for the application, see specially §§ 136, 142.

Contact Transformations.

128. The distinctive idea of contact transformations is derived from geometrical considerations applied to hyperspace. An element of surface at any position in a space of $n+1$ dimensions is determined by means of $z, x_1, \ldots, x_n$, the coordinates of the position, and of $p_1, \ldots, p_n$, the coordinates of the orientation, of the element; thus, if $z, x_1, \ldots, x_n, p_1, \ldots, p_n$ be regarded as $2n+1$ independent magnitudes in general, the element of surface will be given by an equation

$$dz - p_1 dx_1 - \ldots - p_n dx_n = 0.$$

Transformations of the variables are conceived by means of relations between $z, x_1, \ldots, x_n, p_1, \ldots, p_n$ and new variables $z', x_1', \ldots, x_n', p_1', \ldots, p_n'$; if these relations are such that

$$dz' - \sum_{i=1}^{n} p_i' dx_i' = \rho \left(dz - \sum_{i=1}^{n} p_i dx_i \right),$$

where ρ is a non-vanishing quantity independent of differential elements, the transformation is said to be a *contact transformation*; it obviously transforms two elements of surface, that touch one another, into two other elements of surface that also touch one another. Accordingly, Lie's definition* of the most general contact transformation is:—

Let $Z, X_1, \ldots, X_n, P_1, \ldots, P_n$ *be* $2n+1$ *independent functions of* $2n+1$ *independent variables* $z, x_1, \ldots, x_n, p_1, \ldots, p_n$, *such that the relation*

$$dZ - \sum_{i=1}^{n} P_i dX_i = \rho \left(dz - \sum_{i=1}^{n} p_i dx_i \right)$$

is satisfied identically, when ρ *is a non-vanishing quantity independent of differential elements: then the equations*

$$z' = Z, \quad x_i' = X_i, \quad p_i' = P_i,$$

for $i = 1, \ldots, n$, *define a contact transformation.*

The most general contact transformation will be given when the most general functions $Z, X_1, \ldots, X_n, P_1, \ldots, P_n$, satisfying the preceding relations, are known. These are given† by the theorem:—

* *Math. Ann.*, t. VIII (1875), p. 220.

† The analysis establishing the theorem is set out in §§ 134, 135 of Part I of this work. A difference in sign should be noted: it is due to the fact that the quantity $[P, Q]$ is used in this volume in its usual sense, and has the reverse sign of $[P, Q]$ as given in Part I.

When the $n+1$ quantities $Z, X_1, \ldots, X_n$ are obtained as $n+1$ functionally independent integrals of the equations

$$[Z, X_i]=0, \quad [X_i, X_j]=0,$$

for i and $j=1, \ldots, n$; when the quantities $P_1, \ldots, P_n$ are determined, either from the n equations

$$\frac{\partial Z}{\partial x_i}+p_i\frac{\partial Z}{\partial z}=\sum_{r=1}^{n} P_r\left(\frac{\partial X_r}{\partial x_i}+p_i\frac{\partial X_r}{\partial z}\right),$$

for $i=1, \ldots, n$, or from the n equations

$$\frac{\partial Z}{\partial p_i}=\sum_{r=1}^{n} P_r\frac{\partial X_r}{\partial p_i},$$

for $i=1, \ldots, n$, the two sets of n equations being equivalent to one another; and when ρ denotes the non-vanishing quantity

$$\frac{\partial Z}{\partial z}-\sum_{r=1}^{n} P_r\frac{\partial X_r}{\partial z}:$$

then the relation

$$dZ-\sum_{i=1}^{n} P_i dX_i=\rho\left(dz-\sum_{i=1}^{n} p_i dx_i\right)$$

is satisfied identically. The conditions are both necessary and sufficient to secure the property: and other equations satisfied by $Z, X_1, \ldots, X_n, P_1, \ldots, P_n, \rho$, are

$$[Z, P_i]=\rho P_i, \quad [P_i, X_i]=-\rho,$$
$$[P_i, X_j]=0 \quad , \quad [P_i, P_j]= \quad 0,$$

for $i=1, \ldots, n$, and values of j unequal to i.

This is Mayer's form* of Lie's theorem relating to the determination of the most general contact transformation. As regards the conditions, it is to be noted that ρ is known as soon as $Z, X_1, \ldots, X_n, P_1, \ldots, P_n$ have been obtained, and $P_1, \ldots, P_n$ are known as soon as $Z, X_1, \ldots, X_n$ have been obtained.

These quantities are subject to the equations

$$[Z, X_i]=0, \quad [X_i, X_j]=0;$$

and as (§ 53) we have

$$[[Z, X_i], X_j]+[[X_i, X_j], Z]-[[Z, X_j], X_i]$$
$$=-\frac{\partial Z}{\partial z}[X_i, X_j]+\frac{\partial X_i}{\partial z}[Z, X_j]-\frac{\partial X_j}{\partial z}[Z, X_i],$$

for any functions whatever of $z, x_1, \ldots, x_n, p_1, \ldots, p_n$, the equations are consistent with one another and coexist. But the conditions

* *Math. Ann.*, t. VIII (1875), p. 309.

for the existence of the contact transformation do not require any equation to be satisfied by any one of the quantities $Z, X_1, \ldots, X_n$ alone; and the preceding relation for coexistence holds for all functions which satisfy the equations. Hence there is an arbitrary element in the equations that define the contact transformation: thus we could choose any one of the quantities $Z, X_1, \ldots, X_n$ arbitrarily: or any other arbitrary relation could be chosen that is not inconsistent with the aggregate of conditions and equations.

129. There is one most important form of contact transformation, viz. that in which the difference between the old variables and the new variables is small. Such transformations are called *infinitesimal*: they can be represented by

$$Z = z + \epsilon\zeta, \quad X_i = x_i + \epsilon\xi_i, \quad P_i = p_i + \epsilon\pi_i, \qquad (i = 1, \ldots, n),$$

where $\zeta, \xi_1, \ldots, \xi_n, \pi_1, \ldots, \pi_n$ are functions of $z, x_1, \ldots, x_n, p_1, \ldots, p_n$, and ϵ is a quantity so small that its square and higher powers may be neglected. Moreover, when we have the identical transformation, then $\rho = 1$; hence, for an infinitesimal transformation, we have

$$\rho = 1 + \epsilon\sigma,$$

where σ will be a quantity to be determined. Then the equations

$$[Z, X_i] = 0, \quad [X_i, X_j] = 0, \quad [P_i, X_j] = 0, \quad [P_i, P_j] = 0,$$

for unequal values of i and j, respectively give

$$-\frac{\partial\zeta}{\partial p_i} + \sum_{r=1}^{n} p_r \frac{\partial\xi_i}{\partial p_r} = 0,$$

$$-\frac{\partial\xi_i}{\partial p_j} + \frac{\partial\xi_j}{\partial p_i} = 0,$$

$$-\left(\frac{\partial\xi_j}{\partial x_i} + p_i \frac{\partial\xi_j}{\partial z}\right) - \frac{\partial\pi_i}{\partial p_j} = 0,$$

$$\frac{\partial\pi_j}{\partial x_i} + p_i \frac{\partial\pi_j}{\partial z} - \left(\frac{\partial\pi_i}{\partial x_j} + p_j \frac{\partial\pi_i}{\partial z}\right) = 0.$$

The last three equations, taken for all values of i and j, shew that a function U of $z, x_1, \ldots, x_n, p_1, \ldots, p_n$ exists, such that

$$\xi_i = \quad \frac{\partial U}{\partial p_i},$$

$$\pi_i = -\frac{\partial U}{\partial x_i} - p_i \frac{\partial U}{\partial z},$$

for all values of $i=1, \ldots, n$. The equation

$$[P_i, X_i] = -\rho = -1 - \epsilon\sigma$$

then gives

$$\sigma = \frac{\partial \pi_i}{\partial p_i} - \left(\frac{\partial \xi_i}{\partial x_i} + p_i \frac{\partial \xi_i}{\partial z}\right)$$

$$= -\frac{\partial U}{\partial z};$$

and the equations

$$[Z, P_i] = \rho P_i$$

then give

$$\frac{\partial \zeta}{\partial x_i} + p_i \frac{\partial \zeta}{\partial z} = -p_i \frac{\partial U}{\partial z} + \pi_i - \sum_{r=1}^{n} p_r \frac{\partial \pi_i}{\partial p_r}.$$

When these are taken in conjunction with the equations

$$\frac{\partial \zeta}{\partial p_i} = \sum_{r=1}^{n} p_r \frac{\partial \xi_i}{\partial p_r},$$

we find the value of ζ to be

$$\zeta = \sum_{r=1}^{n} \left(p_r \frac{\partial U}{\partial p_r}\right) - U;$$

consequently *the most general infinitesimal transformation is given by the equations*

$$Z = z + \epsilon\zeta, \quad X_i = x_i + \epsilon\xi_i, \quad P_i = p_i + \epsilon\pi_i, \quad \rho = 1 + \epsilon\sigma,$$

where

$$\zeta = \sum_{r=1}^{n} \left(p_r \frac{\partial U}{\partial p_r}\right) - U, \quad \xi_i = \frac{\partial U}{\partial p_i}, \quad -\pi_i = \frac{\partial U}{\partial x_i} + p_i \frac{\partial U}{\partial z}, \quad \sigma = -\frac{\partial U}{\partial z},$$

for $i=1, \ldots, n$, *and* U *denotes any arbitrary function of* $z, x_1, \ldots, x_n, p_1, \ldots, p_n$.

130. Sometimes it is necessary to know the contact transformations for which $X_1, \ldots, X_n, P_1, \ldots, P_n$ are *explicitly independent* of z, so that Z is the only quantity that involves z. The results* are as follows:—

The quantity ρ *is constant and may be made unity; then* Z *is given by*

$$Z = Az + \Pi,$$

where Π *is a function of* $x_1, \ldots, x_n, p_1, \ldots, p_n$ *that does not involve* z, *and* A *is a constant. The quantities* $X_1, \ldots, X_n$ *are functionally independent integrals of the equations*

$$(X_i, X_j) = 0,$$

* See Part I of this work, § 137. The same remark as to difference of sign from the results in Part I applies here as in the foot-note on p. 315 of this volume.

for i and $j=1, \ldots, n$, and Π is an integral of the equations

$$(\Pi, X_i) = -A \sum_{r=1}^{n} p_r \frac{\partial X_i}{\partial p_r},$$

which is functionally independent of $X_1, \ldots, X_n$. The quantities $P_1, \ldots, P_n$ are then given by any n independent equations of the set

$$\frac{\partial \Pi}{\partial x_i} - \sum_{r=1}^{n} P_r \frac{\partial X_r}{\partial x_i} = -Ap_i, \quad \frac{\partial \Pi}{\partial p_i} - \sum_{r=1}^{n} P_r \frac{\partial X_r}{\partial p_i} = 0,$$

for $i=1, \ldots, n$. The functions thus determined satisfy the relation

$$d\Pi - \sum_{i=1}^{n} P_i dX_i = -A \sum_{i=1}^{n} p_i dx_i$$

identically: and other equations satisfied are

$$(X_i, P_i) = A, \quad (\Pi, P_i) = A\left(P_i - \sum_{r=1}^{n} p_r \frac{\partial P_i}{\partial p_r}\right),$$

$$(P_i, X_j) = 0, \quad (P_i, P_j) = 0,$$

for all values of i and j unequal to one another. The constant A is usually taken to be unity.

The determination of Π can be effected by a quadrature, when $X_1, \ldots, X_n$ are known. To see this, let the variables be changed to $y_1, \ldots, y_n, v_1, \ldots, v_n$, where

$$y_1 = X_1, \ldots, y_n = X_n,$$

and $v_1, \ldots, v_n$ are n functions of $x_1, \ldots, x_n, p_1, \ldots, p_n$, chosen so that the new variables make up an aggregate of $2n$ independent functions. Now, because

$$(X_i, X_j) = 0,$$

the quantity (Π, X_i) vanishes when Π is made equal to any of the quantities $X_1, \ldots, X_n$; and therefore, when the new variables are taken, no one of the derivatives

$$\frac{\partial \Pi}{\partial y_1}, \ldots, \frac{\partial \Pi}{\partial y_n}$$

occurs in (Π, X_i). Hence the n equations

$$(\Pi, X_i) = -A \sum_{r=1}^{n} p_r \frac{\partial X_i}{\partial p_r}$$

can be resolved with respect to

$$\frac{\partial \Pi}{\partial v_1}, \ldots, \frac{\partial \Pi}{\partial v_n},$$

giving these quantities in terms of the variables: a quadrature then determines Π.

Conversely, *if we regard the contact transformation as changing the variables from* $Z, X_1, \ldots, X_n, P_1, \ldots, P_n$, *the results are similar.* We then have

$$z = aZ + \Pi,$$

where Π is a function of $X_1, \ldots, X_n, P_1, \ldots, P_n$, and a is a constant; the differential relation is

$$d\Pi - \sum_{i=1}^{n} p_i dx_i = -a \sum_{i=1}^{n} P_i dX_i;$$

the quantities $x_1, \ldots, x_n$ are functionally independent integrals of the equations

$$(x_i, x_j) = 0,$$

for i and $j = 1, \ldots, n$, where the independent variables are $X_1, \ldots, X_n, P_1, \ldots, P_n$; and Π is an integral of the equations

$$(\Pi, x_i) = -a \sum_{r=1}^{n} P_r \frac{\partial x_i}{\partial P_r},$$

which is functionally independent of $x_1, \ldots, x_n$. The quantities $p_1, \ldots, p_n$ are given by any n independent equations of the set

$$\frac{\partial \Pi}{\partial X_i} - \sum_{r=1}^{n} p_r \frac{\partial x_r}{\partial X_i} = -aP_i, \quad \frac{\partial \Pi}{\partial P_i} - \sum_{r=1}^{n} p_r \frac{\partial x_r}{\partial P_i} = 0,$$

for $i = 1, \ldots, n$; and other equations satisfied are

$$(x_i, p_i) = a, \quad (\Pi, p_i) = a\left(p_i - \sum_{r=1}^{n} P_r \frac{\partial p_i}{\partial P_r}\right),$$

for $i = 1, \ldots, n$, as well as

$$(p_i, x_j) = 0, \quad (p_i, p_j) = 0,$$

for unequal values of i, j from the series $1, \ldots, n$.

And $aA = 1$: so that, as A is usually unity, so also is a.

As regards these results, it is to be noted that ρ has become a constant which has justifiably been made unity. The quantities $P_1, \ldots, P_n$ are known as soon as $\Pi, X_1, \ldots, X_n$ are known: and the quantity Π is to be constructed after $X_1, \ldots, X_n$ are known. These quantities are subject to the equations

$$(X_i, X_j) = 0:$$

so that, as (§ 52) the relation

$$((X_i, X_j)\, X_k) + ((X_j, X_k)\, X_i) + ((X_k, X_i)\, X_j) = 0$$

is satisfied for any functions whatever of $x_1, \ldots, x_n, p_1, \ldots, p_n$, the equations are consistent with one another and coexist. The conditions for the existence of the contact transformation do not impose any equation involving only a single one of the quantities $X_1, \ldots, X_n$; and the preceding conditions for coexistence of the equations are satisfied identically, whatever be the functions $X_1, \ldots, X_n$. Hence the equations defining the contact transformation under consideration contain an arbitrary element: thus any one of the quantities $X_1, \ldots, X_n$ can be assigned arbitrarily, or any other arbitrary relation can be chosen that is not inconsistent with the aggregate of conditions and equations.

COROLLARY 1. There is one special case of this contact transformation, usually called the *infinitesimal* transformation of this type.

It is characterised by the properties

$$Z = z + \epsilon\zeta, \quad X_i = x_i + \epsilon\xi_i, \quad P_i = p_i + \epsilon\pi_i, \qquad (i = 1, \ldots, n),$$

where ϵ is a small quantity of such a magnitude that squares and higher powers may be neglected. The critical equations impose limitations upon the forms of ζ, ξ_i, π_i, for $i = 1, \ldots, n$; thus the equations

$$(X_i, X_j) = 0, \quad (X_i, P_j) = 0, \quad (P_i, P_j) = 0, \quad (X_i, P_i) = 1,$$

give the conditions

$$\frac{\partial \xi_i}{\partial p_j} - \frac{\partial \xi_j}{\partial p_i} = 0,$$

$$\frac{\partial \pi_j}{\partial p_i} + \frac{\partial \xi_i}{\partial x_j} = 0,$$

$$\frac{\partial \pi_i}{\partial x_j} - \frac{\partial \pi_j}{\partial x_i} = 0,$$

$$\frac{\partial \pi_i}{\partial x_i} + \frac{\partial \xi_i}{\partial x_i} = 0,$$

respectively. Hence there is some function U of $x_1, \ldots, x_n, p_1, \ldots, p_n$ such that

$$\xi_i = \frac{\partial U}{\partial p_i}, \quad \pi_i = -\frac{\partial U}{\partial x_i}, \qquad (i = 1, \ldots, n);$$

and any function U of these variables will enable all the conditions to be satisfied. Also writing

$$\Pi = \epsilon\zeta,$$

we find the equations for ζ to be

$$\frac{\partial \zeta}{\partial x_i} = -\frac{\partial U}{\partial x_i} + \sum_{r=1}^{n} p_r \frac{\partial^2 U}{\partial p_r \partial x_i},$$

$$\frac{\partial \zeta}{\partial p_i} = \sum_{r=1}^{n} p_r \frac{\partial^2 U}{\partial p_i \partial p_r},$$

so that

$$\zeta = -U + \sum_{r=1}^{n} p_r \frac{\partial U}{\partial p_r}.$$

Hence an infinitesimal contact transformation is given by the equations

$$X_i - x_i = \epsilon \frac{\partial U}{\partial p_i},$$

$$P_i - p_i = -\epsilon \frac{\partial U}{\partial x_i},$$

$$Z - z = \epsilon \left\{ \sum_{r=1}^{n} \left(p_r \frac{\partial U}{\partial p_r} \right) - U \right\}.$$

The equations

$$dz = p_1 dx_1 + \ldots + p_n dx_n,$$

$$dZ = P_1 dX_1 + \ldots + P_n dX_n,$$

are simultaneously satisfied, when either is satisfied: it is easy to verify this property from the foregoing values. Hence an infinitesimal contact transformation, *in which the changes of* x_1, ..., x_n, p_1, ..., p_n *do not involve* z, *is given by the equations*

$$\delta x_i = \epsilon \frac{\partial U}{\partial p_i}, \quad \delta p_i = -\epsilon \frac{\partial U}{\partial x_i}, \qquad (i = 1, \ldots, n),$$

$$\delta z = \epsilon \left\{ \sum_{r=1}^{n} \left(p_r \frac{\partial U}{\partial p_r} \right) - U \right\},$$

where U *is any function of the variables* $x_1, \ldots, x_n, p_1, \ldots, p_n$.

COROLLARY 2. There is another special case of this type of contact transformation, in which X_1, ..., X_n *are homogeneous functions of zero dimension in the variables* p_1, ..., p_n: *the quantity* Z *is given by*

$$Z = z + c,$$

where c *is a constant; and* P_1, ..., P_n *are then homogeneous, of one dimension in the variables* p_1, ..., p_n, *given by*

$$\sum_{r=1}^{n} P_r \frac{\partial X_r}{\partial x_i} = p_i,$$

for $i=1, \dots, n$, *the quantities* $X_1, \dots, X_n$ *still satisfying the equations*

$$(X_i, X_j) = 0,$$

for i *and* $j=1, \dots, n$; *and the differential relation*

$$\sum_{i=1}^{n} P_i dX_i = \sum_{i=1}^{n} p_i dx_i$$

is then satisfied identically. Other equations satisfied are

$$(X_i, P_i) = 1, \quad (P_i, X_j) = 0, \quad (P_i, P_j) = 0,$$

for all values of i *and* j *unequal to one another.*

Such transformations* are called *homogeneous.* As before, any one of the quantities $X_1, \dots, X_n$ may be chosen arbitrarily: or some other arbitrary relation may be assigned that is not inconsistent with the other relations and equations.

COROLLARY 3. The corresponding *infinitesimal* homogeneous contact transformation has already† been given. It affects only the values of $x_1, \dots, x_n, p_1, \dots, p_n$: and *it can be taken in the form*

$$\left. \begin{aligned} X_i &= x_i + \epsilon \frac{\partial H}{\partial p_i} \\ P_i &= p_i - \epsilon \frac{\partial H}{\partial x_i} \end{aligned} \right\}, \qquad (i=1, \dots, n),$$

where H *is a function of* $x_1, \dots, x_n, p_1, \dots, p_n$, *which is homogeneous of one dimension in* $p_1, \dots, p_n$ *and is otherwise arbitrary.*

If we write

$$X_i - x_i = dx_i, \quad P_i - p_i = dp_i, \quad \epsilon = dt,$$

these equations take the form

$$\frac{dx_i}{dt} = \frac{\partial H}{\partial p_i}, \quad \frac{dp_i}{dt} = -\frac{\partial H}{\partial x_i},$$

for $i=1, \dots, n$. These equations, exactly in this form, will occur later as a canonical system of equations in theoretical dynamics.

131. One remark, indicating a relation between these infinitesimal transformations and the integration of a partial differential equation

$$f(x_1, \dots, x_n, p_1, \dots, p_n) = 0,$$

* *l.c.*, § 139. The remark, in the foot-note, p. 315, as to change of sign applies here also.

† See vol. I of this work, § 140.

may at once be made*. If it be required to find the infinitesimal contact transformations

$$\delta z = \epsilon\zeta, \quad \delta x_i = \epsilon\xi_i, \quad \delta p_i = \epsilon\pi_i, \quad (i = 1, \ldots, n),$$

which transform f into itself, we clearly must have

$$\sum_{i=1}^{n} \left(\frac{\partial f}{\partial x_i}\xi_i + \frac{\partial f}{\partial p_i}\pi_i\right) = 0.$$

Denoting by U any arbitrary function of $x_1, \ldots, x_n, p_1, \ldots, p_n$ (thus omitting z from its arguments), we know that the quantities ξ and π can be taken

$$\xi_i = \frac{\partial U}{\partial p_i}, \quad \pi_i = -\frac{\partial U}{\partial x_i};$$

thus the appropriate infinitesimal contact transformations will arise through a function U such that

$$(f, U) = 0.$$

Consequently, the determination of all such transformations is equivalent to the integration of the original equation

$$f = 0.$$

Applications of Contact Transformations to the Integration of an Equation or Equations.

132. The application of the properties of finite contact transformations to the integration of a single partial differential equation is immediate.

First, suppose that the dependent variable z occurs explicitly in the equation so that the given equation may be written in the form

$$f(x_1, \ldots, x_n, z, p_1, \ldots, p_n) = 0.$$

We then take

$$Z = f(x_1, \ldots, x_n, z, p_1, \ldots, p_n),$$

and, applying the preceding results, we assume that we have $2n+1$ independent functions of the $2n+1$ variables $z, p_1, \ldots, p_n, x_1, \ldots, x_n$, being $Z, P_1, \ldots, P_n, X_1, \ldots, X_n$, such that the relation

$$Z - \sum_{i=1}^{n} P_i dX_i = \rho\left(dz - \sum_{i=1}^{n} p_i dx_i\right)$$

* Lie, *Math. Ann.*, t. VIII (1875), p. 240.

is satisfied identically for a non-vanishing quantity ρ that does not involve the differential elements. All the required quantities $P_1, \ldots, P_n$ are known, if $X_1, \ldots, X_n$ are known, by equations characteristic of a contact transformation; and these quantities $X_1, \ldots, X_n$ are such that the equations

$$[Z, X_i]=0, \quad [X_i, X_j]=0,$$

for $i=1, \ldots, n$, and $j=1, \ldots, n$ but unequal to i, are satisfied. It has been seen that, even when Z is arbitrarily assigned, these equations are consistent and coexist. Accordingly, assuming the general results of the theory of contact transformations, we may assume that quantities $X_1, \ldots, X_n$ are determined by these equations and that the quantities $P_1, \ldots, P_n$ have subsequently been obtained.

Now what is desired is an integral of the equation

$$Z=f(x_1, \ldots, x_n, z, p_1, \ldots, p_n)=0,$$

so that, as $p_1, \ldots, p_n$ are derivatives of z with regard to $x_1, \ldots, x_n$ respectively, we have

$$dz-p_1 dx_1-\ldots-p_n dx_n=0;$$

hence the quantities, defining the contact transformation, must satisfy the relation

$$dZ-P_1 dX_1-\ldots-P_n dX_n=0,$$

or, since Z is a permanent zero, we must have

$$P_1 dX_1+\ldots+P_n dX_n=0.$$

This relation is the only relation, except $Z=0$, that needs to be satisfied in order to secure the existence of the relation

$$dz-p_1 dx_1-\ldots-p_n dx_n=0;$$

and, subject to the specified exception, $P_1, \ldots, P_n, X_1, \ldots, X_n$ are independent functions of the variables involved. We therefore require an integral equivalent of the relation

$$P_1 dX_1+\ldots+P_n dX_n=0,$$

where $P_1, \ldots, P_n, X_1, \ldots, X_n, Z$ are $2n+1$ independent functions of the quantities $z, x_1, \ldots, x_n, p_1, \ldots, p_n$, the relation being satisfied identically.

Such a relation possesses* three types of integral equivalents.

(i) It is satisfied identically if

$$X_1 = a_1, \ldots, X_n = a_n,$$

where $a_1, \ldots, a_n$ are n arbitrary constants: these n equations give n differential relations

$$dX_1 = 0, \ldots, dX_n = 0,$$

in virtue of which the differential relation obviously is satisfied.

(ii) It is satisfied identically, if μ equations of the type

$$X_i = g_i(X_{\mu+1}, \ldots, X_n),$$

for $i = 1, \ldots, \mu$ (where $\mu < n$), are postulated, provided the equations

$$\sum_{i=1}^{\mu} P_i \frac{\partial g_i}{\partial X_j} + P_j = 0,$$

for $j = \mu + 1, \ldots, n$, also are satisfied: for these equations give μ differential relations

$$dX_i = \sum_{j=1}^{n-\mu} \frac{\partial g_i}{\partial X_j} dX_j,$$

for $i = 1, \ldots, \mu$ which, in connection with the other $n - \mu$ relations, obviously satisfy the required differential relation.

(iii) It is satisfied identically if

$$P_1 = 0, \ldots, P_n = 0;$$

these relations, however, do not possess (and do not necessarily imply) any differential character.

We consider the significance of these three types of integral equivalent in succession.

In the first place, we have

$$X_1 = a_1, \ldots, X_n = a_n,$$

concurrently with the equation $Z = 0$; and these relations are sufficient to secure the differential relation

$$dz = p_1 dx_1 + \ldots + p_n dx_n.$$

As $Z, X_1, \ldots, X_n$ are independent functions of $z, x_1, \ldots, x_n, p_1, \ldots, p_n$, it is possible to eliminate $p_1, \ldots, p_n$ among the n equations

$$Z = 0, \quad X_1 = a_1, \ldots, X_n = a_n,$$

the eliminant being an equation involving $z, x_1, \ldots, x_n, a_1, \ldots, a_n$. The differential relation shews that the value of z thus provided is an integral of $Z = 0$: manifestly, it is the *complete integral*.

* See Part I of this work, § 142, *foot-note*.

In the second place, we have μ relations

$$X_i = g_i(X_{\mu+1}, \ldots, X_n),$$

for $i = 1, \ldots, \mu$, and $n - \mu$ relations

$$\sum_{i=1}^{\mu} P_i \frac{\partial g_i}{\partial X_j} + P_j = 0,$$

for $j = \mu + 1, \ldots, n$, concurrently with the equation $Z = 0$: and these equations are sufficient to secure the differential relation

$$dz = p_1 dx_1 + \ldots + p_n dx_n.$$

As $Z, X_1, \ldots, X_n, P_1, \ldots, P_n$ are independent functions of their arguments, the n relations are independent of one another and of $Z = 0$: thus it is possible to eliminate $p_1, \ldots, p_n$ among the n relations and $Z = 0$, and the eliminant is an equation involving $z, x_1, \ldots, x_n$ and the functional forms. The differential relation shews that the value of z thus provided is an integral of $Z = 0$: manifestly, it is a *general integral*. Clearly, there will be a number of classes of such integrals, a class being determined by the number of functional relations between the quantities $X_1, \ldots, X_n$ initially postulated: as before, the most comprehensive general integral occurs when only one such relation of the most unrestricted type is postulated.

The equations are expressed* in another form by Lie, as follows. Let H denote a function of $P_1, \ldots, P_\mu, X_{\mu+1}, \ldots, X_n$, which is homogeneous and linear in $P_1, \ldots, P_\mu$ and otherwise is quite arbitrary; then the equations are given by

$$X_i = \frac{\partial H}{\partial P_i}, \quad P_j = -\frac{\partial H}{\partial X_j},$$

for $i = 1, \ldots, \mu$; $j = \mu + 1, \ldots, n$.

For, since H is homogeneous and linear in $P_1, \ldots, P_\mu$, we have

$$H = P_1 \frac{\partial H}{\partial P_1} + \ldots + P_\mu \frac{\partial H}{\partial P_\mu}$$
$$= P_1 X_1 + \ldots + P_\mu X_\mu;$$

hence

$$\sum_{i=1}^{\mu} P_i dX_i + \sum_{i=1}^{\mu} X_i dP_i = dH$$
$$= \sum_{i=1}^{\mu} \frac{\partial H}{\partial P_i} dP_i + \sum_{j=\mu+1}^{n} \frac{\partial H}{\partial X_j} dX_j$$
$$= \sum_{i=1}^{\mu} X_i dP_i - \sum_{j=\mu+1}^{n} P_j dX_j,$$

and therefore

$$\sum_{r=1}^{n} P_r dX_r = 0,$$

which is the equation to be satisfied.

* *Math. Ann.*, t. IX (1876), p. 250.

In the third place, we have n equations

$$P_1 = 0, \ldots, P_n = 0,$$

concurrently with $Z=0$: but as pointed out, they do not definitely secure the differential relation

$$dz - p_1 dx_1 - \ldots - p_n dx_n = 0,$$

for they have no differential character. If, however, they do secure it, we then have

$$dZ = \rho\,(dz - p_1 dx_1 - \ldots - p_n dx_n):$$

consequently

$$\frac{\partial Z}{\partial z} = \rho, \quad \frac{\partial Z}{\partial x_i} = -\rho p_i, \quad \frac{\partial Z}{\partial p_i} = 0,$$

that is,

$$\frac{\partial Z}{\partial p_i} = 0, \quad \frac{\partial Z}{\partial x_i} + p_i \frac{\partial Z}{\partial z} = 0,$$

for $i = 1, \ldots, n$. Thus $2n$ equations must be satisfied, in addition to

$$Z = 0, \quad P_1 = 0, \ldots, P_n = 0;$$

and as ρ is not to vanish, $\dfrac{\partial Z}{\partial z}$ is not zero. Assuming that all the equations coexist, and that it is possible to eliminate $p_1, \ldots, p_n$ so as to leave an equation expressing z in terms of $x_1, \ldots, x_n$, the value of z thus provided is an integral: it is the *singular integral*.

Ex. 1. Let

$$Z = p_1 x_1 + p_2 x_2 + p_3 x_3 - z = 0.$$

The quantities X_1, X_2, X_3 are subject to the equation

$$[Z, X_1] = 0, \quad [Z, X_2] = 0, \quad [Z, X_3] = 0,$$
$$[X_2, X_3] = 0, \quad [X_1, X_3] = 0, \quad [X_1, X_2] = 0;$$

and it is easy to verify that these are satisfied by

$$X_1 = p_1, \quad X_2 = p_2, \quad X_3 = p_3.$$

The quantities P_1, P_2, P_3 are then given by the equations

$$\frac{\partial Z}{\partial p_i} = \sum_{r=1}^{3} P_r \frac{\partial X_r}{\partial p_i},$$

for $i = 1, 2, 3$: evidently

$$P_1 = x_1, \quad P_2 = x_2, \quad P_3 = x_3.$$

It is clear that the quantities Z, X_1, X_2, X_3, P_1, P_2, P_3 are independent of one another: also

$$dZ - P_1 dX_1 - P_2 dX_2 - P_3 dX_3 = -(dz - p_1 dx_1 - p_2 dx_2 - p_3 dx_3):$$

the value of ρ is -1.

The complete integral is given by

$$p_1 = X_1 = a_1, \quad p_2 = X_2 = a_2, \quad p_3 = X_3 = a_3, \quad Z = 0,$$

where a_1, a_2, a_3 are arbitrary constants: it is

$$z = a_1 x_1 + a_2 x_2 + a_3 x_3.$$

One class of general integrals is given by

$$p_1 = g(p_2, p_3),$$

together with

$$x_1 \frac{\partial g}{\partial p_2} + x_2 = 0, \quad x_1 \frac{\partial g}{\partial p_3} + x_3 = 0, \quad Z = 0.$$

Clearly p_2, p_3, and therefore also p_1, are functions of $\frac{x_2}{x_1}$ and $\frac{x_3}{x_1}$; and when g is as unrestricted as possible, they will be arbitrary functions of these quantities. Thus

$$\begin{aligned} z &= p_1 x_1 + p_2 x_2 + p_3 x_3 \\ &= x_1 \left(p_1 + \frac{x_2}{x_1} p_2 + \frac{x_3}{x_1} p_3 \right) \\ &= x_1 G\left(\frac{x_2}{x_1}, \frac{x_3}{x_1} \right), \end{aligned}$$

where G is an arbitrary function of its arguments: the integral represents the first class of general integrals.

Another class of general integrals is given by

$$p_1 = g_1(p_3), \quad p_2 = g_2(p_3),$$

together with

$$x_1 \frac{\partial g_1}{\partial p_3} + x_2 \frac{\partial g_2}{\partial p_3} + x_3 = 0, \quad Z = 0.$$

Manifestly, it is given by the elimination of p_3 between the equations

$$\left. \begin{aligned} z &= x_1 g_1(p_3) + x_2 g_2(p_3) + x_3 p_3 \\ 0 &= x_1 \frac{\partial g_1}{\partial p_3} + x_2 \frac{\partial g_2}{\partial p_3} + x_3 \end{aligned} \right\};$$

it represents the second class of general integrals.

The equations

$$x_1 = P_1 = 0, \quad x_2 = P_2 = 0, \quad x_3 = P_3 = 0,$$

with $Z = 0$, clearly do not provide an integral: there is no singular integral.

Ex. 2. Integrate in the same way the equations:—

(i) $z = p_1 x_1 + p_2 x_2 + p_3 x_3 + p_1 p_2 p_3$;

(ii) $z = p_1 x_2 + p_2 x_3 + p_3 x_1 + p_1 p_2 p_3$;

(iii) $z^2 = (x_1 - a p_1)^2 + (x_2 - a p_2)^2 + (x_3 - a p_3)^2$.

133. Next, suppose that the dependent variable z does not occur explicitly in the differential equation to be solved, which therefore is of the form

$$f(x_1, \ldots, x_n, p_1, \ldots, p_n) = 0.$$

We then take

$$X_n = f(x_1, \ldots, x_n, p_1, \ldots, p_n),$$

and, applying the results of § 130, we assume that we have functions $\Pi, X_1, \ldots, X_{n-1}, P_1, \ldots, P_n$ of $x_1, \ldots, x_n, p_1, \ldots, p_n$, such that the relation

$$d\Pi - \sum_{i=1}^{n} P_i dX_i = -\sum_{i=1}^{n} p_i dx_i$$

is satisfied identically. The quantities $P_1, \ldots, P_n$ are known by means of equations characteristic of the contact transformation, when once $\Pi, X_1, \ldots, X_n$ are known: and Π is determined by a number of equations, as soon as $X_1, \ldots, X_n$ are known. These quantities are subject to the equations

$$(X_i, X_j) = 0,$$

for i and $j = 1, \ldots, n$, the values of i and j being unequal. As has been seen, any one of these quantities can be arbitrarily assumed: accordingly, we assign $f(x_1, \ldots, x_n, p_1, \ldots, p_n)$ as the value of X_n. And then, adopting the general results of the theory of contact transformations, we may suppose that the quantities $X_1, \ldots, X_{n-1}, \Pi, P_1, \ldots, P_n$ are known.

What is desired is an integral of the equation

$$X_n = f(x_1, \ldots, x_n, p_1, \ldots, p_n) = 0.$$

For that purpose, $p_1, \ldots, p_n$ are derivatives of z with regard to $x_1, \ldots, x_n$ respectively, so that

$$dz - p_1 dx_1 - \ldots - p_n dx_n = 0;$$

hence the quantities, defining the contact transformation in the present circumstances, must satisfy the relation

$$d\Pi - \sum_{i=1}^{n} P_i dX_i = -dz,$$

or, since X_n is a permanent zero, we must have

$$d(z + \Pi) - P_1 dX_1 - \ldots - P_{n-1} dX_{n-1} = 0.$$

This is the only relation, other than $X_n = 0$, which needs to be satisfied in order to secure the existence of the relation

$$dz - p_1 dx_1 - \ldots - p_n dx_n = 0;$$

we therefore require an integral equivalent of the differential relation

$$d(z + \Pi) - P_1 dX_1 - \ldots - P_{n-1} dX_{n-1} = 0,$$

so that it may be satisfied identically. As in the former case, there are varieties of integral equivalents: but, as will be seen, the singular integral does not occur.

(i) The differential relation is satisfied identically if

$$z + \Pi = c, \quad X_1 = a_1, \ldots, X_{n-1} = a_{n-1},$$

where $c, a_1, \ldots, a_{n-1}$ are arbitrary constants. These equations are to be taken concurrently with $X_n = 0$; and the quantities Π, $X_1, \ldots, X_{n-1}, X_n$ are functionally independent of one another. Hence we may consider the equations

$$X_1 = a_1, \ldots, X_{n-1} = a_{n-1}, \quad X_n = 0$$

resolved* with regard to $p_1, \ldots, p_n$, and the deduced values substituted in Π: the result is of the form

$$z + h(x_1, \ldots, x_n, a_1, \ldots, a_{n-1}) = c.$$

The integral clearly is the *complete integral.*

(ii) The differential relation is satisfied identically if a number of relations

$$z + \Pi = g(X_{\mu+1}, \ldots, X_{n-1}),$$
$$X_i = g_i(X_{\mu+1}, \ldots, X_{n-1}),$$

for $i = 1, \ldots, \mu$, hold, provided the further relations

$$\frac{\partial g}{\partial X_j} - \sum_{i=1}^{\mu} P_i \frac{\partial g_i}{\partial X_j} - P_j = 0,$$

for $j = \mu + 1, \ldots, n - 1$, also hold. Eliminating $p_1, \ldots, p_n$ among these n relations and $X_n = 0$, we have an integral: it is a *general integral.* Obviously there will be a number of classes of general integrals, a class being determined by the number of functional relations postulated. The most comprehensive general integral is given by the equations

$$z + \Pi = F(X_1, \ldots, X_{n-1}),$$
$$\frac{\partial F}{\partial X_r} = P_r, \quad X_n = 0,$$

for $r = 1, \ldots, n - 1$, the function F being completely arbitrary.

(iii) The differential relation is satisfied identically if

$$P_1 = 0, \ldots, P_{n-1} = 0, \quad z + \Pi = a,$$

* Exceptions might arise, when resolution with regard to $p_1, \ldots, p_n$ is either inconvenient or impossible: we should then proceed as in §§ 58, 59.

where a is a constant that may be zero. Eliminating $p_1, \ldots, p_n$ among these n relations and $X_n = 0$, we have an integral which is an exceedingly special case of the foregoing general integral obtained by taking a as the expression of $F(X_1, \ldots, X_{n-1})$. The relation

$$dz + d\Pi = 0$$

gives

$$d\Pi = -\sum_{i=1}^{n} p_i dx_i,$$

so that

$$\frac{\partial \Pi}{\partial p_i} = 0, \quad \frac{\partial \Pi}{\partial x_i} = -p_i,$$

for $i = 1, \ldots, n$, which are in agreement with the equations characteristic of the contact transformation in this case.

And these exhaust the modes of satisfying the differential relation: thus a singular integral does not arise.

It appears, from the results of § 132 and from the results just obtained, that the construction of the various integrals of a partial differential equation

$$U = 0$$

can immediately be effected, if a contact transformation of which U is an element is known; hence this method of proceeding requires either a knowledge, or the determination, of such a contact transformation. If it has to be determined, then a number of simultaneous partial equations have to be solved.

Ex. As an illustration, consider the equation

$$p_1^2 + p_2^2 + p_3^2 = 1.$$

We take

$$X_3 = p_1^2 + p_2^2 + p_3^2 - 1;$$

a contact transformation of which X_3 is an element, is given by

$$X_1 = p_1, \quad X_2 = p_2,$$

$$\Pi = -p_1 x_1 - p_2 x_2 - p_3 x_3,$$

$$P_1 = -x_1 + \frac{p_1}{p_3} x_3, \quad P_2 = -x_2 + \frac{p_2}{p_3} x_3, \quad P_3 = -\frac{x_3}{2p_3}.$$

The complete integral of the equation is given by the elimination of p_1, p_2, p_3 between

$$z + \Pi = c, \quad X_1 = a_1, \quad X_2 = a_2, \quad X_3 = 0:$$

that is, the complete integral is given by

$$z - a_1 x_1 - a_2 x_2 - a_3 x_3 = c,$$

where

$$a_3^2 = 1 - a_1^2 - a_2^2.$$

There are two classes of general integrals. A general integral of the first class is given by

$$z+\Pi=f(X_1, X_2), \quad X_3=0,$$
$$P_1=\frac{\partial f}{\partial X_1}, \quad P_2=\frac{\partial f}{\partial X_2},$$

where f is an arbitrary function: that is, it is given by the elimination of p_1, p_2, p_3 among the equations

$$\left.\begin{array}{r} p_1^2+p_2^2+p_3^2=1 \\ z-p_1x_1-p_2x_2-p_3x_3=f(p_1, p_2) \\ -x_1+\dfrac{p_1}{p_3}x_3=\dfrac{\partial f}{\partial p_1} \\ -x_2+\dfrac{p_2}{p_3}x_3=\dfrac{\partial f}{\partial p_2} \end{array}\right\}.$$

A general integral of the second class is given by

$$z+\Pi=f(X_2), \quad X_1=g(X_2), \quad X_3=0,$$
$$\frac{df}{dX_2}-P_1\frac{dg}{dX_2}-P_2=0,$$

that is, by the elimination of p_1, p_2, p_3 among the equations

$$\left.\begin{array}{r} p_1^2+p_2^2+p_3^2=1 \\ z-p_1x_1-p_2x_2-p_3x_3=f(p_2) \\ p_1=g(p_2) \\ \dfrac{df}{dp_2}+\left(x_1-\dfrac{p_1}{p_3}x_3\right)\dfrac{dg}{dp_2}+x_2-\dfrac{p_2}{p_3}x_3=0 \end{array}\right\}.$$

There is no singular integral.

134. In the same manner, the properties of contact transformations can be applied to obtain the integrals of a system of equations. We may assume that the system is complete, that is, that all the relations

$$[X_i, X_j]=0, \quad (X_i, X_j)=0,$$

according as the dependent variable does or does not occur, are satisfied, either identically or in virtue of the equations of the system: and we may also assume that the number of such equations is less than $n+1$ or less than n, in the two cases respectively.

In the first place, suppose that the dependent variable does occur: and let the complete system be

$$Z=0, \quad X_1=0, \ldots, X_m=0,$$

where $m<n$, the relations

$$[Z, X_i]=0, \quad [X_i, X_j]=0,$$

being satisfied. If these relations are satisfied identically*, the quantities $Z, X_1, \ldots, X_m$ can be constituents of a contact transformation $Z, X_1, \ldots, X_n, P_1, \ldots, P_n$: for the quantities $Z, X_1, \ldots, X_n$ are determined by equations

$$[Z, X_r] = 0, \quad [X_r, X_s] = 0,$$

for r and $s = 1, \ldots, n$: and the quantities $P_1, \ldots, P_n$ are then given by linear algebraic equations. This contact transformation leads to the differential relation

$$dZ - P_1 dX_1 - \ldots - P_n dX_n = \rho(dz - p_1 dx_1 - \ldots - p_n dx_n),$$

where ρ is a non-vanishing quantity.

When the quantity z is an integral of the given system, we must have

$$dz = p_1 dx_1 + \ldots + p_n dx_n:$$

also

$$dZ = 0, \quad dX_1 = 0, \ldots, dX_m = 0,$$

because $Z = 0$, $X_1 = 0, \ldots, X_m = 0$ permanently: hence the above relation becomes

$$P_{m+1} dX_{m+1} + \ldots + P_n dX_n = 0,$$

and when this relation is satisfied, we have

$$dz - p_1 dx_1 - \ldots - p_n dx_n = 0,$$

so that an integral of the equation is provided. As before, the relation can be satisfied in three kinds of ways.

(i) The relation will be satisfied if

$$X_{m+1} = a_{m+1}, \ldots, X_n = a_n,$$

where $a_{m+1}, \ldots, a_n$ are constants which may be taken arbitrarily. The quantities $Z, X_1, \ldots, X_n$ are independent of one another: hence, eliminating $p_1, \ldots, p_n$ between these $n - m$ equations and the $m + 1$ equations of the complete system, we have a relation between $z, x_1, \ldots, x_n$, which involves $n - m$ arbitrary constants and provides an integral of the system. The integral thus provided is clearly the *complete integral*.

(ii) The relation will be satisfied if the quantities $X_{m+1}, \ldots,$

* The alternative, when these relations are satisfied only in virtue of the members of the complete system, will be a matter for subsequent consideration.

X_n are connected by q equations (where q is less than $n-m$) which may be taken in the form

$$g_1(X_{m+1}, \ldots, X_n) = 0,$$
$$\ldots\ldots\ldots\ldots\ldots\ldots\ldots\ldots$$
$$g_q(X_{m+1}, \ldots, X_n) = 0;$$

but in addition, the equations

$$P_{m+r} = \lambda_1 \frac{\partial g_1}{\partial X_{m+r}} + \ldots + \lambda_q \frac{\partial g_q}{\partial X_{m+r}},$$

for $r = 1, \ldots, n-m$, must be satisfied. We now have the $m+1$ equations of the original system, the q relations of condition among the variables $X_{m+1}, \ldots, X_n$, and the deduced $n-m$ equations connecting the quantities $P_{m+1}, \ldots, P_n$ with the variables $X_{m+1}, \ldots, X_n$: that is, there are $n+1+q$ equations in all. Eliminating $p_1, \ldots, p_n, \lambda_1, \ldots, \lambda_q$ among this aggregate, we obtain a single equation as the eliminant: the form of the equation is affected by the form of the q postulated relations of condition. The value of z thus provided is an integral: obviously it is a *general integral*.

Manifestly there are $n-m-1$ classes of general integrals, determined by the assignment of $1, 2, \ldots, n-m-1$ relations of condition among the variables $X_{m+1}, \ldots, X_n$. The most comprehensive class is that determined by the assignment of only a single relation: if this relation be taken in the form

$$X_{m+1} = h(X_{m+2}, \ldots, X_n),$$

then

$$P_{m+r} + \frac{\partial h}{\partial X_{m+r}} P_{m+1} = 0,$$

for $r = 2, \ldots, n-m-2$: and, as before, we obtain the integral by elimination.

(iii) The relation will be satisfied if

$$P_{m+1} = 0, \ \ldots, \ P_n = 0.$$

We then have

$$dZ - P_1 dX_1 - \ldots - P_m dX_m = \rho(dz - p_1 dx_1 - \ldots - p_n dx_n),$$

so that

$$\frac{\partial Z}{\partial z} - P_1 \frac{\partial X_1}{\partial z} - \ldots - P_m \frac{\partial X_m}{\partial z} = \rho,$$
$$\frac{\partial Z}{\partial x_i} - P_1 \frac{\partial X_1}{\partial x_i} - \ldots - P_m \frac{\partial X_m}{\partial x_i} = -\rho p_i,$$
$$\frac{\partial Z}{\partial p_j} - P_1 \frac{\partial X_1}{\partial p_j} - \ldots - P_m \frac{\partial X_m}{\partial p_j} = 0,$$

for i and $j=1, \ldots, n$. Eliminating $P_1, \ldots, P_m, \rho$ among these $2n+1$ equations, we have $2n-m$ conditions to be satisfied among the derivatives of $Z, X_1, \ldots, X_m$, which can be expressed in the form

$$\left\| \begin{array}{cccccc} \dfrac{dZ}{dx_1}, & \ldots, & \dfrac{dZ}{dx_n}, & \dfrac{\partial Z}{\partial p_1}, & \ldots, & \dfrac{\partial Z}{\partial p_n} \\ \dfrac{dX_1}{dx_1}, & \ldots, & \dfrac{dX_1}{dx_n}, & \dfrac{\partial X_1}{\partial p_1}, & \ldots, & \dfrac{\partial X_1}{\partial p_n} \\ \ldots & \ldots & \ldots & \ldots & \ldots & \ldots \\ \dfrac{dX_m}{dx_1}, & \ldots, & \dfrac{dX_m}{dx_n}, & \dfrac{\partial X_m}{\partial p_1}, & \ldots, & \dfrac{\partial X_m}{\partial p_n} \end{array} \right\| = 0.$$

When these are satisfied, either identically or in virtue of

$$Z=0, \quad X_1=0, \ldots, X_m=0, \quad P_{m+1}=0, \ldots, P_n=0,$$

then the elimination of $p_1, \ldots, p_n$ among these n equations leads to an equation connecting $z, x_1, \ldots, x_n$. This value of z is an integral of the equation: clearly it is the *singular* integral.

Ex. Obtain integrals of the equations

$$\left.\begin{array}{r} Z=z-p_1x_1-p_2x_2-p_3x_3-a=0 \\ X_1=p_1^2+p_2^2+p_3^2-c=0 \end{array}\right\},$$

which satisfy the condition

$$[Z, X_1]=0$$

of coexistence: the quantities a and c are given constants.

The quantities Z and X_1 are constituents of a contact transformation, represented in full by

$$Z, \quad X_1, \quad X_2=p_1, \quad X_3=p_2,$$

$$P_1=\tfrac{1}{2}\frac{x_3}{p_3}, \quad P_2=x_1-x_3\frac{p_1}{p_3}, \quad P_3=x_2-x_3\frac{p_2}{p_3}.$$

The complete integral is represented by

$$X_2=a_2, \quad X_3=a_3,$$

where a_2 and a_3 are arbitrary constants: combining these with

$$Z=0, \quad X_1=0,$$

we find the complete integral given by

$$z=a_1x_1+a_2x_2+a_3x_3+a,$$

where a_1 and a_2 are arbitrary, and a_3 is given by

$$a_1^2+a_2^2+a_3^2=1:$$

or, expressed solely in terms of independent constants, the integral is

$$(z-a_1x_1-a_2x_2-a)^2=x_3^2(1-a_1^2-a_2^2).$$

To find the general integral, we take

$$X_2 = g(X_3),$$
$$P_2 g'(X_3) + P_3 = 0;$$

that is, the general integral is given by the elimination of p_1, p_2, p_3 among the equations

$$\left.\begin{aligned} z - p_1 x_1 - p_2 x_2 - p_3 x_3 - a &= 0 \\ p_1^2 + p_2^2 + p_3^2 - c &= 0 \\ p_1 &= g(p_2) \\ \left(x_1 - x_3 \frac{p_1}{p_3}\right) g'(p_2) + x_2 - x_3 \frac{p_2}{p_3} &= 0 \end{aligned}\right\}.$$

The singular integral, if it exists, is given by

$$Z = 0, \quad X_1 = 0, \quad P_2 = 0, \quad P_3 = 0.$$

The further necessary conditions are satisfied. The elimination of p_1, p_2, p_3 among these four equations leads to a relation

$$(z - a)^2 = c(x_1^2 + x_2^2 + x_3^2),$$

which is therefore the singular integral.

135. Next, suppose that the dependent variable does not occur explicitly and that the complete system is

$$X_1 = 0, \ \ldots, \ X_m = 0,$$

where the relations

$$(X_i, X_j) = 0,$$

for i and $j = 1, \ldots, m$, are satisfied, the integer m being less than n. As these relations are satisfied, the quantities $X_1, \ldots, X_m$ are constituents of a contact transformation such that

$$Z = z + \Pi,$$

and quantities $\Pi, X_{m+1}, \ldots, X_n, P_1, \ldots, P_n$, being functions of $x_1, \ldots, x_n, p_1, \ldots, p_n$, exist such that the differential relation

$$d\Pi - \sum_{i=1}^{n} P_i dX_i = - \sum_{i=1}^{n} p_i dx_i$$

is satisfied identically.

What is desired is an integral of the system

$$X_1 = 0, \ \ldots, \ X_m = 0,$$

so that we must have

$$dX_1 = 0, \ \ldots, \ dX_m = 0;$$

also, as the value of z is to be an integral of this system, we have

$$dz = p_1 dx_1 + \ldots + p_n dx_n.$$

Thus the differential relation can be taken in the form

$$d(z+\Pi) - P_{m+1}dX_{m+1} - \dots - P_n dX_n = 0.$$

When this relation is satisfied concurrently with the m equations of the system, then the relation

$$dz = p_1 dx_1 + \dots + p_n dx_n$$

also is satisfied: so that the value of z, expressed in terms of $x_1, \dots, x_n$, is an integral.

As in the case of a single equation, there are two distinct kinds of integral equivalents of the differential relation, leading to a complete integral and to the general integrals respectively: there is no singular integral.

(i) The differential relation is satisfied identically if

$$z + \Pi = c, \quad X_{m+1} = a_{m+1}, \; \dots, \; X_n = a_n,$$

where $c, a_{m+1}, \dots, a_n$ are arbitrary constants. These equations exist concurrently with the equations

$$X_1 = 0, \; \dots, \; X_m = 0;$$

and the quantities $\Pi, X_1, \dots, X_m, X_{m+1}, \dots, X_n$ are functionally independent of one another. We may therefore resolve the set of equations

$$X_1 = 0, \; \dots, \; X_m = 0, \quad X_{m+1} = a_{m+1}, \; \dots, \; X_n = a_n$$

with regard to the variables* $p_1, \dots, p_n$: when we substitute their values in

$$z + \Pi = c,$$

we have an integral of the system of equations in a form

$$z + H(x_1, \dots, x_n, a_{m+1}, \dots, a_n) = c,$$

involving $n - m + 1$ variables. The integral is clearly the *complete integral*.

(ii) The differential relation is satisfied identically if a number of relations

$$z + \Pi = g\,(X_{\mu+1}, \; \dots, \; X_n),$$
$$X_{m+r} = g_r\,(X_{\mu+1}, \; \dots, \; X_n),$$

* Exceptions may arise when the resolution with regard to another set of n of the variables $x_1, \dots, x_n, p_1, \dots, p_n$ is more convenient: we then proceed as in §§ 58, 59.

for $r=1, \ldots, \mu-m$, hold provided the further relations

$$\frac{\partial g}{\partial X_j} - \sum_{r=1}^{\mu-m} P_{m+r} \frac{\partial g_r}{\partial X_j} - P_j = 0,$$

for $j=\mu+1, \ldots, n$, also hold. Eliminating $p_1, \ldots, p_n$ among these $n-m+1$ relations and the equations of the original system, we have an integral of that system. It is a *general integral.*

Obviously, there will be a number of classes of general integrals, a class being determined by the number of functional relations postulated. The most comprehensive general integral is given by the elimination of $p_1, \ldots, p_n$ among the equations

$$z + \Pi = h(X_{m+1}, \ldots, X_n),$$

$$\frac{\partial h}{\partial X_r} = P_r, \quad X_1 = 0, \ldots, X_m = 0,$$

for $r = m+1, \ldots, n$.

(iii) The differential relation is satisfied identically if

$$P_{m+1} = 0, \ \ldots, \ P_n = 0, \quad z + \Pi = a,$$

where a is a constant that may be zero. Eliminating $p_1, \ldots, p_n$ among these $n-m+1$ relations and the m equations of the complete system, we obtain an integral which is an exceedingly special case of the foregoing comprehensive general integral, the case arising by taking a as the expression of the function $h(X_{m+1}, \ldots, X_n)$. The relation

$$dz + d\Pi = 0$$

gives

$$d\Pi = -\sum_{i=1}^{n} p_i dx_i,$$

so that

$$\frac{\partial \Pi}{\partial p_i} = 0, \quad \frac{\partial \Pi}{\partial x_i} = -p_i,$$

for $i=1, \ldots, n$; these equations are characteristic of the contact transformation for the present case.

The modes of satisfying the differential relation are exhausted: hence a singular integral does not arise for the complete system in the supposed circumstance that the complete system does not explicitly involve the dependent variable.

Ex. 1. Integrate the simultaneous equations

$$\left.\begin{aligned} F=x_2p_1+x_1p_2+ap_3(p_1-p_2)-c=0 \\ G=(p_1+p_2)(x_1+x_2)-b=0 \end{aligned}\right\},$$

where a, b, and c are constants.

It is easy to verify that the condition

$$(F, G)=0$$

is satisfied identically, and that, if

$$H=p_3,$$

then the equations

$$(F, H)=0, \quad (G, H)=0,$$

also are satisfied identically. Hence, taking

$$\begin{aligned} X_1&=x_2p_1+x_1p_2+ap_3(p_1-p_2)-c, \\ X_2&=(p_1+p_2)(x_1+x_2)-b, \\ X_3&=p_3, \end{aligned}$$

the quantities X_1, X_2, X_3 can be constituents of a contact transformation.

To complete the contact transformation, the quantities Π, P_1, P_2, P_3 are required. Of these, Π satisfies the three equations

$$(\Pi, X_3)=-\sum_{r=1}^{3} p_r\frac{\partial X_3}{\partial p_r}=-p_3,$$

$$(\Pi, X_2)=-\sum_{r=1}^{3} p_r\frac{\partial X_2}{\partial p_r}=-(p_1+p_2)(x_1+x_2),$$

$$\begin{aligned}(\Pi, X_1)=-\sum_{r=1}^{3} p_r\frac{\partial X_1}{\partial p_r}&=-p_1(x_2+ap_3)-p_2(x_1-ap_3)-ap_3(p_1-p_2) \\ &=-p_1x_2-p_2x_1-2ap_3(p_1-p_2).\end{aligned}$$

From the first of these, we have

$$\Pi+p_3x_3=u,$$

where u is any function of x_1, x_2, p_1, p_2, p_3. From the second of them, we have

$$u+\tfrac{1}{2}(X_2+b)\log(x_1+x_2)=v,$$

where v is any function of

$$X_2+b, \quad x_2-x_1, \quad p_2-p_1, \quad p_3,$$

or say of X_2+b, a, β, p_3, where

$$a=x_2-x_1, \quad \beta=p_2-p_1.$$

From the third of the equations, we have

$$\frac{2v}{\beta(a+2ap_3)+b}+\log(a+2ap_3)=w,$$

where w is any function of X_2+b, $\beta(a+2ap_3)+b$, p_3. Now for the integration of our equations, $X_1=0$, $X_2=0$; and

$$\begin{aligned}\beta(a+2ap_3)&=X_2-2X_1+b-2c \\ &=b-2c,\end{aligned}$$

so that effectively w is a function of p_3: consequently,

$$\Pi+p_3x_3+\tfrac{1}{2}(X_2+b)\log(x_1+x_2)$$
$$=\tfrac{1}{2}(X_2-2X_1+2b-2c)\{-\log(x_2-x_1+2ap_3)+\phi(p_3)\}.$$

The complete integral is given by

$$z+\Pi=A,\quad X_1=0,\quad X_2=0,\quad X_3=C,$$

where A and C are arbitrary constants; consequently, it is

$$A=z-Cx_3-\tfrac{1}{2}b\log(x_1+x_2)-(b-c)\{-\log(x_2-x_1+2aC)+\phi(C)\},$$

or as $(b-c)\phi(C)$ can be absorbed into A, this complete integral is

$$z-Cx_3-\tfrac{1}{2}b\log(x_1+x_2)-(b-c)\log(x_2-x_1+2aC)=A.$$

The general integral is given by

$$z+\Pi=g(X_3),\quad X_1=0,\quad X_2=0,$$

together with the relation

$$\frac{dg}{dX_3}=P_3.$$

Also, the quantities P_1, P_2, P_3 are given by any three of the six equations

$$\frac{\partial\Pi}{\partial x_i}-\sum_{r=1}^{3}P_r\frac{\partial X_r}{\partial x_i}=-p_i,\qquad \frac{\partial\Pi}{\partial p_i}-\sum_{r=1}^{3}P_r\frac{\partial X_r}{\partial p_i}=0$$

which are independent of one another: in particular,

$$P_3=\frac{\partial\Pi}{\partial p_3}+\frac{a(p_1-p_2)}{x_2-x_1+2ap_3}\left(\frac{\partial\Pi}{\partial p_1}-\frac{\partial\Pi}{\partial p_2}\right),$$

which, on substitution, gives

$$P_3=-x_3+(b-c)\left\{\phi'(p_3)-\frac{2a}{x_2-x_1+2ap_3}\right\}.$$

Hence the general integral is given by the elimination of p_3 between the equations

$$z-p_3x_3-\tfrac{1}{2}b\log(x_1+x_2)-(b-c)\{\log(x_2-x_1+2ap_3)-\phi(p_3)\}=g(p_3),$$
$$-x_3+(b-c)\left\{\phi'(p_3)-\frac{2a}{x_2-x_1+2ap_3}\right\}=g'(p_3),$$

or, writing

$$g(p_3)-(b-c)\phi(p_3)=h(p_3),$$

so that $h(p_3)$ is a new arbitrary function, we have the general integral given by the elimination of p_3 between

$$\left.\begin{aligned} z-p_3x_3-\tfrac{1}{2}b\log(x_1+x_2)-(b-c)\log(x_2-x_1+2ap_3)&=h(p_3)\\ -x_3-\frac{2a(b-c)}{x_2-x_1+2ap_3}&=h'(p_3)\end{aligned}\right\}.$$

And there is no singular integral.

Note. The example is worked out in order to illustrate the relation between the constituents of the contact transformation and the construction of the integral.

If the Jacobian method (Chap. IV) were adopted for the integration of

$$F=0,\quad G=0,$$

we should find a third equation to associate with these: proceeding in the usual way, we could take either

$$p_3 = a_3,$$

or

$$a(p_1 - p_2) - x_3 = c_3,$$

where a_3 and c_3 are arbitrary constants. We then resolve the equations

$$F=0, \quad G=0,$$

with one of the equations

$$p_3 = a_3, \quad a(p_1 - p_2) - x_3 = c_3,$$

for p_1, p_2, p_3; we substitute in

$$dz = p_1 dx_1 + p_2 dx_2 + p_3 dx_3,$$

and effect a quadrature. The resulting equation gives the complete integral: and the general integral would be deduced by the customary process.

Writing

$$p_3 = H, \quad a(p_1 - p_2) - x_3 = K,$$

we have

$$(F, G) = 0, \quad (F, H) = 0, \quad (G, H) = 0,$$

all satisfied identically, so that F, G, H can be the constituents X_1, X_2, X_3 of a contact transformation. Also

$$(F, K) = 0, \quad (G, K) = 0,$$

satisfied identically, so that F, G, K can be the constituents X_1, X_2, X_3 of another contact transformation. The integration of

$$F=0, \quad G=0,$$

by means of the latter contact transformation is left as an exercise.

But

$$(H, K) = 1;$$

hence F, H, K cannot be the constituents X_1, X_2, X_3 of a contact transformation, nor can G, H, K be those constituents.

Ex. 2. Integrate by this method the equations

$$\left.\begin{array}{r} x_2 p_1 + x_1 p_2 + x_3 p_3 - c = 0 \\ a(p_1 - p_2) - x_3 = 0 \end{array}\right\},$$

where a and c are constants.

Ex. 3. Similarly integrate the equations

$$\left.\begin{array}{r} x_1 p_1 - x_2 p_2 + x_3 p_3 - x_4 p_4 = a \\ x_3 p_1 + x_4 p_2 - x_1 p_3 - x_2 p_4 = c \end{array}\right\},$$

where a and c are constants.

136. It thus appears that, when a single differential equation is given in a form

$$U = 0,$$

the quantity U can always be made a constituent of a contact transformation, the explicit expression of which adopts one or other of two forms according as U does, or does not, explicitly involve the dependent variable : and the various kinds of integrals* can be deduced when the full expression of the transformation is known.

Again, when a complete system of equations is given in a form

$$X_1 = 0, \ldots, X_m = 0,$$

(where m is less than the number of independent variables), so that the relations

$$[X_i, X_j] = 0, \text{ or } (X_i, X_j) = 0,$$

(according as the equations in the system do, or do not, involve the dependent variable explicitly) are satisfied, it has been proved that, if the relations are satisfied identically, the theory of contact transformations can be applied† immediately by making $X_1, \ldots, X_m$ constituents of the transformation in the mode indicated. The various kinds of integrals‡ could be deduced when the full expression of the transformation is known.

Consequently, in order that the method of contact transformations may be made effective for the practical integration of a single equation or of a complete system of equations, it is necessary to devise a practical process for the completion of a contact transformation when one constituent can be assumed or when several constituents can be assumed. These constituents have been supposed, in every case thus far considered, to belong to the X-type and not to the P-type in a relation

$$dZ - \sum_{i=1}^{n} P_i dX_i = \rho \left(dz - \sum_{i=1}^{n} p_i dx_i \right);$$

alternative suppositions have not been considered.

* No indication of the occurrence of *special* integrals has been given : these, however, usually depend on the details of form of particular equations, and such details have been ignored. They would, of course, have to be taken into account if a complete theory were being based upon contact transformations and their properties, and upon these alone : but the present chapter has no such purpose. What is intended, is the exposition of the more important general methods of integration, in a sufficient amount of detail to make their scope clear : among such methods, contact transformations must find a place.

† No implication has been made that the theory cannot be applied unless the relations are satisfied identically : the condition was requisite merely for the immediate use of the propositions quoted in §§ 128—130.

‡ With the same exception of special integrals as occurs sometimes when a single equation is propounded for integration.

Moreover, it is quite possible (examples, indeed, have occurred freely in illustration of preceding discussions) that a system $X_1 = 0, \ldots, X_m = 0$ should be complete, even though the relations

$$[X_i, X_j] = 0, \text{ or } (X_i, X_j) = 0$$

should not be satisfied identically; all that is necessary, is that the relations should be satisfied simultaneously with the equations of the system without the intervention of any other equation. The question thus is raised as to the use (if any) that may be made of the property, when the relations of coexistence are satisfied only in virtue of the equations of the system; and the answer to the question is to be found in the properties of groups of functions.

Groups of Functions.

137. Accordingly, we proceed to the consideration of the simpler properties of *groups of functions**: they are based upon the properties of contact transformations, and their development is sufficiently distinct to cause their application to be regarded as a distinct method for the integration of systems of equations.

Also, partly for the sake of some simplicity in the formulæ, and partly because there is no intention of developing the full theory, it will be assumed that we are dealing with the limited contact transformations of § 130, so that we may take

$$z' = z + \Pi, \quad p_i' = P_i, \quad x_i' = X_i,$$

where $i = 1, \ldots, n$, and $\Pi, P_1, \ldots, P_n, X_1, \ldots, X_n$ are functions of $x_1, \ldots, x_n, p_1, \ldots, p_n$, as the type of transformation to be discussed.

The general notion of a group is of exceedingly comprehensive range: for the immediate purpose, it may be described by saying that the aggregate of a number of entities is called a group when, if those entities are compounded in all possible ways which conform to assigned laws, the results of the composition can be expressed in terms of those entities by forms which are subject to other assigned laws.

* This theory is, of course, only a part of Lie's comprehensive theory of transformation-groups. A full exposition is given in his treatise *Theorie der Transformationsgruppen*, vol. II, pp. 178 *et seq.*: see also *Math. Ann.*, t. VIII (1875), pp. 248 *et seq.*, *ib.*, t. XI (1877), pp. 465 *et seq.*

The arguments of the functions are $2n$ in number, taken to be $x_1, \ldots, x_n, p_1, \ldots, p_n$ in two complementary sets, and to be independent of one another, so far as the properties of the sets of functions are concerned. When two functions u and v of the group are known, the assigned law of composition is that they shall be combined in the form

$$\sum_{i=1}^{n}\left(\frac{\partial u}{\partial x_i}\frac{\partial v}{\partial p_i}-\frac{\partial u}{\partial p_i}\frac{\partial v}{\partial x_i}\right);$$

in accordance with the notation already used, this combination is represented by

$$(u, v).$$

Then r functions $u_1, \ldots, u_r$ of the $2n$ variables are, for our purpose, said to be a *group* when the following conditions are satisfied:—

(i) the functions $u_1, \ldots, u_r$ are algebraically independent of one another:

(ii) every combination (u_i, u_j), for i and $j = 1, \ldots, r$, is expressible by means of the r functions $u_1, \ldots, u_r$, the expressions not requiring the introduction of other functions of the variables.

When the conditions are satisfied, the group is said to be of *order* r: and every function of the variables which, when combined with the functions $u_1, \ldots, u_r$ in turn, satisfies the condition of expression in terms of the members $u_1, \ldots, u_r$ of the group, is said to *belong to* the group.

Further, it will appear possible that, in a group of order r, there may be a set of r' functions which, taken by themselves, satisfy the conditions for a group: naturally, in this case, $r' < r$. The set of r' functions is then called a *sub-group* of the group of order r.

When all the combinations (u_i, u_j), for i and $j = 1, \ldots, r$, vanish, the group is called a *system in involution*.

A function v of the $2n$ variables, such that $(v, u_i) = 0$ for $i = 1, \ldots, n$, is said to be *in involution with the group* $u_1, \ldots, u_r$.

Some simple properties of groups, that are practically obvious, may at once be noted.

(i) Every function of the members of a group belongs to the group.

(ii) If r functions $v_1, \ldots, v_r$ of the members of $u_1, \ldots, u_r$ of a group of order r be algebraically independent of one another, they constitute a group. This group, composed of $v_1, \ldots, v_r$, is regarded merely as another form of the original group: thus a group can have an unlimited number of forms.

When any form of a group is in involution, every form of the group is also in involution.

(iii) When r functions $u_1, \ldots, u_r$ are connected by s, and by not more than s, relations, and when all the combinations (u_i, u_j) are expressible in terms of $u_1, \ldots, u_r$, the functions belong to a group of order $r - s$.

(iv) When two different groups have members in common with one another, the aggregate of those common members constitutes a group; for if the aggregate be $t_1, \ldots, t_m$, the combinations (t_i, t_j), for i and $j = 1, \ldots, m$, are expressible in terms of members of the first group and also in terms of members of the second group, that is, they are expressible in terms of the members common to the two groups.

(v) A group of functions, involving $2n$ variables in two complementary sets, is of order not greater than $2n$; for if the number of members were greater than $2n$, they could not be algebraically independent.

We shall see later that the order of a group in involution cannot be greater than n.

(vi) A contact transformation changes a group of order r into another group of order r.

Let the variables introduced by the transformation be $x_1', \ldots, x_n', p_1', \ldots, p_n'$, so that the equations characteristic of the transformation are

$$(x_i', x_j') = 0, \quad (p_i', p_j') = 0, \quad (p_i', x_j') = 0, \quad (p_i', x_i') = -1,$$

where i and $j = 1, \ldots, n$, and are unequal to one another, the independent variables being $x_1, \ldots, x_n, p_1, \ldots, p_n$: also

$$(x_i, x_j) = 0, \quad (p_i, p_j) = 0, \quad (p_i, x_j) = 0, \quad (p_i, x_i) = -1,$$

the independent variables being $x_1', \ldots, x_n', p_1', \ldots, p_n'$.

Let $u_1, \ldots, u_r$ be the group to be transformed, and let v_i be the value of u_i (for $i = 1, \ldots, r$) after the transformation has been effected; as $u_1, \ldots, u_r$ are algebraically independent of one another, so also are $v_1, \ldots, v_r$. Then, since

$$\frac{\partial v_i}{\partial x_s'} = \sum_{\mu=1}^{n} \left(\frac{\partial u_i}{\partial x_\mu} \frac{\partial x_\mu}{\partial x_s'} + \frac{\partial u_i}{\partial p_\mu} \frac{\partial p_\mu}{\partial x_s'} \right),$$

$$\frac{\partial v_i}{\partial p_s'} = \sum_{\mu=1}^{n} \left(\frac{\partial u_i}{\partial x_\mu} \frac{\partial x_\mu}{\partial p_s'} + \frac{\partial u_i}{\partial p_\mu} \frac{\partial p_\mu}{\partial p_s'} \right),$$

we have

$$\begin{aligned}(v_i, v_j) &= \sum_{s=1}^{n} \left(\frac{\partial v_i}{\partial x_s'} \frac{\partial v_j}{\partial p_s'} - \frac{\partial v_i}{\partial p_s'} \frac{\partial v_j}{\partial x_s'} \right) \\ &= \sum_{\mu=1}^{n} \left\{ \left(\frac{\partial u_i}{\partial x_\mu} \frac{\partial u_j}{\partial p_\mu} - \frac{\partial u_i}{\partial p_\mu} \frac{\partial u_j}{\partial x_\mu} \right) \sum_{s=1}^{n} \left(\frac{\partial x_\mu}{\partial x_s'} \frac{\partial p_\mu}{\partial p_s'} - \frac{\partial x_\mu}{\partial p_s'} \frac{\partial p_\mu}{\partial x_s'} \right) \right\} \\ &= \sum_{\mu=1}^{n} \left(\frac{\partial u_i}{\partial x_\mu} \frac{\partial u_j}{\partial p_\mu} - \frac{\partial u_i}{\partial p_\mu} \frac{\partial u_j}{\partial x_\mu} \right) (x_\mu, p_\mu) \\ &= \sum_{\mu=1}^{n} \left(\frac{\partial u_i}{\partial x_\mu} \frac{\partial u_j}{\partial p_\mu} - \frac{\partial u_i}{\partial p_\mu} \frac{\partial u_j}{\partial x_\mu} \right) \\ &= (u_i, u_j).\end{aligned}$$

But (u_i, u_j) is expressible in terms of $u_1, \ldots, u_r$ alone, so that

$$\begin{aligned}(u_i, u_j) &= \theta_{ij}(u_1, \ldots, u_r) \\ &= \theta_{ij}(v_1, \ldots, v_r);\end{aligned}$$

hence

$$(v_i, v_j) = \theta_{ij}(v_1, \ldots, v_r),$$

and therefore the functions $v_1, \ldots, v_r$ form a group of order r.

A question is thus suggested, as follows. Given two groups of order r, each involving $2n$ variables in complementary sets: what are the contact transformations, if any, which transform one of the groups into the other? An answer to the question will be obtained through the determination of a canonical form to represent a group.

138. Passing now to properties of groups that are less obvious than those just given, we have the following theorem, due* to Lie:—

Let $u_1, \ldots, u_r$ be a group of order r in $2n$ variables, composed of two complementary sets; then the r linear partial differential equations

$$(u_1, f) = 0, \ldots, (u_r, f) = 0$$

are a complete system.

* It appears to have been the earliest result in Lie's researches, and was first published in 1872: see *Forh. af Christ.*, (1872), pp. 133—135, *ib.* (1873), pp. 237—262.

In the first place, the r equations are linearly independent of one another: for otherwise a set of determinants involving the derivatives of $u_1, \ldots, u_r$ would vanish and so would imply identical relations among the quantities $u_1, \ldots, u_r$, contrary to the hypothesis that the r functions constitute a group of order r.

When, for convenience, we write

$$(u_i, f) = A_i f,$$

for $i = 1, \ldots, r$, the Poisson-Jacobi identity

$$((u_i, u_j) f) + ((u_j, f) u_i) + ((f, u_i) u_j) = 0$$

becomes

$$((u_i, u_j) f) - A_i (A_j f) + A_j (A_i f) = 0,$$

that is,

$$\begin{aligned} A_i (A_j f) - A_j (A_i f) &= ((u_i, u_j) f) \\ &= \frac{\partial \theta_{ij}}{\partial u_1} A_1 f + \ldots + \frac{\partial \theta_{ij}}{\partial u_r} A_r f, \end{aligned}$$

where, as before,

$$(u_i, u_j) = \theta_{ij} (u_1, \ldots, u_r).$$

Hence all the equations

$$A_i (A_j f) - A_j (A_i f) = 0,$$

for i and $j = 1, \ldots, r$, are satisfied in virtue of the r equations

$$A_1 f = 0, \ldots, A_r f = 0,$$

which therefore (§ 37) are a complete system. The proposition is thus established.

As the system of r equations in $2n$ variables is complete, it possesses $2n - r$ distinct integrals: let these be $v_1, \ldots, v_{2n-r}$. It is known (§ 41) that every other integral of the complete system is expressible in terms of these $2n - r$ functionally distinct integrals; and by the Poisson-Jacobi theorem, (v_i, v_j) is an integral of the system: hence

$$(v_i, v_j) = \phi_{ij} (v_1, \ldots, v_{2n-r}),$$

and therefore the $2n - r$ functions $v_1, \ldots, v_{2n-r}$ form a group.

Applying Lie's theorem to the group $v_1, \ldots, v_{2n-r}$, we see that the system of $2n - r$ equations

$$(v_1, g) = 0, \ldots, (v_{2n-r}, g) = 0$$

is complete, so that it possesses r fundamentally distinct integrals. Evidently the members of the group $u_1, \ldots, u_r$ can be taken as these integrals; and all the equations

$$(u_i, v_j) = 0$$

are satisfied identically, owing to the property (§ 41) of integrals of homogeneous linear partial differential equations of the first order. We thus have the further theorem:—

A group of functions of order r in $2n$ variables determines another group of functions of order $2n-r$ in those variables, and conversely: every function of either group is in involution with the other group.

The two groups thus associated are called, sometimes *reciprocal* to one another, sometimes *polars* of one another.

COROLLARY I. *When a contact transformation is effected upon two reciprocal groups, the transformed groups are reciprocal to one another.* Let $u_1, \ldots, u_r$, and $v_1, \ldots, v_{2n-r}$, be the reciprocal groups, transformed into $u_1', \ldots, u_r'$, and $v_1', \ldots, v'_{2n-r}$, by the contact transformation: then, as before (§ 137),

$$\begin{aligned}(u_i', v_j') &= (u_i, v_j)\\ &= 0,\end{aligned}$$

for all values of i and j. This result is the analytical expression of the property stated.

COROLLARY II. *The order of a system in involution cannot be greater than n,* the total number of variables being $2n$ in two complementary sets. Let $u_1, \ldots, u_r$ be a system in involution, so that the equations

$$(u_i, u_j) = 0$$

are satisfied for all values of i and j. The equations

$$(u_1, f) = 0, \ \ldots, \ (u_r, f) = 0$$

are a complete Jacobian system, possessing $2n-r$ functionally independent integrals. Owing to the equations which express the involution, it is clear that $u_1, \ldots, u_r$ are integrals of the complete Jacobian system, and they are independent of one another: hence

$$r \leqslant 2n - r,$$

so that r cannot be greater than n.

COROLLARY III. The converse of Lie's theorem also is true: that is, if

$$(u_1, f) = 0, \ \ldots, \ (u_r, f) = 0,$$

are a complete system, and if $u_1, \ldots, u_r$ be functionally distinct from one another, then $u_1, \ldots, u_r$ form a group. For, as before, the equation

$$((u_i, u_j)\, f) = 0$$

is satisfied in virtue of the system. Now (u_i, u_j), being some function of the variables, can be expressed in terms of $u_1, \ldots, u_r$, and of $2n - r$ of the variables, say $x_{r+1}, \ldots, x_n, p_1, \ldots, p_n$: hence, if

$$(u_i, u_j) = \theta_{ij}(u_1, \ldots, u_r, x_{r+1}, \ldots, x_n, p_1, \ldots, p_n),$$

we have

$$((u_i, u_j) f) = \sum_{s=1}^{r} \frac{\partial \theta_{ij}}{\partial u_s}(u_s, f) + \sum_{\mu=r+1}^{n} \frac{\partial \theta_{ij}}{\partial x_\mu}(x_\mu, f) + \sum_{\sigma=1}^{n} \frac{\partial \theta_{ij}}{\partial p_\sigma}(p_\sigma, f),$$

and therefore

$$0 = \sum_{\mu=r+1}^{n} \frac{\partial \theta_{ij}}{\partial x_\mu}(x_\mu, f) + \sum_{\sigma=1}^{n} \frac{\partial \theta_{ij}}{\partial p_\sigma}(p_\sigma, f),$$

that is,

$$0 = \sum_{\mu=r+1}^{n} \frac{\partial \theta_{ij}}{\partial x_\mu} \frac{\partial f}{\partial p_\mu} - \sum_{\sigma=1}^{n} \frac{\partial \theta_{ij}}{\partial p_\sigma} \frac{\partial f}{\partial x_\sigma}.$$

Now the given system is complete, so that f satisfies no other equation: hence the last equation must be evanescent, and therefore

$$\frac{\partial \theta_{ij}}{\partial x_\mu} = 0, \quad \frac{\partial \theta_{ij}}{\partial p_\sigma} = 0,$$

for $\mu = r + 1, \ldots, n$, and $\sigma = 1, \ldots, n$. Thus

$$(u_i, u_j) = \theta_{ij}(u_1, \ldots, u_r),$$

and $u_1, \ldots, u_r$ are supposed to be functionally distinct from one another: hence they form a group.

Indicial Functions of a Group.

139. A function, which belongs to a group $u_1, \ldots, u_r$ and is in involution with all the functions of the group, is called* *indicial.* Each indicial function f satisfies the equations

$$(u_1, f) = 0, \ \ldots, \ (u_r, f) = 0;$$

and therefore the number of indicial functions belonging to the group is the number of integrals of these equations, which are independent of one another and can be expressed in terms of $u_1, \ldots, u_r$. These equations are

$$A_1 f = (u_1, u_1) \frac{\partial f}{\partial u_1} + (u_1, u_2) \frac{\partial f}{\partial u_2} + \ldots + (u_1, u_r) \frac{\partial f}{\partial u_r} = 0,$$

....

$$A_r f = (u_r, u_1) \frac{\partial f}{\partial u_1} + (u_r, u_2) \frac{\partial f}{\partial u_2} + \ldots + (u_r, u_r) \frac{\partial f}{\partial u_r} = 0;$$

* The German title is *ausgezeichnete Function*; the French title is *fonction distinguée.*

the coefficients of the derivatives of f are expressible in terms of $u_1, \ldots, u_r$, so that apparently there are r equations in r variables.

Now these equations are not linearly independent of one another: one obvious linear relation is

$$(A_1 f)\frac{\partial f}{\partial u_1} + \ldots + (A_r f)\frac{\partial f}{\partial u_r} = 0\,;$$

and other linear relations may exist. Accordingly, suppose that there are m such relations, so that the r equations reduce to $r-m$ linearly independent equations; the conditions, necessary and sufficient to secure this reduction, are that the determinant

$$\begin{vmatrix} (u_1, u_1), & (u_1, u_2), & \ldots, & (u_1, u_r) \\ (u_2, u_1), & (u_2, u_2), & \ldots, & (u_2, u_r) \\ \ldots & \ldots & \ldots & \ldots \\ (u_r, u_1), & (u_r, u_2), & \ldots, & (u_r, u_r) \end{vmatrix},$$

and its minors of all orders up to $m-1$ inclusive, but not all the minors of order m, shall vanish.

We may assume that the linearly independent equations are the first $r-m$ of the set. As

$$A_i(A_j f) - A_j(A_i f) = \frac{\partial f_{ij}}{\partial u_1} A_1 f + \ldots + \frac{\partial f_{ij}}{\partial u_r} A_r f,$$

where $(u_i, u_j) = f_{ij}(u_1, \ldots, u_r)$, and as each of the quantities $A_{r-m+1}f, \ldots, A_r f$ is linearly expressible in terms of $A_1 f, \ldots, A_{r-m} f$, it follows that the equations

$$A_i(A_j f) - A_j(A_i f) = 0,$$

for i and $j = 1, \ldots, r-m$, are satisfied in virtue of the set of $r-m$ retained equations; hence these $r-m$ equations are a complete system. Any integral is an indicial function, and the complete system of $r-m$ equations in r variables possesses m independent integrals; hence, *under the conditions associated with the preceding determinant, the group possesses m indicial functions.*

Note 1. The number of indicial functions of a group subjected to a contact transformation is unaffected by the transformation. Let the transformed group be $u_1', \ldots, u_r'$: then, as in § 137, we have

$$(u_i', u_j') = (u_i, u_j),$$

so that the critical determinant and its minors are unaltered, and therefore the number of indicial functions is unaffected.

Note 2. The number of indicial functions of a group is independent of the form of the group.

Let $U_1, \ldots, U_r$ be a form of the group $u_1, \ldots, u_r$, so that the Jacobian

$$J\left(\frac{U_1, \ldots, U_r}{u_1, \ldots, u_r}\right)$$

does not vanish, because the functions $U_1, \ldots, U_r$ are independent. Now

$$(U_i, f) = \sum_{s=1}^{r} \frac{\partial U_i}{\partial u_s}(u_s, f),$$

for $i=1, \ldots, r$; hence every function, which is indicial for the form $u_1, \ldots, u_r$, is indicial for the form $U_1, \ldots, U_r$: and, as the determinant of the coefficients of $(u_1, f), \ldots, (u_r, f)$ is the non-vanishing Jacobian, every function indicial for the form $U_1, \ldots, U_r$ is indicial also for the form $u_1, \ldots, u_r$.

Note 3. The critical determinant is obviously skew: it vanishes if r be odd. Hence every group of odd order possesses at least one indicial function.

Note 4. When there are m indicial functions, they are the independent integrals of $r-m$ homogeneous linear equations in r variables. The actual expression of the indicial functions may therefore be assumed known, on constructing the integrals of those equations by any of the methods explained in Chapter III.

Ex. The functions

$$u_1 = p_1 p_2 - x_3 x_4, \quad u_2 = p_3 p_4 - x_1 x_2, \quad u_3 = p_1 x_1 + p_2 x_2 - p_3 x_3 - p_4 x_4,$$

form a group of order three, because

$$(u_1, u_2) = u_3, \quad (u_1, u_3) = -2u_1, \quad (u_2, u_3) = 2u_2.$$

The critical determinant is

$$\begin{vmatrix} 0, & u_3, & -2u_1 \\ -u_3, & 0, & 2u_2 \\ 2u_1, & -2u_2, & 0 \end{vmatrix};$$

it is easy to see that $m=1$, so that the group possesses one indicial function.

This indicial function satisfies the equations

$$\left.\begin{aligned} u_3\frac{\partial f}{\partial u_2} - 2u_1\frac{\partial f}{\partial u_3} &= 0 \\ -u_3\frac{\partial f}{\partial u_1} \qquad\qquad + 2u_2\frac{\partial f}{\partial u_3} &= 0 \\ 2u_1\frac{\partial f}{\partial u_1} - 2u_2\frac{\partial f}{\partial u_2} \qquad\qquad &= 0 \end{aligned}\right\},$$

which are equivalent to two independent equations. These possess one integral; it is easily found to be

$$4u_1u_2+u_3^2,$$

which accordingly is the indicial function of the group.

It has been seen that a group $u_1, \ldots, u_r$ determines a reciprocal group $v_1, \ldots, v_{2n-r}$, the members of the latter being the $2n-r$ independent integrals of

$$(u_1, f)=0, \ldots, (u_r, f)=0;$$

and any integral of these equations is expressible in terms of $v_1, \ldots, v_{2n-r}$. Now the indicial functions of the original group are integrals of these equations; hence there are m relations between the members of the two groups of the form

$$U_i(u_1, \ldots, u_r)=\phi_i(v_1, \ldots, v_{2n-r}),$$

for $i=1, \ldots, m$.

Conversely, a relation of this character implies the existence of an indicial function: because, as the equations

$$(u_1, f)=0, \ldots, (u_r, f)=0,$$

are satisfied for $f=v_1, \ldots, v_{2n-r}$, we have

$$(u_1, \phi_i)=0, \ldots, (u_r, \phi_i)=0,$$

that is,

$$(u_1, U_i)=0, \ldots, (u_r, U_i)=0,$$

shewing that U_i is an indicial function. Denote the value of this indicial function by w_i, so that w_i can be expressed in terms of $u_1, \ldots, u_r$ alone, and therefore w_i belongs to the original group: it can be expressed in terms of $v_1, \ldots, v_{2n-r}$ alone, and therefore w_i belongs to the reciprocal group. Owing to the reciprocity between two polar groups, we know that r independent integrals of the equations

$$(v_1, g)=0, \ldots, (v_{2n-r}, g)=0$$

are $u_1, \ldots, u_r$. As

$$w_i=U_i(u_1, \ldots, u_r),$$

it follows that

$$(v_1, U_i)=0, \ldots, (v_{2n-r}, U_i)=0,$$

and therefore

$$(v_1, w_i)=0, \ldots, (v_{2n-r}, w_i)=0.$$

Now w_i can be expressed in terms of $v_1, \ldots, v_{2n-r}$; hence w_i is an indicial function of the reciprocal group.

Gathering together these results, we can enunciate them as follows:—

Two reciprocal groups $u_1, \ldots, u_r$, and $v_1, \ldots, v_{2n-r}$ have the same indicial functions $w_1, \ldots, w_m$: and the existence of each indicial function implies the existence of a relation between functions of the two groups of the type

$$U_i(u_1, \ldots, u_r) = w_i = \phi_i(v_1, \ldots, v_{2n-r}),$$

for $i = 1, \ldots, m$.

COROLLARY. *If a function belongs to two reciprocal groups, it is an indicial function for each of them.*

Ex. In an earlier example (p. 352), it was seen that

$$u_1 = p_1p_2 - x_3x_4, \quad u_2 = p_3p_4 - x_1x_2, \quad u_3 = p_1x_1 + p_2x_2 - p_3x_3 - p_4x_4,$$

form a group of order three. The reciprocal group is of order five, and is composed of five independent integrals of the equations

$$(u_1, f) = 0, \quad (u_2, f) = 0, \quad (u_3, f) = 0.$$

Five such integrals can be taken in the form

$$v_1 = p_1p_3 - x_2x_4,$$
$$v_2 = p_2p_4 - x_1x_3,$$
$$v_3 = p_1p_4 - x_2x_3,$$
$$v_4 = p_2p_3 - x_1x_4,$$
$$v_5 = p_1x_1 - p_2x_2 + p_3x_3 - p_4x_4,$$

which accordingly constitute the reciprocal group.

It was seen that $u_3^2 + 4u_1u_2$ is the indicial function of the original group: it is therefore, by the preceding theorem, the indicial function of the reciprocal group, and one relation must subsist between the functions u_1, u_2, u_3 and the functions v_1, v_2, v_3, v_4, v_5. It is easy to verify that

$$u_3^2 + 4u_1u_2 = v_5^2 + 4v_1v_2,$$

the common value of these quantities being the indicial function for the two groups.

CANONICAL FORM OF A GROUP.

140. When a group is a system in involution, it is obvious, from the characteristic equations

$$(u_i, u_j) = 0,$$

that every member of the group is an indicial function.

Also, if a group of order r certainly possesses $r-1$ indicial functions, it is a system in involution; for if $u_1, \dots, u_{r-1}$ are the indicial functions, and if v is the other member,

$$(v, u_1)=0, \dots, (v, u_{r-1})=0,$$

that is,

$$(u_1, v)=0, \dots, (u_{r-1}, v)=0;$$

therefore v is an indicial function also, and the group is a system in involution.

Hence, if a group of order r is not a system in involution, it cannot possess more than $r-2$ indicial functions. We now proceed to obtain, for groups that are not systems in involution, a canonical form which shall obviously shew the number and the incidence of the indicial functions of the group*: the main result is contained in the theorem:—

The order of a group, that is not a system in involution, exceeds the number of its indicial functions by an even integer; and a group of order $2q+m$, which possesses m indicial functions, can be transformed into a group $X_1, \dots, X_{q+m}, P_1, \dots, P_q$, such that

$$(X_i, X_j)=0, \quad (P_\mu, P_\kappa)=0, \quad (X_i, P_\mu)=0,$$
$$(P_\mu, X_\mu)=1,$$

for which i and $j=1, \dots, q+m$; μ and $\kappa=1, \dots, q$; while i and μ are unequal to one another.

It will appear that the unity, as the value of (P_μ, X_μ), is merely a determinate constant: and it is clear that $X_{q+1}, \dots, X_{q+m}$ are the m indicial functions of the transformed group. The form, thus selected for the group, is usually called the *canonical form*.

The proposition is established as follows. Let the group be of order r and let its members be $u_1, \dots, u_r$; also, let m be the number of indicial functions so that, after the earlier explanations, $m \leqslant r-2$. Suppose that u_1 is a member of the group which is not an indicial function: then not all the quantities (u_1, u_2), $(u_1, u_3), \dots, (u_1, u_r)$ vanish, and so the equation

$$(u_1, \theta)=(u_1, u_2)\frac{\partial\theta}{\partial u_2}+\dots+(u_1, u_r)\frac{\partial\theta}{\partial u_r}=1$$

* It is unnecessary to deal with a system in involution: every function is an indicial function: and every form of the group is a system in involution.

is possible and possesses integrals which, being functions of $u_2, \ldots, u_r$ or of some of them, belong to the group. Take one of these integrals, and let it be denoted by u_2; thus

$$(u_1, u_2) = 1.$$

Now consider the two equations

$$(u_1, v) = 0, \quad (u_2, v) = 0;$$

they are a complete system, because

$$(u_1(u_2, v)) - (u_2(u_1, v)) = ((u_1, u_2)v) = (1, v) = 0.$$

In full expression, they are

$$\left.\begin{aligned} \frac{\partial v}{\partial u_2} + (u_1, u_3)\frac{\partial v}{\partial u_3} + \ldots + (u_1, u_r)\frac{\partial v}{\partial u_r} &= 0 \\ -\frac{\partial v}{\partial u_1} \quad + (u_2, u_3)\frac{\partial v}{\partial u_3} + \ldots + (u_2, u_r)\frac{\partial v}{\partial u_r} &= 0 \end{aligned}\right\};$$

they determine v in terms of $u_1, \ldots, u_r$; and being a complete system of two equations in r variables, they possess $r-2$ functionally independent integrals. Let these be $v_1, \ldots, v_{r-2}$, which accordingly are independent of one another; also they are functions of $u_1, \ldots, u_r$.

Then $u_1, u_2, v_1, \ldots, v_{r-2}$ are a set of r independent functions: for otherwise, some relation

$$g(u_1, u_2, v_1, \ldots, v_{r-2}) = 0$$

would be satisfied identically, and then

$$\begin{aligned} 0 = (u_1, g) &= (u_1, u_2)\frac{\partial g}{\partial u_2} + (u_1, v_1)\frac{\partial g}{\partial v_1} + \ldots + (u_1, v_{r-2})\frac{\partial g}{\partial v_{r-2}} \\ &= \frac{\partial g}{\partial u_2}, \\ 0 = (u_2, g) &= -\frac{\partial g}{\partial u_1}, \end{aligned}$$

so that g would not involve u_1 or u_2, and the relation would subsist between $v_1, \ldots, v_{r-2}$, which are known to be functionally independent. Hence our original group can be replaced by a group $u_1, u_2, v_1, \ldots, v_{r-2}$ such that

$$(u_1, u_2) = 1, \quad (u_1, v_i) = 0, \quad (u_2, v_i) = 0,$$

for $i = 1, \ldots, r-2$. We have seen (§ 139, Note 2) that the number of indicial functions is independent of the form of the group; and it is clear that neither u_1 nor u_2 is an indicial function.

Let $w_1, \ldots, w_{2n-r}$ be the polar of the original group, so that

$$(u_1, w_i) = 0, \quad (u_2, w_i) = 0,$$

for $i = 1, \ldots, 2n - r$; and as the $2n - r$ quantities form a group, all the quantities (w_i, w_j) are expressible in terms of $w_1, \ldots, w_{2n-r}$. Then $u_1, u_2, w_1, \ldots, w_{2n-r}$ form a group: for they are $2n - r + 2$ independent functions and, as

$$(u_1, u_2) = 1, \quad (u_1, w_i) = 0, \quad (u_2, w_i) = 0,$$

all the combinations of members of the group are expressible in terms of those members. Now

$$(v_i, u_1) = 0, \quad (v_i, u_2) = 0, \quad (v_i, w_j) = 0,$$

for $i = 1, \ldots, r - 2$, and $j = 1, \ldots, 2n - r$: the first two are the equations defining v_i, and the rest are satisfied because v_i belongs to the group that is reciprocal to $w_1, \ldots, w_{2n-r}$. Hence $v_1, \ldots, v_{r-2}$ is a group reciprocal to $u_1, u_2, w_1, \ldots, w_{2n-r}$: so that $v_1, \ldots, v_{r-2}$ is a sub-group of the group $u_1, u_2, v_1, \ldots, v_{r-2}$. Moreover, the indicial functions of our group are functions of $u_1, u_2, v_1, \ldots, v_{r-2}$: denoting any one of them by $\theta(u_1, u_2, v_1, \ldots, v_{r-2})$, we know that the equations

$$(u_1, \theta) = 0, \quad (u_2, \theta) = 0, \quad (v_1, \theta) = 0, \ldots, (v_{r-2}, \theta) = 0$$

must be satisfied. The first of these equations is

$$\frac{\partial \theta}{\partial u_2} = 0,$$

so that θ does not involve u_2; the second is

$$-\frac{\partial \theta}{\partial u_1} = 0,$$

so that θ does not involve u_1; and so θ is a function of $v_1, \ldots, v_{r-2}$.

Thus the group of order r possessing m indicial functions has been transformed into another group

$$u_1, u_2, v_1, \ldots, v_{r-2};$$

the quantity $(u_1, u_2) = 1$; and the $r - 2$ quantities $v_1, \ldots, v_{r-2}$ constitute a group of order $r - 2$, possessing m indicial functions.

If the group $v_1, \ldots, v_{r-2}$ is a system in involution, then, as it possesses m indicial functions, we have

$$m = r - 2;$$

and writing

$$P_1 = u_1, \quad X_1 = u_2, \quad X_2 = v_1, \ldots, X_{r-1} = v_{r-2},$$

we have

$$(P_1, X_1) = 1, \quad (P_1, X_i) = 0, \quad (X_1, X_i) = 0, \quad (X_i, X_j) = 0,$$

for i and $j = 2, \ldots, r-1$. The reduction, indicated in the theorem, has been made.

If the group $v_1, \ldots, v_{r-2}$ possessing m indicial functions is not a system in involution, so that $m \leqslant r-4$, then we transform it in the same manner as the original group was transformed. It can be made to take a form

$$X_2, P_2, w_1, \ldots, w_{r-4},$$

where

$$(P_2, X_2) = 1, \quad (P_2, w_i) = 0, \quad (X_2, w_i) = 0,$$

for $i = 1, \ldots, r-4$; and then $w_1, \ldots, w_{r-4}$ constitute a group of order $r-4$, possessing m indicial functions. The original group has thus been changed to the form

$$X_1, P_1, X_2, P_2, w_1, \ldots, w_{r-4},$$

such that

$$\begin{gathered}(P_1, X_1) = 1, \quad (P_2, X_2) = 1, \quad (X_1, X_2) = 0, \quad (X_1, P_2) = 0, \\ (X_2, P_1) = 0, \quad (P_1, P_2) = 0, \\ (X_1, w_i) = 0, \quad (X_2, w_i) = 0, \quad (P_1, w_i) = 0, \quad (P_2, w_i) = 0.\end{gathered}$$

If the group $w_1, \ldots, w_{r-4}$ is a system in involution, the required reduction has been effected: and, as the group possesses m indicial functions, we then have

$$m = r - 4.$$

If the group $w_1, \ldots, w_{r-4}$ is not a system in involution, we proceed as before. At each stage in the successive changes, we isolate two functions X_i and P_i such that

$$(P_i, X_i) = 1,$$

and we are left with a group of order $r - 2i$ possessing m indicial functions. Ultimately, we shall reach a stage when this remaining group is a system in involution, so that

$$r - 2q = m;$$

the isolated pairs of functions are

$$X_1, P_1; \ X_2, P_2; \ \ldots; \ X_q, P_q;$$

the remaining functions, being (as stated) a system in involution, may be represented by $X_{q+1}, \ldots, X_{q+m}$, such that

$$(X_{q+i}, X_{q+j}) = 0.$$

The original group of order r, possessing m indicial functions, has been replaced by

$$X_1, \ldots, X_{q+m}, \quad P_1, \ldots, P_q,$$

such that

$$(P_i, X_i) = 1,$$

for $i = 1, \ldots, q$; also

$$(P_\mu, P_\kappa) = 0, \quad (X_i, X_j) = 0, \quad (P_\mu, X_i) = 0,$$

for μ and $\kappa = 1, \ldots, q$, and for i and $j = 1, \ldots, q + m$, the values of μ and i being unequal. Also

$$r - m = 2q.$$

The proposition is thus established.

Note 1. Obviously $X_1, \ldots, X_{m+q}$, a group of order $m + q$, is a system in involution: hence (§ 138)

$$m + q \leqslant n.$$

Note 2. It is obvious that, on account of the relations

$$(P_1, X_1) = 1, \ldots, (P_q, X_q) = 1,$$

no one of the functions $P_1, X_1, \ldots, P_q, X_q$ can itself be an indicial function. It is equally obvious that, on account of the relations

$$(P_\mu, X_{q+i}) = 0, \quad (X_\mu, X_{q+i}) = 0,$$
$$(X_{q+i}, X_{q+j}) = 0,$$

for $\mu = 1, \ldots, q$, and i and $j = 1, \ldots, m$, the quantities $X_{q+1}, \ldots, X_{q+m}$ are indicial functions.

If it were possible to have any other indicial function, say $g(P_1, \ldots, P_q, X_1, \ldots, X_{q+m})$, it would have to satisfy the equations

$$(P_\mu, g) = 0, \quad (X_\mu, g) = 0, \quad (X_{q+i}, g) = 0,$$

for $\mu = 1, \ldots, q$, and $i = 1, \ldots, m$. The first set of these equations is

$$\frac{\partial g}{\partial X_\mu} = 0;$$

the second set is

$$-\frac{\partial g}{\partial P_\mu} = 0;$$

and the third is identically satisfied. Thus g does not involve $P_1, \ldots, P_q, X_1, \ldots, X_q$: and every indicial function is expressible* in terms of $X_{q+1}, \ldots, X_{q+m}$.

* This is only another way of stating that, as the group $X_{q+1}, \ldots, X_{q+m}$ is in involution, any form of the group is in involution.

Note 3. The relation

$$r - m = 2q,$$

enables us to verify the former result (§ 139, Note 3) that a group of odd order cannot be devoid of indicial forms. If

$$r = 2p + 1,$$

then

$$m = 2p - 2q + 1,$$

so that the number of indicial functions possessed by a group of odd order is certainly odd.

Ex. 1. On p. 352, it was seen that the functions

$$u_1 = p_1 p_2 - x_3 x_4, \quad u_2 = p_3 p_4 - x_1 x_2, \quad u_3 = p_1 x_1 + p_2 x_2 - p_3 x_3 - p_4 x_4,$$

form a group of order 3: and, as

$$(u_1, u_2) = u_3, \quad (u_1, u_3) = -2u_1, \quad (u_2, u_3) = 2u_2,$$

this group is not a system in involution.

The canonical form can be obtained as in the text. Obviously u_1 is not an indicial function: consequently, we require an integral of the equation

$$(u_1, \theta) = 1,$$

that is, of

$$u_3 \frac{\partial \theta}{\partial u_2} - 2u_1 \frac{\partial \theta}{\partial u_3} = 1.$$

An integral is given by

$$\theta = -\frac{u_3}{2u_1},$$

so that two functions for the canonical form are

$$P_1 = u_1, \quad X_1 = -\frac{u_3}{2u_1}.$$

One other function is required: it must be a common integral of

$$(P_1, \phi) = 0, \quad (X_1, \phi) = 0.$$

The former equation is

$$u_3 \frac{\partial \phi}{\partial u_2} - 2u_1 \frac{\partial \phi}{\partial u_3} = 0;$$

and the latter, after reduction in conjunction with this equation, is

$$u_1 \frac{\partial \phi}{\partial u_1} - u_2 \frac{\partial \phi}{\partial u_2} = 0.$$

A common integral is given by

$$\phi = u_3^2 + 4u_1 u_2.$$

Accordingly, we take

$$P_1 = u_1, \quad X_1 = -\frac{u_3}{2u_1}, \quad X_2 = u_3^2 + 4u_1 u_2:$$

we have

$$(P_1, X_1) = 1, \quad (P_1, X_2) = 0, \quad (X_1, X_2) = 0,$$

which is a canonical form. Obviously, X_2 is the one indicial function of the group.

Another transformation arises by taking

$$P_1=u_2,\quad X_1=-\frac{u_1}{u_3},\quad X_2=u_3^2+4u_1u_2\,;$$

and again, another, by taking

$$P_1=u_3,\quad X_1=\tfrac{1}{2}\log u_1,\quad X_2=u_3^2+4u_1u_2.$$

Both of these are canonical: they satisfy the equations

$$(P_1,\,X_1)=1,\quad (P_1,\,X_2)=0,\quad (X_1,\,X_2)=0.$$

Ex. 2. On p. 354, it was seen that the functions

$$v_1=p_1p_3-x_2x_4,\quad v_2=p_2p_4-x_1x_3,$$
$$v_3=p_1p_4-x_2x_3,\quad v_4=p_2p_3-x_1x_4,$$
$$v_5=p_1x_1-p_2x_2+p_3x_3-p_4x_4,$$

form a group, being the reciprocal of the group u_1, u_2, u_3 in the preceding example.

Introducing a quantity t, where

$$t=p_1x_1-p_2x_2-p_3x_3+p_4x_4,$$

so that

$$t^2=v_5^2+4v_1v_2-4v_3v_4,$$

obtain a canonical form for the group, given by

$$P_1=v_1,$$
$$X_1=-\frac{v_5}{2v_1},$$
$$P_2=v_3,$$
$$X_2=-\frac{t}{2v_3},$$
$$X_3=v_5^2+4v_1v_2,$$

so that

$$(P_1,\,X_1)=1,\quad (P_2,\,X_2)=1,$$

and all other combinations of the functions in this form of the group vanish.

141. Next, *when a group of order $m+2q$, having m indicial functions, is given in canonical form, it can be amplified, by the association of $2n-(m+2q)$ appropriately determined functions, so that the $2n$ functions are a group of order $2n$ in canonical form.*

Let the group be

$$X_1,\ P_1,\ \ldots,\ X_q,\ P_q,\ X_{q+1},\ \ldots,\ X_{q+m},$$

with the preceding notation, so that $X_{q+1},\ \ldots,\ X_{q+m}$ are the indicial functions of the group.

Suppose that X_{q+1} is omitted; the surviving $m+2q-1$ functions form a group having $m-1$ indicial functions. Form the reciprocal of this diminished group; this reciprocal contains X_{q+1} and other $2n-(m+2q)$ functions. Now X_{q+1} does not belong to the diminished group and therefore it cannot be an indicial function of the reciprocal group; and therefore, as in the preceding investigation, we can determine a function P_{q+1} of the other $2n-(m+2q)$ functions in the reciprocal group such that

$$(P_{q+1},\ X_{q+1}) = 1.$$

Moreover, P_{q+1} cannot belong to the original unmodified group: if it were expressible in terms of the members of that group, we should have

$$(X_{q+1},\ P_{q+1}) = 0,$$

because X_{q+1} is an indicial function of the unmodified group. Also, as P_{q+1} belongs to the reciprocal of the diminished group, we have

$$(P_{q+1},\ P_i) = 0,\quad (P_{q+1},\ X_i) = 0,\quad (P_{q+1},\ X_{q+s}) = 0,$$

for $i = 1, \ldots, q$, and $s = 2, \ldots, m$. Hence, when P_{q+1} is associated with the original unmodified group, we have a new group

$$X_1,\ P_1,\ \ldots,\ X_q,\ P_q,\ X_{q+1},\ P_{q+1},\ X_{q+2},\ \ldots,\ X_{q+m},$$

which is in a canonical form and possesses $m-1$ indicial functions.

Repeating this process $m-1$ times so as, on each occasion, to associate a new function P with the group and to diminish the number of indicial functions by one unit, we ultimately obtain a group

$$X_1,\ P_1,\ \ldots,\ X_{q+m},\ P_{q+m},$$

which is in a canonical form and possesses no indicial functions. We have seen (§ 140, Note 1) that

$$q + m \leqslant n.$$

If $q+m$ is equal to n, the required amplification of the original group has been effected.

If $q+m$ is less than n, take any member of the group reciprocal to

$$X_1,\ P_1,\ \ldots,\ X_{q+m},\ P_{q+m},$$

and denote it by X_{q+m+1}. Associating it with this group of order $2q+2m$, we have a new group of order $2q+2m+1$, possessing X_{q+m+1} as its one indicial function. We then apply the earlier process so as to determine a new function P_{q+m+1}: and we have a new group

$$X_1,\ P_1,\ \ldots,\ X_{q+m+1},\ P_{q+m+1},$$

which is in a canonical form and possesses no indicial function.

Proceeding in this way, by associating alternately a function X and a function P until the amplified group is of order $2n$, we ultimately obtain a group

$$X_1,\ P_1,\ \ldots,\ X_n,\ P_n,$$

which is of order $2n$, is in a canonical form, and possesses no indicial functions.

Ex. Let it be required to achieve the first stage in the completion of the group v_1, v_2, v_3, v_4, v_5 which is given in § 140, Ex. 2, so that it shall have a canonical form.

As there indicated, we take

$$P_1 = v_1,\quad X_1 = -\frac{v_5}{2v_1},\quad P_2 = v_3,\quad X_2 = -\frac{t}{2v_3},\quad X_3 = v_5{}^2 + 4v_1 v_2;$$

the required first step towards completing the group is to determine a function P_3 such that

$$(P_3,\ X_3) = 1.$$

For this purpose, we need a set of four independent integrals of

$$(P_1,\ \theta) = 0,\quad (X_1,\ \theta) = 0,\quad (P_2,\ \theta) = 0,\quad (X_2,\ \theta) = 0,$$

or, what is the same thing, four independent integrals of

$$(v_1,\ \theta) = 0,\quad (v_5,\ \theta) = 0,\quad (v_3,\ \theta) = 0,\quad (t,\ \theta) = 0.$$

Also, we know that the group v_1, v_2, v_3, v_4, v_5 (or, what is the same thing, the group v_1, v_5, v_3, t, v_2) is the reciprocal of u_1, u_2, u_3; so that three independent integrals of the preceding complete system of four equations are given by u_1, u_2, u_3. Moreover,

$$X_3 = v_5{}^2 + 4v_1 v_2 = u_3{}^2 + 4u_1 u_2;$$

so that what is needed is an integral of those four equations, independent of u_1, u_2, u_3.

Expanding the equations in full, resolving them so as to express $\dfrac{\partial\theta}{\partial x_1}$, $\dfrac{\partial\theta}{\partial x_2}$, $\dfrac{\partial\theta}{\partial x_3}$, $\dfrac{\partial\theta}{\partial x_4}$ linearly in terms of $\dfrac{\partial\theta}{\partial p_1}$, $\dfrac{\partial\theta}{\partial p_2}$, $\dfrac{\partial\theta}{\partial p_3}$, $\dfrac{\partial\theta}{\partial p_4}$, and integrating them either by Jacobi's method or by Mayer's method (Chap. IV), we find

$$u_1,\ u_2,\ u_3,\ \frac{p_1}{x_2},$$

as four independent integrals, so that $\dfrac{p_1}{x_2}$ is the fourth integral required. The quantity P_3 is to be a function of $u_1, u_2, u_3, \dfrac{p_1}{x_2}$, such that $(P_3, X_3) = 1$.

Now

$$(u_1,\ v) = v^2,\quad (u_2,\ v) = -1,\quad (u_3,\ v) = 2v,$$

so that

$$(X_3,\ v) = 4\,(u_2 v^2 + u_3 v - u_1):$$

and we know that

$$(X_3,\ u_1) = 0,\quad (X_3,\ u_2) = 0,\quad (X_3,\ u_3) = 0.$$

Hence P_3 is given by

$$P_3 = \int \frac{dv}{4(u_1 - u_3 v - u_2 v^2)},$$

where u_1, u_2, u_3 are constant in the quadrature: that is,

$$P_3 = \tfrac{1}{4} X_3^{-\frac{1}{2}} \log\left(\frac{X_3^{\frac{1}{2}} + u_3 + 2u_2 v}{X_3^{\frac{1}{2}} - u_3 - 2u_2 v}\right).$$

Two more steps are needed in order to obtain the complete group expressed in canonical form. We first require the group of two members which is reciprocal to X_1, P_1, X_2, P_2, X_3, P_3; or, as u_1, u_2, u_3 is the group reciprocal to X_1, P_1, X_2, P_2, X_3, we require a function θ of u_1, u_2, u_3 such that

$$(P_3, \theta) = 0.$$

There are two such functions, independent of one another: let them be w_1 and w_2. As

$$(P_3, X_3) = 1,$$

w_1, w_2, X_3 are three independent functions of u_1, u_2, u_3. We then take

$$X_4 = w_1.$$

The last step is the determination of P_4 as a function of w_2 and the other functions w_1, P_3, X_3, P_2, X_2, P_1, X_1, such that

$$(P_4, X_4) = 1:$$

or since

$$(X_4, w_1) = 0, \quad (X_4, X_i) = 0, \quad (X_4, P_i) = 0,$$

for $i = 1, 2, 3$, we have

$$P_4 = \int \frac{dw_2}{(w_2, w_1)},$$

where (w_2, w_1) should be expressed in terms of w_2, w_1, P_3, X_3, P_2, X_2, P_1, X_1 and, for the quadratures, w_2 should be regarded as the only variable.

Groups of Functions and Contact Transformations.

142. The preceding results can be used to establish an important property of groups, viz. *when a group of functions is subjected to a contact transformation, there are two invariants, being the order of the group and the number of indicial functions; and when two groups in the same variables have the same invariants, they can be transformed into one another by a contact transformation.*

The first part of this proposition is merely a restatement of two results already established. It was seen, in § 137, (vi), that a contact transformation does not alter the order of a group; and, in § 139, Note 1, that a contact transformation does not alter the number of indicial functions.

For the second part, we express each of the groups in a canonical form: as the orders are the same, say $2q+m$, and as the numbers of indicial functions are the same, say m, the canonical forms of the groups may be expressed by

$$X_1, P_1, \ldots, X_q, P_q, X_{q+1}, \ldots, X_{q+m},$$
$$Y_1, Q_1, \ldots, Y_q, Q_q, Y_{q+1}, \ldots, Y_{q+m},$$

respectively. The former group can be amplified into

$$X_1, P_1, \ldots, X_n, P_n,$$

and the latter can be amplified into

$$Y_1, Q_1, \ldots, Y_n, Q_n.$$

Now, on account of the relations

$$(P_i, X_i)=1, \quad (P_i, P_j)=0, \quad (P_i, X_j)=0, \quad (X_i, X_j)=0,$$

for i and $j=1, \ldots, n$, with unequal values for i and j, the equations

$$x_\mu'=X_\mu, \quad p_\mu'=P_\mu,$$

for $\mu=1, \ldots, n$, determine a contact transformation; and the equations

$$x_\mu'=Y_\mu, \quad p_\mu'=Q_\mu,$$

for the same values of μ, similarly determine a contact transformation. Consequently, the equations

$$X_\mu=Y_\mu, \quad P_\mu=Q_\mu,$$

for $\mu=1, \ldots, n$, determine a contact transformation, which manifestly transforms the one group into the other.

143. We have seen that, when a group of order $2q+m$ possessing m indicial functions is expressed in a canonical form

$$X_1, P_1, \ldots, X_q, P_q, X_{q+1}, \ldots, X_{q+m},$$

the $q+m$ quantities $X_1, \ldots, X_{q+m}$ are such that

$$(X_i, X_j)=0:$$

that is, the group contains a sub-group of order $q+m$ which is a system in involution. It will now be proved that *any sub-group which is a system in involution is of order not greater than* $q+m$.

Let a sub-group, being a system in involution, be

$$Z_1, \ldots, Z_\mu.$$

Conceive the original group amplified so as to be of order $2n$, expressed in canonical form by the association of $n-q$ functions

$P_{q+1}, \ldots, P_n$ and of $n-q-m$ functions $X_{q+m+1}, \ldots, X_n$. Now these $n-q-m$ new functions X are themselves a system in involution: they are in involution with every member of the original group, and therefore with $Z_1, \ldots, Z_\mu$: and therefore

$$Z_1, \ldots, Z_\mu, \quad X_{q+m+1}, \ldots, X_n$$

is a system in involution. The order of a system in involution cannot be greater than n (§ 138): hence

$$\mu + n - q - m \leqslant n,$$

that is,

$$\mu \leqslant q + m.$$

Further, this result can be used to obtain an upper limit for the tale of indicial functions, when the group is of order greater than n and therefore is not a system in involution. Let

$$r = 2q + m = n + k,$$

where k is positive: then as

$$q + m \leqslant n,$$

we have

$$\begin{aligned} 2n &\geqslant 2q + 2m \\ &\geqslant m + n + k, \end{aligned}$$

and therefore

$$m \leqslant n - k,$$

so that *a group of order* $n+k$ *cannot possess more than* $n-k$ *indicial functions.*

144. It is of importance to be able to construct the sub-group of greatest order $q+m$ which is a system in involution. Denoting the group by

$$u_1, \ldots, u_{2q+m},$$

we first determine the m indicial functions as the m functionally independent integrals of the equations

$$(u_1, \theta) = 0, \ldots, (u_{2q+m}, \theta) = 0,$$

making $u_1, \ldots, u_{2q+m}$ the independent variables. This system of equations is equivalent to $2q$ linearly independent equations and is a complete system: it can be integrated by any of the methods in Chapter III. Let the m independent integrals be

$$v_1, \ldots, v_m,$$

which are therefore the m indicial functions and can be taken as m members of the required sub-group.

Now let u_1 denote a function of the group, and suppose that it is not an indicial function so that it cannot be expressed in terms of $v_1, \ldots, v_m$. Then, for the equation

$$(u_1, \theta) = 0,$$

where θ is regarded as a function of the members of the group, we know $m+1$ independent integrals, viz. $u_1, v_1, \ldots, v_m$, and the number of variables is $2q+m$; hence it possesses $2q-2$ other independent integrals. Assuming that q is greater than unity, let w_2 be one of these other $2q-2$ integrals: then $u_1, w_2, v_1, \ldots, v_m$ are in involution with one another.

Again, for the equations

$$(u_1, \theta) = 0, \quad (w_2, \theta) = 0,$$

where θ is regarded as a function of the members of the group, we know that it is a complete system in the $2q+m$ variables; it therefore possesses $2q+m-2$ independent integrals. We already know $m+2$ of these integrals, in the form $u_1, w_2, v_1, \ldots, v_m$; hence there are $2q-4$ other independent integrals. Assuming that $q>2$, let w_3 be one of these $2q-4$ integrals; then $u_1, w_2, w_3, v_1, \ldots, v_m$ are in involution with one another.

Proceeding in this way, we shall (after $q-1$ similar stages) have obtained $q+m$ functions, independent of one another and in involution with one another; the aggregate is a sub-group of the greatest order that permits it to be a system in involution.

Application of Groups of Functions to the Integration of Systems of Equations.

145. As our main purpose, in connection with these groups of functions, is their application to the integration of a system of differential equations in one dependent variable, we shall not pursue the further development of their properties which will be found in Lie's treatise already quoted (p. 344): we proceed to apply them for the purpose of integration.

Accordingly, let the equations

$$f_1 = 0, \ldots, f_\mu = 0$$

be a system in involution: they may be a system initially given, in which case there is no question of arbitrary constants occurring in them: or they may be a system in a stage of gradual construction as in Jacobi's method, in which case some at least of the quantities $f_1, \ldots, f_\mu$ will contain additive arbitrary constants. Suppose also that independent integrals $\phi_1, \ldots, \phi_r$ of the complete system of equations

$$(f_1, \phi) = 0, \ldots, (f_\mu, \phi) = 0$$

have been obtained, and that the Poisson-Jacobi theorem has been applied so as to give all the integrals of the type (ϕ_i, ϕ_j) that can thus be constructed. Then the set of functions in the aggregate $f_1, \ldots, f_\mu, \phi_1, \ldots, \phi_r$ constitute a group: and $f_1, \ldots, f_\mu$ certainly are indicial functions of this group. It may happen that $f_1, \ldots, f_\mu$ do not complete the tale of indicial functions of the group: if they do not, let $f_{\mu+1}, \ldots, f_m$ be the other independent indicial functions of the group, so that

$$f_1, \ldots, f_\mu, f_{\mu+1}, \ldots, f_m$$

constitute a system in involution. Moreover, as $r + \mu$ is the order of a group which possesses m indicial functions, we have

$$\begin{aligned} r + \mu - m &= \text{even integer} \\ &= 2q, \end{aligned}$$

say, where q is a whole number.

It has been proved that a group of order $m + 2q$, possessing m indicial functions, contains a sub-group of order $m + q$ which is a system in involution; and consequently, our group of order $r + \mu$ contains a sub-group of order $m + q$, which is in involution and of which m members are given by

$$f_1, \ldots, f_m:$$

let the other members of this sub-group be $f_{m+1}, \ldots, f_{m+q}$. Then the integration of the original system of μ equations in involution is reduced to the integration of the modified system of

$$m + q \;(= \mu + m - \mu + q)$$

equations in involution: as $m \geqslant \mu$, $q \geqslant 0$, the modified form of the problem is usually simpler than the original form.

To complete the integration, we need integrals of the complete system

$$(f_1, \phi) = 0, \ldots, (f_{m+q}, \phi) = 0;$$

by the earlier theory, this is known to possess $2n - m - q$ independent integrals. Of this aggregate $m + q$ integrals are known, being $f_1, \ldots, f_{m+q}$: so that other $2n-2m-2q$ integrals are required.

If $q+m=n$, the complete aggregate of integrals is possessed. If $q+m<n$, then we can conceive that the proper number of integrals necessary to complete the aggregate has been determined, e.g. by Jacobi's method as amplified by Mayer. Let this aggregate be denoted by

$$f_1=0, \ldots, f_\mu=0, \quad f_{\mu+1}=a_1, \ldots, f_n=a_{n-\mu},$$

where we can take $a_1, \ldots, a_{n-\mu}$ to be arbitrary constants.

The construction of the integral of the system of equations now proceeds as before. If the n equations can be resolved so as to express $p_1, \ldots, p_n$ in terms of $x_1, \ldots, x_n, a_1, \ldots, a_{n-\mu}$, then, when the resolved values are substituted in

$$dz = p_1 dx_1 + \ldots + p_n dx_n,$$

a single quadrature leads to an equation of the form

$$z + c = F(x_1, \ldots, x_n, a_1, \ldots, a_{n-\mu}),$$

which is the complete integral of the system: and the remaining integrals can be deduced by the known general theory. If, however, the n equations cannot be conveniently resolved for $p_1, \ldots, p_n$, we determine a function Π from the equations

$$(\Pi, f_i) = -\sum_{r=1}^{n} p_r \frac{\partial f_i}{\partial p_r},$$

for $i=1, \ldots, n$, by a quadrature (§ 130): the complete integral of the system of equations is then given by

$$z - c = \Pi,$$

and the other integrals can be deduced as before.

A sufficient indication of the method has been given: for further developments, reference may be made to the authorities quoted (p. 314) at the beginning of this chapter*.

* Special reference should be made to two memoirs by Lie, *Math. Ann.*, t. IX (1876), pp. 245—296, *ib.*, t. XI (1877), pp. 464—557.

CHAPTER X.

THE EQUATIONS OF THEORETICAL DYNAMICS.

THE present chapter is devoted, more to matters cognate with the theory of the integration of partial differential equations than to the theory itself or to processes of integration.

The only process of integration included is that which is commonly called the Jacobi-Hamilton process: in order to make it more easily comprehended and to shew the source of its inspiration, a brief account of Hamilton's investigations in theoretical dynamics is prefixed.

The analysis shews once more, as so often before in the processes already explained, the close relation between the integration of a partial differential equation and the integrals of the set of ordinary equations, sometimes called subsidiary equations, sometimes the equations of the characteristics, here a canonical system. Some properties of canonical systems are given: but there is not an attempt to deal with them exhaustively because, as every property of such a system can be expressed as a result in theoretical dynamics, they really belong to the subject of theoretical dynamics.

The older development of theoretical dynamics was due mainly to Lagrange, Poisson, Hamilton, Jacobi, Donkin, Bertrand; and expositions of that theory will be found in Jacobi's *Vorlesungen über Dynamik*, in Imschenetsky's memoir* *Sur l'intégration des équations aux dérivées partielles du premier ordre*, and in Graindorge's treatise *Intégration des équations de la mécanique*. Further developments have been effected by Routh and are expounded in his *Treatise on the Dynamics of Rigid Bodies*.

The subject has developed in a different direction, since the application of Lie's theory of contact transformations to a quite general canonical system and the discovery of his important proposition that such transformations at once conserve the form of a general canonical system and are the only transformations which do conserve that form—a result that enables many older properties to be seen in an entirely new relation. An account of this theory and of the mode of development will be found in Dziobek's treatise†

* Translated from the original Russian by Hoüel, and published in *Grunert's Archiv*, t. L (1869), pp. 278—474; see, in particular, chapter VII of the memoir.

† An English translation was published in 1892 (The Inland Press, Ann Arbor).

Die mathematischen Theorien der Planetenbewegungen (Leipzig, 1888), Section II, and in Whittaker's *Analytical Dynamics* (Cambridge, 1904). Reference may also be made to an interesting paper by E. O. Lovett*, which gives a critical and historical account of the subject.

While these results hold of quite general systems, they are an incomplete statement of the case (particularly as to contact transformations being the only transformations which conserve the form) for particular given systems. Lie's original memoir† discusses the whole matter. The present chapter purports to give indications of the theory as connected with the theory of partial differential equations; it does not aim at being an introductory account of theoretical dynamics, as developed on the lines of Lie's theory.

Hamilton's Characteristic Equations.

146. We have seen that Cauchy's method of integration introduces the notion of initial values of the variables and utilises them in the expression of an integral. The same idea was used by Jacobi in developing some researches of Hamilton on theoretical dynamics where such initial values had been used: and in connection with the idea, he devised a method of integration, which is sometimes called *Jacobi's first method* and more often the *Jacobi-Hamiltonian method.* The details of the method differ from those in Cauchy's method: but on account of the ideas and the results, both Lie and Mansion claim‡ the method as Cauchy's. Some account of the method will be given here, partly because of its close association with methods and results obtained in the region of theoretical dynamics when the equations are taken in their canonical form. Later researches in some branches of this subject have diverged from the earlier course, mainly because of the application of Lie's theory of contact transformations.

In treatises concerned with the dynamics of systems of bodies§, it is shewn that the equations of motion of a holonomic system can be expressed in a form

$$\frac{d\theta_i}{dt}=\frac{\partial H}{\partial u_i},\quad \frac{du_i}{dt}=-\frac{\partial H}{\partial \theta_i},$$

* "The theory of perturbations and Lie's theory of contact transformations," *Quart. Journ. Math.*, t. XXX (1899), pp. 47—149.

† "Die Störungstheorie und die Berührungstransformationen," *Arch. f. Math. og Nat.*, t. II (1877), pp. 10—38.

‡ See Part I of this Treatise, p. 188, foot note.

§ Such as Routh's *Treatise on Rigid Dynamics*; see vol. I, ch. VIII.

for $i=1, \ldots, m$, the quantity H being (in the simplest case) the total energy of the system expressed in terms of t and of the variables $\theta_1, u_1, \ldots$. The form is usually associated with the name of Hamilton, as having been obtained by him*.

The following derivation of this result in the simplest case will give some indication as to the source of the transformation adopted by Jacobi in his method of integration. Denoting, as usual, the kinetic energy of the system by T and its potential energy by V, by $\theta_1, \ldots, \theta_m$ the m independent coordinates of the system, and by $\theta_1', \ldots, \theta_m'$ their derivatives with regard to the time t, we have Lagrange's equations of motion in the form

$$\frac{d}{dt}\left(\frac{\partial T}{\partial \theta_i'}\right) - \frac{\partial T}{\partial \theta_i} + \frac{\partial V}{\partial \theta_i} = 0.$$

Introducing a function L, such that

$$L = T - V,$$

and noting that V does not involve $\theta_1', \ldots, \theta_m'$, we have the equations in the form

$$\frac{d}{dt}\left(\frac{\partial L}{\partial \theta_i'}\right) = \frac{\partial L}{\partial \theta_i}.$$

The function H is defined by the equation

$$H = \theta_1' \frac{\partial L}{\partial \theta_1'} + \ldots + \theta_m' \frac{\partial L}{\partial \theta_m'} - L,$$

the form of which has analogies with Legendre's contact transformation; and it is convenient to introduce variables $u_1, \ldots, u_m$ such that

$$\frac{\partial L}{\partial \theta_i'} = u_i,$$

for $i=1, \ldots, m$. Thus

$$H = \theta_1' u_1 + \ldots + \theta_m' u_m - L,$$

and therefore

$$\begin{aligned} dH = {} & u_1 d\theta_1' + \ldots + u_m d\theta_m' + \theta_1' du_1 + \ldots + \theta_m' du_m \\ & - \left(u_1 d\theta_1' + \ldots + u_m d\theta_m' + \frac{\partial L}{\partial \theta_1} d\theta_1 + \ldots + \frac{\partial L}{\partial \theta_m} d\theta_m + \frac{\partial L}{\partial t} dt\right) \\ = {} & \theta_1' du_1 + \ldots + \theta_m' du_m - \frac{\partial L}{\partial \theta_1} d\theta_1 - \ldots - \frac{\partial L}{\partial \theta_m} d\theta_m - \frac{\partial L}{\partial t} dt. \end{aligned}$$

* *Phil. Trans.*, (1834), pp. 247—308, (1835), pp. 95—144.

In its initially defined form, H is a function of t, $\theta_1, \dots, \theta_m$, $\theta_1', \dots, \theta_m'$; let its expression be supposed changed, by means of the m equations

$$\frac{\partial L}{\partial \theta_1'} = u_1, \ \dots, \ \frac{\partial L}{\partial \theta_m'} = u_m,$$

so that it becomes a function of $u_1, \dots, u_m$, $\theta_1, \dots, \theta_m$, t, the variables $\theta_1', \dots, \theta_m'$ being replaced by $u_1, \dots, u_m$. Then the foregoing differential relation gives

$$\frac{\partial H}{\partial u_i} = \theta_i' = \frac{d\theta_i}{dt},$$

$$-\frac{\partial H}{\partial \theta_i} = \frac{\partial L}{\partial \theta_i}$$

$$= \frac{d}{dt}\left(\frac{\partial L}{\partial \theta_i'}\right)$$

$$= \frac{du_i}{dt},$$

for $i = 1, \dots, m$: and these equations are frequently called the *canonical form* of the equations of motion.

Now

$$H = \theta_1' u_1 + \dots + \theta_m' u_m - L,$$

so that, as

$$\frac{\partial L}{\partial \theta_i'} = u_i,$$

for $i = 1, \dots, m$, we have

$$dH = \theta_1' du_1 + \dots + \theta_m' du_m - \frac{\partial L}{\partial \theta_1} d\theta_1 - \dots - \frac{\partial L}{\partial \theta_m} d\theta_m - \frac{\partial L}{\partial t} dt.$$

Hence, when H is expressed as a function of t, $\theta_1, \dots, \theta_m$, $u_1, \dots, u_m$ through the removal of $\theta_1', \dots, \theta_m'$ by means of the equations

$$\frac{\partial L}{\partial \theta_i'} = u_i,$$

we have

$$\frac{\partial H}{\partial u_r} = \theta_r',$$

for $r = 1, \dots, m$; thus

$$H = u_1 \frac{\partial H}{\partial u_1} + \dots + u_r \frac{\partial H}{\partial u_r} - L,$$

and therefore

$$L = u_1 \frac{\partial H}{\partial u_1} + \dots + u_r \frac{\partial H}{\partial u_r} - H,$$

so that the relation between L and H is reciprocal.

This is true when t occurs explicitly in L and in H: it can easily be verified, in the simplest case, when neither T nor V involves t explicitly. Then T is a homogeneous quadratic function of the second order in $\theta_1', \ldots, \theta_m'$, so that

$$\theta_1' \frac{\partial T}{\partial \theta_1'} + \ldots + \theta_m' \frac{\partial T}{\partial \theta_1'} = 2T,$$

that is,

$$\theta_1' \frac{\partial L}{\partial \theta_1'} + \ldots + \theta_m' \frac{\partial L}{\partial \theta_1'} = 2T,$$

that is,

$$\theta_1' u_1 + \ldots + \theta_m' u_m = 2T,$$

and therefore

$$\begin{aligned} H &= 2T - L \\ &= T + V, \end{aligned}$$

so that, in this case, H is the total energy of the system. Also, as the equations

$$\frac{\partial L}{\partial \theta_i'} = u_i$$

in this case determine $\theta_1', \ldots, \theta_m'$ as quantities linear and homogeneous in $u_1, \ldots, u_m$, the quantity T is a homogeneous quadratic function of the second order in $u_1, \ldots, u_m$ after the change of variables is effected, so that

$$u_1 \frac{\partial T}{\partial u_1} + \ldots + u_m \frac{\partial T}{\partial u_m} = 2T,$$

and therefore

$$u_1 \frac{\partial H}{\partial u_1} + \ldots + u_m \frac{\partial H}{\partial u_m} = 2T:$$

consequently,

$$\begin{aligned} L &= 2T - H \\ &= u_1 \frac{\partial H}{\partial u_1} + \ldots + u_m \frac{\partial H}{\partial u_m} - H, \end{aligned}$$

so that, as before, H and L are reciprocal to one another in form. The preceding analysis shews that, when L is derived thus from the function H, and when it is expressed as a function of $\theta_1, \ldots, \theta_m, \theta_1', \ldots, \theta_m'$ by means of the equations

$$\frac{\partial H}{\partial u_1} = \theta_1', \ \ldots, \ \frac{\partial H}{\partial u_m} = \theta_m',$$

the equations

$$\frac{d}{dt}\left(\frac{\partial L}{\partial \theta_i'}\right) = \frac{\partial L}{\partial \theta_i},$$

for $i = 1, \ldots, m$, are satisfied.

The canonical equations

$$\frac{d\theta_i}{dt} = \frac{\partial H}{\partial u_i}, \quad -\frac{\partial H}{\partial \theta_i} = \frac{du_i}{dt},$$

are the general equations of motion, whether H involves t explicitly or not; but there is a substantial difference, as regards the relation of H to the equations, according as t does or does not occur in H.

When t does not occur explicitly in H, we have

$$\begin{aligned}\frac{dH}{dt} &= \sum_{r=1}^{m} \left(\frac{\partial H}{\partial u_r}\frac{du_r}{dt} + \frac{\partial H}{\partial \theta_r}\frac{d\theta_r}{dt}\right)\\ &= 0,\end{aligned}$$

that is, H is constant throughout the motion: or

$$H = \text{constant}$$

is an integral of the system, being of course the energy integral. Also, if

$$f = f(u_1, \ldots, u_m, \theta_1, \ldots, \theta_m) = \text{constant},$$

be any other integral of the system, we have

$$\begin{aligned}0 = \frac{df}{dt} &= \sum_{r=1}^{m} \left(\frac{\partial f}{\partial \theta_r}\frac{d\theta_r}{dt} + \frac{\partial f}{\partial u_r}\frac{du_r}{dt}\right)\\ &= \sum_{r=1}^{m} \left(\frac{\partial f}{\partial \theta_r}\frac{\partial H}{\partial u_r} - \frac{\partial f}{\partial u_r}\frac{\partial H}{\partial \theta_r}\right)\\ &= (f, H),\end{aligned}$$

in the earlier notation. Conversely, any quantity f, distinct from H, involving the variables but not involving t, and satisfying this equation, is an integral of the canonical system.

But, if t does occur explicitly in H, then in connection with the system of equations we have

$$\begin{aligned}\frac{dH}{dt} &= \frac{\partial H}{\partial t} + \sum_{r=1}^{m} \left(\frac{\partial H}{\partial u_r}\frac{du_r}{dt} + \frac{\partial H}{\partial \theta_r}\frac{d\theta_r}{dt}\right)\\ &= \frac{\partial H}{\partial t},\end{aligned}$$

which does not vanish: so that $H = \text{constant}$ is not then an integral of the system. If

$$g = g(t, \theta_1, \ldots, \theta_m, u_1, \ldots, u_m) = \text{constant},$$

be an integral of the system, we have

$$0 = \frac{dg}{dt} = \frac{\partial g}{\partial t} + \sum_{r=1}^{m} \left(\frac{\partial g}{\partial \theta_r} \frac{d\theta_r}{dt} + \frac{\partial g}{\partial u_r} \frac{du_r}{dt} \right)$$
$$= \frac{\partial g}{\partial t} + \sum_{r=1}^{m} \left(\frac{\partial g}{\partial \theta_r} \frac{\partial H}{\partial u_r} - \frac{\partial g}{\partial u_r} \frac{\partial H}{\partial \theta_r} \right)$$
$$= \frac{\partial g}{\partial t} + (g, H);$$

and this is the partial differential equation which is characteristic of every function g leading to an integral of the system. It is one of the equations that occur in Hamilton's theory: and (g, H) is homogeneous and linear in the derivatives of g.

147. Another characteristic partial differential equation is derived by Hamilton through the consideration of the integral

$$S = \int_{t_0}^{t} L dt,$$

so that, in the most general case, S is a function of t and t_0 and also of the values* of $\theta_1, \ldots, \theta_m, \theta_1', \ldots, \theta_m'$ at t and at t_0. To obtain some of its properties, imagine a quite general variation and (in order the more simply to allow t also to undergo this variation) introduce a new variable s, so that

$$\theta_i' = \frac{d\theta_i}{dt} = \frac{1}{t'} \frac{d\theta_i}{ds} = \frac{1}{t'} \phi_i,$$

say, for $i = 1, \ldots, m$, where $t' = \dfrac{dt}{ds}$; thus

$$S = \int_{s_0}^{s} t' L \left(t, \theta_1, \ldots, \theta_m, \frac{1}{t'} \phi_1, \ldots, \frac{1}{t'} \phi_m \right) ds$$
$$= \int_{s_0}^{s} \lambda (t, t', \theta_1, \phi_1, \ldots, \theta_m, \phi_m) \, ds,$$

where now all the arguments in λ are assumed functions of s, and s itself is not subject to variation. Taking a variation $t + \delta t$, $\theta_1 + \delta\theta_1, \ldots, \theta_m + \delta\theta_m$, we find, as usual,

$$\delta S = \left[\frac{\partial \lambda}{\partial t'} \delta t + \sum_{r=1}^{m} \frac{\partial \lambda}{\partial \phi_r} \delta\theta_r \right]_{s_0}^{s}$$
$$+ \int_{s_0}^{s} \left\{ \frac{\partial \lambda}{\partial t} - \frac{d}{ds} \left(\frac{\partial \lambda}{\partial t'} \right) \right\} \delta t \, ds$$
$$+ \sum_{\mu=1}^{m} \int_{s_0}^{s} \left\{ \frac{\partial \lambda}{\partial \theta_\mu} - \frac{d}{ds} \left(\frac{\partial \lambda}{\partial \phi_\mu} \right) \right\} \delta\theta_\mu ds.$$

* Only half of these $4m$ quantities can be taken as independent variables.

As s is at our choice, we shall choose it so that the equation

$$\frac{\partial \lambda}{\partial t} - \frac{d}{ds}\left(\frac{\partial \lambda}{\partial t'}\right) = 0$$

is satisfied: the actual value of s is not required. Also

$$\frac{\partial \lambda}{\partial \phi_\mu} = t' \left\{\frac{1}{t'} \frac{\partial L}{\partial \theta_\mu'}\right\} = \frac{\partial L}{\partial \theta_\mu'},$$

$$\frac{\partial \lambda}{\partial \theta_\mu} = t' \frac{\partial L}{\partial \theta_\mu},$$

so that

$$\begin{aligned}
\frac{\partial \lambda}{\partial \theta_\mu} - \frac{d}{ds}\left(\frac{\partial \lambda}{\partial \phi_\mu}\right) &= t' \frac{\partial L}{\partial \theta_\mu} - \frac{d}{ds}\left(\frac{\partial L}{\partial \theta_\mu'}\right) \\
&= t' \left\{\frac{\partial L}{\partial \theta_\mu} - \frac{d}{dt}\left(\frac{\partial L}{\partial \theta_\mu'}\right)\right\} \\
&= 0,
\end{aligned}$$

for all values of μ. Hence

$$\begin{aligned}
\delta S &= \left[\frac{\partial \lambda}{\partial t'} \delta t + \sum_{r=1}^{m} \frac{\partial \lambda}{\partial \phi_r} \delta \theta_r\right]_{s_0}^{s} \\
&= \left[\frac{\partial \lambda}{\partial t'} \delta t + \sum_{r=1}^{m} \frac{\partial \lambda}{\partial \phi_r} \delta \theta_r\right]_{t_0}^{t}.
\end{aligned}$$

Now, as above,

$$\frac{\partial \lambda}{\partial \phi_r} = \frac{\partial L}{\partial \theta_r'},$$

and

$$\frac{\partial \lambda}{\partial t'} = L - \Sigma \theta_r' \frac{\partial L}{\partial \theta_r'};$$

consequently

$$\begin{aligned}
\delta S &= \left[\left(L - \Sigma \theta_r' \frac{\partial L}{\partial \theta_r'}\right) dt + \sum_{r=1}^{m} \frac{\partial L}{\partial \theta_r'} \delta \theta_r\right]_{t_0}^{t} \\
&= \left[-H \delta t + \sum_{r=1}^{m} \frac{\partial L}{\partial \theta_r'} \delta \theta_r\right]_{t_0}^{t}.
\end{aligned}$$

It is an immediate consequence that in any configuration, as developed from assigned initial conditions, the value of S at any time depends only upon the configuration at that moment and upon the initial conditions.

To make these initial conditions precise, let

$$\begin{aligned}
&\theta_\mu = \beta_\mu, \text{ when } t = t_0, \text{ for } \mu = 1, \ldots, m, \\
&L = L_0, \ \ldots\ldots\ldots\ldots, \\
&H = H_0, \ \ldots\ldots\ldots\ldots, \\
&\frac{\partial L_0}{\partial \theta_r'} = c_r, \text{ when } t = t_0, \text{ for } r = 1, \ldots, m:
\end{aligned}$$

then we have

$$\left.\begin{aligned} \frac{\partial S}{\partial t} &= -H, & \frac{\partial S}{\partial t_0} &= -H_0 \\ \frac{\partial S}{\partial \theta_r} &= u_r, & \frac{\partial S}{\partial \beta_r} &= -c_r \end{aligned}\right\},$$

for $r = 1, \ldots, m$. These are the values of the derivatives of S with regard to each of the quantities it involves.

As our purpose is not the discussion of the organic significance of any property or group of properties of the quantities concerned, but only to indicate so much of the analysis connected with the equations of theoretical dynamics as will throw some light upon the analysis introduced into what is commonly called the Jacobi-Hamilton method, we shall indicate only one inference from the preceding equations. The quantity H can be expressed as a function of $t, \theta_1, \ldots, \theta_m, u_1, \ldots, u_m$, say

$$H = \psi(t, \theta_1, \ldots, \theta_m, u_1, \ldots, u_m);$$

hence S satisfies the equation

$$\frac{\partial S}{\partial t} + \psi\left(t, \theta_1, \ldots, \theta_m, \frac{\partial S}{\partial \theta_1}, \ldots, \frac{\partial S}{\partial \theta_m}\right) = 0,$$

and this is another of the characteristic equations in Hamilton's theory. Moreover, when H involves t explicitly, H is not homogeneous in the quantities u, so that the equation satisfied by S is not homogeneous in the derivatives.

If, however, H is independent of any explicit occurrence of t, we know that

$$H = h,$$

where h is a constant; and then

$$S = -h(t - t_0) + S_1,$$

so that

$$\frac{\partial S}{\partial \theta_r} = u_r = \frac{\partial S_1}{\partial \theta_r};$$

and the equation satisfied by S_1 is

$$\begin{aligned} h &= H(\theta_1, \ldots, \theta_m, u_1, \ldots, u_m) \\ &= H\left(\theta_1, \ldots, \theta_m, \frac{\partial S_1}{\partial \theta_1}, \ldots, \frac{\partial S_1}{\partial \theta_m}\right). \end{aligned}$$

This is the modified form of Hamilton's characteristic equation when H does not explicitly involve t: the right-hand is homogeneous in the derivatives of S_1.

148. These characteristic equations have been derived from the initial set of equations in the canonical form: the relation between them can be exhibited in another light. By the existence-theorem of a set of ordinary simultaneous equations of the first order*, the canonical equations

$$\frac{d\theta_i}{dt}=\frac{\partial H}{\partial u_i},\quad \frac{du_i}{dt}=-\frac{\partial H}{\partial \theta_i},$$

determine the $2m$ quantities $\theta_1, \ldots, \theta_m, u_1, \ldots, u_m$ as functions of t and of parameters, which are the values of those quantities when $t=t_0$: these are

$$\theta_i=\beta_i,\quad u_i=c_i,$$

when $t=t_0$. Now L is a function of $t, \theta_1, \ldots, \theta_m, \theta_1', \ldots, \theta_m'$: as, by means of the equations

$$\theta_i'=\frac{\partial H}{\partial u_i},$$

the quantities $\theta_1', \ldots, \theta_m'$ are expressible in terms of $u_1, \ldots, u_m$, it follows that L can be expressed as a function of $t, \theta_1, \ldots, \theta_m, u_1, \ldots, u_m$. When the integrals of the canonical system are used, they can express $u_1, \ldots, u_m, c_1, \ldots, c_m$ in terms of the other quantities: thus L can be expressed in terms of $t, t_0, \theta_1, \ldots, \theta_m, \beta_1, \ldots, \beta_m$, and therefore also S, which is

$$\int_{t_0}^{t} L\,dt,$$

can be expressed in terms of $t, t_0, \theta_1, \ldots, \theta_m, \beta_1, \ldots, \beta_m$. Assuming this expression effected, we have

$$\frac{\partial S}{\partial \theta_r}=u_r,\quad \frac{\partial S}{\partial \beta_r}=-c_r,$$

which are $2m$ equations expressing $u_1, \ldots, u_m$ and $c_1, \ldots, c_m$ in terms of $t, t_0, \theta_1, \ldots, \theta_m, \beta_1, \ldots, \beta_m$: that is, they are equivalent to the $2m$ relations which are the integrals of the canonical system. Combining these results, we have the following theorem:—

The system of ordinary canonical equations

$$\frac{d\theta_i}{dt}=\frac{\partial H}{\partial u_i},\quad \frac{du_i}{dt}=-\frac{\partial H}{\partial \theta_i},$$

* See vol. II of this work, § 10.

where $H = H(t, \theta_1, \ldots, \theta_m, u_1, \ldots, u_m)$, *is connected in the following manner with the partial differential equation*

$$\frac{\partial S}{\partial t} + H(t, \theta_1, \ldots, \theta_m, u_1, \ldots, u_m) = 0,$$

where

$$u_1 = \frac{\partial S}{\partial \theta_1}, \ldots, u_m = \frac{\partial S}{\partial \theta_m}.$$

Obtain the complete set of integrals of the system of ordinary equations, determining the arbitrary constants so that

$$\theta_1, \ldots, \theta_m = \beta_1, \ldots, \beta_m,$$
$$u_1, \ldots, u_m = c_1, \ldots, c_m,$$

when $t = t_0$. *Construct the function*

$$L = u_1 \frac{\partial H}{\partial u_1} + \ldots + u_m \frac{\partial H}{\partial u_m} - H,$$

and, expressing it in terms of $t, \beta_1, \ldots, \beta_m, c_1, \ldots, c_m$ *by means of the foregoing integrals, obtain the function*

$$I = \int_{t_0}^{t} L\,dt - \sum_{r=1}^{m} c_r \beta_r;$$

and when this is obtained, use the integrals to express I *in terms of* $t, \beta_1, \ldots, \beta_m, \theta_1, \ldots, \theta_m$, *denoting the resulting expression by* s. *Then the relation*

$$S = a + s,$$

when a *is an arbitrary constant, is an integral of the partial differential equation: as it contains* $m+1$ *arbitrary constants* $a, \beta_1, \ldots, \beta_m$, *it is a complete integral. Moreover, the* $2m$ *equations*

$$\frac{\partial S}{\partial \theta_i} = u_i, \quad \frac{\partial S}{\partial \beta_i} = -c_i, \qquad (i = 1, \ldots, m),$$

can be regarded as an integral equivalent of the system of ordinary equations.

This process has been derived through the general equations of motion of a dynamical system, so that there are limitations on the form of H, quà function of $u_1, \ldots, u_m$, when the theorem is thus obtained. The result, however, is not subject in fact to these limitations: and it is this extension and generalisation of Hamilton's investigation which constitute the method of integration. As it was published* by Jacobi, it is often called the

* *Crelle*, t. XVII (1837), pp. 136 et seq.: see also Part I of this treatise, § 109.

Jacobi-Hamiltonian method; but, as already pointed out, the use of the initial values had been introduced earlier by Cauchy.

149. Other characteristic functions are introduced into the study of theoretical dynamics: among them, one of the most important is Hamilton's function often denoted by A, which represents the dynamical Action when H does not explicitly involve t, and which in general is defined by the equation

$$\begin{aligned} A &= S + \int_{t_0}^{t} \frac{d}{dt}(Ht)\,dt \\ &= S + Ht - H_0 t_0. \end{aligned}$$

With this value, we have

$$\begin{aligned} \delta A &= \delta S + \Big[H\delta t + t\delta H \Big]_{t_0}^{t} \\ &= \left[t\delta H + \sum_{r=1}^{m} \frac{\partial L}{\partial \theta_r'}\delta\theta_r \right]_{t_0}^{t} \\ &= \left[t\delta H + \sum_{r=1}^{m} u_r \delta\theta_r \right]_{t_0}^{t}. \end{aligned}$$

Now by means of the integrals of the canonical system as associated with initial values, H can be expressed in terms of $t, t_0, \theta_1, \ldots, \theta_m, \beta_1, \ldots, \beta_m$, and H_0 can be expressed in terms of $t_0, \beta_1, \ldots, \beta_m$. Also A can be expressed in terms of $t, t_0, \theta_1, \ldots, \theta_m, \beta_1, \ldots, \beta_m$: when t and t_0 are eliminated from its expression by means of the expressions for H and H_0, it comes to be a function of $H, H_0, \theta_1, \ldots, \theta_m, \beta_1, \ldots, \beta_m$. In this form, the equations

$$\frac{\partial A}{\partial H} = t, \quad \frac{\partial A}{\partial \theta_r} = u_r, \quad \frac{\partial A}{\partial \beta_r} = -c_r, \quad \frac{\partial A}{\partial H_0} = -t_0,$$

are satisfied; they must be equivalent to the integrals of the canonical system.

Ex. 1. Prove that, when the expressions for the kinetic energy T and the potential energy V do not explicitly involve the time, the function A satisfies the partial differential equation

$$T\left(\theta_1, \ldots, \theta_m, \frac{\partial A}{\partial \theta_1}, \ldots, \frac{\partial A}{\partial \theta_m}\right) = h - V(\theta_1, \ldots, \theta_m),$$

where h is a constant. (Jacobi.)

Ex. 2. Shew that, if a complete integral of the partial differential equation satisfied by A be obtained in the form

$$A = g(\theta_1, \ldots, \theta_m, h, a_1, \ldots, a_{m-1}) + a_m,$$

where $a_1, \ldots, a_{m-1}, a_m$ are the arbitrary constants, then a set of integrals of the canonical system is given by

$$\frac{\partial g}{\partial a_1}=b_1, \ldots, \frac{\partial g}{\partial a_{m-1}}=b_{m-1}, \frac{\partial g}{\partial h}=t+\tau,$$

where $a_1, \ldots, a_{m-1}, h, b_1, \ldots, b_{m-1}, \tau$ are the $2m$ arbitrary constants in the integrals. (Jacobi.)

Ex. 3. Let m integrals (in involution) of the canonical system be supposed known, involving $\theta_1, \ldots, \theta_m, u_1, \ldots, u_m$ in such a way that $u_1, \ldots, u_m$ can be expressed in terms of $\theta_1, \ldots, \theta_m, t$, and the m arbitrary constants of the integrals; and let these values be substituted in H, the resulting value being denoted by $\overline{H}$. Prove that

$$u_1 d\theta_1 + \ldots + u_m d\theta_m - \overline{H} dt$$

is an exact differential; and shew how the remaining integrals of the canonical system can be obtained. (Liouville.)

Ex. 4. When the expression for the energy of the system does not involve the time and when $m-1$ integrals (other than $H=h$) of the canonical system have been obtained, so that $u_1, \ldots, u_m$ can be expressed in terms of $\theta_1, \ldots, \theta_m$ by means of those $m-1$ integrals and $H=h$, prove that

$$u_1 d\theta_1 + \ldots + u_m d\theta_m$$

is an exact differential $d\Sigma$. Obtain the other integrals of the canonical system: and shew that the variables in the integrals are connected with t by the relation

$$\frac{\partial \Sigma}{\partial h}=t+\tau.$$

(Liouville.)

Ex. 5. Integrate the equation

$$(p_1^2+p_2^2+p_3^2)(x_1^2+x_2^2+x_3^2)^2=a^4,$$

where a is a constant, by using the theorems in any of the preceding examples.

JACOBI'S GENERALISATION OF HAMILTON'S RESULTS.

150. The preceding brief discussion will sufficiently illustrate the connection between a partial differential equation and a canonical system of ordinary equations, as it arises in the discussion of theoretical dynamics: and each of the methods of integration, which have been expounded in the preceding chapters, shews a similar organic relation. The detailed application of the method, suggested by the processes of theoretical dynamics, differs from the use made in other methods; and though it is somewhat more cumbrous than those methods, its association with the results of theoretical dynamics seems ample justification for its retention among the principal methods of integration.

We proceed from the generalised form of the equation satisfied by S and, making a change in the notation, we suppose that a given partial differential equation has been resolved with regard to one of the derivatives into the form

$$p + H(x, x_1, \ldots, x_n, p_1, \ldots, p_n) = 0,$$

where, as usual

$$p = \frac{\partial z}{\partial x}, \quad p_i = \frac{\partial z}{\partial x_i},$$

for $i = 1, \ldots, n$. It will be noticed that z does not occur explicitly: the alternative forms, when z does occur, will be given later.

The equations of the characteristics are

$$\frac{dx}{1} = \frac{dx_1}{\frac{\partial H}{\partial p_1}} = \ldots = \frac{dx_n}{\frac{\partial H}{\partial p_n}} = \frac{dp_1}{-\frac{\partial H}{\partial x_1}} = \ldots = \frac{dp_n}{-\frac{\partial H}{\partial x_n}} = \frac{dp}{-\frac{\partial H}{\partial x}}:$$

but the last fraction can be omitted, because p occurs there only and we have the permanent equation

$$p = -H.$$

Thus the equations can be taken in the form

$$\frac{dx_i}{dx} = \frac{\partial H}{\partial p_i}, \quad \frac{dp_i}{dx} = -\frac{\partial H}{\partial x_i},$$

for $i = 1, \ldots, n$: and these agree with the canonical form of the equations of theoretical dynamics. Assume that these equations have been completely integrated, the arbitrary constants being determined by the conditions that

$$x_1, \ldots, x_n, p_1, \ldots, p_n = a_1, \ldots, a_n, b_1, \ldots, b_n,$$

when $x = a$; and let the results be expressible in the form

$$x_i = \xi_i(x, a_1, \ldots, a_n, b_1, \ldots, b_n),$$
$$p_i = \pi_i(x, a_1, \ldots, a_n, b_1, \ldots, b_n),$$

for $i = 1, \ldots, n$. The determinant

$$J\left(\frac{\xi_1, \ldots, \xi_n}{a_1, \ldots, a_n}\right)$$

is unity when $x = a$, so that it cannot vanish identically; hence the n equations $x_i = \xi_i$ can be resolved for $a_1, \ldots, a_n$.

Now the quantity

$$p_1 dx_1 + \dots + p_n dx_n + p dx$$

is an exact differential: substituting for $dx_1, \dots, dx_n$ from the ordinary equations, and also $-H$ for p, it takes the form

$$\left(p_1 \frac{\partial H}{\partial p_1} + \dots + p_n \frac{\partial H}{\partial p_n} - H\right) dx,$$

which also is a form suggested by the analysis connected with the equations of theoretical dynamics. We therefore take a quantity

$$\zeta = \int_a^x \left(\sum_{i=1}^n p_i \frac{\partial H}{\partial p_i} - H\right) dx + \sum_{i=1}^n a_i b_i,$$

on the analogy of the dynamical results and, substituting for the variables $x_1, \dots, x_n, p_1, \dots, p_n$ their values as given by the integrals of the canonical system, we effect the quadrature which then gives ζ as a function of $x, a_1, \dots, a_n, b_1, \dots, b_n$. Let c denote any one of these $2n$ constants that occur in ζ; then, taking account of the fact that the values of the variables have been substituted in the initial form of ζ, we have

$$\begin{aligned}
\frac{\partial \zeta}{\partial c} - \frac{\partial}{\partial c}\left(\sum_{i=1}^n a_i b_i\right)
&= \int_a^x \sum_{i=1}^n \left\{\pi_i \frac{\partial}{\partial c}\left(\frac{\partial H}{\partial p_i}\right) + \frac{\partial H}{\partial p_i}\frac{\partial \pi_i}{\partial c} - \frac{\partial H}{\partial x_i}\frac{\partial \xi_i}{\partial c} - \frac{\partial H}{\partial p_i}\frac{\partial \pi_i}{\partial c}\right\} dx \\
&= \int_a^x \sum_{i=1}^n \left\{\pi_i \frac{\partial}{\partial c}\left(\frac{\partial \xi_i}{\partial x}\right) + \frac{\partial \xi_i}{\partial c}\frac{\partial \pi_i}{\partial x}\right\} dx \\
&= \int_a^x \sum_{i=1}^n \left\{\pi_i \frac{\partial}{\partial x}\left(\frac{\partial \xi_i}{\partial c}\right) + \frac{\partial \xi_i}{\partial c}\frac{\partial \pi_i}{\partial x}\right\} dx \\
&= \sum_{i=1}^n \left[\pi_i \frac{\partial \xi_i}{\partial c}\right]_a^x \\
&= \sum_{i=1}^n \left(\pi_i \frac{\partial \xi_i}{\partial c} - b_i \frac{\partial a_i}{\partial c}\right);
\end{aligned}$$

consequently

$$\frac{\partial \zeta}{\partial a_s} = \sum_{i=1}^n \left(\pi_i \frac{\partial \xi_i}{\partial a_s}\right),$$

$$\frac{\partial \zeta}{\partial b_s} = \sum_{i=1}^n \left(\pi_i \frac{\partial \xi_i}{\partial b_s}\right) + a_s.$$

We have seen that the n equations $x_1 = \xi_1, \dots, x_n = \xi_n$ can be resolved so as to express $a_1, \dots, a_n$ in terms of $x_1, \dots, x_n, x, b_1, \dots,$

b_n; let the values thus obtained be substituted in ζ, and denote the resulting value by Z. Then, as

$$\frac{\partial \zeta}{\partial a_s} = \sum_{i=1}^{n} \frac{\partial Z}{\partial x_i} \frac{\partial \xi_i}{\partial a_s},$$

$$\frac{\partial \zeta}{\partial b_s} = \frac{\partial Z}{\partial b_s} + \sum_{i=1}^{n} \frac{\partial Z}{\partial x_i} \frac{\partial \xi_i}{\partial b_s},$$

we have, on substituting the preceding values for the derivatives of ζ, the equations

$$\sum_{i=1}^{n} \left(\frac{\partial Z}{\partial x_i} - \pi_i\right) \frac{\partial \xi_i}{\partial a_s} = 0,$$

$$\sum_{i=1}^{n} \left(\frac{\partial Z}{\partial x_i} - \pi_i\right) \frac{\partial \xi_i}{\partial b_s} = a_s - \frac{\partial Z}{\partial b_s},$$

for $s = 1, \ldots, n$. The former set of n equations is linear and homogeneous in the quantities $\frac{\partial Z}{\partial x_i} - \pi_i$, and the determinant of the coefficients of these quantities does not vanish; hence

$$\frac{\partial Z}{\partial x_i} - \pi_i = 0,$$

for $i = 1, \ldots, n$, and therefore from the remaining equations

$$\frac{\partial Z}{\partial b_i} - a_i = 0,$$

for $i = 1, \ldots, n$. These relations are not identities, because $\pi_1, \ldots, \pi_n, a_1, \ldots, a_n$ do not occur in Z; and they clearly are independent of one another. Moreover, they are satisfied in connection with the equations $x_1 = \xi_1, \ldots, x_n = \xi_n, p_1 = \pi_1, \ldots, p_n = \pi_n$; hence they are a general integral equivalent of the $2n$ differential equations

$$\frac{dx_i}{dx} = \frac{\partial H}{\partial p_i}, \quad \frac{dp_i}{dx} = -\frac{\partial H}{\partial x_i}.$$

Again, we have

$$\frac{d\zeta}{dx} = \sum_{i=1}^{n} \left(\pi_i \frac{\partial H}{\partial p_i} - H\right),$$

on replacing $x_1, \ldots, x_n, p_1, \ldots, p_n$ by their values: and

$$\frac{dZ}{dx} = \frac{\partial Z}{\partial x} + \sum_{i=1}^{n} \frac{\partial Z}{\partial x_i} \frac{\partial \xi_i}{\partial x}$$

$$= \frac{\partial Z}{\partial x} + \sum_{i=1}^{n} \pi_i \frac{\partial H}{\partial p_i},$$

on using the preceding equations. Hence the equality

$$\frac{d\zeta}{dx} = \frac{dZ}{dx},$$

leads to the relation

$$\begin{aligned}\frac{\partial Z}{\partial x} &= -H(x, \xi_1, \ldots, \xi_n, \pi_1, \ldots, \pi_n)\\ &= -H\left(x, x_1, \ldots, x_n, \frac{\partial Z}{\partial x_1}, \ldots, \frac{\partial Z}{\partial x_n}\right);\end{aligned}$$

and therefore the equation

$$z = Z + c,$$

where c is an arbitrary constant, is an integral of the equation

$$p + H(x, x_1, \ldots, x_n, p_1, \ldots, p_n) = 0.$$

Accordingly, the process may be stated as follows:—

To obtain an integral of the equation

$$p + H(x, x_1, \ldots, x_n, p_1, \ldots, p_n) = 0,$$

form the canonical system

$$\frac{dx_i}{dx} = \frac{\partial H}{\partial p_i}, \quad \frac{dp_i}{dx} = -\frac{\partial H}{\partial x_i}, \qquad (i = 1, \ldots, n),$$

of ordinary equations, and construct their complete set of integrals

$$\left.\begin{aligned} x_i &= \xi_i(x, a_1, \ldots, a_n, b_1, \ldots, b_n)\\ p_i &= \pi_i(x, a_1, \ldots, a_n, b_1, \ldots, b_n)\end{aligned}\right\}, \quad (i = 1, \ldots, n),$$

such that $x_1, \ldots, x_n, p_1, \ldots, p_n = a_1, \ldots, a_n, b_1, \ldots, b_n$ *respectively, when* $x = a$. *Take a quantity* ζ *defined by the relation*

$$\zeta = \int_a^x \left(\sum_{i=1}^n p_i \frac{\partial H}{\partial p_i} - H\right) dx + \sum_{i=1}^n a_i b_i;$$

and, substituting $\xi_1, \ldots, \xi_n, \pi_1, \ldots, \pi_n$ *for the variables under the sign of integration, effect the quadrature which gives* ζ *as a function of* $x, a_1, \ldots, a_n, b_1, \ldots, b_n$. *From* ζ *eliminate* $a_1, \ldots, a_n$ *by means of the equations* $x_1 = \xi_1, \ldots, x_n = \xi_n$, *and let the resulting function of* $x, x_1, \ldots, x_n, b_1, \ldots, b_n$ *be denoted by* Z; *then*

$$z = Z + c,$$

where c *is an arbitrary constant, is an integral of the partial differential equation, and manifestly it is a complete integral. Moreover, a complete set of integrals of the canonical system is given by*

$$\frac{\partial Z}{\partial x_i} = p_i, \quad \frac{\partial Z}{\partial b_i} = a_i,$$

for $i = 1, \ldots, n$, *the constants* $a_1, \ldots, a_n$ *being arbitrary.*

151. It is easy to see that the complete integral thus obtained is an integral which, when $x=a$, acquires the value

$$b_1x_1 + \dots + b_nx_n + c.$$

For Z is the value of ζ when $a_1, \dots, a_n$ are eliminated from ζ by means of $x_1 = \xi_1, \dots, x_n = \xi_n$; and therefore the value of Z, when $x=a$, is obtained from that of ζ when $x=a$ (which value is $a_1b_1 + \dots + a_nb_n$) by eliminating $a_1, \dots, a_n$ through the forms of the equations $x_1 = \xi_1, \dots, x_n = \xi_n$, when $x=a$: and these forms are $x_1 = a_1, \dots, x_n = a_n$.

The complete integral is therefore somewhat restricted, though it contains the appropriate number of arbitrary constants: its relation to any other complete integral, say

$$z = \phi(x, x_1, \dots, x_n, k_1, \dots, k_n) + c,$$

can be simply obtained. In the case of this complete integral, a set of integrals of the canonical system is given by

$$\frac{\partial \phi}{\partial x_i} = p_i, \quad \frac{\partial \phi}{\partial k_i} = \kappa_i,$$

for $i = 1, \dots, n$. Let

$$\phi_0 = \phi(a, a_1, \dots, a_n, k_1, \dots, k_n);$$

then as $a_1, \dots, a_n, b_1, \dots, b_n$ are the values of $x_1, \dots, x_n, p_1, \dots, p_n$, when $x=a$, we must have

$$\frac{\partial \phi_0}{\partial a_i} = b_i, \quad \frac{\partial \phi_0}{\partial k_i} = \kappa_i.$$

Now

$$\frac{\partial H}{\partial p_i} = \frac{dx_i}{dx}, \quad \frac{\partial \phi}{\partial x_i} = p_i, \quad H = -\frac{\partial \phi}{\partial x};$$

hence

$$\begin{aligned}\zeta &= \sum_{i=1}^{n} a_ib_i + \int_a^x \left(\sum_{i=1}^{n} p_i \frac{\partial H}{\partial p_i} - H\right) dx \\ &= \sum_{i=1}^{n} a_ib_i + \int_a^x \left(\frac{\partial \phi}{\partial x_i}\frac{dx_i}{dx} + \frac{\partial \phi}{\partial x}\right) dx \\ &= \sum_{i=1}^{n} a_ib_i + \phi - \phi_0.\end{aligned}$$

The constants $k_1, \dots, k_n$ are such that

$$\frac{\partial \phi}{\partial k_i} = \kappa_i, \quad \frac{\partial \phi_0}{\partial k_i} = \kappa_i,$$

that is, such that

$$\frac{\partial\phi}{\partial k_i}=\frac{\partial\phi_0}{\partial k_i},$$

for $i=1, \ldots, n$: let their values be determined and substituted in ζ. In order to obtain Z from ζ, the constants $a_1, \ldots, a_n$ must be eliminated; their values are given by

$$\frac{\partial\phi_0}{\partial a_i}=b_i,$$

in connection with the preceding values of $k_1, \ldots, k_n$: and so Z is the value of ζ when, from the equation

$$\zeta=\sum_{i=1}^{n} a_i b_i+\phi-\phi_0,$$

the quantities $a_1, \ldots, a_n, k_1, \ldots, k_n$ are removed by means of the equations

$$\frac{\partial\phi}{\partial k_i}=\frac{\partial\phi_0}{\partial k_i},\quad \frac{\partial\phi_0}{\partial a_i}=b_i,$$

for $i=1, \ldots, n$. Hence the complete integral in the theorem can be derived from any given complete integral*.

Ex. 1. The detailed working can be shewn by thus solving the equation

$$\frac{x}{p}+\frac{x_1}{p_1}+\frac{x_2}{p_2}=1,$$

which clearly has an integral

$$z=kx^2+k_1x_1^2+k_2x_2^2+c,$$

where

$$\frac{1}{k}+\frac{1}{k_1}+\frac{1}{k_2}=2.$$

The value of H for the form $p+H=0$ is

$$H=\frac{xp_1p_2}{p_1x_2+p_2x_1-p_1p_2}.$$

The canonical system is

$$\frac{dx_1}{dx}=\frac{xx_1p_2^2}{(p_1x_2+p_2x_1-p_1p_2)^2},\quad \frac{dx_2}{dx}=\frac{xx_2p_1^2}{(p_1x_2+p_2x_1-p_1p_2)^2},$$

$$\frac{dp_1}{dx}=\frac{xp_1p_2^2}{(p_1x_2+p_2x_1-p_1p_2)^2},\quad \frac{dp_2}{dx}=\frac{xp_1^2p_2}{(p_1x_2+p_2x_1-p_1p_2)^2};$$

and integrals are

$$p_1=ax_1,\quad p_2=bx_2,\quad x_1^2=\frac{c^2}{a^2}x^2+a',\quad x_2^2=\frac{c^2}{b^2}x^2+b',$$

* The whole of the preceding exposition follows that which is given by Mayer, *Math. Ann.*, t. III (1871), pp. 434—452.

where a, b, a', b' are arbitrary constants, and

$$c=\frac{ab}{a+b-ab}.$$

Hence, taking a_1, a_2, b_1, b_2 as initial values when $x=a$, we have

$$a=\frac{b_1}{a_1}, \quad b=\frac{b_2}{a_2}, \quad \frac{c}{a}=\frac{b_1 a_2}{b_2 a_1+b_1 a_2-b_1 b_2}, \quad \frac{c}{b}=\frac{b_2 a_1}{b_2 a_1+b_1 a_2-b_1 b_2};$$

and the integrals are

$$p_1=\frac{b_1}{a_1}x_1, \quad p_2=\frac{b_2}{a_2}x_2,$$

$$x_1^2-a_1^2=\frac{c^2}{a^2}(x^2-a^2), \quad x_2^2-a_2^2=\frac{c^2}{b^2}(x^2-a^2).$$

If we require the integral which becomes $b_1x_1+b_2x_2+\gamma$, when $x=a$, we take

$$\begin{aligned}\zeta&=a_1b_1+a_2b_2+\int_a^x\left(p_1\frac{\partial H}{\partial p_1}+p_2\frac{\partial H}{\partial p_2}-H\right)dx\\&=a_1b_1+a_2b_2+\tfrac{1}{2}c^2(x^2-a^2);\end{aligned}$$

and the required integral is given by

$$z=Z+\gamma,$$

where Z is given as a function of x, x_1, x_2, by the elimination of a_1 and a_2 between the three equations

$$\left.\begin{aligned}Z&=a_1b_1+a_2b_2+\tfrac{1}{2}c^2(x^2-a^2)\\x_1^2-a_1^2&=\frac{c^2}{a^2}(x^2-a^2)\\x_2^2-a_2^2&=\frac{c^2}{b^2}(x^2-a^2)\end{aligned}\right\},$$

account being taken of the values of a, b, c in terms of a_1, a_2, b_1, b_2.

To derive this integral from the integral

$$z=kx^2+k_1x_1^2+k_2x_2^2+c,$$

where

$$\frac{1}{k}+\frac{1}{k_1}+\frac{1}{k_2}=2,$$

we write

$$\phi=kx^2+k_1x_1^2+k_2x_2^2,$$
$$\phi_0=ka^2+k_1a_1^2+k_2a_2^2;$$

and then we take

$$\zeta=a_1b_1+a_2b_2+\phi-\phi_0.$$

The relations between the quantities a_1, a_2, b_1, b_2, k, k_1, k_2 (other than the single relation between k, k_1, k_2) are

$$\frac{\partial\phi_0}{\partial a_1}=b_1, \quad \frac{\partial\phi_0}{\partial a_2}=b_2, \quad \frac{\partial\phi}{\partial k_1}=\frac{\partial\phi_0}{\partial k_1}, \quad \frac{\partial\phi}{\partial k_2}=\frac{\partial\phi_0}{\partial k_2},$$

account being taken of the relation between k, k_1, k_2. These give

$$b_1=2k_1a_1, \quad b_2=2k_2a_2,$$
$$k_1^2(x_1^2-a_1^2)=k^2(x^2-a^2), \quad k_2^2(x_2^2-a_2^2)=k^2(x^2-a^2);$$

hence

$$\begin{aligned}\zeta &= a_1 b_1 + a_2 b_2 + \phi - \phi_0\\ &= \frac{b_1^2}{2k_1} + \frac{b_2^2}{2k_2} + k(x^2 - a^2) + \frac{k^2}{k_1}(x^2 - a^2) + \frac{k^2}{k_2}(x^2 - a^2)\\ &= \frac{b_1^2}{2k_1} + \frac{b_2^2}{2k_2} + 2k^2(x^2 - a^2).\end{aligned}$$

We eliminate k, k_1, k_2 between the equations

$$\left.\begin{aligned} Z &= \frac{b_1^2}{2k_1} + \frac{b_2^2}{2k_2} + 2k^2(x^2 - a^2)\\ k^2 x_1^2 - \tfrac{1}{4} b_1^2 &= k^2(x^2 - a^2)\\ k^2 x_2^2 - \tfrac{1}{4} b_2^2 &= k^2(x^2 - a^2)\\ 2 &= \frac{1}{k} + \frac{1}{k_1} + \frac{1}{k_2}\end{aligned}\right\};$$

and then the integral is

$$z = Z + \gamma.$$

The verification that Z becomes $b_1 x_1 + b_2 x_2$, when $x = a$, is immediate.

Ex. 2. Obtain, in the preceding manner, the integral of the equation

$$pq - px - qy = 0,$$

such that $q = c$ when $x = a$, in the form

$$z = xy + y(x^2 - 2ac + c^2)^{\frac{1}{2}} + \gamma;$$

where γ is an arbitrary constant.

Deduce it also from the complete integral

$$z = \frac{1}{2a}(y + ax)^2 + \gamma.$$

Another integral is given by

$$z = xy + \{(x^2 - a^2)(y^2 - b^2)\}^{\frac{1}{2}};$$

is there any relation between this integral and the first integral?

Ex. 3. Let

$$z = \phi(x, x_1, \ldots, x_n, b_1, \ldots, b_n) + \gamma$$

be a complete integral of the differential equation in the text. Shew that, if

$$\phi_0 = \phi(a, a_1, \ldots, a_n, b_1, \ldots, b_n),$$

and

$$f_0 = f(a_1, \ldots, a_n),$$

and if $a_1, \ldots, a_n, b_1, \ldots, b_n$ be eliminated between the equations

$$\left.\begin{aligned} Z &= \phi - \phi_0 + f_0\\ \frac{\partial \phi}{\partial b_i} &= \frac{\partial \phi_0}{\partial b_i}, \quad \frac{\partial f_0}{\partial a_i} = \frac{\partial \phi_0}{\partial a_i}\end{aligned}\right\},$$

for $i = 1, \ldots, n$, the resulting value of Z is also an integral of the equation and that, when $x = a$, it acquires the value $f(x_1, \ldots, x_n)$. (Mayer.)

152. The discussion in §§ 150, 151 related to a resolved equation in which the dependent variable does not occur explicitly: and the inverse operations required consisted of the integration of a system of $2n$ ordinary equations, followed by a quadrature.

When the partial differential equation involves the dependent variable explicitly, and when it can be resolved so that it may be taken in the form

$$p + f(z, x, x_1, \ldots, x_n, p_1, \ldots, p_n) = 0,$$

then the corresponding result is as follows:—

Form and integrate the equations

$$\frac{dx_i}{dx} = \frac{\partial f}{\partial p_i}, \quad \frac{dp_i}{dx} = -\frac{\partial f}{\partial x_i} - p_i \frac{\partial f}{\partial z}, \quad \frac{dz}{dx} = \sum_{i=1}^{n} p_i \frac{\partial f}{\partial p_i} - f,$$

determining the arbitrary constants by the conditions that $x_1, \ldots, x_n, p_1, \ldots, p_n, z$ *acquire the values*

$$a_1, \ldots, a_n, b_1, \ldots, b_n, c + a_1 b_1 + \ldots + a_n b_n$$

respectively, when $x = a$. *Among this integral system of* $2n+1$ *equations, eliminate* $a_1, \ldots, a_n, p_1, \ldots, p_n$; *and let* Z *denote the resulting value of* z, *which is a function of* $x, x_1, \ldots, x_n, b_1, \ldots, b_n, c$. *Then*

$$z = Z$$

is a complete integral of the partial differential equation; and

$$\frac{\partial Z}{\partial x_i} = p_i, \quad \frac{\partial Z}{\partial b_i} = a_i \frac{\partial Z}{\partial c}, \quad z = Z,$$

for $i = 1, \ldots, n$, *the constants* $a_1, \ldots, a_n$ *being arbitrary, are a set of integrals of the* $2n+1$ *ordinary equations.*

This result may be deduced from the former case, or it may be obtained directly; we shall leave the establishment as an exercise. It will be noticed that, in the present case, the inverse operations required are the integration of a system of $2n+1$ ordinary equations, as contrasted with the slightly simpler inverse operations in the former case constituted by the integration of a system of $2n$ ordinary equations and a quadrature.

Lastly, it may happen that the partial differential equation contains the dependent variable explicitly but that it cannot be resolved, or cannot conveniently be resolved, in terms of any of the derivatives. In such a case, a similar process exists: and the result has already been stated*.

The Poisson-Jacobi Combinant (ϕ, ψ).

153. The determination of a complete integral of a partial differential equation

$$p + H(x, x_1, \ldots, x_n, p_1, \ldots, p_n) = 0,$$

and the determination of a full set of integrals of the associated canonical system

$$\frac{dx_i}{dx} = \frac{\partial H}{\partial p_i}, \quad \frac{dp_i}{dx} = -\frac{\partial H}{\partial x_i},$$

for $i = 1, \ldots, n$, have been shewn to be practically equivalent problems. It is known that, if two equations compatible with the original differential equation have been obtained, the Poisson-Jacobi combination of those equations provides another equation (which may be insignificant or may be evanescent) also compatible with the equation: and naturally therefore a question arises whether the same combination can similarly be effective in assisting the construction of the integrals of the canonical system.

Let

$$\phi = \phi(x, x_1, \ldots, x_n, p_1, \ldots, p_n) = \text{constant}$$

be an integral of the canonical system: then, in connection with that system, we have

$$\frac{d\phi}{dx} = \frac{\partial \phi}{\partial x} + \sum_{i=1}^{n} \frac{\partial \phi}{\partial x_i}\frac{dx_i}{dx} + \sum_{i=1}^{n} \frac{\partial \phi}{\partial p_i}\frac{dp_i}{dx} = 0,$$

and therefore

$$\frac{\partial \phi}{\partial x} + \sum_{i=1}^{n} \left(\frac{\partial \phi}{\partial x_i}\frac{\partial H}{\partial p_i} - \frac{\partial \phi}{\partial p_i}\frac{\partial H}{\partial x_i}\right) = 0,$$

that is, using the Poisson-Jacobi symbol, we have

$$\frac{\partial \phi}{\partial x} + (\phi, H) = 0,$$

analogous with a corresponding equation (§ 146) in theoretical dynamics. Similarly, if

$$\psi = \psi(x, x_1, \ldots, x_n, p_1, \ldots, p_n) = \text{constant}$$

* In Part I, § 109, of the present work. All the results are given in Mayer's memoir quoted on p. 388, foot-note.

be an integral of the canonical system, we have

$$\frac{\partial \psi}{\partial x} + (\psi, H) = 0.$$

It is natural to inquire whether (ϕ, ψ) also is an integral of the canonical system: we have

$$\begin{aligned}\frac{d}{dx}(\phi, \psi) &= \frac{\partial}{\partial x}(\phi, \psi) + ((\phi, \psi), H) \\ &= \left(\frac{\partial \phi}{\partial x}, \psi\right) + \left(\phi, \frac{\partial \psi}{\partial x}\right) + ((\phi, \psi), H) \\ &= -((\phi, H), \psi) - (\phi, (\psi, H)) + ((\phi, \psi), H) \\ &= ((H, \phi), \psi) + ((\psi, H), \phi) + ((\phi, \psi), H) \\ &= 0,\end{aligned}$$

on account of the identically satisfied relation of § 52. Hence

$$(\phi, \psi) = \text{constant},$$

as an integral equation, is compatible with the canonical system. Various cases may arise, as in the former investigation.

It may happen that (ϕ, ψ) vanishes identically: no new integral is provided.

It may happen that (ϕ, ψ) is a pure constant not zero; instances have occurred in which (ϕ, ψ) is equal to unity: no new integral is provided.

It may happen that (ϕ, ψ), while a function of the variables, can be expressed in terms of ϕ and ψ alone (and possibly in terms of previously known integrals, if any): no new integral is provided.

And it may happen that (ϕ, ψ) is a function of the variables which cannot be expressed in terms of ϕ and ψ alone (or in terms of these and of previously known integrals, if any): a new integral of the canonical system is then provided.

Note. It may happen that, when the partial differential equation is

$$H(x_1, \ldots, x_n, p_1, \ldots, p_n) = \text{constant} = h,$$

so that p and x have disappeared, care has to be exercised concerning new integrals of the canonical system: such new integrals do not necessarily provide equations compatible with $H=h$ and with equations which coexist with it. The canonical system is effectively the same as before, for it is

$$\frac{dx_i}{\dfrac{\partial H}{\partial p_i}} = \frac{dp_i}{-\dfrac{\partial H}{\partial x_i}} = dx,$$

say, for $i=1, \ldots, n$; and if

$$\phi=\phi(x_1, \ldots, x_n, p_1, \ldots, p_n, x)=\text{constant},$$
$$\psi=\psi(x_1, \ldots, x_n, p_1, \ldots, p_n, x)=\text{constant},$$

are two integrals of the system, then

$$(\phi, \psi)=\text{constant}$$

is also an integral of the canonical system; and we have

$$(H, \phi)=0, \quad (H, \psi)=0, \quad (H, (\phi, \psi))=0.$$

But if we are proceeding on the lines of Jacobi's second method, as explained in Chapter IV, for the integration of the partial equation $H=0$, and if we have associated the equations

$$H=0, \quad \phi=a_1,$$

where a_1 is arbitrary, then we can associate

$$\psi=a_2$$

with these, only if

$$(\phi, \psi)=0;$$

and we can associate

$$(\phi, \psi)=a_3$$

with $H=0$, $\phi=a_1$, only if

$$(\phi, (\phi, \psi))=0;$$

and these conditions are not always satisfied.

Thus if the equation be

$$H=p_1^2+p_2^2+p_3^2+p_4^2-f(x_1^2+x_2^2+x_3^2, x_4)=0,$$

integrals of the canonical system are given by

$$\phi=x_2p_3-x_3p_2=a_1,$$
$$\psi=x_3p_1-x_1p_3=a_2;$$

and then

$$\chi=(\phi, \psi)=x_1p_2-x_2p_1,$$

which leads to a new integral of the canonical system. Now

$$H=0, \quad \phi=a_1,$$

can be associated, because

$$(H, \phi)=0.$$

But $\psi=a_2$ cannot be associated with these two equations; for though $(H, \psi)=0$, we have

$$(\phi, \psi)=\chi,$$

different from zero. Again, the equations

$$H=0, \quad \psi=a_2,$$

can be associated, because

$$(H, \psi)=0.$$

But $\phi=a_1$ cannot be associated with these equations because

$$(\psi, \phi)=-\chi,$$

which is not zero. Moreover,

$$\chi=a_3$$

cannot be associated with either pair, because

$$(\chi, \phi)=\psi, \quad (\chi, \psi)=-\phi.$$

But

$$(H, \psi^2+\chi^2)=0, \quad (\phi, \psi^2+\chi^2)=0:$$

and so

$$H=0, \quad \phi=a_1, \quad \psi^2+\chi^2=c,$$

can be associated.

It thus appears that, while the Poisson-Jacobi combination of two integrals of the canonical system can provide a new integral of that system, the new integral cannot necessarily be associated with a retained system compatible with the original equation.

154. It will be convenient, for the sake of brevity, to call the Poisson-Jacobi combination (ϕ, ψ), of two functions ϕ and ψ, their *combinant*. From the preceding results, it is clear that any integral ϕ of the canonical system, which does not involve x, satisfies the equation

$$(\phi, H)=0$$

identically: hence the combinant of H with any such integral ϕ leads to no new integral.

Moreover, when the function H in the canonical system does not explicitly involve x (which corresponds to the case in theoretical dynamics when the total energy of the dynamical system is constant), the combinant of H and of any integral ϕ of the canonical system, that does not explicitly involve x, vanishes identically: for the equation

$$(\phi, H)=0$$

is then satisfied identically, so that (ϕ, H) provides no new integral. Also, with the same supposition concerning H, the integrals of the canonical system can be so taken that $2n-1$ of them are relations among the variables $x_1, \ldots, x_n, p_1, \ldots, p_n$, and the remaining integral can be taken in the form

$$\theta = x - x_0 + \vartheta(x_1, \ldots, x_n, p_1, \ldots, p_n) = 0,$$

where x_0 is an arbitrary quantity. In that case, the equation

$$(\vartheta, H)+1=0,$$

is satisfied identically: that is, (θ, H) provides no new integral. It therefore follows that, when the quantity H in the canonical system does not involve the variable x, no new integral can be derived by combining H with any other integral of the system: in fact, the quantity H is useless for any combinant construction with any integral ϕ of the system with a view to the derivation of new integrals.

It thus appears that, when an integral of the canonical system has been obtained, no new integral is furnished by the combinant of H with that integral. Clearly, when H is explicitly independent of the variable x, it furnishes an integral of the system: but it is not the only integral of the system which, under the combinant construction, leads to evanescent or unfruitful results. Indeed, earlier results obtained in connection with the development of Jacobi's second method in Chapter IV shew that, when any integral of the canonical system of equations has been obtained, other integrals exist such that their combinants with the given integral provide no new integral but only an evanescent result. For let ϕ be a given integral of the system: and let ψ denote some other integral, distinct from H in case H should not involve x explicitly: then (ϕ, ψ) also satisfies the equations of the system. If (ϕ, ψ) vanishes, or is equal to a pure constant, or is not functionally independent of ϕ and ψ, then (ϕ, ψ) is illusory as providing a new integral. But if no one of these alternatives is valid, we write

$$(\phi, \psi) = \psi_1,$$

and we proceed (as in § 62) to form the series of functions

$$(\phi, \psi_1) = \psi_2, \quad (\phi, \psi_2) = \psi_3, \quad \ldots, \quad (\phi, \psi_{i-1}) = \psi_i.$$

Each of the functions ψ_1, ψ_2, ... is an integral of the canonical system; and the set of such functions, that are independent of one another, is limited in number because the canonical system is of finite order. Accordingly, we may assume that the series of functions, derived through combination with ϕ, terminates with ψ_i: the termination can come (§ 62) in one of three ways.

(i) If ψ_i vanishes identically, then ψ_{i-1} is such that

$$(\phi, \psi_{i-1}) = 0,$$

identically, that is, ψ_{i-1} is an integral of the type indicated.

(ii) If ψ_i is a pure constant, say c, then

$$(\phi, \psi^2_{i-1} - 2c\psi_{i-2}) = 0,$$

so that as i is greater than unity (for otherwise ψ_1 would be an integral of the type indicated), then $\psi^2_{i-1} - 2c\psi_{i-2}$ is an integral of the type indicated.

(iii) If ψ_i is a function of the integrals that occur earlier in the series, say θ, then any integral (say Ψ) of the system

$$\frac{d\psi}{\psi_1} = \frac{d\psi_1}{\psi_2} = \dots = \frac{d\psi_{i-1}}{\theta}$$

is such that

$$(\phi, \Psi) = 0,$$

identically; and so Ψ is an integral of the indicated type belonging to the canonical system.

The process of combination is thus seen to provide a number of integrals: but, account being taken only of integrals that are independent of one another, the process cannot lead to all the independent integrals because, as has been seen, there are integrals which, when combined with a given integral, lead to an illusory result*. We shall not pursue this subject further, and shall be content with referring the reader to an important memoir† by Bertrand.

Ex. Prove that, when the function H in the canonical system does not explicitly involve x and when an integral ϕ other than H is given, the complete set of integrals of the canonical system is given by

$$H = h, \quad \phi = \text{constant},$$

and by

(i) an integral χ, not explicitly involving x and such that

$$(\phi, \chi) = 1,$$

(ii) an integral ψ, where

$$\psi = g(x_1, \dots, x_n, p_1, \dots, p_n) - x,$$

such that

$$(\phi, \psi) = 0,$$

(iii) other $2n-4$ integrals $a_1, \dots, a_{2n-4}$, explicitly independent of x, such that

$$(\phi, a_i) = 0,$$

for $i = 1, \dots, 2n-4$. (Bertrand.)

* Contrary to the opinion formed by Jacobi according to which it can be seen "in omnibus problematibus mechanicis in quibus virium vivarum conservatio "locum habet, *generaliter* e duobus integralibus præter principium illud inventis "reliqua omnia absque ulla ulteriore integratione inveniri posse": *Ges. Werke*, t. v, p. 49. The original theorem due to Poisson was published at the end of the year 1809: it seems that Jacobi's application and development of Poisson's theorem were made about 1838.

The frequently illusory character of the combinant is one of the causes which limit the number of the general algebraic integrals of the dynamical problem of n bodies to the classical integrals: see vol. III of this work, chapter XVII.

† *Liouville's Journal*, t. XVII (1852), pp. 393–436. Other references will be found in Graindorge's treatise, already quoted.

Contact Transformations and Canonical Systems.

155. In the last chapter, it was seen that the theory of contact transformations could be applied to the integration of a partial differential equation: and it has also been seen, from various points of view, that the integration of such an equation is bound up with the integration of a canonical system of ordinary equations. It is therefore natural to suppose that the theory of contact transformations can be brought into relation with the integration of a canonical system.

Let the canonical system be

$$\frac{dx_i}{dx} = \frac{\partial H}{\partial p_i}, \quad \frac{dp_i}{dx} = -\frac{\partial H}{\partial x_i},$$

for $i = 1, \ldots, n$; and suppose that H does not explicitly involve x. Let a contact transformation be given which passes from $x_1, \ldots, x_n, p_1, \ldots, p_n$ to $X_1, \ldots, X_n, P_1, \ldots, P_n$, such that

$$(X_\mu, P_\mu) = 1,$$
$$(X_m, X_\mu) = 0, \quad (P_m, P_\mu) = 0, \quad (X_m, P_\mu) = 0,$$

for μ and $m = 1, \ldots, n$, with unequal values of m and μ; and let it be applied to transform the canonical system. We have

$$\begin{aligned} \frac{dX_i}{dx} &= \sum_{m=1}^{n} \left(\frac{\partial X_i}{\partial x_m}\frac{dx_m}{dx} + \frac{\partial X_i}{\partial p_m}\frac{dp_m}{dx} \right) \\ &= \sum_{m=1}^{n} \left(\frac{\partial X_i}{\partial x_m}\frac{\partial H}{\partial p_m} - \frac{\partial X_i}{\partial p_m}\frac{\partial H}{\partial x_m} \right) \\ &= (X_i, H). \end{aligned}$$

When the variables in H are transformed, let the resulting quantity be denoted by K; then

$$\begin{aligned} (X_i, H) &= \sum_{m=1}^{n} \left\{ (X_i, X_m)\frac{\partial K}{\partial X_m} + (X_i, P_m)\frac{\partial K}{\partial P_m} \right\} \\ &= \frac{\partial K}{\partial P_i}, \end{aligned}$$

on account of the properties of the contact transformation. Consequently, we have

$$\frac{dX_i}{dx} = \frac{\partial K}{\partial P_i};$$

and, by similar analysis, we also have

$$\frac{dP_i}{dx} = -\frac{\partial K}{\partial X_i}.$$

Hence the contact transformation leaves the form of the canonical system unchanged. But this is not the limit of the property: it is easy to see that any transformation, which leaves the form of the canonical system (supposed perfectly general) unchanged, is of the contact type. For taking any transformation from $x_1, \ldots, x_n$, $p_1, \ldots, p_n$ to $X_1, \ldots, X_n, P_1, \ldots, P_n$, we have

$$\frac{dX_i}{dx} = (X_i, H), \quad \frac{dP_i}{dx} = (P_i, H);$$

if H' be the value of H after transformation has been effected, then

$$(X_i, H) = \sum_{m=1}^{n} \left\{ (X_i, X_m) \frac{\partial H'}{\partial X_m} + (X_i, P_m) \frac{\partial H'}{\partial P_m} \right\},$$

$$(P_i, H) = \sum_{m=1}^{n} \left\{ (P_i, X_m) \frac{\partial H'}{\partial X_m} + (P_i, P_m) \frac{\partial H'}{\partial P_m} \right\}.$$

If the new form of the equations is still canonical, the former of these must be $\frac{\partial H'}{\partial P_i}$, and the latter must be $-\frac{\partial H'}{\partial X_i}$, for all values of i: hence, as H and H' are supposed quite general functions, we must have

$$(X_m, P_m) = 1,$$

for all values $1, \ldots, n$ of m, and

$$(X_i, X_m) = 0, \quad (X_i, P_m) = -(P_m, X_i) = 0, \quad (P_i, P_m) = 0,$$

for all unequal values $1, \ldots, n$ of i and m. These are the equations which define a contact transformation. Therefore *a canonical system is unchanged in form by a contact transformation; and every transformation, which conserves the form of a quite general canonical system, is of the contact type.*

There is an immediate practical advantage in such a transformation, whenever the form of H' is simpler: the equations may be simpler to integrate.

156. In the next place, suppose that the canonical system is of the form

$$\frac{dx_i}{dx} = \frac{\partial H}{\partial p_i}, \quad \frac{dp_i}{dx} = -\frac{\partial H}{\partial x_i},$$

where H now involves x, as well as $x_1, \ldots, x_n, p_1, \ldots, p_n$. In this case, we have

$$\frac{dH}{dx} = \frac{\partial H}{\partial x} + (H, H)$$

$$= \frac{\partial H}{\partial x};$$

hence introducing a new variable p, such that

$$\Theta = H + p = \text{constant} = 0,$$

we have

$$\frac{dp}{dx} = -\frac{\partial H}{\partial x} = -\frac{\partial \Theta}{\partial x},$$

$$\frac{dx}{dx} = 1 \quad = \frac{\partial \Theta}{\partial p},$$

$$\frac{dx_i}{dx} = \frac{\partial H}{\partial p_i} = \frac{\partial \Theta}{\partial p_i},$$

$$\frac{dp_i}{dx} = -\frac{\partial H}{\partial x_i} = -\frac{\partial \Theta}{\partial x_i},$$

so that the canonical system may be replaced by the amplified system

$$\frac{dx}{\frac{\partial \Theta}{\partial p}} = \frac{dp}{-\frac{\partial \Theta}{\partial x}} = \frac{dx_i}{\frac{\partial \Theta}{\partial p_i}} = \frac{dp_i}{-\frac{\partial \Theta}{\partial x_i}},$$

for $i = 1, \ldots, n$. Now take a contact transformation changing the variables from $x, x_1, \ldots, x_n, p, p_1, \ldots, p_n$ to $X, X_1, \ldots, X_n, P, P_1, \ldots, P_n$: then denoting x, p, X, P by x_0, p_0, X_0, P_0 for convenience, we must have

$$(X_i, P_i) = 1,$$

for $i = 0, 1, \ldots, n$, and

$$(X_i, X_j) = 0, \quad (P_i, P_j) = 0, \quad (X_i, P_j) = 0,$$

for unequal values of i and j from the series $0, 1, \ldots, n$. As in the earlier case, if Φ be the transformed value of Θ, the amplified canonical system can be replaced by

$$\frac{dX}{\frac{\partial \Phi}{\partial P}} = \frac{dP}{-\frac{\partial \Phi}{\partial X}} = \frac{dX_i}{\frac{\partial \Phi}{\partial P_i}} = \frac{dP_i}{-\frac{\partial \Phi}{\partial X_i}},$$

for $i = 1, \ldots, m$. Let $\Phi = 0$ be resolved so as to express P in terms of $X, X_1, \ldots, X_n, P_1, \ldots, P_n$; and let the resolved form be

$$K(X, X_1, \ldots, X_n, P_1, \ldots, P_n) + P = 0.$$

Then

$$\frac{\partial \Phi}{\partial X} - \frac{\partial \Phi}{\partial P}\frac{\partial K}{\partial X} = 0,$$

$$\frac{\partial \Phi}{\partial X_i} - \frac{\partial \Phi}{\partial P}\frac{\partial K}{\partial X_i} = 0,$$

$$\frac{\partial \Phi}{\partial P_i} - \frac{\partial \Phi}{\partial P}\frac{\partial K}{\partial P_i} = 0,$$

for $i = 1, \ldots, n$. The preceding system can now, in its turn, be replaced by the equations

$$\frac{dX_i}{dX} = \frac{\frac{\partial \Phi}{\partial P_i}}{\frac{\partial \Phi}{\partial P}} = \frac{\partial K}{\partial P_i},$$

$$\frac{dP_i}{dX} = -\frac{\frac{\partial \Phi}{\partial X_i}}{\frac{\partial \Phi}{\partial P}} = -\frac{\partial K}{\partial X_i},$$

for $i = 1, \ldots, n$, which again are a canonical system: and we also have

$$\frac{dP}{dX} = -\frac{\partial K}{\partial X}.$$

It thus appears that a contact transformation, applied to a canonical system even when the function H involves the variable x, changes the system into another canonical system.

Conversely, any transformation between $x, x_1, \ldots, x_n, p_1, \ldots, p_n$ and $X, X_1, \ldots, X_n, P_1, \ldots, P_n$, which changes a canonical system

$$\frac{dx_i}{dx} = \frac{\partial H}{\partial p_i}, \quad \frac{dp_i}{dx} = -\frac{\partial H}{\partial x_i}, \qquad (i = 1, \ldots, n),$$

where H involves x as well as $x_1, \ldots, x_n, p_1, \ldots, p_n$, into another canonical system

$$\frac{dX_i}{dX} = \frac{\partial K}{\partial P_i}, \quad \frac{dP_i}{dX} = -\frac{\partial K}{\partial X_i}, \qquad (i = 1, \ldots, n),$$

is a contact transformation in an increased number of variables. To establish the result, we introduce a variable p such that

$$\Theta = p + H = 0;$$

hence

$$\frac{dp}{dx} = -\frac{\partial H}{\partial x}.$$

We then consider any transformation which changes the variables from $x, p, x_1, \ldots, x_n, p_1, \ldots, p_n$ to $X, P, X_1, \ldots, X_n, P_1, \ldots, P_n$; and, for convenience, we write x_0, p_0, X_0, P_0 for x, p, X, P respectively. Then

$$\begin{aligned}\frac{dX_i}{dx} &= \sum_{m=0}^{n} \frac{\partial X_i}{\partial x_m}\frac{dx_m}{dx} + \frac{\partial X_i}{\partial p_m}\frac{dp_m}{dx} \\ &= \sum_{m=0}^{n} \left(\frac{\partial X_i}{\partial x_m}\frac{\partial \Theta}{\partial p_m} - \frac{\partial X_i}{\partial p_m}\frac{\partial \Theta}{\partial x_m}\right) \\ &= (X_i, \Theta) \\ &= \sum_{m=0}^{n} \left\{(X_i, X_m)\frac{\partial \Theta}{\partial X_m} + (X_i, P_m)\frac{\partial \Theta}{\partial P_m}\right\},\end{aligned}$$

for $i = 0, 1, \ldots, m$; and similarly

$$\frac{dP_i}{dx} = \sum_{m=0}^{n} \left\{(P_i, X_m)\frac{\partial \Theta}{\partial X_m} + (P_i, P_m)\frac{\partial \Theta}{\partial P_m}\right\}.$$

Let $\Theta = 0$, after substitution has been made for p and in H, be resolved so as to express P_0 in terms of $X_0, X_1, \ldots, X_n, P_1, \ldots, P_n$; let the resolved equivalent be

$$P_0 + K = 0,$$

where K is a function of $X_0, X_1, \ldots, X_n, P_1, \ldots, P_n$; then

$$\frac{\partial \Theta}{\partial X_\mu} - \frac{\partial \Theta}{\partial P_0}\frac{\partial K}{\partial X_\mu} = 0,$$

for $\mu = 0, 1, \ldots, m$, and

$$\frac{\partial \Theta}{\partial P_\mu} - \frac{\partial \Theta}{\partial P_0}\frac{\partial K}{\partial P_\mu} = 0,$$

for $\mu = 1, \ldots, m$. Then, for $i = 1, \ldots, m$, we have

$$\begin{aligned}\frac{dX_i}{dX} &= \frac{dX_i}{dx} \div \frac{dX_0}{dx} \\ &= \frac{\sum_{m=0}^{n} \left\{(X_i, X_m)\frac{\partial \Theta}{\partial X_m} + (X_i, P_m)\frac{\partial \Theta}{\partial P_m}\right\}}{\sum_{m=0}^{n} \left\{(X_0, X_m)\frac{\partial \Theta}{\partial X_m} + (X_0, P_m)\frac{\partial \Theta}{\partial P_m}\right\}} \\ &= \frac{(X_i, P_0) + (X_i, X_0)\frac{\partial K}{\partial X} + \sum_{m=1}^{n} \left\{(X_i, X_m)\frac{\partial K}{\partial X_m} + (X_i, P_m)\frac{\partial K}{\partial P_m}\right\}}{(X_0, P_0) + (X_0, X_0)\frac{\partial K}{\partial X} + \sum_{m=1}^{n} \left\{(X_0, X_m)\frac{\partial K}{\partial X_m} + (X_0, P_m)\frac{\partial K}{\partial P_m}\right\}},\end{aligned}$$

and

$$\frac{dP_i}{dX}=\frac{(P_i, P_0)+(P_i, X_0)\dfrac{\partial K}{\partial X}+\sum\limits_{m=1}^{n}\left\{(P_i, X_m)\dfrac{\partial K}{\partial X_m}+(P_i, P_m)\dfrac{\partial K}{\partial P_m}\right\}}{(X_0, P_0)+(X_0, X_0)\dfrac{\partial K}{\partial X}+\sum\limits_{m=1}^{n}\left\{(X_0, X_m)\dfrac{\partial K}{\partial X_m}+(X_0, X_m)\dfrac{\partial K}{\partial P_m}\right\}}.$$

We are in quest of transformations which will make the new system canonical, and therefore the transformed equations should be of the form

$$\frac{dX_i}{dX}=\frac{\partial \bar{K}}{\partial P_i},\quad \frac{dP_i}{dX}=-\frac{\partial \bar{K}}{\partial X_i},$$

for $i=1, \ldots, m$. Hence, in order that the preceding equations may be of this type, we take

$$\bar{K}=K+\alpha,$$

where α is a constant: and the conditions, necessary and sufficient for the purpose when the system is of the most general type, are

$$(X_0, P_0)=(X_1, P_1)=\ldots=(X_n, P_n),$$
$$(X_i, X_j)=0,\quad (P_i, P_j)=0,\quad (P_i, X_j)=0,$$

for unequal values of i and j from the series $0, 1, \ldots, m$. These equations are characteristic of, and define, a contact transformation in the increased aggregate of variables. Moreover,

$$\frac{dP_0}{dX}=\frac{dP_0}{dx}\div\frac{dX_0}{dx}$$
$$=-\frac{\partial K}{\partial X}.$$

Consequently, *even when the function H in the general canonical system*

$$\frac{dx_i}{dx}=\frac{\partial H}{\partial p_i},\quad \frac{dp_i}{dx}=-\frac{\partial H}{\partial x_i},$$

involves the variable x, any contact transformation of the amplified system leads to a new canonical system; and every transformation, which transforms one canonical system of the most general type into another, is a contact transformation in the increased number of variables.

157. But it may be asked whether a contact transformation, involving only the variables $x_1, \ldots, x_n, p_1, \ldots, p_n$, will transform one canonical system into another when H involves the variable x; it is easy to see that such a transformation is possible and that it is, in fact, a special case of the contact transformation in the amplified number of variables. To verify this statement, we make

$$X_0 = x_0 = x, \quad P_0 = p_0 = p;$$

then, as (X_0, P_0) is unity, we have

$$(X_1, P_1) = 1, \ldots, (X_n, P_n) = 1.$$

Now the equations

$$(X_0, X_i) = 0, \quad (X_0, P_i) = 0, \quad (P_0, X_i) = 0, \quad (P_0, P_i) = 0,$$

give

$$\frac{\partial X_i}{\partial p_0} = 0, \quad \frac{\partial P_i}{\partial p_0} = 0, \quad \frac{\partial X_i}{\partial x_0} = 0, \quad \frac{\partial P_i}{\partial x_0} = 0;$$

and therefore the other equations are

$$(X_i, P_i) = \sum_{m=1}^{n} \left(\frac{\partial X_i}{\partial x_m} \frac{\partial P_i}{\partial p_m} - \frac{\partial X_i}{\partial p_m} \frac{\partial P_i}{\partial x_m} \right) = 1,$$

for $i = 1, \ldots, n$: also

$$(X_i, X_m) = \sum_{j=1}^{n} \left(\frac{\partial X_i}{\partial x_j} \frac{\partial X_m}{\partial p_j} - \frac{\partial X_i}{\partial p_j} \frac{\partial X_m}{\partial x_j} \right) = 0,$$

$$(X_i, P_m) = 0, \quad (P_i, P_m) = 0.$$

These equations clearly define a contact transformation between $X_1, \ldots, X_n, P_1, \ldots, P_n$ and $x_1, \ldots, x_n, p_1, \ldots, p_n$ alone: and they give a special case of the contact transformation in the amplified number of variables conserving the form of the canonical system.

158. Returning now to the canonical system of equations in the simpler form in which the quantity H does not involve the independent variable of the system explicitly, and denoting that variable by t, we have the system in the form

$$\frac{dx_i}{dt} = \frac{\partial H}{\partial p_i}, \qquad \frac{dp_i}{dt} = -\frac{\partial H}{\partial x_i},$$

for $i = 1, \ldots, n$. Here, H is the total energy of the system and it remains constant throughout the motion; and, with the variables adopted for the construction of the canonical system, H is a function of $x_1, \ldots, x_n, p_1, \ldots, p_n$ alone.

We have seen that the most general form of infinitesimal contact transformation is given (§ 129) by

$$\delta z = \epsilon \zeta, \quad \delta x_i = \epsilon \xi_i, \quad \delta p_i = \epsilon \pi_i, \quad (i = 1, \ldots, n),$$

where

$$\xi_i = \frac{\partial U}{\partial p_i}, \quad -\pi_i = \frac{\partial U}{\partial x_i} + p_i \frac{\partial U}{\partial z},$$

U denoting any arbitrary function of $x_1, \ldots, x_n, p_1, \ldots, p_n, z$. Let U be chosen so as not to involve z explicitly: then the equations become

$$\xi_i = \frac{\partial U}{\partial p_i}, \quad -\pi_i = \frac{\partial U}{\partial x_i};$$

or, on writing $\epsilon = \delta t$, the equations of the infinitesimal contact transformation may be taken in the form

$$\frac{dx_i}{dt} = \frac{\partial U}{\partial p_i}, \quad \frac{dp_i}{dt} = -\frac{\partial U}{\partial x_i},$$

for $i = 1, \ldots, n$.

It therefore follows that the equations of the canonical system are the equations of an infinitesimal contact transformation, applied to the variables of the system and derived from the energy H of the originating system; and therefore the changes in the variables of the system can be regarded as the changes caused by the continued application of the infinitesimal contact transformation derived from the energy of the system. It is known, from the theory of groups of transformations, that the infinitesimal contact transformations determine uniquely the finite contact transformations of which they are the infinitesimal expression: moreover, what is the equivalent of this proposition for the present purpose, we have shewn that a finite contact transformation conserves the form of the canonical system. Hence, if we denote the values of the variables of the canonical system at any epoch t_0 by $\alpha_1, \ldots, \alpha_n, \beta_1, \ldots, \beta_n$, and their values at the epoch t by $X_1, \ldots, X_n, P_1, \ldots, P_n$, there is a contact transformation between $X_1, \ldots, X_n, P_1, \ldots, P_n$ and $\alpha_1, \ldots, \alpha_n, \beta_1, \ldots, \beta_n$; and therefore the variables of the canonical system change continuously from their initial values under the continuous domination of the infinitesimal contact transformation determined by the energy.

This result includes the properties established by Bertrand* as regards canonical constants; for the equations defining these canonical constants are the equations expressing the contact trans-

* *Liouville*, t. XVII (1852), pp. 393 et seq.

formation between $\alpha_1, \ldots, \alpha_n, \beta_1, \ldots, \beta_n$ and $x_1, \ldots, x_n, p_1, \ldots, p_n$, viz.

$$(\alpha_i, \beta_i) = 1,$$

$$(\alpha_i, \alpha_j) = 0, \quad (\alpha_i, \beta_j) = 0, \quad (\beta_i, \beta_j) = 0,$$

for i and $j = 1, \ldots n$, with unequal values of i and j.

This stage will mark the limit of our discussion of the canonical equations of theoretical dynamics. Their detailed properties constitute a subject, distinct in many of its developments from the theory of partial differential equations; for a fuller discussion, reference may be made to the authorities quoted at the beginning of the chapter.

CHAPTER XI.

SIMULTANEOUS EQUATIONS OF THE FIRST ORDER.

THE present chapter is a discussion of systems of simultaneous partial equations of the first order, the number of equations being the same as the number of dependent variables. The operation of integrating such equations is an inverse operation of class greater than unity in general, that is, it cannot generally be resolved into operations of the first order such as the integration of a number of ordinary equations each of the first order. General inverse operations of class greater than unity cannot be performed in finite terms, in the present state of analysis; those particular inverse operations, which can be resolved into operations of the first order, can however be performed, in the sense of the methods given in some of the preceding chapters. Naturally, the simultaneous partial equations involving several dependent variables, which can be integrated by these resoluble operations, are subject to corresponding limitations as regards generality of form: and consequently, owing to this somewhat particularised character, the theory of these equations is not so fully discussed here as has been the theory of equations in a single dependent variable.

The subject appears to have been considered first* by Jacobi: as presented in this form, further developments of Jacobi's theory are given by Natani†, and Zajaçrkowski‡.

A different presentation, and a completely different class of equations, occur in Hamburger's treatment§; cognate investigations have been effected by Königsberger||, who also deals with the existence-theorem for a set of equations, the number of which is equal to the number of dependent variables; and Hamburger's method has been extended by von Weber¶ to the case, when the number of equations is greater than the number of dependent variables.

* *Ges. Werke*, t. IV, pp. 3—15.

† *Die höhere Analysis*, pp. 339—341.

‡ *Grunert's Archiv*, t. LVI (1874), pp. 163—174.

§ *Crelle*, t. LXXXI (1876), pp. 243—280, *ib.*, t. XCIII (1882), pp. 188—214.

|| *Crelle*, t. CIX (1892), pp. 261—340; *Math. Ann.*, t. XLI (1893), pp. 260—285.

¶ *Crelle*, t. CXVIII (1897), pp. 123—157.

A class of equations, the number of which is an exact multiple of the number of dependent variables involved, has been considered by König*: an account of his method and of his results is given in the course of the chapter.

Hamburger has shewn that it is possible to apply the method to partial differential equations of the second order and of higher orders in one dependent variable and two independent variables. The subsidiary equations obtained as ancillary to the integration have substantial similarity with those obtained in the method, devised by Darboux for the integration of such equations and developed by Speckman and others. Accordingly, an account of Hamburger's application of his method to the integration of equations of the second order and of higher orders will not be considered in this chapter but will be deferred until the stage when such equations are being generally considered.

It may be added that some of the geometrical properties that can be associated with the simplest case, viz. when there are two dependent variables and two independent variables, are considered by Bäcklund†. As the processes of integration in this chapter are only applicable to limited classes of equations, these geometrical associations are not discussed in this connection: moreover, they belong more properly to the theory of equations of higher orders and, like the extension of Hamburger's method to such equations, they also will be deferred for consideration in connection with that theory.

159. The investigations in the preceding chapters have been concerned with the integration and the general theory of partial differential equations involving only a single dependent variable; no restriction was laid upon the number of independent variables; and, when more than a single equation occurred, the conditions necessary and sufficient to secure coexistence were obtained. It was shewn how to deduce, from a complete integral, other classes of integrals of various types: the aggregate of these classes was completely comprehensive for some types of equations and largely so (the exceptions being the special integrals) for the remainder. The construction of the complete integral was made to depend upon the integration, complete or incomplete, of a simultaneous system of ordinary equations, formed from the partial differential equations: the integration required depends, in practice, solely upon the possibility of actually effecting general inverse processes of the first order. Speaking broadly, we may say that the theory of partial differential equations of the first order in a single dependent variable can be considered a known theory.

* *Math. Ann.*, t. XXIII (1884), pp. 520—526.

† *Math. Ann.*, t. XVII (1880), pp. 285—328; *ib.* t. XIX (1882), pp. 387—422.

The problems which, in scale of difficulty, lie next to that of partial differential equations of the first order in a single dependent variable, are obtained, on the one hand, by increasing the number of dependent variables and keeping the equations still of the first order, and on the other hand, by taking equations of a higher order than the first, still in a single dependent variable.

As concerns partial differential equations of the second order (and of higher orders) in a single dependent variable, there is a considerable body of theory: moreover, the frequent occurrence of such differential equations, in subjects such as geometry and many of the developed branches of mathematical physics, has led to the discussion of detailed properties of particular equations which, once known, have pointed the way to further developments of the general theory.

But as concerns sets of partial differential equations of the first order in several dependent variables, when these sets are not the equivalent of a single equation of higher order in a single dependent variable, the amount of finished theory that has been obtained is comparatively slight. Thus, when the number of equations is equal to the number of dependent variables and when these equations have a special form which, among other limitations, is linear in the derivatives, it is known (§§ 9—14) that integrals of the equations do exist, satisfying assigned conditions of a given type. But when there is a question of constructing an integral in some form other than a multiple power-series as it occurs in the establishment of the existence-theorem, methods even only theoretically effective for the purpose have been devised solely for very restricted classes of systems of equations. Accordingly, before passing to equations of higher order in a single dependent variable, we shall deal with systems of equations of the first order in several variables, so as to indicate such general methods and results as have been obtained.

As in the early stages of the development of the theory of equations of the first order in a single dependent variable, some indications of results, which may be expected to hold frequently in simple cases though far from universally, can be obtained by proceeding from a set of integral equations. Let the independent variables be

$$x_1, \ldots, x_n,$$

and the dependent variables be

$$z_1, \ldots, z_m;$$

then the m dependent variables will be given by m integral equations. These equations may contain a number of arbitrary constants: let this number be N, and suppose that these are essential constants, so that they cannot be expressed by a number smaller than N.

When the first derivatives of these equations are formed, by differentiating with respect to the independent variables in turn, and are associated with the integral system, the total number of equations then possessed is $m(n+1)$. Suppose that all the arbitrary constants can be eliminated and that no peculiarities* occur during the processes of elimination; then the number of differential equations of the first order, emerging after the elimination, is $m(n+1)-N$. If these differential equations are to be conceived as capable of determining the m dependent variables, their number cannot be less than m; hence

$$m(n+1)-N \geqslant m,$$

that is,

$$N \leqslant mn,$$

thus giving an upper limit for N.

If $N = m(n+1-r)$, where $1 \leqslant r \leqslant n$, and if the same suppositions be made concerning the integral system in the passage to the differential equations, the number of emerging differential equations is rm.

But conversely, unless conditions equivalent to the reversibility of the preceding process are satisfied by a given system of simultaneous equations, it does not follow that their integral is of the assumed initial form: indeed, if the number of equations in the simultaneous system be greater than the number of dependent variables, it does not follow that the system possesses any integral at all. In order that the equations in such a system may coexist, conditions will have to be satisfied.

* Such, for instance, as occur when a partial differential equation in a single dependent variable is thus constructed from its general integral which may contain any number of arbitrary constants. The supposition, adopted in the face of such an instance, is enough to destroy any confidence as to more than possibility in the inferences that can be drawn.

König's completely integrable Equations.

160. The conditions just indicated can be set out in the case of certain classes of equations of simple types: one such class is discussed* by König. Let

$$p_{ij} = \frac{\partial z_i}{\partial x_j},$$

for $i = 1, \ldots, m$, and $j = 1, \ldots, n$: suppose that the system contains rm equations and that they can be resolved so as to express the derivatives of the m dependent variables with regard to one and the same set of r independent variables, in terms of the remaining quantities of the system. Let these r independent variables be $x_1, \ldots, x_r$; then the rm equations may be taken in the form

$$p_{ij} = f_{ij},$$

for $i = 1, \ldots, m$, and $j = 1, \ldots, r$; the arguments of f_{ij} are the variables $x_1, \ldots, x_n, z_1, \ldots, z_m$, and also the derivatives $p_{\lambda\mu}$, where

$$\lambda = 1, \ldots, m, \quad \mu = r + 1, \ldots, n.$$

Then König's theorem is as follows:—

When appropriate formal conditions are satisfied, the system of equations

$$p_{ij} = f_{ij}$$

possesses an integral equivalent $z_1, \ldots, z_m$ *such that, when initial values* $c_1, \ldots, c_r$ *are assigned to* $x_1, \ldots, x_r$ *respectively, the functions* $z_1, \ldots, z_m$ *become functions of* $x_{r+1}, \ldots, x_n$, *which are regular functions in a certain domain and otherwise can be arbitrarily assigned.*

Denote by S_j the aggregate of the m differential equations in which the second suffix is j; and let Σ_j denote the set made up of the aggregates

$$S_j, S_{j+1}, \ldots, S_r.$$

Consider the aggregate S_r: it contains no derivatives with regard to $x_1, \ldots, x_{r-1}$, which therefore may be regarded as parameters during processes of integration. It thus is a system of m equations involving the m dependent variables and the $n - r + 1$ independent variables $x_r, x_{r+1}, \ldots, x_n$; and it is resolved with respect to

$$\frac{\partial z_1}{\partial x_r}, \ldots, \frac{\partial z_m}{\partial x_r}.$$

* *Math. Ann.*, t. XXIII (1884), pp. 520—526.

It thus is of the class to which Madame Kowalevsky's proof of Cauchy's existence-theorem can be applied, if the formal conditions imposed for the theorem are satisfied. Assuming that these conditions are satisfied, the system possesses a set of integrals $z_1, \ldots, z_m$ which, when $x_r = c_r$, become assigned functions of $x_{r+1}, \ldots, x_n$, taken to be regular in a definite domain and otherwise arbitrarily assigned. The variables $x_1, \ldots, x_{r-1}$ are parametric throughout; the integrals of the aggregate S_r are a set of functions satisfying the conditions assigned in the theorem as stated, when we make $x_1, \ldots, x_{r-1}$ equal to $c_1, \ldots, c_{r-1}$.

Next, consider the aggregate of equations represented by S_{r-1}: it contains no derivatives with regard to $x_1, \ldots, x_{r-2}, x_r$, which therefore may be regarded as parameters during processes of integration. It is a system of m equations in the m dependent variables and the $n-r+1$ independent variables $x_{r-1}, x_{r+1}, \ldots, x_n$; and it is resolved with respect to

$$\frac{\partial z_1}{\partial x_{r-1}}, \ldots, \frac{\partial z_m}{\partial x_{r-1}}.$$

Applying Cauchy's existence-theorem to this system, on the assumption that the formal conditions are satisfied, we infer that the system possesses a set of integrals $z_1, \ldots, z_m$ which, when $x_{r-1} = c_{r-1}$, become the assigned functions of $x_{r+1}, \ldots, x_n$. The variables $x_1, \ldots, x_{r-2}, x_r$ are parametric throughout: the integrals of the aggregate S_{r-1} are a set of functions satisfying the conditions assigned in the theorem as stated, when we make $x_1, \ldots, x_{r-2}, x_r$ equal to $c_1, \ldots, c_{r-2}, c_r$.

And so on, for each of the aggregates in turn: in the case of each of them, we obtain a set of integrals which satisfy the initial conditions assigned in König's theorem as stated.

But though the integrals of the aggregate S_{r-1} satisfy the same initial conditions as the integrals of the aggregate S_r, it does not follow that they are the same functions of the variables; and, *à fortiori*, it does not follow that the integrals of the aggregate S_j are integrals of all the succeeding aggregates, that is, are integrals of the set Σ_j.

It may however happen that the integrals determined for the aggregate S_j are integrals for the set Σ_j. When this is the case for all values of j and, in particular, for $j = 1$, it is clear that the original system of equations possesses a set of integrals with the properties stated in the theorem. In that case, the system is said to be *completely integrable*: it will therefore be necessary to determine the conditions which are necessary and sufficient to secure the complete integrability of the system.

161. We have assumed that the formal conditions, which justify the application of Cauchy's theorem, are satisfied. These conditions relate to the form of the functions f_{ij}, and require those functions to be regular within a domain, which belongs to the values $c_1, \ldots, c_r$ of $x_1, \ldots, x_r$ and to initial values assignable at will to $x_{r+1}, \ldots, x_n$: it is within such a domain that the functions, postulated in the initial conditions, are regular. Moreover, the determination of $z_1, \ldots, z_m$, for any aggregate S_j, as regular functions of the variables is unique under the assigned initial conditions; so that integrals are, or are not, possessed by the system in accordance with the initial conditions according as, for all values of j, the integrals of the aggregate S_j are, or are not, integrals of the set Σ_{j+1}. And, in particular, it is sufficient, in order to secure that the integrals of the aggregate S_j are the same as the integrals (if any) of the set Σ_{j+1}, that the integrals of S_j should satisfy the equations in the set Σ_{j+1}.

The conditions of complete integrability are therefore such that the integrals of S_j should satisfy the equations in Σ_{j+1}, for all values $1, \ldots, r-1$ of j. In order that the integrals of S_j may satisfy the equations in the set Σ_{j+1}, it is necessary and sufficient that, when they are substituted in those equations, they should make each of the equations an identity. Let $E=0$ be any one of the equations in the set Σ_{j+1}, thus made an identity: then we have

$$\frac{\partial E}{\partial x_1}=0, \; \ldots, \; \frac{\partial E}{\partial x_n}=0,$$

in virtue of those integrals and of the equations of the system. Now in S_j and Σ_{j+1} there are no derivatives with regard to $x_1, \ldots, x_{j-1}$; consequently, the conditions

$$\frac{\partial E}{\partial x_{j-1}}=0, \quad \frac{\partial E}{\partial x_{j-2}}=0, \; \ldots, \; \frac{\partial E}{\partial x_1}=0,$$

can be held over for consideration with the aggregates S_{j-1}, $S_{j-2}, \ldots, S_1$ respectively. Also, the conditions

$$\frac{\partial E}{\partial x_{j+1}}=0, \quad \frac{\partial E}{\partial x_{j+2}}=0, \; \ldots, \; \frac{\partial E}{\partial x_{n-1}}=0, \quad \frac{\partial E}{\partial x_n}=0,$$

can be regarded as satisfied, on the hypothesis that Σ_{j+1} possesses integrals; and therefore, at the stage of considering whether the integrals of S_j satisfy the equations in the set Σ_{j+1}, it is sufficient to take the conditions

$$\frac{\partial E}{\partial x_j}=0,$$

where E is any equation of the set Σ_{j+1}. These conditions must be satisfied in virtue of the equations of the system: and they are

$$\frac{\partial (p_{i\alpha}-f_{i\alpha})}{\partial x_j}=0,$$

for $i=1, \ldots, n$, and $\alpha=j+1, \ldots, r$.

Now, when the integrals are substituted, we have

$$\begin{aligned}0=\frac{\partial (p_{i\alpha}-f_{i\alpha})}{\partial x_j}&=\frac{\partial p_{i\alpha}}{\partial x_j}-\frac{df_{i\alpha}}{dx_j}\\&=\frac{\partial p_{ij}}{\partial x_\alpha}-\frac{df_{i\alpha}}{dx_j}\\&=\frac{df_{ij}}{dx_\alpha}-\frac{df_{i\alpha}}{dx_j}.\end{aligned}$$

Here

$$\frac{df_{ij}}{dx_\alpha}=\frac{\partial f_{ij}}{\partial x_\alpha}+\sum_{\lambda=1}^{m}\frac{\partial f_{ij}}{\partial z_\lambda}p_{\lambda\alpha}+\sum_{\rho=1}^{m}\sum_{\mu=r+1}^{n}\frac{\partial f_{ij}}{\partial p_{\rho\mu}}\frac{\partial p_{\rho\mu}}{\partial x_\alpha},$$

and

$$\begin{aligned}\frac{\partial p_{\rho\mu}}{\partial x_\alpha}&=\frac{\partial p_{\rho\alpha}}{\partial x_\mu}\\&=\frac{df_{\rho\alpha}}{dx_\mu}\\&=\frac{\partial f_{\rho\alpha}}{\partial x_\mu}+\sum_{\lambda=1}^{m}\frac{\partial f_{\rho\alpha}}{\partial z_\lambda}p_{\lambda\mu}+\sum_{\sigma=1}^{m}\sum_{\tau=r+1}^{n}\frac{\partial f_{\rho\alpha}}{\partial p_{\sigma\tau}}\frac{\partial p_{\sigma\tau}}{\partial x_\mu},\end{aligned}$$

while

$$p_{\lambda\alpha}=f_{\lambda\alpha};$$

hence

$$\begin{aligned}\frac{df_{ij}}{dx_\alpha}=\frac{\partial f_{ij}}{\partial x_\alpha}&+\sum_{\lambda=1}^{m}\frac{\partial f_{ij}}{\partial z_\lambda}f_{\lambda\alpha}+\sum_{\rho=1}^{m}\sum_{\mu=r+1}^{n}\frac{\partial f_{ij}}{\partial p_{\rho\mu}}\frac{\partial f_{\rho\alpha}}{\partial x_\mu}\\&+\sum_{\lambda=1}^{m}\sum_{\rho=1}^{m}\sum_{\mu=r+1}^{n}\frac{\partial f_{ij}}{\partial p_{\rho\mu}}\frac{\partial f_{\rho\alpha}}{\partial z_\lambda}p_{\lambda\mu}\\&+\sum_{\rho=1}^{m}\sum_{\mu=r+1}^{n}\sum_{\sigma=1}^{m}\sum_{\tau=r+1}^{n}\frac{\partial f_{ij}}{\partial p_{\rho\mu}}\frac{\partial f_{\rho\alpha}}{\partial p_{\sigma\tau}}\frac{\partial p_{\sigma\tau}}{\partial x_\mu}.\end{aligned}$$

Similarly,

$$\frac{df_{i\alpha}}{dx_j} = \frac{\partial f_{i\alpha}}{\partial x_j} + \sum_{\lambda=1}^{m} \frac{\partial f_{i\alpha}}{\partial z_\lambda} f_{\lambda j} + \sum_{\rho=1}^{m} \sum_{\mu=r+1}^{n} \frac{\partial f_{i\alpha}}{\partial p_{\rho\mu}} \frac{\partial f_{\rho j}}{\partial x_\mu}$$

$$+ \sum_{\lambda=1}^{m} \sum_{\rho=1}^{n} \sum_{\mu=r+1}^{n} \frac{\partial f_{i\alpha}}{\partial p_{\rho\mu}} \frac{\partial f_{\rho j}}{\partial z_\lambda} p_{\lambda\mu}$$

$$+ \sum_{\rho=1}^{m} \sum_{\mu=r+1}^{n} \sum_{\sigma=1}^{m} \sum_{\tau=r+1}^{n} \frac{\partial f_{i\alpha}}{\partial p_{\rho\mu}} \frac{\partial f_{\rho j}}{\partial p_{\sigma\tau}} \frac{\partial p_{\sigma\tau}}{\partial x_\mu}.$$

Consequently, as our condition is

$$\frac{df_{ij}}{dx_\alpha} - \frac{df_{i\alpha}}{dx_j} = 0,$$

we have

$$\frac{\partial f_{ij}}{\partial x_\alpha} - \frac{\partial f_{i\alpha}}{\partial x_j} + \sum_{\lambda=1}^{m} \left(f_{\lambda\alpha} \frac{\partial f_{ij}}{\partial z_\lambda} - f_{\lambda j} \frac{\partial f_{i\alpha}}{\partial z_\lambda} \right)$$

$$+ \sum_{\rho=1}^{m} \sum_{\mu=r+1}^{n} \left(\frac{\partial f_{ij}}{\partial p_{\rho\mu}} \frac{\partial f_{\rho\alpha}}{\partial x_\mu} - \frac{\partial f_{i\alpha}}{\partial p_{\rho\mu}} \frac{\partial f_{\rho j}}{\partial x_\mu} \right)$$

$$+ \sum_{\rho=1}^{m} \sum_{\mu=r+1}^{n} \sum_{\lambda=1}^{m} \left\{ \left(\frac{\partial f_{ij}}{\partial p_{\rho\mu}} \frac{\partial f_{\rho\alpha}}{\partial z_\lambda} - \frac{\partial f_{i\alpha}}{\partial p_{\rho\mu}} \frac{\partial f_{\rho j}}{\partial z_\lambda} \right) p_{\lambda\mu} \right\}$$

$$+ \sum_{\rho=1}^{m} \sum_{\mu=r+1}^{n} \sum_{\sigma=1}^{m} \sum_{\tau=r+1}^{n} \left\{ \left(\frac{\partial f_{ij}}{\partial p_{\rho\mu}} \frac{\partial f_{\rho\alpha}}{\partial p_{\sigma\tau}} - \frac{\partial f_{i\alpha}}{\partial p_{\rho\mu}} \frac{\partial f_{\rho j}}{\partial p_{\sigma\tau}} \right) \frac{\partial p_{\sigma\tau}}{\partial x_\mu} \right\} = 0.$$

Now

$$\frac{\partial p_{\sigma\tau}}{\partial x_\mu} = \frac{\partial p_{\sigma\mu}}{\partial x_\tau},$$

while μ and $\tau = r+1, \ldots, n$ in the last summation; and the preceding condition is to be satisfied, either identically or in connection with the equations of the original series

$$p_{ij} = f_{ij},$$

for $i = 1, \ldots, m$, and $j = 1, \ldots, r$. The quantities f_{ij} involve the quantities $p_{\lambda\mu}$, $p_{\sigma\tau}$, but they do not involve $\dfrac{\partial p_{\sigma\tau}}{\partial x_\mu}$; and no derivatives of the equations in the original system involve only derivatives of $p_{\sigma\tau}$ for $\tau = r+1, \ldots, n$, with respect to x_μ or only derivatives of $p_{\sigma\mu}$, for $\mu = r+1, \ldots, n$, with respect to x_τ, because such derivatives of the equations would introduce derivatives of $p_{\sigma\alpha}$, where α is less than $r+1$. The preceding condition therefore must be satisfied without the assistance of the equations of the original system: and therefore the relations

$$\frac{\partial f_{ij}}{\partial x_\alpha} - \frac{\partial f_{i\alpha}}{\partial x_j} + \sum_{\lambda=1}^{m} \left(f_{\lambda\alpha} \frac{\partial f_{ij}}{\partial z_\lambda} - f_{\lambda j} \frac{\partial f_{i\alpha}}{\partial z_\lambda} \right)$$
$$+ \sum_{\rho=1}^{m} \sum_{\mu=r+1}^{n} \left(\frac{\partial f_{ij}}{\partial p_{\rho\mu}} \frac{\partial f_{\rho\alpha}}{\partial x_\mu} - \frac{\partial f_{i\alpha}}{\partial p_{\rho\mu}} \frac{\partial f_{\rho j}}{\partial x_\mu} \right)$$
$$+ \sum_{\rho=1}^{m} \sum_{\mu=r+1}^{n} \sum_{\lambda=1}^{m} \left\{ \left(\frac{\partial f_{ij}}{\partial p_{\rho\mu}} \frac{\partial f_{\rho\alpha}}{\partial z_\lambda} - \frac{\partial f_{i\alpha}}{\partial p_{\rho\mu}} \frac{\partial f_{\rho j}}{\partial z_\lambda} \right) p_{\lambda\mu} \right\} = 0,$$

and

$$\sum_{\rho=1}^{m} \left(\frac{\partial f_{ij}}{\partial p_{\rho\mu}} \frac{\partial f_{\rho\alpha}}{\partial p_{\sigma\tau}} - \frac{\partial f_{i\alpha}}{\partial p_{\rho\mu}} \frac{\partial f_{\rho j}}{\partial p_{\sigma\tau}} + \frac{\partial f_{ij}}{\partial p_{\rho\tau}} \frac{\partial f_{\rho\alpha}}{\partial p_{\sigma\mu}} - \frac{\partial f_{i\alpha}}{\partial p_{\rho\tau}} \frac{\partial f_{\rho j}}{\partial p_{\sigma\mu}} \right) = 0,$$

the latter relation arising from the combination of the coefficients of the equal quantities $\frac{\partial p_{\sigma\tau}}{\partial x_\mu}$ and $\frac{\partial p_{\sigma\mu}}{\partial x_\tau}$, must be satisfied identically.

The first of these identical relations holds for

$$\alpha = j+1, \ldots, r; \quad i = 1, \ldots, m.$$

The second of these identical relations holds for

$$\alpha = j+1, \ldots, r; \quad i \text{ and } \sigma = 1, \ldots, m;$$
$$\mu \text{ and } \tau = r+1, \ldots, n;$$

the subscripts μ and τ may have the same value, in which case there is a superfluous factor 2; or they may have different values, and then only the pair of values from the series $r+1, \ldots, n$ need be taken. Lastly, as α is greater than j, the preceding tale of relations holds for

$$j = 1, \ldots, r-1.$$

Note 1. There are three extreme cases.

(i) Let $r=1$: there is no possible value of j, and so there are no conditions. In this case, we have a system of m equations in m dependent variables: they are of the form

$$\frac{\partial z_1}{\partial x_1} = \phi_1, \quad \frac{\partial z_2}{\partial x_1} = \phi_2, \ldots, \frac{\partial z_m}{\partial x_1} = \phi_m,$$

where $\phi_1, \ldots, \phi_m$ involve all the variables and all the derivatives except those on the left-hand sides of the equations. It is clear that such equations can coexist without the necessity of submitting $\phi_1, \ldots, \phi_m$ to conditions.

(ii) Let $m=1$, so that there is a single dependent variable: the system of equations is

$$p_j = f_j(z, x_1, \ldots, x_n, p_{r+1}, \ldots, p_n),$$

for $j=1, \ldots, r$, and thus it belongs to the type of Jacobian systems.

The first set of conditions is

$$\frac{\partial f_j}{\partial x_a} - \frac{\partial f_a}{\partial x_j} + f_a \frac{\partial f_j}{\partial z} - f_j \frac{\partial f_a}{\partial z}$$
$$+ \sum_{\mu=r+1}^{n} \left\{ \frac{\partial f_j}{\partial p_\mu} \left(\frac{\partial f_a}{\partial x_\mu} + p_\mu \frac{\partial f_a}{\partial z} \right) - \frac{\partial f_a}{\partial p_\mu} \left(\frac{\partial f_j}{\partial x_\mu} + p_\mu \frac{\partial f_j}{\partial z} \right) \right\} = 0,$$

for $\alpha = 1, \ldots, j-1$, and $j=1, \ldots, r$; and the second set of conditions, being

$$\frac{\partial (f_j, f_a)}{\partial (p_\mu, p_\tau)} + \frac{\partial (f_j, f_a)}{\partial (p_\tau, p_\mu)} = 0,$$

is evanescent. The aggregate of these conditions constitutes the aggregate for a complete Jacobian system.

(iii) Let $r=n$, so that there is a system of mn equations of the form

$$p_{ij} = f_{ij}(z_1, \ldots, z_m, x_1, \ldots, x_n),$$

for $i=1, \ldots, m$, and $j=1, \ldots, n$. As the functions f_{ij} involve no derivatives, the second set of conditions does not appear; and the first set becomes

$$\frac{\partial f_{ij}}{\partial x_a} - \frac{\partial f_{ia}}{\partial x_j} + \sum_{\lambda=1}^{m} \left(f_{\lambda a} \frac{\partial f_{ij}}{\partial z_\lambda} - f_{\lambda j} \frac{\partial f_{ia}}{\partial z_\lambda} \right) = 0,$$

for $i=1, \ldots, m$; $\alpha = 1, \ldots, j-1$; and $j=1, \ldots, n$. This is Mayer's system of completely integrable equations*.

Note 2. The first set of conditions is sufficient to secure that

$$\frac{\partial z_i}{\partial x_j} = p_{ij}, \text{ for } i=1, \ldots, m, \text{ and } j=1, \ldots, n;$$

$$\frac{\partial p_{i\rho}}{\partial x_\beta} = \frac{\partial p_{i\beta}}{\partial x_\rho}, \text{ for } i=1, \ldots, m;\ \beta=1, \ldots, n;\ \rho=1, \ldots, r;$$

and the second set of conditions is sufficient to secure that

$$\frac{\partial p_{i\tau}}{\partial x_\mu} = \frac{\partial p_{i\mu}}{\partial x_\tau},$$

for $i=1, \ldots, m$; μ and $\tau = r+1, \ldots, n$.

* They are discussed fully in vol. I of this work, §§ 34—42.

162. Suppose that all the conditions for complete integrability are satisfied; then the theorem is established, according to which the completely integrable system of rm equations possesses a set of integrals $z_1, \ldots, z_m$; when $x_1, \ldots, x_r$ are made equal to $c_1, \ldots, c_r$, these integrals become functions of $x_{r+1}, \ldots, x_n$ which, subject solely to the condition of being regular within an assigned domain, may be arbitrarily assumed. As already stated, it is necessary, in addition to the conditions for complete integrability, that the quantities f_{ij} should be regular functions of their arguments within the domains considered.

Moreover, the argument shews that, in order to obtain the integrals required, it is necessary to integrate an aggregate of m equations. In practice, instead of beginning with a selected aggregate, it is convenient to effect Mayer's transformation adopted (§ 43, Note 1) for a complete Jacobian system in a single dependent variable. For this purpose, we write

$$x_1 = y_1,$$
$$x_2 - a_2 = (y_1 - a_1)\, y_2,$$
$$\ldots\ldots\ldots\ldots\ldots\ldots$$
$$x_r - a_r = (y_1 - a_1)\, y_r,$$

leaving the other variables unaltered: then, taking

$$\frac{\partial z_i}{\partial y_j} = p'_{ij},$$

for $i = 1, \ldots, m$, and $j = 1, \ldots, r$, we have an equivalent set of equations in the form

$$p'_{i1} = f_{i1} + y_2 f_{i2} + \ldots + y_r f_{ir} = g_{i1},$$
$$p'_{i\rho} = (y_1 - a_1) f_{i\rho},$$

for $\rho = 2, \ldots, r$, and $i = 1, \ldots, n$. The first aggregate is

$$p'_{i1} = g_{i1},$$

for $i = 1, \ldots, n$: suppose it possible to obtain a set of integrals of this set of m equations such that, when $y_1 = a_1$, the integrals become functions of $x_{r+1}, \ldots, x_n$ only. These integrals satisfy the other equations, by the preceding argument: and as regards initial conditions for those equations, we see that

$$p'_{i\rho} = 0, \qquad (\rho = 2, \ldots, r),$$

when $y_1 = a_1$, that is, when $y_1 = a_1$, the integrals are not to involve $y_2, \ldots, y_n$—a set of conditions actually satisfied by the form of the functions assigned to the integrals when $y_1 = a_1$.

Hence the integration of a set of rm completely integrable equations of the resolved type indicated can be effected, if the integration of a set of only m equations of that resolved type can be effected.

Different kinds of Integrals: their Relations.

163. Before proceeding to the discussion of systematic attempts at the integration of simultaneous equations involving more than one dependent variable, it is worth noting that two kinds of integrals of simultaneous equations have been indicated. In one kind, the arbitrary element consists of arbitrary constants; in the other, it consists of arbitrary functions introduced through initial conditions. We have seen that, in the case of equations in a single dependent variable, it is possible to connect the two kinds of integrals organically; and it is natural to inquire whether the organic relation can be extended to the integrals of systems of equations involving several dependent variables.

It has appeared that, in a general sense, mn is the greatest number of arbitrary constants that can be eliminated from a system of m integral equations, in m dependent and n independent variables, so as to lead to m partial differential equations of the first order: and conversely, it is natural to expect, also in a general sense, that mn is the greatest number of arbitrary constants that can occur in the integral equivalent of the system of m partial differential equations. On the analogy of the case when $m=1$, such an integral involving this greatest number mn of arbitrary constants is called the *complete integral.* Let mn be denoted by μ, and suppose that the integral equations are resolved so as to express each of the dependent variables explicitly in terms of the independent variables and the arbitrary constants: the complete integral then* is of the form

$$\left.\begin{array}{c} z_1 = g_1(x_1, \dots, x_n, a_1, \dots, a_\mu) \\ \dots\dots\dots\dots\dots\dots\dots\dots \\ z_m = g_m(x_1, \dots, x_n, a_1, \dots, a_\mu) \end{array}\right\}.$$

To deduce other integrals, if possible, from the complete integral, let the customary Lagrangian method be adopted. The

* The following discussion is partly based upon that which is given by Königsberger, *Crelle*, t. CIX (1892), pp. 303 et seq.

quantities $a_1, \ldots, a_\mu$ are made variable, subject to the condition that the m differential equations are unaltered in form; and this condition will be satisfied if the expressions for the values of the derivatives p_{ij}, for $i=1, \ldots, m$, and $j=1, \ldots, n$, are unaltered. Hence we must have

$$\sum_{r=1}^{\mu} \frac{\partial g_i}{\partial a_r} \frac{\partial a_r}{\partial x_j} = 0,$$

for the specified values of i and j; and these are the equations, mn in number, to be satisfied by the mn quantities a. As these quantities $a_1, \ldots, a_\mu$ are variable, being functions of $x_1, \ldots, x_n$, we may take n of them as equivalent to $x_1, \ldots, x_n$; let these be $a_1, \ldots, a_n$, and let the remainder $a_{n+1}, \ldots, a_\mu$ be regarded as functions of $a_1, \ldots, a_n$. The preceding equations now become

$$\sum_{r=1}^{n} \left[\frac{\partial a_r}{\partial x_j} \left\{ \frac{\partial g_i}{\partial a_r} + \sum_{s=n+1}^{\mu} \left(\frac{\partial g_i}{\partial a_s} \frac{\partial a_s}{\partial a_r} \right) \right\} \right] = 0,$$

for $i = 1, \ldots, m$, and $j = 1, \ldots, n$.

Out of this set of mn equations, let that aggregate be selected which is obtained by taking one value of i and all the n values of j; it is

$$\left.\begin{array}{l} P_{i1} \dfrac{\partial a_1}{\partial x_1} + P_{i2} \dfrac{\partial a_2}{\partial x_1} + \ldots + P_{in} \dfrac{\partial a_n}{\partial x_1} = 0 \\ P_{i1} \dfrac{\partial a_1}{\partial x_2} + P_{i2} \dfrac{\partial a_2}{\partial x_2} + \ldots + P_{in} \dfrac{\partial a_n}{\partial x_2} = 0 \\ \ldots\ldots\ldots\ldots\ldots\ldots\ldots\ldots\ldots\ldots \\ P_{i1} \dfrac{\partial a_1}{\partial x_n} + P_{i2} \dfrac{\partial a_2}{\partial x_n} + \ldots + P_{in} \dfrac{\partial a_n}{\partial x_n} = 0 \end{array}\right\},$$

where

$$P_{ij} = \frac{\partial g_i}{\partial a_j} + \sum_{s=n+1}^{\mu} \left(\frac{\partial g_i}{\partial a_s} \frac{\partial a_s}{\partial a_j} \right).$$

The n equations, contained in this aggregate, are homogeneous and linear in the quantities $P_{i1}, P_{i2}, \ldots, P_{in}$; hence, either

$$P_{i1} = 0, \quad P_{i2} = 0, \ldots, P_{in} = 0,$$

or else

$$\frac{\partial (a_1, \ldots, a_n)}{\partial (x_1, \ldots, x_n)} = 0.$$

The latter result implies a functional relation between $a_1, \ldots, a_n$ without the intervention of quantities other than pure constants: let it be

$$a_n = F(a_1, \ldots, a_{n-1}).$$

Multiply the n equations in the aggregate by $dx_1, \ldots, dx_n$ respectively, and add: then

$$P_{i1}da_1 + P_{i2}da_2 + \ldots + P_{in}da_n = 0,$$

so that, in conjunction with $a_n = F(a_1, \ldots, a_{n-1})$, we must have

$$P_{ij} + P_{in}\frac{\partial F}{\partial a_j} = 0,$$

for $j = 1, \ldots, n-1$, and $i = 1, \ldots, m$. There are thus two ways of satisfying the selected aggregate of equations: either

$$P_{i1} = 0, \quad P_{i2} = 0, \ldots, P_{in} = 0,$$

or

$$\left.\begin{aligned} a_n &= F(a_1, \ldots, a_{n-1}) \\ 0 &= P_{ij} + P_{in}\frac{\partial F}{\partial a_j} \end{aligned}\right\},$$

where $j = 1, \ldots, n-1$ in the latter set. And then, taking all the aggregates which can thus be selected so that we use the full set of mn equations, we see that the two sets of equations in virtue of which the mn equations can be satisfied are (i) the system of equations

$$P_{ij} = 0,$$

for $i = 1, \ldots, m$, and $j = 1, \ldots, n$; and (ii) the system of equations

$$a_n = F(a_1, \ldots, a_{n-1}),$$

$$0 = P_{ij} + P_{in}\frac{\partial F}{\partial a_j},$$

for $i = 1, \ldots, m$, and $j = 1, \ldots, n$.

The alternatives must be considered separately.

164. I. For the first alternative, we have the system of mn equations

$$P_{ij} = 0,$$

where

$$P_{ij} = \frac{\partial g_i}{\partial a_j} + \sum_{s=n+1}^{\mu}\left(\frac{\partial g_i}{\partial a_s}\frac{\partial a_s}{\partial a_j}\right).$$

These equations contain the variables $x_1, \ldots, x_n$ and the mn parameters $a_1, \ldots, a_\mu$. When the variables are eliminated, $n(m-1)$ equations survive as the eliminant; and these are partial differential equations of the first order, in which $a_{n+1}, \ldots, a_\mu$ are $n(m-1)$ dependent variables and $a_1, \ldots, a_n$ are n independent variables. Now

$$n(m-1) > m,$$

except in three cases; in the first case of exception, $m=1$, and then we have the usual Jacobian theory of partial differential equations in a single dependent variable; in the second case of exception, $n=1$, and then we have a system of ordinary equations; the third case of exception is given by the equality of the numbers $n(m-1)$ and m, and then the only possible values are $n=2$, $m=2$.

Hence, when $m \geqslant 2$ and $n \geqslant 2$, the derivation of the values of $a_{n+1}, \ldots, a_\mu$ through $P_{ij}=0$ exacts the integration of a system of simultaneous partial equations in a number of dependent variables greater than the number in the original system of equations; and therefore the process of deducing new integrals from the complete integral by means of the equations $P_{ij}=0$ is of more elaborate extent than the process of integrating the original system. There is one exception to this result, and it is given by $m=2$, $n=2$; in that case, the two processes are of the same degree of difficulty.

II. For the second alternative, we have the $m(n-1)$ equations

$$P_{ij} + P_{in}\frac{\partial F}{\partial a_j} = 0,$$

for $j=1, \ldots, n-1$, and $i=1, \ldots, m$, together with the relation

$$a_n = F(a_1, \ldots, a_{n-1}).$$

As $a_1, \ldots, a_n$ are now connected by a relation, they are no longer eligible as a set of n independent quantities. We therefore choose some other set, say $a_{n+1}, a_1, \ldots, a_{n-1}$, as the n independent quantities equivalent to $x_1, \ldots, x_n$; and, instead of using the $m(n-1)$ equations associated with $a_n=F$, we return to the equations, which secure the absence of change of form in the derivatives and therefore conserve the form of the differential equations. The quantities $a_n, a_{n+2}, a_{n+3}, \ldots, a_\mu$ are now functions of $a_1, \ldots, a_{n-1}, a_{n+1}$: so that, writing

$$Q_{ij} = \frac{\partial g_i}{\partial a_j} + \frac{\partial g_i}{\partial a_n}\frac{\partial F}{\partial a_j} + \sum_{s=n+2}^{\mu} \frac{\partial g_i}{\partial a_s}\frac{\partial a_s}{\partial a_j},$$

for $j=1, \ldots, n-1$, and

$$Q_{ij'} = \frac{\partial g_i}{\partial a_{j'}} + \sum_{s=n+2}^{\mu} \frac{\partial g_i}{\partial a_s}\frac{\partial a_s}{\partial a_{j'}},$$

for $j' = n+1$, we have these equations in the form

$$\sum_{j=1}^{n-1} Q_{ij} \frac{\partial a_j}{\partial x_k} + Q_{ij'} \frac{\partial a_{j'}}{\partial x_k} = 0,$$

for $i = 1, \ldots, m$, and $k = 1, \ldots, n$.

Proceeding as before, these equations can be satisfied in two different ways. We may have

$$Q_{ij} = 0,$$

for $i = 1, \ldots, m$, and $j = 1, \ldots, n-1, n+1$, being a system of mn equations; or we may have a relation of the form

$$a_{n+1} = G(a_1, \ldots, a_{n-1}),$$

coupled with the equations

$$Q_{ij} + Q_{ij'} \frac{\partial G}{\partial a_j} = 0,$$

where $j' = n+1$; $j = 1, \ldots, n-1$; $i = 1, \ldots, m$.

In the former case, we eliminate $x_1, \ldots, x_n$ from the system of mn equations: when the elimination has been effected, there remain $n(m-1)$ differential equations in the n independent variables $a_1, \ldots, a_{n-1}, a_{n+1}$ and the $n(m-1)-1$ dependent variables $a_{n+2}, a_{n+3}, \ldots, a_\mu$, the value of a_n being already known. As the number of equations is greater than the number of dependent variables by unity, and as the equations are formally independent of one another, the system can coexist only if conditions are satisfied: it will not unconditionally determine the dependent variables.

In the latter case, the quantities $a_1, \ldots, a_{n-1}, a_{n+1}$, being connected by a relation, are not eligible as independent variables; we proceed to choose a set of quantities independent of one another as equivalent to $x_1, \ldots, x_n$, say $a_1, \ldots, a_{n-1}, a_{n+2}$, and construct the corresponding equations. The equations can be satisfied, as before, in two ways: either by a system of mn equations which, on the elimination of $x_1, \ldots, x_n$, give a set of $n(m-1)$ differential equations, involving $n(m-1)-2$ dependent variables and therefore not unconditionally determining those variables: or by a relation

$$a_{n+2} = H(a_1, \ldots, a_{n-1}),$$

with associated equations.

Pursuing the latter alternative to the extreme end, we have, at that end, a number of relations

$$a_s = F_s(a_1, \ldots, a_{n-1}),$$

for $s = n, n+1, \ldots, \mu$. Writing

$$T_{ij} = \frac{\partial g_i}{\partial a_j} + \sum_{s=n}^{\mu} \frac{\partial g_i}{\partial a_s}\frac{\partial F_s}{\partial a_j},$$

for $i = 1, \ldots, m$, and $j = 1, \ldots, n-1$, the equations to be associated with the $n(m-1)+1$ relations are

$$T_{ij} = 0.$$

These $m(n-1)$ equations, together with the $n(m-1)+1$ relations, make up $2mn - m - n + 1$ equations to be associated with the original m equations: they are more in number than the mn unknown quantities a; and therefore they cannot unconditionally be satisfied, for they are formally independent.

Summing up the various results we see that, *except in the single case $m = 2$, $n = 2$, it is not possible with unconditioned equations to use the Lagrangian process of variation of constants for the derivation of integrals from the complete integral without integrating a set of differential equations of more elaborate extent than the original system: when this set can be integrated, the integrals thus provided are, in general, the only integrals that can be derived from the complete integral. In the case of exception, the equations to be integrated are of the same order of difficulty as the original system.*

165. It is perhaps superfluous to point out that such integrals as can be obtained may belong to various classes. When they are derived through the $n(m-1)$ partial differential equations which determine the $n(m-1)$ quantities $a_{n+1}, \ldots, a_\mu$ in terms of a and b, there will be as many kinds of integrals of the original equations thus provided as there are different types of integrals of these new equations. Thus some integrals will involve arbitrary functional forms: these will correspond to one or other of the classes of general integrals. We know already that all the demands are satisfied by having $a_1, \ldots, a_\mu$ constant: we then have the complete integral. There may be equations of intermediate types in which some arbitrary functional forms occur and, at the same time, some of the varied arbitrary constants may survive merely as constants, by arising as trivial constant integrals satisfying the partial differential equations.

Ex. 1. Such systematic processes as have been devised for the integration of particular systems of simultaneous equations will be discussed almost immediately. Meanwhile, the special case can be illustrated by an example given* by Königsberger in the form

$$yp_1 - xq_2 = 0, \quad xq_1 - yp_2 = 0,$$

where z_1, z_2 are the dependent variables, x, y are the independent variables, and

$$\frac{\partial z_1}{\partial x} = p_1, \quad \frac{\partial z_1}{\partial y} = q_1, \quad \frac{\partial z_2}{\partial x} = p_2, \quad \frac{\partial z_2}{\partial y} = q_2.$$

The actual integration of the equations happens to be easy. We have

$$\begin{aligned} dz_1 &= p_1 dx + q_1 dy \\ &= \frac{x}{y} q_2 dx + \frac{y}{x} p_2 dy ; \end{aligned}$$

the right hand side must be a perfect differential, so that

$$\frac{\partial}{\partial x}\left(\frac{y}{x} p_2\right) = \frac{\partial}{\partial y}\left(\frac{x}{y} q_2\right),$$

that is,

$$\frac{1}{x}\frac{\partial}{\partial x}\left(\frac{1}{x}\frac{\partial z_2}{\partial x}\right) = \frac{1}{y}\frac{\partial}{\partial y}\left(\frac{1}{y}\frac{\partial z_2}{\partial y}\right),$$

and therefore

$$z_2 = F(x^2 + y^2) - G(x^2 - y^2),$$

where f and g are arbitrary functions of their arguments. It is then easy to deduce, by means of the values of p_1 and q_1, that the value of z_1 is

$$z_1 = F(x^2 + y^2) + G(x^2 - y^2).$$

Now a set of integrals containing four arbitrary constants (and therefore constituting a complete integral) is

$$\left.\begin{aligned} z_1 &= \alpha + ax^2 + by^2 \\ z_2 &= \beta + bx^2 + ay^2 \end{aligned}\right\},$$

where a, b, α, β are arbitrary constants. To deduce other integrals if possible, we make α, β, a, b variable quantities, being functions of x and y; and, in the first place, we make a and b equivalent to x and y, and can therefore use them as independent variables. Then if

$$A_1 = \frac{\partial \alpha}{\partial a} + x^2, \quad A_2 = \frac{\partial \beta}{\partial a} + y^2,$$

$$B_1 = \frac{\partial \alpha}{\partial b} + y^2, \quad B_2 = \frac{\partial \beta}{\partial b} + x^2,$$

the values of p_1, q_1, p_2, q_2 will be unaltered if

$$A_1 \frac{\partial a}{\partial x} + B_1 \frac{\partial b}{\partial x} = 0, \quad A_2 \frac{\partial a}{\partial x} + B_2 \frac{\partial b}{\partial x} = 0,$$

$$A_1 \frac{\partial a}{\partial y} + B_1 \frac{\partial b}{\partial y} = 0, \quad A_2 \frac{\partial a}{\partial y} + B_2 \frac{\partial b}{\partial y} = 0,$$

* *Crelle*, t. CIX, p. 319: the integral selected is simpler than Königsberger's.

which are equations for the determination of a and b. They can be satisfied in various ways.

(i) They are satisfied if

$$\frac{\partial a}{\partial x}=0,\quad \frac{\partial a}{\partial y}=0,\quad \frac{\partial b}{\partial x}=0,\quad \frac{\partial b}{\partial y}=0:$$

then a, b (and therefore α and β, which are functions of a and b) are constants. We return to the complete integral.

(ii) They are satisfied if

$$A_1=0,\quad B_1=0,\quad A_2=0,\quad B_2=0.$$

From $A_1=0$, $B_2=0$, we have

$$\frac{\partial \alpha}{\partial a}=\frac{\partial \beta}{\partial b};$$

and from $B_1=0$, $A_2=0$, we have

$$\frac{\partial \alpha}{\partial b}=\frac{\partial \beta}{\partial a}.$$

Hence

$$\frac{\partial^2 \alpha}{\partial a^2}-\frac{\partial^2 \alpha}{\partial b^2}=0,$$

so that

$$\alpha=f(a+b)+g(a-b),$$

where f and g are arbitrary functions of their arguments; and then it easily follows that

$$\beta=f(a+b)-g(a-b).$$

Also from $A_1=0$, $B_1=0$, we now have

$$-x^2=\frac{\partial \alpha}{\partial a}=f'(a+b)+g'(a-b),$$

$$-y^2=\frac{\partial \alpha}{\partial b}=f'(a+b)-g'(a-b),$$

so that

$$-(x^2+y^2)=2f'(a+b),\quad -(x^2-y^2)=2g'(a-b).$$

Hence $a+b$ is an arbitrary function of x^2+y^2, and $a-b$ is an arbitrary function of x^2-y^2, say

$$a+b=\theta(x^2+y^2),\quad a-b=\phi(x^2-y^2).$$

Thus

$$\begin{aligned}
z_1&=\alpha+ax^2+by^2\\
&=f\{\theta(x^2+y^2)\}+g\{\phi(x^2-y^2)\}+\tfrac{1}{2}(x^2+y^2)\,\theta(x^2+y^2)+\tfrac{1}{2}(x^2-y^2)\,\phi(x^2-y^2)\\
&=\Theta(x^2+y^2)+\Phi(x^2-y^2),\\
z_2&=\beta+bx^2+ay^2\\
&=f\{\theta(x^2+y^2)\}-g\{\phi(x^2-y^2)\}+\tfrac{1}{2}(x^2+y^2)\,\theta(x^2+y^2)-\tfrac{1}{2}(x^2-y^2)\,\phi(x^2-y^2)\\
&=\Theta(x^2+y^2)-\Phi(x^2-y^2),
\end{aligned}$$

where Θ and Φ are arbitrary functions of their arguments.

(iii) The equations necessary in order that the forms of p_1, q_1, p_2, q_2 may be conserved can be satisfied by

$$\frac{\partial(a, b)}{\partial(x, y)} = 0,$$

that is, by

$$b = h(a),$$

where h is any function of its argument, together with

$$A_1 + B_1 \frac{db}{da} = 0,$$

$$A_2 + B_2 \frac{db}{da} = 0.$$

As a and b are not now independent, we return to the equations; and we make α and a the independent variables.

Proceeding as before, we find that the alternative to the integral obtained in the last case is

$$\alpha = k(a),$$

where k is any function of its argument, with associated equations.

To deal with the latter alternative, we again return to the initial equations; and we make β and a the independent variables. A similar process leads to the result that the alternative to the integral already obtained is given by

$$\beta = l(a),$$

with the associated equations.

We thus have

$$b = h(a), \quad \alpha = k(a), \quad \beta = l(a);$$

and the associated equations are

$$\left. \begin{array}{l} k'(a) + x^2 + y^2 h'(a) = 0 \\ l'(a) + x^2 h'(a) + y^2 = 0 \end{array} \right\}.$$

These can coexist only in two cases. In the first case,

$$h'(a) = 1,$$

and

$$k'(a) = l'(a) = -(x^2 + y^2);$$

the corresponding integrals are

$$z_1 = \Omega(x^2 + y^2), \quad z_2 = \Omega(x^2 + y^2),$$

being particular forms of the integrals already obtained. In the second case,

$$h'(a) = -1,$$

$$k'(a) = -l'(a) = -(x^2 - y^2);$$

the corresponding integrals are

$$z_1 = \Psi(x^2 - y^2), \quad z_2 = -\Psi(x^2 - y^2),$$

being particular forms of the integrals already obtained.

Hence the most general integrals that can thus be derived from the complete integrals are

$$z_1 = \Theta(x^2+y^2) + \Phi(x^2-y^2),$$
$$z_2 = \Theta(x^2+y^2) - \Phi(x^2-y^2).$$

Ex. 2. Generalise similarly the integrals

$$z_1 = \quad \alpha + (\alpha+\gamma)x^2 + (\alpha-\gamma)y^2 + \beta(x^2+y^2)^2,$$
$$z_2 = -\alpha + (\alpha-\gamma)x^2 + (\alpha+\gamma)y^2 + \beta(x^2+y^2)^2,$$

of the same equations

$$yp_1 - xq_2 = 0, \quad xq_1 - yp_2 = 0.$$

(Königsberger.)

Ex. 3. Construct the differential equations of the first order satisfied by

$$z_1 = a^2x + by + c^2xy + kx^2,$$
$$z_2 = ay + b^2x + cx + k^2y,$$

where a, b, c, k are arbitrary constants.

Generalise the integrals so as to deduce others from this complete integral. In particular, shew that another integral is given by keeping c and k constant, and by making a and b functions of x and y such that

$$\left.\begin{aligned} 4abx^2 &= y^2 \\ 8a^3 - \alpha a^{\frac{3}{2}} + \frac{y^3}{x^3} &= 0 \end{aligned}\right\},$$

where α is an arbitrary constant. (Königsberger.)

Hamburger's Linear Equations.

166. It has appeared, from the discussion of the classes of simultaneous equations already considered, that the construction of a system of integrals can be made to depend on the construction of the integrals of a set of simultaneous equations the number of which is the same as the number of dependent variables involved. The only indication of any systematic method of obtaining the integrals is furnished in the proof of the existence theorem; they are obtained in the form of converging power-series in the independent variables. What is usually desired for the purpose is an expression for the integrals in some form more compact than multiple power-series.

A method, which has been found effective for a limited number of classes of equations, has been devised* by Hamburger.

* *Crelle*, t. LXXXI (1876), pp. 243—280, the equations being linear in the derivatives of the dependent variables; *ib.*, t. XCIII (1882), pp. 188—214, the equations not being necessarily linear in those derivatives.

See also a paper by Königsberger, *Math. Ann.*, t. XLI (1893), pp. 260—285.

Denoting the independent variables by $x_1, \ldots, x_n$, the dependent variables by $z_1, \ldots, z_m$, and the derivatives of the dependent variables by p_{ij} as before, where

$$p_{ij} = \frac{\partial z_i}{\partial x_j},$$

we first consider a set of m algebraically independent equations which are linear in the derivatives. We also assume that they can be resolved so as to express the derivatives of the m dependent variables with regard to one and the same independent variable: let this variable be x_1, so that the system may be taken in the form

$$p_{i1} = \pi_i + \sum_{j=1}^{m} \sum_{s=2}^{n} p_{js}\theta_{jsi},$$

for $i = 1, \ldots, m$: the quantities π_i and θ_{jsi}, for all values of i, j, s, are functions of the variables $z_1, \ldots, z_m, x_1, \ldots, x_n$. Multiplying the equations by $\lambda_1, \ldots, \lambda_m$, a set of provisionally indeterminate multipliers, and adding, we have

$$\sum_{i=1}^{m} \lambda_i\pi_i - \sum_{i=1}^{m} \lambda_i p_{i1} + \sum_{j=1}^{m} \sum_{s=2}^{n} p_{js}\left(\sum_{i=1}^{m} \lambda_i\theta_{jsi}\right) = 0.$$

The values of the derivatives must be such that the differential relations

$$dz_i - p_{i1}dx_1 - p_{i2}dx_2 - \ldots - p_{in}dx_n = 0,$$

for $i = 1, \ldots, m$, must be satisfied: consequently, the relation

$$\sum_{i=1}^{m} \lambda_i dz_i - dx_1 \sum_{i=1}^{n} \lambda_i p_{i1} - \sum_{j=1}^{m} \sum_{s=2}^{n} p_{js}\lambda_j dx_s = 0$$

also must be satisfied. Comparing this differential relation with the preceding composite equation and having regard to the ordinary subsidiary equations constructed in connection with a single partial differential equation, we construct the set

$$\frac{\sum_{i=1}^{m} \lambda_i dz_i}{\sum_{i=1}^{m} \lambda_i \pi_i} = dx_1 = \frac{-\lambda_j dx_s}{\sum_{i=1}^{m} \lambda_i \theta_{jsi}}$$

of ordinary equations, to hold for all values of j and s.

In this system of ordinary equations, let

$$\frac{dx_s}{dx_1} = \mu_s,$$

so that

$$\sum_{i=1}^{m} \lambda_i \theta_{jsi} + \lambda_j \mu_s = 0,$$

for all values of j and s. Selecting those of the equations given by one value of s and the m values of j, we can eliminate $\lambda_1, \ldots, \lambda_m$ determinantally; and we obtain an equation satisfied by μ_s. For each root μ_s of this equation we obtain a set of ratios $\lambda_1 : \lambda_2 : \ldots : \lambda_m$, the values of these ratios depending upon the coefficients θ_{jsi}, where s is the same for all the coefficients in the tableau.

If there be more than one value of s, say if σ be another value, then certain combinations of the coefficients $\theta_{j\sigma i}$ must be the same as those combinations of the coefficients θ_{jsi}, in order to secure the same values for the ratios $\lambda_1 : \lambda_2 : \ldots : \lambda_m$. This requirement would impose conditions upon the equations which would not, in general, be satisfied: though it might be of interest to construct classes of equations for which the appropriate conditions are satisfied, we shall assume that our equations are not thus conditioned. Accordingly, there will be only one value of s, say $s = 2$.

167. Thus for the present purpose, we restrict ourselves to the consideration of equations in two independent variables, which will be denoted by x and y. Writing

$$p_i = \frac{\partial z_i}{\partial x}, \quad q_i = \frac{\partial z_i}{\partial y},$$

we may take the equations in the form

$$p_i = \pi_i + \sum_{s=1}^{m} a_{is} q_s,$$

for $i = 1, \ldots, m$. In connection with the differential relations

$$dz_i - p_i dx - q_i dy = 0,$$

we form the set of ordinary equations

$$\frac{\sum_{i=1}^{m} \lambda_i dz_i}{\sum_{i=1}^{m} \lambda_i \pi_i} = dx = -\frac{\lambda_s dy}{\sum_{i=1}^{m} \lambda_i a_{is}},$$

for $s = 1, \ldots, m$. Take

$$dy = \mu dx,$$

so that

$$\sum_{i=1}^{m} \lambda_i a_{is} + \mu\lambda_s = 0,$$

for $s = 1, \ldots, m$, and eliminate the quantities $\lambda_1, \ldots, \lambda_m$ among these equations: then μ satisfies the equation

$$\Theta = \begin{vmatrix} a_{11}+\mu, & a_{21}\;, & a_{31}\;, & \ldots, & a_{m1} \\ a_{12}\;, & a_{22}+\mu, & a_{32}\;, & \ldots, & a_{m2} \\ a_{13}\;, & a_{23}\;, & a_{33}+\mu, & \ldots, & a_{m3} \\ \ldots & \ldots & \ldots & \ldots & \ldots \\ a_{1m}\;, & a_{2m}\;, & a_{3m}\;, & \ldots, & a_{mm}+\mu \end{vmatrix} = 0,$$

which is of order m.

When μ is a simple root of this equation $\Theta = 0$, the preceding equations determine a unique set of values of $\lambda_1 : \lambda_2 : \ldots : \lambda_m$; then, when we take

$$\frac{\lambda_i}{\sum_{i=1}^{m} \lambda_i \pi_i} = \alpha_i, \qquad (i = 1, \ldots, m),$$

the quantities $\alpha_1, \ldots, \alpha_m$ are uniquely determinate. Hence we have

$$\left.\begin{aligned} \alpha_1 dz_1 + \ldots + \alpha_m dz_m &= dx \\ dy &= \mu dx \end{aligned}\right\};$$

that is, there are two linear equations for each simple root of $\Theta = 0$.

Next, let μ be a multiple root of the equation $\Theta = 0$, and suppose that it occurs θ times; then t of the m equations that lead to $\Theta = 0$ are deducible from the remainder, where

$$1 \leqslant t \leqslant \theta \leqslant m.$$

The ratios $\lambda_1 : \lambda_2 : \ldots : \lambda_m$ are no longer determinate: the quantities $\lambda_1, \lambda_2, \ldots, \lambda_m$ are expressible in terms of t arbitrary quantities $\kappa_1, \ldots, \kappa_t$ by equations of the form

$$\lambda_r = \gamma_{r1}\kappa_1 + \ldots + \gamma_{rt}\kappa_t,$$

for $r = 1, \ldots, m$, the quantities $\gamma_{r1}, \ldots, \gamma_{rt}$ being determinate. The equation

$$\sum_{r=1}^{m} \lambda_r dz_r = \left(\sum_{r=1}^{m} \lambda_r \pi_r\right) dx$$

now becomes

$$\sum_{r=1}^{m}\sum_{i=1}^{t}(\gamma_{ri}dz_r)\,\kappa_i=\sum_{r=1}^{m}\sum_{i=1}^{t}(\gamma_{ri}\pi_r)\,\kappa_i dx\,;$$

and therefore, as the quantities $\kappa_1, \ldots, \kappa_t$ are arbitrary, we have the $t+1$ linear equations

$$dy=\mu\,dx,$$

$$\sum_{r=1}^{m}\gamma_{ri}dz_r=\left(\sum_{r=1}^{m}\gamma_{ri}\pi_r\right)dx,$$

for $i=1, \ldots, t$.

168. The extreme case among multiple roots is that in which

$$\theta=m:$$

the quantity Θ is then the mth power of a linear factor: the coefficients $a_{11}, a_{22}, \ldots, a_{mm}$ have a common value, say α; and all the coefficients a_{ij}, for unequal values of i and j, are zero. The equations are

$$p_i=\pi_i+\alpha q_i\,;$$

and the associated subsidiary equations are

$$\frac{dx}{1}=\frac{dy}{-\alpha}=\frac{dz_1}{\pi_1}=\ldots=\frac{dz_m}{\pi_m}.$$

The equations in this extreme case were considered by Jacobi*, independently of the preceding mode of origin: he approached them from the stage of integral relations with any number of dependent variables and any number of independent variables, and the particular set just given are his equations when there are two independent variables.

When μ is a multiple root of order θ, the most important case is that in which $t=\theta$; the conditions† are connected with the elementary divisors (or elementary factors) of the determinant Θ. They will not be set out in detail because the method devised by Hamburger will be sufficiently illustrated for the equations, to which it can be applied, by a discussion of the simplest cases.

The integral of the Jacobian set can easily be obtained. Let $u_1, \ldots, u_{m+1}$ be a complete set of independent integrals of the $m+1$ ordinary equations

$$\frac{dx}{1}=\frac{dy}{-\alpha}=\frac{dz_1}{\pi_1}=\ldots=\frac{dz_m}{\pi_m},$$

* *Ges. Werke*, t. IV, p. 7; *Crelle*, t. II (1827), p. 321.

† For this theory, see Weierstrass, *Ges. Werke*, t. II, pp. 19—44. Other references are given in vol. IV of the present work, p. 42, foot-note.

which are subsidiary to the system

$$p_i = \pi_i + \alpha q_i,$$

for $i = 1, \ldots, m$: then the differential relation

$$\frac{\partial u_r}{\partial x} dx + \frac{\partial u_r}{\partial y} dy + \frac{\partial u_r}{\partial z_1} dz_1 + \ldots + \frac{\partial u_r}{\partial z_m} dz_m = 0$$

is satisfied in virtue of the above set of ordinary equations, and consequently

$$\frac{\partial u_r}{\partial x} - \alpha \frac{\partial u_r}{\partial y} + \pi_1 \frac{\partial u_r}{\partial z_1} + \ldots + \pi_m \frac{\partial u_r}{\partial z_m} = 0,$$

holding for $r = 1, \ldots, m+1$.

Now take any equation

$$\psi_i(u_1, \ldots, u_{m+1}) = 0,$$

regarded as one of a system of m equations to express m dependent variables $z_1, \ldots, z_m$, in terms of x and y. We have

$$\sum_{r=1}^{m+1} \frac{\partial \psi_i}{\partial u_r} \left\{ \frac{\partial u_r}{\partial x} + p_1 \frac{\partial u_r}{\partial z_1} + \ldots + p_m \frac{\partial u_r}{\partial z_m} \right\} = 0,$$

$$\sum_{r=1}^{m+1} \frac{\partial \psi_i}{\partial u_r} \left\{ \frac{\partial u_r}{\partial y} + q_1 \frac{\partial u_r}{\partial z_1} + \ldots + q_m \frac{\partial u_r}{\partial z_m} \right\} = 0;$$

multiplying the latter by α, subtracting from the former, and using the partial equation satisfied by the quantity u_r, we have

$$\sum_{r=1}^{m+1} \frac{\partial \psi_i}{\partial u_r} \left\{ (p_1 - \pi_1 - \alpha q_1) \frac{\partial u_r}{\partial z_1} + \ldots + (p_m - \pi_m - \alpha q_m) \frac{\partial u_r}{\partial z_m} \right\} = 0.$$

Now, when we write

$$\frac{d\psi_i}{dz_s} = \sum_{r=1}^{m+1} \frac{\partial \psi_i}{\partial u_r} \frac{\partial u_r}{\partial z_s},$$

so that $\frac{d\psi_i}{dz_s}$ is the complete derivative of ψ_i with regard to z_s, this equation is

$$(p_1 - \pi_1 - \alpha q_1) \frac{d\psi_i}{dz_1} + \ldots + (p_m - \pi_m - \alpha q_m) \frac{d\psi_i}{dz_m} = 0.$$

Accordingly, take m equations

$$\psi_i(u_1, \ldots, u_{m+1}) = 0,$$

for $i=1, \dots, m$, independent of one another and determining* the m quantities $z_1, \dots, z_m$; then the equation

$$(p_1-\pi_1-\alpha q_1)\frac{d\psi_i}{dz_1}+\dots+(p_m-\pi_m-\alpha q_m)\frac{d\psi_i}{dz_m}=0$$

is satisfied for each of the values $i=1, \dots, m$. Also as the equations are independent of one another and determine the quantities $z_1, \dots, z_m$, the determinant of the quantities $\frac{d\psi_i}{dz_j}$, for i and $j=1, \dots, m$, does not vanish: consequently

$$p_j-\pi_i-\alpha q_j=0,$$

for $j=1, \dots, m$.

Hence the m equations

$$\psi_i(u_1, \dots, u_{m+1})=0,$$

for $i=1, \dots, m$, give integrals of the Jacobian set; and this is true however arbitrary the functions ψ may be, provided only that they are independent of one another.

It can be proved, as in the case of a single dependent variable (§§ 31—33), that the preceding integrals (when $\psi_1, \dots, \psi_m$ are kept as arbitrary as possible) include all integrals that are not of the type called *special* in the simpler case. Such special integrals could occur in connection with zeros, with branch-values, and with singularities of the quantities $\alpha, \pi_1, \dots, \pi_m$.

* There cannot be an identical relation between $u_1, \dots, u_{m+1}$ which leads to an equation independent of $z_1, \dots, z_m$. If such an equation were possible in a form

$$\theta(u_1, \dots, u_{m+1})=0,$$

then as

$$\sum_{r=1}^{m+1}\frac{\partial\theta}{\partial u_r}\frac{\partial u_r}{\partial x}+p_1\frac{d\theta}{dz_1}+\dots+p_m\frac{d\theta}{dz_m}=0,$$

$$\sum_{r=1}^{m+1}\frac{\partial\theta}{\partial u_r}\frac{\partial u_r}{\partial y}+q_1\frac{d\theta}{dz_1}+\dots+q_m\frac{d\theta}{dz_m}=0,$$

we should then have

$$\sum_{r=1}^{m+1}\frac{\partial\theta}{\partial u_r}\frac{\partial u_r}{\partial x}=0, \quad \sum_{r=1}^{m+1}\frac{\partial\theta}{\partial u_r}\frac{\partial u_r}{\partial y}=0.$$

Also

$$0=\frac{d\theta}{dz_i}=\sum_{r=1}^{m+1}\frac{\partial\theta}{\partial u_r}\frac{\partial u_r}{\partial z_i}=0,$$

for $i=1, \dots, m$: consequently, if some of the derivatives $\frac{\partial\theta}{\partial u_1}, \dots, \frac{\partial\theta}{\partial u_{m+1}}$ do not vanish,

$$\frac{\partial(u_1, \dots, u_{m+1})}{\partial(z_1, \dots, z_m, x)}=0, \quad \frac{\partial(u_1, \dots, u_{m+1})}{\partial(z_1, \dots, z_m, y)}=0,$$

which are not true because $u_1, \dots, u_{m+1}$ are a set of independent integrals of the subsidiary system.

Ex. 1. Integrate the equations

$$\left.\begin{aligned} p_1+q_1&=\frac{1-z_1(x+y)}{z_1-z_2} \\ p_2+q_2&=\frac{1-z_2(x+y)}{z_2-z_1} \end{aligned}\right\}.$$

The subsidiary equations, taken according to the preceding explanations, are

$$\frac{dx}{1}=\frac{dy}{1}=\frac{dz_1}{\dfrac{1-z_1(x+y)}{z_1-z_2}}=\frac{dz_2}{\dfrac{1-z_2(x+y)}{z_2-z_1}}.$$

Three independent integrals of this system are easily obtained in the form

$$\begin{aligned} u_1&=x-y=a, \\ u_2&=z_1+z_2+xy=b, \\ u_3&=z_1z_2+x=c, \end{aligned}$$

where a, b, c are arbitrary constants; and therefore a set of integrals of the partial differential equations is given by the equations

$$\left.\begin{aligned} z_1+z_2+xy&=\phi(x-y) \\ z_1z_2+x&=\psi(x-y) \end{aligned}\right\},$$

where ϕ and ψ are arbitrary functions. These equations constitute a general integral.

Ex. 2. Integrate the equations

$$\left.\begin{aligned} p_1-q_1&=-\tfrac{1}{2}z_2\left(\frac{y-x}{z_2^2}+\frac{y+x}{y^2}\right) \\ p_2-q_2&=\tfrac{1}{2}\frac{z_2^2}{z_1}\left(\frac{y+x}{y^2}-\frac{y-x}{z_2^2}\right) \end{aligned}\right\},$$

obtaining a general integral in the form

$$\frac{z_1}{z_2}+\frac{x}{y}=f(x+y), \quad z_1z_2+xy=g(x+y),$$

where f and g are arbitrary functions.

State also a complete integral: and from it deduce the general integral.

Hamburger's Equations in Two Dependent Variables.

169. The simplest case of the general problem occurs when there are two dependent variables and two independent variables, so that the equations may be taken to be

$$\left.\begin{aligned} p_1&=\gamma_1+a_1q_1+b_1q_2 \\ p_2&=\gamma_2+a_2q_1+b_2q_2 \end{aligned}\right\}.$$

The subsidiary ordinary equations are

$$\frac{\lambda_1dz_1+\lambda_2dz_2}{\lambda_1\gamma_1+\lambda_2\gamma_2}=dx=\frac{-\lambda_1dy}{\lambda_1a_1+\lambda_2a_2}=\frac{-\lambda_2dy}{\lambda_1b_1+\lambda_2b_2}.$$

Hence, if

$$dy = \mu dx,$$

we have

$$\lambda_1 (a_1 + \mu) + \lambda_2 a_2 = 0,$$

$$\lambda_1 b_1 + \lambda_2 (b_2 + \mu) = 0:$$

consequently

$$\mu^2 + \mu (a_1 + b_2) + a_1 b_2 - a_2 b_1 = 0,$$

and therefore

$$2\mu + a_1 + b_2 = \{(a_1 - b_2)^2 + 4a_2 b_1\}^{\frac{1}{2}}.$$

Thus, if the radical does not vanish, there are two values of μ. Denoting either of these values by μ, we have the subsidiary equations in the form

$$\left.\begin{aligned} \alpha_1 dz_1 + \alpha_2 dz_2 &= \quad dx \\ dy &= \mu dx \end{aligned}\right\},$$

where

$$\frac{\alpha_1}{a_2} = \frac{\alpha_2}{-(a_1 + \mu)} = \frac{1}{a_2\gamma_1 - (a_1 + \mu)\gamma_2}.$$

Let

$$u(x, y, z_1, z_2) = \text{constant}$$

be an integral of these differential relations: then the relation

$$\frac{\partial u}{\partial x} dx + \frac{\partial u}{\partial y} dy + \frac{\partial u}{\partial z_1} dz_1 + \frac{\partial u}{\partial z_2} dz_2$$
$$= \rho (\alpha_1 dz_1 + \alpha_2 dz_2 - dx) + \sigma (dy - \mu dx)$$

must be identically satisfied, so that

$$\left.\begin{aligned} \frac{\partial u}{\partial z_1} + \alpha_1 \frac{\partial u}{\partial x} + \alpha_1 \mu \frac{\partial u}{\partial y} = 0 \\ \frac{\partial u}{\partial z_2} + \alpha_2 \frac{\partial u}{\partial x} + \alpha_2 \mu \frac{\partial u}{\partial y} = 0 \end{aligned}\right\}.$$

It is easy to see that these two equations are not generally a complete system: for if they were, and if

$$u(x, y, z_1, z_2) = \text{constant}$$

were an integral, we should have

$$\frac{\partial u}{\partial x} + \frac{\partial u}{\partial z_1} p_1 + \frac{\partial u}{\partial z_2} p_2 = 0,$$

that is,

$$(1 - p_1\alpha_1 - p_2\alpha_2)\frac{\partial u}{\partial x} - \mu (p_1\alpha_1 + p_2\alpha_2)\frac{\partial u}{\partial y} = 0,$$

an equation which, in general, cannot be satisfied identically. Writing

$$A_1 u = \frac{\partial u}{\partial z_1} + \alpha_1 \frac{\partial u}{\partial x} + \alpha_1 \mu \frac{\partial u}{\partial y},$$

$$A_2 u = \frac{\partial u}{\partial z_2} + \alpha_2 \frac{\partial u}{\partial x} + \alpha_2 \mu \frac{\partial u}{\partial y},$$

we have

$$A_1(A_2 u) - A_2(A_1 u)$$
$$= \{A_1(\alpha_2) - A_2(\alpha_1)\} \frac{\partial u}{\partial x} + \{A_1(\alpha_2 \mu) - A_2(\alpha_1 \mu)\} \frac{\partial u}{\partial y};$$

and therefore, if a common integral of the two partial equations for u exists, we must have

$$\{A_1(\alpha_2) - A_2(\alpha_1)\} \frac{\partial u}{\partial x} + \{A_1(\alpha_2 \mu) - A_2(\alpha_1 \mu)\} \frac{\partial u}{\partial y} = 0.$$

When this is associated with the other two, the three may make a complete system: in that case, there is one integral of the complete system of three equations in four variables, which may be denoted by

$$u = u(x, y, z_1, z_2).$$

Similarly, from the other value of μ, there may be an integral: let it be denoted by

$$v = v(x, y, z_1, z_2).$$

Then the equations

$$u = \text{constant}, \quad v = \text{constant},$$

give an integral system of the original equations.

It may however happen that only one of the values of μ may lead to an integral equation of the form

$$u = u(x, y, z_1, z_2) = \text{constant}.$$

In that case, we can use the equation thus obtained to eliminate one of the dependent variables and its derivatives from the original equations: and it appears as follows that, if one of the original equations is then satisfied, the other also is satisfied so that, in fact, the integral can be used to reduce the two original equations to one only.

The integral $u = \text{constant}$ is the one integral common to a complete system of three equations, which may be taken in the form

$$\frac{1}{\theta} \frac{\partial u}{\partial y} = \frac{1}{\phi} \frac{\partial u}{\partial z_1} = \frac{1}{\psi} \frac{\partial u}{\partial z_2} = \frac{\partial u}{\partial x}:$$

and it also is an integral of the system

$$\left.\begin{aligned} \alpha_1 dz_1 + \alpha_2 dz_2 &= dx \\ dy &= \mu dx \end{aligned}\right\},$$

so that

$$\frac{\partial u}{\partial z_1} + \alpha_1 \frac{\partial u}{\partial x} + \alpha_1 \mu \frac{\partial u}{\partial y} = 0,$$

$$\frac{\partial u}{\partial z_2} + \alpha_2 \frac{\partial u}{\partial x} + \alpha_2 \mu \frac{\partial u}{\partial y} = 0;$$

and therefore

$$\phi + \alpha_1 (1 + \mu\theta) = 0, \quad \psi + \alpha_2 (1 + \mu\theta) = 0.$$

Now if $u = u(x, y, z_1, z_2)$ is to be used to eliminate z, p_1, q_1 from the original equations, we have

$$\frac{\partial u}{\partial x} + p_1 \frac{\partial u}{\partial z_1} + p_2 \frac{\partial u}{\partial z_2} = 0,$$

$$\frac{\partial u}{\partial y} + q_1 \frac{\partial u}{\partial z_1} + q_2 \frac{\partial u}{\partial z_2} = 0,$$

as the equations giving the values of the derivatives: and therefore

$$1 + \phi p_1 + \psi p_2 = 0,$$

$$\theta + \phi q_1 + \psi q_2 = 0,$$

that is,

$$1 - (\alpha_1 p_1 + \alpha_2 p_2)(1 + \mu\theta) = 0,$$

$$\theta - (\alpha_1 q_1 + \alpha_2 q_2)(1 + \mu\theta) = 0.$$

When by means of these two relations, we eliminate p_1 and q_1 from the original equations, they become

$$\left(\frac{1}{1+\mu\theta} - \alpha_2 p_2\right)\frac{1}{\alpha_1} = \gamma_1 + \frac{a_1}{\alpha_1}\left(\frac{\theta}{1+\mu\theta} - \alpha_2 q_2\right) + b_1 q_2,$$

$$p_2 = \gamma_2 + \frac{a_2}{\alpha_1}\left(\frac{\theta}{1+\mu\theta} - \alpha_2 q_2\right) + b_2 q_2,$$

respectively; and these are easily proved to be one and the same equation, in virtue of the relations between the quantities a, α, γ, μ. Eliminating z_1 from either of them, we then have a single partial equation of the first order involving only z_2 and its derivatives; its integral can be associated with

$$u = u(x, y, z_1, z_2) = \text{constant},$$

and the two equations constitute an integral equivalent of the original equations.

It is an immediate consequence of this analytical investigation that, if the two equations can be combined in any way with

$$u = u(x, y, z_1, z_2) = \text{constant},$$

so as to lead to a new integral equation independent of $u =$ constant, then the new integral can be combined with $u =$ constant as above to provide an integral equivalent of the original system.

Ex. 1. As an example*, consider the equations

$$\left.\begin{aligned} p_1 &= \frac{2x+2y(z_1-z_2)}{z_2-2z_1} - z_1 q_1 - z_2 q_2 \\ p_2 &= \frac{-x+y(5z_1-2z_2)}{z_2-2z_1} - 2z_1 q_1 - 2z_2 q_2 \end{aligned}\right\}.$$

The equation for μ is

$$\begin{vmatrix} -z_1+\mu, & -2z_1 \\ -z_2\ , & -2z_2+\mu \end{vmatrix} = 0,$$

so that there are two values for μ, viz.

$$\mu = z_1 + 2z_2, \quad \mu = 0.$$

Taking the value $\mu = z_1 + 2z_2$, the associated values of a_1 and a_2 are

$$a_1 = \frac{-z_1}{x+y(z_1+2z_2)},$$

$$a_2 = \frac{-z_2}{x+y(z_1+2z_2)};$$

and then the equations for u are

$$A_1 u = \frac{\partial u}{\partial z_1} - \frac{z_1}{x+y(z_1+2z_2)} \left\{\frac{\partial u}{\partial x} + (z_1+2z_2)\frac{\partial u}{\partial y}\right\} = 0,$$

$$A_2 u = \frac{\partial u}{\partial z_2} - \frac{z_2}{x+y(z_1+2z_2)} \left\{\frac{\partial u}{\partial x} + (z_1+2z_2)\frac{\partial u}{\partial y}\right\} = 0.$$

Also

$$A_1(A_2 u) - A_2(A_1 u) = \frac{2z_1 - z_2}{\{x+y(z_1+2z_2)\}^2}\left(-y\frac{\partial u}{\partial x} + x\frac{\partial u}{\partial y}\right),$$

so that

$$\frac{\partial u}{\partial y} = \frac{y}{x}\frac{\partial u}{\partial x};$$

and then the other two equations are

$$\frac{\partial u}{\partial z_1} = \frac{z_1}{x}\frac{\partial u}{\partial x},$$

$$\frac{\partial u}{\partial z_2} = \frac{z_2}{x}\frac{\partial u}{\partial x}.$$

The system for u is complete: it has the single integral

$$u = x^2 + y^2 + z_1^2 + z_2^2.$$

* It is given by Königsberger, *Math. Ann.*, t. XLI (1893), p. 264.

Next, taking the value $\mu=0$, (and this implies $y=$constant, as an integral of the subsidiary system), we have

$$a_1=\frac{2(z_2-2z_1)}{5x-y(z_1+2z_2)},$$

$$a_2=\frac{-(z_2-2z_1)}{5x-y(z_1+2z_2)};$$

and the equations for u are

$$B_1u=\frac{\partial u}{\partial z_1}-\frac{2(z_2-2z_1)}{5x-y(z_1+2z_2)}\frac{\partial u}{\partial x}=0,$$

$$B_2u=\frac{\partial u}{\partial z_2}+\frac{z_2-2z_1}{5x-y(z_1+2z_2)}\frac{\partial u}{\partial x}=0,$$

clearly satisfied by $u=y$, which however is not an effective integral for our purpose. Also, the equation

$$B_1(B_2u)-B_2(B_1u)=0$$

is satisfied only in virtue of

$$\frac{\partial u}{\partial x}=0;$$

and the other two equations then are

$$\frac{\partial u}{\partial z_1}=0,\quad \frac{\partial u}{\partial z_2}=0.$$

Clearly no integral that is effective can be derived through $\mu=0$.

We thus have only one integral of the original system, viz.

$$u=x^2+y^2+z_1{}^2+z_2{}^2=\text{constant}.$$

As explained in the text, this integral can be used to eliminate one of the dependent variables and its derivatives from the two original equations: when this elimination has been effected, the resulting equations are one and the same: and the integral of this last equation will complete the integral equivalent of the original equations. Or, also as explained in the text, it can be used so as, in combination with the original equations, to construct a new integral, independent of $u=$constant. The latter process happens, in the present example, to be the simpler. We have

$$y+z_1q_1+z_2q_2=0.$$

When this is combined with the first of the equations, we find

$$p_1+\frac{yz_2-2x}{z_2-2z_1}=0;$$

and, when it is combined with the second of the equations, we find

$$p_2+\frac{x-yz_1}{z_2-2z_1}=0.$$

These two are equivalent to one another in virtue of

$$x+z_1p_1+z_2p_2=0,$$

as derived from the integral already obtained: and so they can be replaced by any relation which is a combination of the two. Such a relation is

$$p_1+2p_2+y=0,$$

an integral of which will serve to complete the integral system. An integral is

$$z_1+2z_2+xy=\phi(y),$$

where ϕ is an arbitrary function.

Consequently, an integral equivalent of the two original equations is

$$\left.\begin{array}{r}x^2+y^2+z_1^2+z_2^2=a\\ z_1+2z_2+xy=\phi(y)\end{array}\right\},$$

where a is an arbitrary constant, and ϕ is an arbitrary function.

It is also possible to obtain the integral from

$$p_2+\frac{x-yz_1}{z_2-2z_1}=0,$$

by substituting $(a-x^2-y^2-z_2^2)^{\frac{1}{2}}$ for z_1 and integrating.

Ex. 2. Obtain an integral system of the equations

$$\left.\begin{array}{r}p_1-\dfrac{x^2+y}{x+y^2}q_1+x^2q_2=0\\ p_2-\dfrac{1}{x+y^2}q_1+q_2=0\end{array}\right\},$$

in the form

$$z_1-\tfrac{1}{3}x^3-2xy-\tfrac{2}{3}y^3=a,\quad z_2-x-y=b,$$

where a and b are arbitrary constants. (Königsberger.)

Ex. 3. Obtain an integral system of the equations

$$\left.\begin{array}{r}p_1+x^2q_1-\tfrac{1}{4}x(x-y)^2q_2=0\\ p_2+xq_1+xyq_2=0\end{array}\right\},$$

in the form

$$z_1-\tfrac{1}{3}x^3+y=a,\quad z_2-\tfrac{1}{2}x^2=b,$$

where a and b are arbitrary constants. (Königsberger.)

Ex. 4. Shew that, if u_1, u_2, u_3 are any three functions of x, y, z_1, z_2, the differential equations for z_1 and z_2 that correspond to the integral relations

$$\phi(u_1, u_2, u_3)=0,\quad \psi(u_1, u_2, u_3)=0,$$

where ϕ and ψ are arbitrary, are

$$\left.\begin{array}{l}a_1(p_1q_2-p_2q_1)+\beta_1p_1+\gamma_1q_1+\delta_1p_2+\epsilon_1q_2=\zeta_1\\ a_2(p_1q_2-p_2q_1)+\beta_2p_1+\gamma_2q_1+\delta_2p_2+\epsilon_2q_2=\zeta_2\end{array}\right\}.$$

(Hamburger.)

What are the limitations on u_1, u_2, u_3, in order that these equations may reduce to Jacobi's set?

HAMBURGER'S PROCESS WHEN THERE ARE MORE THAN TWO EQUATIONS.

170. When the number of dependent variables is greater than two, and μ is a simple root of the critical equation, we proceed in a similar manner. The subsidiary equations are

$$\left.\begin{aligned} \alpha_1 dz_1 + \ldots + \alpha_m dz_m &= dx \\ dy &= \mu dx \end{aligned}\right\},$$

where the quantities $\alpha_1, \ldots, \alpha_m$ are determinate. Let

$$u(x, y, z_1, \ldots, z_m) = \text{constant}$$

be an integral of these relations; then the differential relation

$$\frac{\partial u}{\partial x} dx + \frac{\partial u}{\partial y} dy + \sum_{i=1}^{m} \frac{\partial u}{\partial z_i} dz_i$$

$$= \rho \left(\sum_{i=1}^{m} \alpha_i dz_i - dx \right) + \sigma (dy - \mu dx)$$

must be satisfied identically, where ρ and σ are independent of the differential elements: hence

$$\frac{\partial u}{\partial z_i} + \alpha_i \frac{\partial u}{\partial x} + \alpha_i \mu \frac{\partial u}{\partial y} = 0,$$

for $i = 1, \ldots, m$. This is a system of m equations in $m+2$ variables; according to its character in respect of completeness, it may possess two independent integrals, or only one, or none. The most general integral, which it possesses and which involves any of the dependent variables, provides an integral of the original system.

It is possible that such an integral may be provided by each simple root of the critical equation. If each root of the critical equation is simple, and if an integral can be determined in association with each of the roots, the aggregate of all the integrals thus obtained is an integral equivalent of the original system of equations.

But an integral equivalent will not thus be provided if it should not be possible to obtain an integral in connection with a simple root of the critical equation. In that case, we take such integrals, say $m - \mu$, as are thus provided: and we use them to eliminate, from the m original equations, $m - \mu$ of the dependent variables with their derivatives; there will then be left a system of μ equations, which are of the same form as before and which involve only μ dependent variables. The problem now is similar to the problem in its initial stage: but it is simpler, because the number of dependent variables has been decreased.

Next, consider a multiple root of the critical equation, and let it give rise to a system of differential relations represented by

$$\left.\begin{array}{c} dy = \mu dx \\ \sum_{i=1}^{m} \gamma_{is} dz_i = \left(\sum_{i=1}^{m} \gamma_{is}\pi_i\right) dx \end{array}\right\},$$

for $s = 1, \ldots, t$, the system thus containing $t+1$ relations.

Let the last t relations be resolved so as to express t of the elements dz, say $dz_1, \ldots, dz_m$, in terms of the remainder, so that the system may be replaced by a system of the form

$$\left.\begin{array}{l} dy = \mu dx \\ dz_s = \pi_s dx + \sum_{\sigma=t+1}^{m} \epsilon_{s\sigma}(\pi_\sigma dx - dz_\sigma) \end{array}\right\},$$

for $s = 1, \ldots, t$. The modified system implies, as the former system implied, that t of the equations

$$E_\rho = \sum_{i=1}^{m} \lambda_i a_{i\rho} + \mu\lambda_\rho = 0,$$

for $\rho = 1, \ldots, m$, are deducible from the remainder, the quantities $a_{i\rho}$ being the coefficients in the original set of equations

$$p_i = \pi_i + \sum_{\rho=1}^{m} a_{i\rho} q_\rho, \qquad (i = 1, \ldots, m),$$

and the quantities $\lambda_1, \ldots, \lambda_m$ are such that the t differential relations arise from

$$\sum_{i=1}^{m} \lambda_i (dz_i - \pi_i dx) = 0.$$

In order to express the interdependence of some of the equations $E_1 = 0, \ldots, E_m = 0$, we write

$$E_i + \sum_{j=t+1}^{m} \epsilon_{ij} E_j = 0,$$

for $i = 1, \ldots, t$; and therefore the quantities ϵ_{ij} are such that

$$a_{i\mu} + \sum_{j=t+1}^{m} \epsilon_{ij} a_{j\mu} = 0,$$

for all values of μ different from $i, t+1, \ldots, m$, together with

$$a_{ii} + \mu + \sum_{j=t+1}^{m} \epsilon_{ij} a_{ji} = 0,$$

$$a_{ij} + \sum_{\rho=t+1}^{m} \epsilon_{i\rho} a_{\rho j} + \epsilon_{ij} (a_{jj} + \mu) = 0,$$

where the summation with regard to ρ excludes $\rho = j$, while i has the values $1, \ldots, t$, and j has the values $t+1, \ldots, m$. With these values, we have

$$\begin{aligned} p_i + \mu q_i - \pi_i + \sum_{j=t+1}^{m} \epsilon_{ij} (p_j + \mu q_j - \pi_j) \\ &= a_{i1} q_1 + a_{i2} q_2 + \ldots + (a_{ii} + \mu) q_i + \ldots + a_{im} q_m \\ &\quad + \sum_{j=t+1}^{m} \epsilon_{ij} \{a_{j1} q_1 + a_{j2} q_2 + \ldots + (a_{jj} + \mu) q_j + \ldots + a_{jm} q_m\} \\ &= 0, \end{aligned}$$

because the coefficient of each of the quantities $q_1, \ldots, q_m$ vanishes on account of the above equations in the quantities ϵ_{ij}.

Thus the t equations

$$p_i + \mu q_i - \pi_i + \sum_{j=t+1}^{m} \epsilon_{ij} (p_i + \mu q_j - \pi_j) = 0$$

can be regarded as replacing t of the equations in the original system: other $m - t$ equations would be required to have a complete equivalent of that system.

Now let

$$\phi (x, y, z_1, \ldots, z_m) = \text{constant}$$

be one of the equations in the integral equivalent of the subsidiary differential relations; and assume that the differential relations are completely integrable*, so that there are $t+1$ such integrals. Then the relation

$$\frac{\partial \phi}{\partial x} dx + \frac{\partial \phi}{\partial y} dy + \sum_{i=1}^{m} \frac{\partial \phi}{\partial z_i} dz_i = 0$$

is consistent with the $t+1$ differential relations

$$dy = \mu dx,$$

$$dz_j = \pi_j dx + \sum_{\sigma=t+1}^{m} \epsilon_{j\sigma} (\pi_\sigma dx - dz_\sigma);$$

* The assumption is a distinct limitation, as its justification requires that conditions should be satisfied. It must be remembered, however, that we are dealing with equations, restricted in form and in the number of independent variables, so that the method does not claim to be general ; it is therefore hardly necessary to deal generally with all the minutiæ of alternatives, when these could be dealt with in any particular case.

and therefore

$$\frac{\partial\phi}{\partial x}+\mu\frac{\partial\phi}{\partial y}+\sum_{s=1}^{t}\left(\pi_s+\sum_{\sigma=t+1}^{m}\epsilon_{s\sigma}\pi_\sigma\right)\frac{\partial\phi}{\partial z_s}=0,$$

$$\frac{\partial\phi}{\partial z_\rho}-\sum_{s=1}^{t}\epsilon_{s\rho}\frac{\partial\phi}{\partial z_s}=0,$$

for $\rho=t+1,\ \ldots,\ m$. And these equations must be satisfied by each of the functions ϕ.

171. Let the $t+1$ functions, supposed to be thus obtained, be denoted by $\phi_1,\ \ldots,\ \phi_{t+1}$; and let t arbitrary combinations of these functions be taken which, when equated to zero, may be regarded as t equations helping to express $z_1,\ \ldots,\ z_m$ in terms of x and y. If

$$f_i(\phi_1,\ \ldots,\ \phi_{t+1})=0$$

be any one of these integral equations, then, on multiplying by $\frac{\partial f}{\partial\phi_i}$ the differential equations which determine ϕ_i and on adding the results, we have

$$\left.\begin{aligned}\frac{\partial f_i}{\partial x}+\mu\frac{\partial f_i}{\partial y}+\sum_{s=1}^{t}\left(\pi_s+\sum_{\sigma=t+1}^{m}\epsilon_{s\sigma}\pi_\sigma\right)\frac{\partial f_i}{\partial z_s}=0\\ \frac{\partial f_i}{\partial z_\rho}-\sum_{s=1}^{t}\epsilon_{s\rho}\frac{\partial f_i}{\partial z_s}=0\end{aligned}\right\},$$

for $\rho=t+1,\ \ldots,\ m$. But when $f_i=0$ is regarded as an integral equation, we have

$$\frac{\partial f_i}{\partial x}+\sum_{s=1}^{m}p_s\frac{\partial f_i}{\partial z_s}=0,$$

$$\frac{\partial f_i}{\partial y}+\sum_{s=1}^{m}q_s\frac{\partial f_i}{\partial z_s}=0,$$

and therefore

$$\frac{\partial f_i}{\partial x}+\mu\frac{\partial f_i}{\partial y}+\sum_{s=1}^{m}(p_s+\mu q_s)\frac{\partial f_i}{\partial z_s}=0.$$

Substituting in this equation for $\frac{\partial f_i}{\partial x}+\mu\frac{\partial f_i}{\partial y}$, and for $\frac{\partial f_i}{\partial z_\rho}$ (for $\rho=t+1,\ \ldots,\ m$), we have

$$\sum_{s=1}^{t}\left\{p_s+\mu q_s-\pi_s+\sum_{\sigma=t+1}^{m}\epsilon_{s\sigma}(p_\sigma+\mu q_\sigma-\pi_\sigma)\right\}\frac{\partial f_i}{\partial z_s}=0.$$

This relation holds for $i=1,\ \ldots,\ t$; and the functions $f_1,\ \ldots,f_t$ are independent, so that the quantity

$$J\left(\frac{f_1,\ \ldots,f_t}{z_1,\ \ldots,\ z_t}\right)$$

is not evanescent; consequently, the equations

$$p_s + \mu q_s - \pi_s + \sum_{\sigma=t+1}^{m} \epsilon_{s\sigma} (p_\sigma + \mu q_\sigma - \pi_\sigma) = 0,$$

for $s = 1, \ldots, t$, are satisfied. And these equations are part of the system equivalent to the original system: the set of integrals thus obtained effectively satisfy t of the equations of the original system.

When we proceed in this way with all the multiple roots of the critical equation and obtain the integrals associated with them in turn, and when we similarly retain all the integrals associated with the simple roots of that equation, we obtain an aggregate of integrals of the differential equations: let the number of these be τ. Then these τ equations can be resolved so as to express τ of the dependent variables, say $z_1, \ldots, z_\tau$, in terms of the remainder and of x, y; and they are such as to satisfy an appropriate set of τ combinations of the original equations. When the values of $z_1, \ldots, z_\tau$ and of their derivatives are substituted in the $m - \tau$ other combinations, which (with the τ just satisfied) constitute an algebraic equivalent of the original system, then we have a simultaneous set of $m - \tau$ equations having $z_{\tau+1}, \ldots, z_\tau$ for the dependent variables, and x, y for the independent variables. The problem of obtaining the integrals of this new set of equations is similar to the initial problem: but it is simpler, because the number of dependent variables has been reduced from m to $m - \tau$.

As already remarked, the easiest case is that in which each root μ of the critical equation is simple. With each such root, two differential relations are satisfied: let u_i=constant, v_i=constant be their integral equivalent. Then

$$g_i (u_i, v_i) = 0,$$

where g_i is arbitrary, is an integral of the original system; and a full set of integrals of the original equations is given by

$$g_1 (u_1, v_1) = 0, \quad g_2 (u_2, v_2) = 0, \ldots, g_m (u_m, v_m) = 0,$$

where $g_1, \ldots, g_m$ are arbitrary functions.

If, connected with a simple root μ_j of the critical equation, only one integral (say u_j) can be obtained, then

$$u_j = \text{constant}$$

takes the place of $g_j(u_j, v_j)=0$. And if no integral can be obtained, then (as already explained) we use the known integrals to reduce the order of the system and adopt the process for the integration of the reduced system.

It is to be noticed that what is wanted at each stage, in connection with the differential relations of the form

$$\left.\begin{aligned} \alpha_1 dz_1 + \ldots + \alpha_m dz_m &= dx \\ dy &= \mu dx \end{aligned}\right\}$$

for a simple root μ of the critical equation, and of similar relations for a multiple root, is not a complete equivalent of each set regarded as a set of Pfaffian equations but only those integral equations (if any), which arise by forming an exactly integrable combination of the differential relations or which can be obtained by some equivalent process.

Further it is clear from the general argument that, if circumstances make the use of an obtained integral convenient at any stage, the integral can be used to reduce the order of the system at once without determining any further integral or integrals connected with the root in question, or with any other root, of the critical equation in μ.

Ex. 1. Integrate the equations

$$\left.\begin{aligned} p_1 &= \frac{1}{2x}(3z_1+2z_2+z_3) - \frac{y}{x}(q_1+q_2+q_3) \\ p_2 &= \frac{1}{2x}(z_1+2z_2-z_3) + \frac{y}{x}q_3 \\ p_3 &= \frac{1}{2x}(-z_1-2z_2+z_3) + \frac{y}{x}q_2 \end{aligned}\right\}.$$

The critical equation for the determination of μ is

$$\begin{vmatrix} \mu-\frac{y}{x}, & -\frac{y}{x}, & -\frac{y}{x} \\ 0\;, & \mu\;, & \frac{y}{x} \\ 0\;, & \frac{y}{x}, & \mu \end{vmatrix} = 0,$$

that is,

$$\left(\mu-\frac{y}{x}\right)\left(\mu^2-\frac{y^2}{x^2}\right)=0,$$

so that $\mu=-\frac{y}{x}$ is a simple root and $\mu=\frac{y}{x}$ is a repeated root.

Taking $\mu=-\frac{y}{x}$, we have

$$-\lambda_1\frac{y}{x}=\frac{y}{x}\lambda_1,$$

$$-\lambda_1\frac{y}{x}+\lambda_3\frac{y}{x}=\frac{y}{x}\lambda_2,$$

$$-\lambda_1\frac{y}{x}+\lambda_2\frac{y}{x}=\frac{y}{x}\lambda_3,$$

so that

$$\lambda_1=0, \quad \lambda_2=\lambda_3.$$

The subsidiary equations are

$$\frac{dz_2+dz_3}{0}=dx,$$

$$dy=-\frac{y}{x}dx:$$

two integrals of these are

$$z_2+z_3=\text{constant}, \quad xy=\text{constant}:$$

hence an integral of the original system of equations is

$$z_2+z_3=f(xy),$$

where f is an arbitrary function.

Next, taking the repeated root $\mu=\frac{y}{x}$, we find that there is only a single relation among the three quantities λ: it is

$$\lambda_2+\lambda_3=\lambda_1.$$

The subsidiary equations, on the substitution of this value of λ_1, take the form

$$\frac{\lambda_2(dz_1+dz_2)+\lambda_3(dz_1+dz_3)}{\frac{\lambda_2}{2x}(4z_1+4z_2)+\frac{\lambda_3}{2x}(2z_1+2z_3)}=dx,$$

$$dy=\frac{y}{x}dx;$$

hence as $\lambda_2:\lambda_3$ is undetermined, we take the subsidiary equations in the form

$$\left.\begin{aligned}\frac{x}{2z_1+2z_2}(dz_1+dz_2)&=dx\\ \frac{x}{z_1+z_3}(dz_1+dz_3)&=dx\\ \frac{x}{y}dy&=dx\end{aligned}\right\}.$$

Three independent integrals of these equations are

$$\frac{z_1+z_2}{x^2}=\text{constant},$$

$$\frac{z_1+z_3}{x}=\text{constant},$$

$$\frac{y}{x}=\text{constant};$$

hence two integrals of the original system are given by

$$\phi\left(\frac{z_1+z_2}{x^2}, \frac{z_1+z_3}{x}, \frac{y}{x}\right)=0,$$

$$\psi\left(\frac{z_1+z_2}{x^2}, \frac{z_1+z_3}{x}, \frac{y}{x}\right)=0,$$

where ϕ and ψ are arbitrary functions or, what is the same thing, two integrals are given by

$$\frac{z_1+z_2}{x^2}=g\left(\frac{y}{x}\right), \quad \frac{z_1+z_3}{x}=h\left(\frac{y}{x}\right),$$

where g and h are arbitrary functions.

Hence an integral equivalent of the system of differential equations is given by the three equations

$$\left.\begin{aligned} z_2+z_3&=f(xy) \\ z_1 \quad +z_3&=xh\left(\frac{y}{x}\right) \\ z_1+z_2 \quad &=x^2g\left(\frac{y}{x}\right) \end{aligned}\right\},$$

where f, g, h are arbitrary functions.

Ex. 2. Integrate the equations

$$\left.\begin{aligned} 7xp_1&=7z_1+z_2+2z_3-y(7q_1+2q_2+4q_3) \\ 7xp_2&=6z_2-2z_3-y(5q_2-4q_3) \\ 7xp_3&=-3z_2+z_3+y(6q_2+5q_3) \end{aligned}\right\}.$$

Alternative Method, with Partial Subsidiary Equations.

172. In the preceding investigation, the construction of the integrals of the system

$$p_i=\pi_i+\sum_{s=1}^{m} a_{is}q_s, \qquad (i=1, \ldots, m)$$

was made to depend upon the integration of a subsidiary set of equations homogeneous and linear in differential elements. As is well known from many discussions in earlier parts of this treatise, the integration of such a set can be replaced by the integration of a system of simultaneous partial differential equations in a single dependent variable: and indeed, in §§ 169, 170, the problem was thus actually transferred from the region of ordinary equations to that of partial equations. The construction of these partial equations can be effected, without the intervention of the subsidiary equations as follows. Let

$$u=u(x, y, z_1, \ldots, z_m)=\text{constant}$$

be a member of the integral equivalent of the above system of partial differential equations: then the equations

$$\frac{\partial u}{\partial x}+\sum_{i=1}^{m}\frac{\partial u}{\partial z_i}p_i=0,$$

$$\frac{\partial u}{\partial y}+\sum_{i=1}^{m}\frac{\partial u}{\partial z_i}q_i=0,$$

not merely coexist with the system of m equations but they are not independent of them, quà equations in the derivatives. Consequently, when substitution in the prior of the last two equations is made for $p_1, \ldots, p_m$, so that the original system has been completely used for purposes of inter-relation of the whole set, the two equations

$$\frac{\partial u}{\partial x}+\sum_{i=1}^{m}\left(\pi_i+\sum_{s=1}^{m}a_{is}q_s\right)\frac{\partial u}{\partial z_i}=0,$$

$$\frac{\partial u}{\partial y}+\sum_{i=1}^{n}q_i\frac{\partial u}{\partial z_i}=0,$$

are not independent of one another so far as concerns the derivatives. Comparing the coefficients of $q_1, \ldots, q_m$, and the quantities not involving these derivatives, in the two equations, we have

$$\sum_{j=1}^{m}a_{ji}\frac{\partial u}{\partial z_j}=-\mu\frac{\partial u}{\partial z_i},$$

$$\frac{\partial u}{\partial x}+\sum_{j=1}^{m}\pi_j\frac{\partial u}{\partial z_j}=-\mu\frac{\partial u}{\partial y},$$

where μ is an unknown quantity and the former equation holds for $i=1, \ldots, m$.

The m equations

$$\sum_{j=1}^{m}a_{ji}\frac{\partial u}{\partial z_j}=-\mu\frac{\partial u}{\partial z_i},$$

for $i=1, \ldots, m$, are homogeneous and linear in the m derivatives of u with regard to $z_1, \ldots, z_m$: and these do not all vanish, for $u=$ constant is postulated as an integral equation. Hence μ is given by the equation

$$\Theta=\begin{vmatrix} a_{11}+\mu, & a_{21}\ , & \ldots, & a_{m1} \\ a_{12}\ , & a_{22}+\mu, & \ldots, & a_{m2} \\ \cdots & \cdots & \cdots & \cdots \\ a_{1m}\ , & a_{2m}\ , & \ldots, & a_{mm}+\mu \end{vmatrix}=0,$$

being the same equation for μ as in the other investigation (§ 167). According to the character of μ as a root of this equation $\Theta = 0$, the form of the system of equations for u alters.

Let $\mu = \sigma$ be a simple root of $\Theta = 0$; then the former set of $m + 1$ equations involving the then unknown quantity μ and the derivatives of u can be replaced by the set of m equations

$$\frac{\partial u}{\partial z_i} + \alpha_i \left(\frac{\partial u}{\partial x} + \sigma \frac{\partial u}{\partial y}\right) = 0,$$

for $i = 1, \ldots, m$, the quantity μ now being known; and the quantities $\alpha_1, \ldots, \alpha_m$ are given by the equation

$$\sum_{j=1}^{m} \pi_j \alpha_j = -1,$$

and by any $m - 1$ of the m equations

$$\sum_{j=1}^{m} a_{ji} \alpha_j = -\sigma \alpha_i,$$

for $i = 1, \ldots, m$. The set of m equations for u involves $m + 2$ variables. It may possess no integral at all or no integral involving any one of the variables $z_1, \ldots, z_m$; in that case, no integral of the original system of equations is derivable through the root σ of $\Theta = 0$. Or it may possess one integral involving at least one of the variables $z_1, \ldots, z_m$; in that case,

$$U = \text{constant},$$

where U is the integral in question, is an integral of the original system. Or it may possess two integrals U and V, one at least of which involves one or more than one of the variables $z_1, \ldots, z_m$; in that case,

$$\phi(U, V) = 0$$

where ϕ is arbitrary, is the most general integral of the original system thus obtainable.

Similarly for each simple root of $\Theta = 0$.

Next, let μ be a multiple root of $\Theta = 0$; then the m equations

$$E_i = \sum_{j=1}^{m} a_{ji} \alpha_j + \mu \alpha_i = 0, \qquad (i = 1, \ldots, m),$$

are not independent of one another. Let them be such that t (and not more than t) of them can be deduced from the remainder, so that there will be t relations of the form

$$E_s + \sum_{j=1}^{m} \epsilon_{sj} E_j = 0,$$

for $s=1, \ldots, t$: thus the quantities ϵ are given by

$$a_{ss} + \mu + \sum_{j=t+1}^{m} \epsilon_{sj} a_{js} = 0,$$

$$\epsilon_{sj}(a_{jj} + \mu) + a_{sj} + \sum_{\rho=t+1}^{m} \epsilon_{s\rho} a_{\rho j} = 0,$$

$$a_{s\mu} + \sum_{j=t+1}^{m} \epsilon_{sj} a_{j\mu} = 0,$$

where the first equation holds for $s=1, \ldots, t$: the second holds for $s=1, \ldots, t$, and $j=t+1, \ldots, m$, and in it the summation with regard to ρ is for all values $t+1, \ldots, m$ except $\rho=j$; and the third holds for the values $s=1, \ldots, t$, and for the values $1, \ldots, t$ except $\mu=s$. Proceeding as for the simple root, we find that the equations for u are

$$\frac{\partial u}{\partial x} + \mu \frac{\partial u}{\partial y} + \sum_{s=1}^{t} \left(\pi_s + \sum_{\sigma=t+1}^{m} \epsilon_{s\sigma} \pi_\sigma \right) \frac{\partial u}{\partial z_s} = 0,$$

$$\frac{\partial u}{\partial z_\rho} - \sum_{s=1}^{t} \epsilon_{s\rho} \frac{\partial u}{\partial z_s} = 0,$$

for $\rho=t+1, \ldots, m$, being the same equations as in § 171. This set of equations, being $m-t+1$ in number and involving $m+2$ variables, can have any number* of integrals from 0 up to $t+1$; let these integrals be

$$U_1, \ldots, U_\kappa,$$

where

$$0 \leqslant \kappa \leqslant t+1.$$

Not more than one of these integrals can be independent of all the variables $z_1, \ldots, z_m$: if there be one such, let it be U_κ.

If $\kappa > 1$, then the equations

$$U_1 = f_1(U_\kappa), \ldots, U_{\kappa-1} = f_{\kappa-1}(U_\kappa),$$

where $f_1, \ldots, f_{\kappa-1}$ are arbitrary functions, constitute $\kappa - 1$ integrals of the original system; they are associated with the multiple root μ of the equation $\Theta = 0$.

If $\kappa = 1$, and if U_1 involves one at least of the variables $z_1, \ldots, z_m$, then

$$U_1 = \text{constant}$$

is an integral of the original system: it is associated with the multiple root μ.

* The conditions as to number depend solely upon the number of equations in the system when rendered complete. If this number be κ', where $\kappa' \leqslant m+2$, then the number of integrals is $m+2-\kappa'$.

When $\kappa=1$ and U_1 does not involve any of the variables $z_1, \ldots, z_m$, and when $\kappa=0$, no integral is provided for the original system in association with the multiple root.

Similarly for each multiple root of the critical equation $\Theta=0$.

As before, the integrals can be used to eliminate some of the dependent variables and so to reduce the order of the original system.

Ex. The preceding process can be illustrated by being applied to the equations in Ex. 1, § 171, viz.

$$\left.\begin{aligned} p_1 &= \frac{1}{2x}(3z_1+2z_2+z_3)-\frac{y}{x}(q_1+q_2+q_3) \\ p_2 &= \frac{1}{2x}(z_1+2z_2-z_3)+\frac{y}{x}q_3 \\ p_3 &= \frac{1}{2x}(-z_1-2z_2+z_3)+\frac{y}{x}q_2 \end{aligned}\right\}.$$

Let it be supposed that

$$u(x, y, z_1, z_2, z_3)=0$$

is an integral of these equations: then the preceding explanations shew that the equations

$$\begin{aligned} \frac{\partial u}{\partial x}+\frac{\partial u}{\partial z_1}&\left\{\frac{1}{2x}(3z_1+2z_2+z_3)-\frac{y}{x}(q_1+q_2+q_3)\right\} \\ &+\frac{\partial u}{\partial z_2}\left\{\frac{1}{2x}(z_1+2z_2-z_3)+\frac{y}{x}q_3\right\} \\ &+\frac{\partial u}{\partial z_3}\left\{\frac{1}{2x}(-z_1-2z_2+z_3)+\frac{y}{x}q_2\right\}=0, \end{aligned}$$

and

$$\frac{\partial u}{\partial y}+\frac{\partial u}{\partial z_1}q_1+\frac{\partial u}{\partial z_2}q_2+\frac{\partial u}{\partial z_3}q_3=0,$$

quà equations in q_1, q_2 and q_3, are one and the same. Hence

$$\begin{aligned} \theta\frac{\partial u}{\partial z_1}&=-\frac{y}{x}\frac{\partial u}{\partial z_1}, \\ \theta\frac{\partial u}{\partial z_2}&=-\frac{y}{x}\frac{\partial u}{\partial z_1}+\frac{y}{x}\frac{\partial u}{\partial z_3}, \\ \theta\frac{\partial u}{\partial z_3}&=-\frac{y}{x}\frac{\partial u}{\partial z_1}+\frac{y}{x}\frac{\partial u}{\partial z_2}, \\ \theta\frac{\partial u}{\partial y}&=\frac{\partial u}{\partial x}+\frac{1}{2x}\left\{(3z_1+2z_2+z_3)\frac{\partial u}{\partial z_1}+(z_1+2z_2-z_3)\left(\frac{\partial u}{\partial z_2}-\frac{\partial u}{\partial z_3}\right)\right\}. \end{aligned}$$

From the first three of these equations, we have

$$\begin{vmatrix} \theta+\frac{y}{x}, & 0, & 0 \\ \frac{y}{x}, & \theta, & -\frac{y}{x} \\ \frac{y}{x}, & -\frac{y}{x}, & \theta \end{vmatrix}=0,$$

that is,

$$\left(\theta+\frac{y}{x}\right)\left(\theta^2-\frac{y^2}{x^2}\right)=0,$$

so that a simple root is given by

$$\theta=\frac{y}{x},$$

and a double root is given by

$$\theta=-\frac{y}{x}.$$

I. Let $\theta=\frac{y}{x}$. The first equation gives

$$\frac{\partial u}{\partial z_1}=0;$$

and the other equations then are

$$\frac{\partial u}{\partial z_2}-\frac{\partial u}{\partial z_3}=0,$$

$$\frac{y}{x}\frac{\partial u}{\partial y}-\frac{\partial u}{\partial x}=0.$$

These three equations are a complete system: they possess two independent integrals, in the form

$$u=z_2+z_3, \quad u=xy;$$

hence the equation

$$z_2+z_3=F(xy),$$

where F is an arbitrary function, is part of an integral of the original equations.

II. Let $\theta=-\frac{y}{x}$. The first equation becomes evanescent: the next two equations both become

$$\frac{\partial u}{\partial z_1}-\frac{\partial u}{\partial z_2}-\frac{\partial u}{\partial z_3}=0,$$

and the last equation is

$$x\frac{\partial u}{\partial x}+y\frac{\partial u}{\partial y}+2(z_1+z_2)\frac{\partial u}{\partial z_2}+(z_1+z_3)\frac{\partial u}{\partial z_3}=0.$$

These two equations are a complete system: they possess three independent integrals, in the form

$$\frac{z_1+z_2}{x^2}, \quad \frac{z_1+z_3}{x}, \quad \frac{y}{x};$$

hence the equations

$$z_1+z_2=x^2G\left(\frac{y}{x}\right),$$

$$z_1+z_3=xH\left(\frac{y}{x}\right),$$

where G and H are arbitrary functions, are part of an integral of the original equations.

The full integral of the original equations is given by combining all the parts obtained: it is

$$z_2+z_3=F(xy),$$

$$z_1+z_2=x^2G\left(\frac{y}{x}\right),$$

$$z_1+z_3=xH\left(\frac{y}{x}\right),$$

being the same as by the other process.

Hamburger's Method applied to Non-linear Equations.

173. The method of constructing the integral of a system of simultaneous equations in a number of dependent variables, as just expounded, depends apparently on the formal property that the equations in question are linear in the derivatives of the dependent variable. It was only natural to expect that the method could be extended so as to apply to equations not restricted to being linear in those derivatives; and this extension, due* initially to Hamburger, was effected by a device, (successful specially in connection with equations of the second order, as will be seen later), which replaces the non-linear system by an amplified linear system.

Adopting the same notation as before for the independent variables, for the dependent variables and for their derivatives, and assuming that the number of partial differential equations algebraically independent of one another is the same as the number of dependent variables, we take these equations in the form

$$f_i(x, y, z_1, \ldots, z_n, p_1, \ldots, p_n, q_1, \ldots, q_n)=0,$$

for $i=1, \ldots, n$. Were the integrals known, and the values of $z_1, \ldots, z_n$ and of their derivatives substituted in the differential equations, the latter would become identities; accordingly, when we take

$$\frac{df_i}{dx}=\frac{\partial f_i}{\partial x}+\sum_{r=1}^{n}\frac{\partial f_i}{\partial z_r}p_r, \quad \frac{df_i}{dy}=\frac{\partial f_i}{\partial y}+\sum_{r=1}^{n}\frac{\partial f_i}{\partial z_r}q_r,$$

the integrals of the equations are in accord with the further equations

$$\sum_{\kappa=1}^{n}\frac{\partial f_i}{\partial p_\kappa}\frac{\partial p_\kappa}{\partial x}+\sum_{\kappa=1}^{n}\frac{\partial f_i}{\partial q_\kappa}\frac{\partial p_\kappa}{\partial y}=-\frac{df_i}{dx},$$

$$\sum_{\kappa=1}^{n}\frac{\partial f_i}{\partial p_\kappa}\frac{\partial q_\kappa}{\partial x}+\sum_{\kappa=1}^{n}\frac{\partial f_i}{\partial q_\kappa}\frac{\partial q_\kappa}{\partial y}=-\frac{df_i}{dy},$$

* For references, see p. 407.

deduced from derivatives of the identities by using the necessary relations

$$\frac{\partial q_\kappa}{\partial x} = \frac{\partial p_\kappa}{\partial y}, \qquad (\kappa = 1, \ldots, n).$$

Also, because of the necessary relations

$$\frac{\partial z_\kappa}{\partial x} = p_\kappa, \quad \frac{\partial z_\kappa}{\partial y} = q_\kappa,$$

we have

$$\sum_{\kappa=1}^{n} \frac{\partial f_i}{\partial p_\kappa} \frac{\partial z_\kappa}{\partial x} + \sum_{\kappa=1}^{n} \frac{\partial f_i}{\partial q_\kappa} \frac{\partial z_\kappa}{\partial y} = \sum_{\kappa=1}^{n} \left(p_\kappa \frac{\partial f_i}{\partial p_\kappa} + q_\kappa \frac{\partial f_i}{\partial q_\kappa} \right).$$

These equations hold for $i=1, \ldots, n$: they thus constitute a system of $3n$ equations, involving $3n$ dependent variables $z_1, \ldots, z_n$, $p_1, \ldots, p_n$, $q_1, \ldots, q_n$; and they are linear in the derivatives of those $3n$ dependent variables. Hence this system of equations is amenable to the Hamburger method for linear equations already expounded.

To apply the method, we introduce n quantities $l_1, \ldots, l_n$, which are functions of all the variables and which (as to their ratios) will be determined subsequently; and we write

$$\sum_{i=1}^{n} l_i \frac{\partial f_i}{\partial p_\kappa} = P_\kappa,$$

$$\sum_{i=1}^{n} l_i \frac{\partial f_i}{\partial q_\kappa} = Q_\kappa,$$

$$\sum_{i=1}^{n} l_i \frac{df_i}{dx} = X,$$

$$\sum_{i=1}^{n} l_i \frac{df_i}{dy} = Y.$$

Then, multiplying the preceding typical equations by l_i and adding for all values of i, we have

$$\sum_{\kappa=1}^{n} P_\kappa \frac{\partial p_\kappa}{\partial x} + \sum_{\kappa=1}^{n} Q_\kappa \frac{\partial p_\kappa}{\partial y} = -X,$$

$$\sum_{\kappa=1}^{n} P_\kappa \frac{\partial q_\kappa}{\partial x} + \sum_{\kappa=1}^{n} Q_\kappa \frac{\partial q_\kappa}{\partial y} = -Y,$$

$$\sum_{\kappa=1}^{n} P_\kappa \frac{\partial z_\kappa}{\partial x} + \sum_{\kappa=1}^{n} Q_\kappa \frac{\partial z_\kappa}{\partial y} = \sum_{\kappa=1}^{n} (P_\kappa p_\kappa + Q_\kappa q_\kappa)$$

Again, we have

$$dp_\kappa = \frac{\partial p_\kappa}{\partial x} dx + \frac{\partial p_\kappa}{\partial y} dy,$$

$$dq_\kappa = \frac{\partial q_\kappa}{\partial x} dx + \frac{\partial q_\kappa}{\partial y} dy,$$

$$dz_\kappa = \frac{\partial z_\kappa}{\partial x} dx + \frac{\partial z_\kappa}{\partial y} dy;$$

and therefore, if $\lambda_1, \ldots, \lambda_n$ denote another series of quantities, which are functions of all the variables and which (also as to their ratios) will be determined subsequently, we have

$$\sum_{\kappa=1}^{n} \lambda_\kappa \frac{\partial p_\kappa}{\partial x} dx + \sum_{\kappa=1}^{n} \lambda_\kappa \frac{\partial p_\kappa}{\partial y} dy = \sum_{\kappa=1}^{n} \lambda_\kappa dp_\kappa,$$

$$\sum_{\kappa=1}^{n} \lambda_\kappa \frac{\partial q_\kappa}{\partial x} dx + \sum_{\kappa=1}^{n} \lambda_\kappa \frac{\partial q_\kappa}{\partial y} dy = \sum_{\kappa=1}^{n} \lambda_\kappa dq_\kappa,$$

$$\sum_{\kappa=1}^{n} \lambda_\kappa \frac{\partial z_\kappa}{\partial x} dx + \sum_{\kappa=1}^{n} \lambda_\kappa \frac{\partial z_\kappa}{\partial y} dy = \sum_{\kappa=1}^{n} \lambda_\kappa dz_\kappa.$$

In connection with these two sets of equations and as a generalisation of the corresponding step in the earlier process, we construct a subsidiary system of equations

$$\frac{\sum_{\kappa=1}^{n} \lambda_\kappa dp_\kappa}{-X} = \frac{\sum_{\kappa=1}^{n} \lambda_\kappa dq_\kappa}{-Y} = \frac{\sum_{\kappa=1}^{n} \lambda_\kappa dz_\kappa}{\sum_{\kappa=1}^{n} (P_\kappa p_\kappa + Q_\kappa q_\kappa)}$$

$$= \frac{\lambda_1 dx}{P_1} = \ldots = \frac{\lambda_n dx}{P_n}$$

$$= \frac{\lambda_1 dy}{Q_1} = \ldots = \frac{\lambda_n dy}{Q_n}.$$

Take

$$dy = \mu dx;$$

then

$$\mu = \frac{Q_r}{P_r},$$

for $r = 1, \ldots, n$, that is,

$$l_1 \left(\frac{\partial f_1}{\partial q_r} - \mu \frac{\partial f_1}{\partial p_r} \right) + \ldots + l_n \left(\frac{\partial f_n}{\partial q_r} - \mu \frac{\partial f_n}{\partial p_r} \right) = 0,$$

for $r=1, \ldots, n$. Hence μ must satisfy the equation

$$\Delta = \begin{vmatrix} \frac{\partial f_1}{\partial q_1} - \mu \frac{\partial f_1}{\partial p_1}, \ldots, \frac{\partial f_n}{\partial q_1} - \mu \frac{\partial f_n}{\partial p_1} \\ \cdots\cdots\cdots\cdots\cdots\cdots \\ \frac{\partial f_1}{\partial q_n} - \mu \frac{\partial f_1}{\partial p_n}, \ldots, \frac{\partial f_n}{\partial q_n} - \mu \frac{\partial f_n}{\partial p_n} \end{vmatrix} = 0.$$

When μ is determined as a simple root of $\Delta = 0$, the ratios $l_1 : l_2 : \ldots : l_n$ are those of first minors of the determinant. When μ is determined as a multiple root of $\Delta = 0$, say of multiplicity θ, then the quantities $l_1, \ldots, l_n$ can be expressed linearly and homogeneously in terms of t independent quantities, where $t \leqslant \theta$. Moreover, when μ and the ratios of the quantities $l_1, \ldots, l_n$ are known, the ratios of the quantities $\lambda_1, \ldots, \lambda_n$ can be considered known; for we have

$$\frac{\lambda_1}{P_1} = \ldots = \frac{\lambda_n}{P_n},$$

and

$$\frac{\lambda_1}{Q_1} = \ldots = \frac{\lambda_n}{Q_n},$$

these two sets being the same in virtue of the relations

$$Q_r = \mu P_r,$$

for $r=1, \ldots, n$. Being concerned only with ratios of $\lambda_1, \ldots, \lambda_n$, for it is only the ratios that occur in the subsidiary system, we take

$$\lambda_1 = P_1, \ldots, \lambda_n = P_n;$$

and then we have the subsidiary system in the form

$$\Delta = 0,$$

$$dy = \mu\, dx,$$

$$\sum_{\kappa=1}^{n} P_\kappa dp_\kappa = -X dx,$$

$$\sum_{\kappa=1}^{n} P_\kappa dq_\kappa = -Y dx,$$

$$\begin{aligned} \sum_{\kappa=1}^{n} P_\kappa dz_\kappa &= \sum_{\kappa=1}^{n} \{P_\kappa p_\kappa + Q_\kappa q_\kappa\} dx \\ &= \sum_{\kappa=1}^{n} P_\kappa (p_\kappa + \mu q_\kappa) dx \\ &= \sum_{\kappa=1}^{n} P_\kappa (p_\kappa dx + q_\kappa dy). \end{aligned}$$

The last equation can evidently be replaced by

$$dz_\kappa = p_\kappa dx + q_\kappa dy,$$

for $\kappa = 1, \ldots, n$: and so we take the final form of the equations of the subsidiary system to be

$$\left.\begin{aligned} &\Delta = 0, \quad dy = \mu dx \\ &dz_\kappa = p_\kappa dx + q_\kappa dy, \quad (\kappa = 1, \ldots, n), \\ &\sum_{\kappa=1}^{n} P_\kappa dp_\kappa = - X dx \\ &\sum_{\kappa=1}^{n} P_\kappa dq_\kappa = - Y dx \end{aligned}\right\}.$$

This set includes $n+3$ total differential relations, which involve the $3n+2$ variables $x, y, z_1, \ldots, z_n, p_1, \ldots, p_n, q_1, \ldots, q_n$. For each simple root of the equation $\Delta = 0$, one such set arises; and there is a corresponding set, similar in form but larger in number, for any multiple root of $\Delta = 0$. The application of the method will be sufficiently indicated for a system of equations such that all the roots of $\Delta = 0$ are simple: it is given by the theorem :—

Assuming that all the n roots of $\Delta = 0$ are simple, let $v = c$ be an integral of the subsidiary system, distinct from $f_1 = 0, \ldots, f_n = 0$, and associated with a root μ; and suppose that, on taking the n roots μ in succession, the successive subsidiary systems give integrals

$$v_1 = c_1, \ \ldots, \ v_n = c_n,$$

these equations being distinct from $f_1 = 0, \ldots, f_n = 0$, quà functions of $p_1, \ldots, p_n, q_1, \ldots, q_n$. When the $2n$ equations

$$f_1 = 0, \ \ldots, f_n = 0, \quad v_1 = c_1, \ \ldots, \ v_n = c_n$$

are resolved for $p_1, \ldots, p_n, q_1, \ldots, q_n$, and the deduced values are substituted in

$$dz_\kappa = p_\kappa dx + q_\kappa dy,$$

for $\kappa = 1, \ldots, n$, the latter n equations become a completely integrable aggregate. The integral of this aggregate is an integral of the original system which, as it contains $2n$ arbitrary constants, is a complete integral.

174. This theorem, which is due to Hamburger, can be established as follows.

When $v=c$ is an integral of the subsidiary system associated with a root μ of $\Delta=0$, then the relation

$$dv=0$$

is satisfied in consequence of that system: that is, the relation

$$\frac{\partial v}{\partial x}dx+\frac{\partial v}{\partial y}dy+\sum_{s=1}^{n}\left(\frac{\partial v}{\partial z_s}dz_s+\frac{\partial v}{\partial p_s}dp_s+\frac{\partial v}{\partial q_s}dq_s\right)=0$$

must be a consequence of the $n+3$ equations. When the $n+3$ equations are used to remove dy, dz_1, ..., dz_n, dp_1, dq_1 from the differential relation $dv=0$, the coefficients of the remaining differential elements must vanish; and therefore

$$\frac{\partial v}{\partial p_s}-\frac{P_s}{P_1}\frac{\partial v}{\partial p_1}=0,$$

$$\frac{\partial v}{\partial q_s}-\frac{P_s}{P_1}\frac{\partial v}{\partial q_1}=0,$$

for $s=2$, ..., n, together with

$$\left(\frac{dv}{dx}+\mu\frac{dv}{dy}\right)P_1-X\frac{\partial v}{\partial p_1}-Y\frac{\partial v}{\partial q_1}=0,$$

where

$$\frac{dv}{dx}=\frac{\partial v}{\partial x}+\sum_{i=1}^{n}\frac{\partial v}{\partial z_i}p_i,$$

$$\frac{dv}{dy}=\frac{\partial v}{\partial y}+\sum_{i=1}^{n}\frac{\partial v}{\partial z_i}q_i.$$

We thus have $2n-1$ equations, homogeneous and linear in the derivatives of v, and the number of arguments occurring is $3n+2$: hence the number of integrals common to the system may be anything from zero to $n+3$, according to the extra number of equations required to make the system complete. We shall assume that the conditions securing the existence of a variable non-trivial integral are satisfied: and we shall make this assumption for each of the roots of $\Delta=0$: so that there thus will arise n equations

$$v_1=c_1, \ \ldots, \ v_n=c_n.$$

Now let the $2n$ equations

$$f_1=0, \ \ldots, \ f_n=0, \quad v_1=c_1, \ \ldots, \ v_n=c_n$$

be resolved so as to express $p_1, \ldots, p_n, q_1, \ldots, q_n$ in terms of $x, y, z_1, \ldots, z_n$. When their values are substituted in the $2n$ equations, each of these becomes an identity; and therefore, from each equation $v_i = c_i$ thus changed, we have

$$\frac{\partial v_i}{\partial x} + \sum_{s=1}^{n} \frac{\partial v_i}{\partial p_s}\frac{\partial p_s}{\partial x} + \sum_{s=1}^{n} \frac{\partial v_i}{\partial q_s}\frac{\partial q_s}{\partial x} = 0,$$

$$\frac{\partial v_i}{\partial y} + \sum_{s=1}^{n} \frac{\partial v_i}{\partial p_s}\frac{\partial p_s}{\partial y} + \sum_{s=1}^{n} \frac{\partial v_i}{\partial q_s}\frac{\partial q_s}{\partial y} = 0,$$

$$\frac{\partial v_i}{\partial z_j} + \sum_{s=1}^{n} \frac{\partial v_i}{\partial p_s}\frac{\partial p_s}{\partial z_j} + \sum_{s=1}^{n} \frac{\partial v_i}{\partial q_s}\frac{\partial q_s}{\partial z_j} = 0,$$

for $j = 1, \ldots, n$. Writing

$$\frac{d}{dx} = \frac{\partial}{\partial x} + \sum_{j=1}^{n} p_j \frac{\partial}{\partial z_j}, \quad \frac{d}{dy} = \frac{\partial}{\partial y} + \sum_{j=1}^{n} q_j \frac{\partial}{\partial z_j},$$

we have

$$\frac{dv_i}{dx} + \sum_{s=1}^{n} \frac{\partial v_i}{\partial p_s}\frac{dp_s}{dx} + \sum_{s=1}^{n} \frac{\partial v_i}{\partial q_s}\frac{dq_s}{dx} = 0,$$

$$\frac{dv_i}{dy} + \sum_{s=1}^{n} \frac{\partial v_i}{\partial p_s}\frac{dp_s}{dy} + \sum_{s=1}^{n} \frac{\partial v_i}{\partial q_s}\frac{dq_s}{dy} = 0,$$

and therefore

$$\frac{dv_i}{dx} + \mu\frac{dv_i}{dy} + \sum_{s=1}^{n} \frac{\partial v_i}{\partial p_s}\left(\frac{dp_s}{dx} + \mu\frac{dp_s}{dy}\right) + \sum_{s=1}^{n} \frac{\partial v_i}{\partial q_s}\left(\frac{dq_s}{dx} + \mu\frac{dq_s}{dy}\right) = 0.$$

When the formal equations satisfied by v_i are used, and the equivalent values of $\frac{\partial v_i}{\partial x} + \mu\frac{\partial v_i}{\partial y}$ and $\frac{\partial v_i}{\partial p_s}, \frac{\partial v_i}{\partial q_s}$, as given by those equations, are substituted in the last relation, it becomes

$$\frac{\partial v_i}{\partial p_1}\left[X + \sum_{s=1}^{n} P_s\left(\frac{dp_s}{dx} + \mu\frac{dp_s}{dy}\right)\right] + \frac{\partial v_i}{\partial q_1}\left[Y + \sum_{s=1}^{n} P_s\left(\frac{dq_s}{dx} + \mu\frac{dq_s}{dy}\right)\right] = 0;$$

and this relation, when regard is paid to the equations

$$\frac{Q_1}{P_1} = \ldots = \frac{Q_n}{P_n} = \mu,$$

can be transformed so that it becomes

$$\frac{\partial v_i}{\partial p_1}\left[X + \sum_{s=1}^{n}\left(P_s\frac{dp_s}{dx} + Q_s\frac{dp_s}{dy}\right)\right] + \frac{\partial v_i}{\partial q_1}\left[Y + \sum_{s=1}^{n}\left(P_s\frac{dq_s}{dx} + Q_s\frac{dq_s}{dy}\right)\right] = 0.$$

Further, each of the equations

$$f_j = 0,$$

for $j=1, \ldots, n$, becomes an identity when the values of $p_1, \ldots, p_n$, $q_1, \ldots, q_n$ are substituted therein. Hence

$$\frac{\partial f_j}{\partial x}+\sum_{s=1}^{n}\frac{\partial f_j}{\partial p_s}\frac{\partial p_s}{\partial x}+\sum_{s=1}^{n}\frac{\partial f_j}{\partial q_s}\frac{\partial q_s}{\partial x}=0,$$

$$\frac{\partial f_j}{\partial y}+\sum_{s=1}^{n}\frac{\partial f_j}{\partial p_s}\frac{\partial p_s}{\partial y}+\sum_{s=1}^{n}\frac{\partial f_j}{\partial q_s}\frac{\partial q_s}{\partial y}=0,$$

$$\frac{\partial f_j}{\partial z_\mu}+\sum_{s=1}^{n}\frac{\partial f_j}{\partial p_s}\frac{\partial p_s}{\partial z_\mu}+\sum_{s=1}^{n}\frac{\partial f_j}{\partial q_s}\frac{\partial q_s}{\partial z_\mu}=0,$$

for $\mu=1, \ldots, n$; and therefore

$$\frac{df_j}{dx}+\sum_{s=1}^{n}\frac{\partial f_j}{\partial p_s}\frac{dp_s}{dx}+\sum_{s=1}^{n}\frac{\partial f_j}{\partial q_s}\frac{dq_s}{dx}=0,$$

$$\frac{df_j}{dy}+\sum_{s=1}^{n}\frac{\partial f_j}{\partial p_s}\frac{dp_s}{dy}+\sum_{s=1}^{n}\frac{\partial f_j}{\partial q_s}\frac{dq_s}{dy}=0.$$

Multiplying these equations by l_j, and adding the respective equations for all values of j, we have

$$X+\sum_{s=1}^{n}P_s\frac{dp_s}{dx}+\sum_{s=1}^{n}Q_s\frac{dq_s}{dx}=0,$$

$$Y+\sum_{s=1}^{n}P_s\frac{dp_s}{dy}+\sum_{s=1}^{n}Q_s\frac{dq_s}{dy}=0.$$

When the values thus given for X and Y are substituted in the earlier equation which is homogeneous and linear in $\frac{\partial v_i}{\partial p_1}$ and $\frac{\partial v_i}{\partial q_1}$, it becomes

$$\frac{\partial v_i}{\partial p_1}\sum_{s=1}^{n}\left\{Q_s\left(\frac{dp_s}{dy}-\frac{dq_s}{dx}\right)\right\}+\frac{\partial v_i}{\partial q_1}\sum_{s=1}^{n}\left\{P_s\left(\frac{dq_s}{dx}-\frac{dp_s}{dy}\right)\right\}=0,$$

and therefore, as

$$Q_s=\mu P_s,$$

for $s=1, \ldots, n$, we have

$$\left(\mu\frac{\partial v_i}{\partial p_1}-\frac{\partial v_i}{\partial q_1}\right)\sum_{s=1}^{n}\left\{P_s\left(\frac{dp_s}{dy}-\frac{dq_s}{dx}\right)\right\}=0.$$

It is impossible, owing to the independence of $v_1, \ldots, v_n$, $f_1, \ldots, f_n$, quà functions of $p_1, \ldots, p_n, q_1, \ldots, q_n$, that the quantity

$$\mu\frac{\partial v_i}{\partial p_1}-\frac{\partial v_i}{\partial q_1}$$

shall be evanescent. For let

$$\frac{1}{P_1}\frac{\partial v_i}{\partial p_1}=M,$$

so that

$$\frac{\partial v_i}{\partial p_1} = MP_1,$$

and therefore, on account of the equations satisfied by v_i,

$$\frac{\partial v_i}{\partial p_s} = MP_s,$$

for $s = 2, \ldots, n$: then, if

$$\frac{\partial v_i}{\partial q_1} - \mu \frac{\partial v_i}{\partial p_1} = 0,$$

we should have

$$\begin{aligned}\frac{\partial v_i}{\partial q_1} &= \mu \frac{\partial v_i}{\partial p_1} \\ &= \mu MP_1 \\ &= MQ_1,\end{aligned}$$

and therefore, on account of the equations satisfied by v_i,

$$\frac{\partial v_i}{\partial q_s} = MQ_s,$$

for $s = 2, \ldots, n$. Now

$$P_s = \sum_{j=1}^{n} l_j \frac{\partial f_j}{\partial p_s}, \quad Q_s = \sum_{j=1}^{n} l_j \frac{\partial f_j}{\partial p_s};$$

consequently, we should have

$$\left\| \begin{matrix} \frac{\partial v_i}{\partial p_1}, & \ldots, & \frac{\partial v_i}{\partial p_n}, & \frac{\partial v_i}{\partial q_1}, & \ldots, & \frac{\partial v_i}{\partial q_n} \\ \frac{\partial f_1}{\partial p_1}, & \ldots, & \frac{\partial f_1}{\partial p_n}, & \frac{\partial f_1}{\partial q_1}, & \ldots, & \frac{\partial f_1}{\partial q_n} \\ \cdots & \cdots & \cdots & \cdots & \cdots & \cdots \\ \frac{\partial f_n}{\partial p_1}, & \ldots, & \frac{\partial f_n}{\partial p_n}, & \frac{\partial f_n}{\partial q_1}, & \ldots, & \frac{\partial f_n}{\partial q_n} \end{matrix} \right\| = 0.$$

Hence v_i, regarded as a function of $p_1, \ldots, p_n, q_1, \ldots, q_n$, would not be functionally distinct from $f_1, \ldots, f_n$, contrary to the hypothesis as to the actual construction of $v_1, \ldots, v_n$. Thus $\mu \frac{\partial v_i}{\partial p_1} - \frac{\partial v_i}{\partial q_1}$ is not evanescent; and therefore

$$\sum_{s=1}^{n} P_s \left(\frac{dp_s}{dy} - \frac{dq_s}{dx} \right) = 0,$$

that is,

$$\sum_{s=1}^{n} \left\{ \left(\frac{dp_s}{dy} - \frac{dq_s}{dx} \right) \sum_{j=1}^{n} \left(l_j \frac{\partial f_j}{\partial p_s} \right) \right\} = 0.$$

The quantities $l_1, \ldots, l_n$ are proportional to the first minors of any row of constituents in Δ, which has n distinct roots; and thus, taking the n sets of quantities $l_1, \ldots, l_n$ associated with the n roots of $\Delta = 0$ in succession, we have n independent sets such that the quantity

$$\begin{vmatrix} l_1^{(1)}, & \ldots, & l_n^{(1)} \\ \cdots & \cdots & \cdots \\ l_1^{(n)}, & \ldots, & l_n^{(n)} \end{vmatrix}$$

does not vanish, where $l_1^{(r)}, \ldots, l_n^{(r)}$ are the set associated with the root μ_r.

We thus have n equations, homogeneous and linear in the quantities

$$\frac{dp_1}{dy} - \frac{dq_1}{dx}, \ \ldots, \ \frac{dp_n}{dy} - \frac{dq_n}{dx}.$$

The determinant of their coefficients is

$$\begin{vmatrix} l_1^{(1)}, & \ldots, & l_n^{(1)} \\ \cdots & \cdots & \cdots \\ l_1^{(n)}, & \ldots, & l_n^{(n)} \end{vmatrix} J\begin{pmatrix} f_1, & \ldots, & f_n \\ p_1, & \ldots, & p_n \end{pmatrix}:$$

neither of the factors in this quantity vanishes: and therefore

$$\frac{dp_s}{dy} - \frac{dq_s}{dx} = 0,$$

for $s = 1, \ldots, n$. Consequently, the n equations

$$\left.\begin{array}{l} dz_1 = p_1 dx + q_1 dy \\ \cdots\cdots\cdots\cdots\cdots\cdots \\ dz_n = p_n dx + q_n dy \end{array}\right\},$$

where the values of $p_1, \ldots, p_n, q_1, \ldots, q_n$ are given by

$$f_1 = 0, \ \ldots, \ f_n = 0, \quad v_1 = c_1, \ \ldots, \ v_n = c_n,$$

are a completely integrable system: their integral equivalent contains $2n$ arbitrary constants: and it constitutes a complete integral of the original system

$$f_1 = 0, \ \ldots, \ f_n = 0.$$

Hamburger's theorem is thus established.

Note 1. When the complete integral has thus been obtained, the customary Lagrangian process of varying the parameters, subject to the conservation of form of the equations, can be used in order to deduce other integrals from the complete integral.

Note 2. If only a number of integrals

$$v_1 = c_1, \ldots, v_m = c_m,$$

where $m < n$, of the various subsidiary systems are known, they can be used to eliminate m of the dependent variables, say $z_1, \ldots, z_m$, and also their derivatives, from $f_1 = 0, \ldots, f_n = 0$. The integration of the surviving equations is a problem of simpler extent than the original problem.

Note 3. The preceding theorem requires an integral

$$v_i = c_i$$

of a subsidiary system. That subsidiary system may have a number of functionally independent integrals

$$v_i^{(1)}, v_i^{(2)}, \ldots,$$

the number not being greater than $n+3$: if the number is greater than unity, we replace the equation

$$v_i = c_i$$

by the equation

$$\phi_i (v_i^{(1)}, v_i^{(2)}, \ldots) = 0,$$

where ϕ_i is arbitrary. The argument then proceeds as before.

Note 4. The case, when the equations are linear in the derivatives and are of the form

$$f_i = -p_i + \pi_i + \sum_{s=1}^{n} a_{is} q_s = 0,$$

for $i = 1, \ldots, n$, being the case treated in the earlier sections of this chapter, is included simply in the general case. The critical equation $\Delta = 0$, being

$$\left\| \frac{\partial f_j}{\partial q_j} - \mu \frac{\partial f_j}{\partial p_j} \right\| = 0,$$

becomes

$$\begin{vmatrix} a_{11} + \mu, & a_{12}, & \ldots \\ a_{21}, & a_{22} + \mu, & \ldots \\ \ldots & \ldots & \ldots \end{vmatrix} = 0,$$

that is, $\Theta = 0$, being the critical equation for the simpler case.

Note 5. When no one of the dependent variables occurs explicitly in the original system, then

$$\frac{df_i}{dx} = \frac{\partial f_i}{\partial x}, \quad \frac{df_i}{dy} = \frac{\partial f_i}{\partial y},$$

for $i = 1, \ldots, n$, so that the equations are simplified. In that case, it may be possible to construct $v_1, \ldots, v_n$, so that no one of them contains any of the dependent variables explicitly: each of the equations

$$dz_r = p_r dx + q_r dy, \qquad (r = 1, \ldots, n),$$

is then completely integrable by itself without reference to the other equations.

175. Sometimes a member of the final integral equivalent can be obtained more directly as follows. Let

$$u(x, y, z_1, \ldots, z_n) = 0$$

be one integral relation in the integral equivalent of a system

$$f_1 = 0, \ldots, f_n = 0;$$

then the equations

$$\frac{\partial u}{\partial x} + \sum_{i=1}^{n} p_i \frac{\partial u}{\partial z_i} = 0,$$

$$\frac{\partial u}{\partial y} + \sum_{i=1}^{n} q_i \frac{\partial u}{\partial z_i} = 0,$$

are satisfied in connection with the system. If then the quantities $p_1, \ldots, p_n$ are eliminated from the equation

$$\frac{\partial u}{\partial x} + \sum_{i=1}^{n} p_i \frac{\partial u}{\partial z_i} = 0,$$

by means of

$$f_1 = 0, \ldots, f_n = 0,$$

the resulting equation must effectively be the same as

$$\frac{\partial u}{\partial y} + \sum_{i=1}^{n} q_i \frac{\partial u}{\partial z_i} = 0.$$

When the latter equation is used to eliminate any of the quantities $q_1, \ldots, q_n$, say q_n (on the supposition that $\frac{\partial u}{\partial z_n}$ is not zero), from the transformed shape of

$$\frac{\partial u}{\partial x} + \sum_{i=1}^{n} p_i \frac{\partial u}{\partial z_i} = 0,$$

the result must be an identity: the necessary conditions that it should reduce to an identity are a number of relations among the derivatives of u, which accordingly are a set of simultaneous partial equations for the determination of u.

It is clear, however, that the process is only of limited application; for instance, if u be a function of x, y, z_1 only, it can be only in the case of equations of exceedingly special form that the transformation of the equation

$$\frac{\partial u}{\partial x} + p_1 \frac{\partial u}{\partial z_1} = 0$$

will lead to the equation

$$\frac{\partial u}{\partial y} + q_1 \frac{\partial u}{\partial z_1} = 0,$$

in a way that gives useful relations between the derivatives of u. Moreover, just as in the classical problem of the three bodies*, it is not a fact that any integral leads to an identically satisfied equation: it may only lead to a relation merely compatible with the others.

Should the method fail, then it is necessary to fall back upon the method given in Hamburger's general theory.

Ex. 1. Let it be required to integrate the equations

$$\left.\begin{aligned} p_1(xq_1+yq_2-xy)+q_1(y^2-yq_1-xq_2)=0 \\ p_2(xq_1+yq_2-xy)+q_2(x^2-yq_1-xq_2)=0 \\ p_3(xq_1+yq_2-xy)+q_3(y^2-yq_1-xq_2)=0 \end{aligned}\right\}.$$

To test the method just suggested in § 175, we assume that

$$u=u(x, y, z_1, z_2, z_3)=0$$

is an equation in the integral equivalent: then we have

$$\frac{\partial u}{\partial x}+p_1\frac{\partial u}{\partial z_1}+p_2\frac{\partial u}{\partial z_2}+p_3\frac{\partial u}{\partial z_3}=0,$$

which must be consistent with the given equations, and therefore

$$(xy-xq_1-yq_2)\frac{\partial u}{\partial x}+q_1\frac{\partial u}{\partial z_1}(y^2-yq_1-xq_2)$$
$$+q_2\frac{\partial u}{\partial z_2}(x^2-yq_1-xq_2)+q_3\frac{\partial u}{\partial z_3}(y^2-yq_1-xq_2)=0.$$

* See vol. III of this work, §§ 265, 266.

This effectively must be the same as

$$\frac{\partial u}{\partial y}+q_1\frac{\partial u}{\partial z_1}+q_2\frac{\partial u}{\partial z_2}+q_3\frac{\partial u}{\partial z_3}=0,$$

and therefore the equation

$$(xy-xq_1-yq_2)\frac{\partial u}{\partial x}+q_2\frac{\partial u}{\partial z_2}(x^2-yq_1-xq_2)$$
$$=(y^2-yq_1-xq_2)\left(\frac{\partial u}{\partial y}+q_2\frac{\partial u}{\partial z_2}\right),$$

quà equation in q_1 and q_2, must be an identity. Hence

$$xy\frac{\partial u}{\partial x}=y^2\frac{\partial u}{\partial y},$$

$$x\frac{\partial u}{\partial x}=y\frac{\partial u}{\partial y},$$

$$-y\frac{\partial u}{\partial x}+x^2\frac{\partial u}{\partial z_2}=y^2\frac{\partial u}{\partial z_2}-x\frac{\partial u}{\partial y},$$

from the terms independent of q_1 and q_2, from the coefficient of q_1, and from the coefficient of q_2, respectively: the other terms in q_1 and q_2 disappear of themselves. These three equations are equivalent to the two

$$x\frac{\partial u}{\partial z_2}+\frac{\partial u}{\partial y}=0,$$

$$y\frac{\partial u}{\partial z_2}+\frac{\partial u}{\partial x}=0,$$

which are a complete Jacobian system: they have three independent integrals, viz.

$$z_1,\quad z_2-xy,\quad z_3,$$

and therefore we should take

$$z_2-xy=g(z_3),$$
$$z_1=f(z_3),$$

where f and g are arbitrary functions.

It is conceivable that there should be an integral in the full integral equivalent which does not involve z_3: let it be

$$v(x, y, z_1, z_2)=0.$$

Then the two equations

$$\frac{\partial v}{\partial x}+p_1\frac{\partial v}{\partial z_1}+p_2\frac{\partial v}{\partial z_2}=0,$$

$$\frac{\partial v}{\partial y}+q_1\frac{\partial v}{\partial z_1}+q_2\frac{\partial v}{\partial z_2}=0,$$

must be treated in a similar manner. We substitute in the first for p_1 and p_2 by means of the original equations and, after substitution, we eliminate q_2

by means of the second: the result should be an identity. The necessary conditions are easily found to be

$$y\frac{\partial v}{\partial y}-x\frac{\partial v}{\partial x}+(y^2-x^2)\frac{\partial v}{\partial z_1}+\frac{1}{\frac{\partial v}{\partial z_2}}\left(y\frac{\partial v}{\partial x}-x\frac{\partial v}{\partial y}\right)\frac{\partial v}{\partial z_1}=0,$$

$$xy\frac{\partial v}{\partial x}-x^2\frac{\partial v}{\partial y}+\frac{1}{\frac{\partial v}{\partial z_2}}\left(y\frac{\partial v}{\partial x}-x\frac{\partial v}{\partial y}\right)\frac{\partial v}{\partial y}=0.$$

The second of these gives either

$$y\frac{\partial v}{\partial x}-x\frac{\partial v}{\partial y}=0,$$

or

$$x\frac{\partial v}{\partial z_2}+\frac{\partial v}{\partial y}=0.$$

Using the latter alternative, the first condition can be transformed to

$$\left(y\frac{\partial v}{\partial z_1}-\frac{\partial v}{\partial y}\right)\left(y\frac{\partial v}{\partial y}-x\frac{\partial v}{\partial x}\right)=0.$$

If we could have

$$y\frac{\partial v}{\partial z_1}-\frac{\partial v}{\partial y}=0,$$

together with

$$x\frac{\partial v}{\partial z_2}+\frac{\partial v}{\partial y}=0,$$

then completing the system, we should have

$$\frac{\partial v}{\partial z_1}=0,$$

and so $\frac{\partial v}{\partial y}=0$, $\frac{\partial v}{\partial z_2}=0$: no integral would be obtained. When we take

$$y\frac{\partial v}{\partial y}-x\frac{\partial v}{\partial x}=0,$$

together with

$$x\frac{\partial v}{\partial z_2}+\frac{\partial v}{\partial y}=0,$$

they become

$$x\frac{\partial v}{\partial z_2}+\frac{\partial v}{\partial y}=0,$$

$$y\frac{\partial v}{\partial z_2}+\frac{\partial v}{\partial x}=0,$$

the system already used. Thus we obtain no new integral from the alternative

$$x\frac{\partial v}{\partial z_2}+\frac{\partial v}{\partial y}=0.$$

Using the prior alternative

$$y\frac{\partial v}{\partial x}-x\frac{\partial v}{\partial y}=0,$$

the first equation becomes

$$y\frac{\partial v}{\partial y}-x\frac{\partial v}{\partial x}+(y^2-x^2)\frac{\partial v}{\partial z_1}=0:$$

the two equations can be replaced by

$$x\frac{\partial v}{\partial z_1}-\frac{\partial v}{\partial x}=0,$$

$$y\frac{\partial v}{\partial z_1}-\frac{\partial v}{\partial y}=0.$$

These are a complete Jacobian system: they have two independent integrals

$$z_1-\tfrac{1}{2}x^2-\tfrac{1}{2}y^2, \quad z_2;$$

accordingly, we take

$$z_1-\tfrac{1}{2}(x^2+y^2)=k(z_2),$$

where k is an arbitrary function.

Thus an integral equivalent of the original equations is

$$z_1-\tfrac{1}{2}(x^2+y^2)=k(z_2), \quad z_2-xy=g(z_3), \quad z_1=f(z_3),$$

where f, g, k are arbitrary functions.

Had it been impossible to obtain a third integral by the preceding process, the known integrals could have been used as follows.

Owing to the relation

$$z_1=f(z_3),$$

we have

$$p_1q_3-p_3q_1=0;$$

so that, when this integral is retained, the third differential equation is a consequence of the first and can therefore be neglected. Again, eliminating z_3 between

$$z_2-xy=g(z_3), \quad z_1=f(z_3),$$

let the result be

$$z_2-xy=\theta(z_1).$$

It is easy to verify that

$$p_2(xq_1+yq_2-xy)+q_2(x^2-yq_1-xq_2)$$
$$=\{p_1(xq_1+yq_2-xy)+q_1(y^2-yq_1-xq_2)\}\,\theta'(z_1),$$

so that the second differential equation is satisfied if the first is satisfied; it need not be retained when the first is retained. Substituting the value of z_2 in the first equation, we have

$$p_1q_1\{x+y\theta'(z_1)\}-q_1^2\{y+x\theta'(z_1)\}+(y^2-x^2)q_1=0$$

as the one equation to be satisfied, or neglecting $q_1=0$, we have

$$p_1\{x+y\theta'(z_1)\}-q_1\{y+x\theta'(z_1)\}=x^2-y^2.$$

This is an equation of Lagrangian form. To integrate it, we construct two integrals of the ordinary equations

$$\frac{dx}{x+y\theta'(z_1)}=\frac{dy}{-y-x\theta'(z_1)}=\frac{dz_1}{x^2-y^2}:$$

these are obtainable in the form

$$z_1 - \tfrac{1}{2}(x^2+y^2) = \text{constant},$$

$$xy + \theta(z_1) = \text{constant};$$

and therefore a general integral is given by

$$z_1 - \tfrac{1}{2}(x^2+y^2) = k\{xy + \theta(z_1)\},$$

where k is an arbitrary functional form. Hence

$$z_1 - \tfrac{1}{2}(x^2+y^2) = k(z_2),$$

agreeing with the former result.

Ex. 2. As an illustration of the general method, we still consider the same system as in the last example. It is easy to see that, as the equation for μ is

$$\begin{vmatrix} p_1x+y^2-2yq_1-xq_2-\mu\theta, & p_2x-yq_2 \quad , & p_3x-yq_3 \\ p_1y-xq_1 \quad , & p_2y+x^2-yq_1-2xq_2-\mu\theta, & p_3y-xq_3 \\ 0 \quad , & 0 \quad , & y^2-yq_1-xq_2-\mu\theta \end{vmatrix} = 0,$$

where θ denotes xq_1+yq_2-xy, one root is given by

$$\mu\theta = y^2 - yq_1 - xq_2.$$

The corresponding sets of quantities l_1, l_2, l_3 are such that

$$l_1(p_1x-yq_1)+l_2(p_2x-yq_2)+l_3(p_3x-yq_3)=0,$$

$$l_1(p_1y-xq_1)+l_2(p_2y-xq_2)+l_3(p_3y-xq_3)=0:$$

hence we may take

$$l_1 = p_2q_3-p_3q_2, \quad l_2 = p_3q_1-p_1q_3, \quad l_3 = p_1q_2-p_2q_1.$$

But by the given equations

$$p_3q_1 - p_1q_3 = 0,$$

so that

$$l_2 = 0;$$

and then

$$\frac{l_3}{l_1} = -\frac{p_1}{p_3} = -\frac{q_1}{q_3}.$$

Again

$$X = l_1\frac{\partial f_1}{\partial x} + l_2\frac{\partial f_2}{\partial x} + l_3\frac{\partial f_3}{\partial x} = 0,$$

$$Y = l_1\frac{\partial f_1}{\partial y} + l_2\frac{\partial f_2}{\partial y} + l_3\frac{\partial f_3}{\partial y} = 0,$$

in the present case. Hence the equations for v are

$$\frac{\partial v}{\partial p_2} = 0, \quad \frac{\partial v}{\partial q_2} = 0,$$

$$\frac{\partial v}{\partial p_3} + \frac{p_1}{p_3}\frac{\partial v}{\partial p_1} = 0,$$

$$\frac{\partial v}{\partial q_3} + \frac{q_1}{q_3}\frac{\partial v}{\partial q_1} = 0,$$

$$(xq_1+yq_2-xy)\frac{dv}{dx} + (y^2-yq_1-xq_2)\frac{dv}{dy} = 0.$$

We make the system complete: it then is

$$\frac{\partial v}{\partial p_2}=0,\ \frac{\partial v}{\partial q_2}=0,\ \frac{\partial v}{\partial x}=0,\ \frac{\partial v}{\partial y}=0,$$

$$\frac{\partial v}{\partial z_1}=0,\ \frac{\partial v}{\partial z_2}=0,\ \frac{\partial v}{\partial z_3}=0,$$

$$\frac{\partial v}{\partial p_3}+\frac{p_1}{p_3}\frac{\partial v}{\partial p_1}=0,\quad \frac{\partial v}{\partial q_3}+\frac{q_1}{q_3}\frac{\partial v}{\partial q_1}=0.$$

The two integrals, independent so far as these equations are concerned, are

$$v=\frac{p_3}{p_1},\quad v=\frac{q_3}{q_1}.$$

But, from the original system of equations, $p_3q_1-p_1q_3=0$: hence we take

$$\frac{p_3}{p_1}=a,\ \frac{q_3}{q_1}=a.$$

Then

$$\begin{aligned} dz_3&=p_3\,dx+q_3\,dy\\ &=a\,(p_1dx+q_1dy)\\ &=a\,dz_1; \end{aligned}$$

and therefore

$$z_3=az_1+b,$$

an integral of the complete type.

The discussion for the other values of μ is left as an exercise.

Ex. 3. Integrate the equations

$$\left.\begin{aligned} x\,(q_1+q_2-y)\,p_1&=yq_1^2+yq_1q_2-y^2q_1+(x^2-y^2)\,q_2\\ x\,(q_1+q_2-y)\,p_2&=yq_1q_2+yq_2^2-x^2q_2 \end{aligned}\right\}.$$

[The equation for μ is

$$\begin{vmatrix} \mu x\,(q_1+q_2-y)-xp_1+2yq_1+yq_2-y^2, & -xp_2+yq_2\\ -xp_1+yq_1+x^2-y^2\quad, & \mu x\,(q_1+q_2-y)-xp_2+yq_1+2yq_2-x^2 \end{vmatrix}=0;$$

and the two values of μ are given by

$$\mu x\,(q_1+q_2-y)=-y\,(q_1+q_2)+x^2,$$

$$\mu x\,(q_1+q_2-y)=-2y\,(q_1+q_2)+x\,(p_1+p_2)+y^2.$$

Integrals of the complete type are

$$\left.\begin{aligned} z_1+az_2&=\tfrac{1}{2}\,(x^2+y^2)+b\\ z_1+z_2&=a'xy+b' \end{aligned}\right\};$$

and integrals of the general type are

$$\left.\begin{aligned} z_1-\tfrac{1}{2}\,(x^2+y^2)&=f\,(z_2)\\ z_1+z_2&=g\,(x,y) \end{aligned}\right\},$$

where f and g are arbitrary functions.]

Ex. 4. Integrate the equations

$$\left.\begin{aligned}\lambda p_1&=q_1q_3(q_1y+q_2)\\ \lambda p_2&=-yq_1q_2q_3+q_2(x-q_2)(1-q_3)\\ \lambda p_3&=-q_3(1-q_3)(q_1y+q_2)\end{aligned}\right\},$$

where λ denotes $q_1(q_2+xq_3-x)$.

Ex. 5. Merely as an indication that the general method is not always effective, consider the comparatively simple pair of equations

$$\left.\begin{aligned}f_1&=xp_1+yq_2-1=0\\ f_2&=yp_2+xq_1-1=0\end{aligned}\right\}$$

The equation for μ becomes

$$\mu^2-1=0;$$

the relation between the quantities l_1 and l_2 is

$$l_2=l_1\mu.$$

Also

$$P_1=l_1x, \qquad P_2=l_2y,$$
$$X=l_1(p_1+q_1), \quad Y=l_2(p_2+q_2).$$

The differential equations for v are

$$\frac{\partial v}{\partial p_2}=\frac{l_2y}{l_1x}\frac{\partial v}{\partial p_1}=\mu\frac{y}{x}\frac{\partial v}{\partial p_1},$$

$$\frac{\partial v}{\partial q_2}=\mu\frac{y}{x}\frac{\partial v}{\partial q_1},$$

$$\frac{dv}{dx}+\mu\frac{dv}{dy}-\frac{p_1+q_1}{x}\frac{\partial v}{\partial p_1}-\frac{p_2+q_2}{x}\frac{\partial v}{\partial q_1}=0.$$

First, let $\mu=1$, so that the equations are

$$\frac{\partial v}{\partial p_2}-\frac{y}{x}\frac{\partial v}{\partial p_1}=0, \quad \frac{\partial v}{\partial q_2}-\frac{y}{x}\frac{\partial v}{\partial q_1}=0,$$

$$\frac{\partial v}{\partial x}+p_1\frac{\partial v}{\partial z_1}+p_2\frac{\partial v}{\partial z_2}+\frac{\partial v}{\partial y}+q_1\frac{\partial v}{\partial z_1}+q_2\frac{\partial v}{\partial z_2}$$
$$-\frac{p_1+q_1}{x}\frac{\partial v}{\partial p_1}-\frac{p_2+q_2}{x}\frac{\partial v}{\partial q_1}=0.$$

When we complete the system, it becomes

$$\frac{\partial v}{\partial p_1}=0, \ \frac{\partial v}{\partial p_2}=0, \ \frac{\partial v}{\partial q_1}=0, \ \frac{\partial v}{\partial q_2}=0, \ \frac{\partial v}{\partial z_1}=0, \ \frac{\partial v}{\partial z_2}=0,$$

$$\frac{\partial v}{\partial x}+\frac{\partial v}{\partial y}=0:$$

the only integral is

$$v=v_1=x-y,$$

and it is useless for the application of the process.

Similarly, when $\mu=-1$, it appears that the only derivable value of v is

$$v=v_2=x+y,$$

and it is useless for the application of the process. In fact, p_1, q_1, p_2, q_2 cannot be deduced from

$$f_1=0, \quad f_2=0, \quad v_1=c_1, \quad v_2=c_2.$$

It may be remarked that, if we write

$$z_1=z+\log(x+y),$$

and eliminate z_2 between the two equations, then z satisfies an equation of the second order which (in the usual notation) is

$$r+\frac{p}{x}=t-\frac{q}{y}.$$

This equation (as may easily be verified) does not possess an intermediate integral. When we take

$$x+y=2u, \quad x-y=2v,$$

the equation becomes

$$\frac{\partial^2 z}{\partial u \partial v}-\frac{u}{u^2-v^2}\frac{\partial z}{\partial u}+\frac{v}{u^2-v^2}\frac{\partial z}{\partial v}=0,$$

which is of Laplace's linear type having equal invariants, hereafter to be considered.

Can the Jacobian process be generalised?

176. The preceding investigations of Hamburger shew that, for limited classes of equations*, it is possible to construct auxiliary systems suggested by the analogy of the subsidiary equations constructed in association with a linear equation. And it is precisely this linear form which has made the process effective for the appropriate equations. Moreover, when the original simultaneous equations propounded for integration are not linear, Hamburger's method is to change the set into an amplified set with an amplified aggregate of dependent variables, the new set being linear.

It is natural to enquire whether, as the Lagrangian subsidiary equations for a single linear equation in a single dependent variable have thus been generalised so as to be associable with a number of linear equations in the same number of dependent variables, there is any corresponding possibility of generalising the Jacobian method of proceeding with a single equation. It is, however, possible to see, from even the simplest case, that the

* The limitations are imposed by the hypothesis that the equations in the auxiliary systems are so far consistent with one another as to possess one or more integrals: also, there are only two independent variables.

generalisation of the Jacobian method requires, in order to be effective, a process which is of too high an order for the applicability of analysis in its present stage of attainment.

Thus let

$$f = f(z_1, z_2, p_1, q_1, p_2, q_2, x, y) = 0,$$

$$g = g(z_1, z_2, p_1, q_1, p_2, q_2, x, y) = 0,$$

be propounded as a couple of equations, algebraically independent of one another; and let

$$h = h(z_1, z_2, p_1, q_1, p_2, q_2, x, y) = \text{constant}$$

be an equation compatible with them. Then, denoting the second derivatives of z_1 and z_2 by r_1, s_1, t_1, and r_2, s_2, t_2 respectively, we have

$$\frac{\partial f}{\partial p_1} r_1 + \frac{\partial f}{\partial q_1} s_1 + \frac{\partial f}{\partial p_2} r_2 + \frac{\partial f}{\partial q_2} s_2 + \frac{df}{dx} = 0,$$

$$\frac{\partial g}{\partial p_1} r_1 + \frac{\partial g}{\partial q_1} s_1 + \frac{\partial g}{\partial p_2} r_2 + \frac{\partial g}{\partial q_2} s_2 + \frac{dg}{dy} = 0,$$

$$\frac{\partial h}{\partial p_1} r_1 + \frac{\partial h}{\partial q_1} s_1 + \frac{\partial h}{\partial p_2} r_2 + \frac{\partial h}{\partial q_2} s_2 + \frac{dh}{dy} = 0;$$

the elimination of r_1 and r_2 leads to the equation

$$\frac{\partial(f, g, h)}{\partial(p_1, p_2, q_1)} s_1 + \frac{\partial(f, g, h)}{\partial(p_1, p_2, q_2)} s_2 + \frac{\partial(f, g, h)}{\partial(p_1, p_2, x)} = 0.$$

Similarly, we have the equation

$$\frac{\partial(f, g, h)}{\partial(q_1, q_2, p_1)} s_1 + \frac{\partial(f, g, h)}{\partial(q_1, q_2, p_2)} s_2 + \frac{\partial(f, g, h)}{\partial(p_1, p_2, x)} = 0.$$

It is generally impossible to eliminate s_1 and s_2 between these two equations. Thus, by associating only a single equation with a given pair, it is not possible to generalise Jacobi's process.

If, however, a second equation, say

$$k = k(z_1, z_2, p_1, q_1, p_2, q_2, x, y) = \text{constant},$$

can be associated with the first pair and with the equation $g = \text{constant}$, so that

$$\frac{\partial k}{\partial p_1} r_1 + \frac{\partial k}{\partial q_1} s_1 + \frac{\partial k}{\partial p_2} r_2 + \frac{\partial k}{\partial q_2} s_2 + \frac{dk}{dx} = 0,$$

we find

$$\frac{\partial(f, g, h, k)}{\partial(p_1, q_1, p_2, q_2)} s_1 + \frac{\partial(f, g, h, k)}{\partial(p_1, x, p_2, q_2)} = 0,$$

$$\frac{\partial(f, g, h, k)}{\partial(p_1, q_1, p_2, q_2)} s_2 + \frac{\partial(f, g, h, k)}{\partial(p_1, q_1, p_2, x)} = 0.$$

Similarly, we should have

$$\frac{\partial(f, g, h, k)}{\partial(p_1, q_1, p_2, q_2)} s_1 + \frac{\partial(f, g, h, k)}{\partial(y, q_1, p_2, q_2)} = 0,$$

$$\frac{\partial(f, g, h, k)}{\partial(p_1, q_1, p_2, q_2)} s_2 + \frac{\partial(f, g, h, k)}{\partial(p_1, q_1, y, q_2)} = 0.$$

Consequently,

$$\frac{\partial(f, g, h, k)}{\partial(x, p_1, p_2, q_2)} + \frac{\partial(f, g, h, k)}{\partial(y, q_1, p_2, q_2)} = 0,$$

$$\frac{\partial(f, g, h, k)}{\partial(x, p_2, p_1, q_1)} + \frac{\partial(f, g, h, k)}{\partial(y, q_2, p_1, p_2)} = 0,$$

which may be regarded as two equations for the determination of h and k. The first of them secures the relation

$$\frac{\partial p_1}{\partial y} = \frac{\partial q_1}{\partial x};$$

the second secures the relation

$$\frac{\partial p_2}{\partial y} = \frac{\partial q_2}{\partial x};$$

and the two equations are thus the necessary and sufficient conditions for integrability.

The two equations are also two equations for the determination of h and k, being lineo-linear in the derivatives of these quantities. They are simpler in form than the original equations $f=0$, $g=0$: yet, even so, the integration of the two equations appears to be an operation of the second order, which is not resoluble into operations of the first order.

Thus the Jacobian process cannot be generalised when there are two, or more than two, dependent variables without requiring for its completion operations of higher order than are required for Hamburger's generalisation of Lagrange's process.

Note. It may happen that three equations

$$f=0, \quad g=0, \quad h=\text{constant},$$

are given: conditions as to coexistence must be satisfied. The two preceding equations can be regarded as simultaneous equations for the determination of one dependent variable: the conditions that they possess a common integral will be the conditions for the coexistence of the three equations. When these are satisfied, k can be determined by the usual processes: and then the equations

$$f=0, \quad g=0, \quad h=\text{constant}, \quad k=\text{constant},$$

give values of p_1, q_1, p_2, q_2 which make

$$dz_1=p_1dx+q_1dy, \quad dz_2=p_2dx+q_2dy,$$

a completely integrable system.

Ex. Shew that, if n dependent variables $z_1, \ldots, z_n$ are to be determined in terms of two independent variables x and y by means of n partial differential equations of the first order

$$f_1=0, \ldots, f_n=0,$$

and if other n equations

$$v_1=c_1, \ldots, v_n=c_n,$$

where $c_1, \ldots, c_n$ are constants, can be associated with them, then the equations

$$\frac{\partial(f_1, \ldots, f_n, v_1, \ldots, v_n)}{\partial(x, p_r, p_1, q_1, \ldots, p_n, q_n)}+\frac{\partial(f_1, \ldots, f_n, v_1, \ldots, v_n)}{\partial(y, q_r, p_1, q_1, \ldots, p_n, q_n)}=0,$$

for $r=1, \ldots, n$, derivatives with regard to q_r not appearing in the first expression and those with regard to p_r not appearing in the second expression, are satisfied.

177. In a preceding example of very simple type (§ 175, Ex. 5), it was seen that the elimination of one of the dependent variables and its derivatives led to an equation of the second order. This result is partly due to the special form of the equations there given: that it is not true in general for two equations of the form

$$f(z_1, z_2, p_1, p_2, q_1, q_2, x, y)=0,$$
$$g(z_1, z_2, p_1, p_2, q_1, q_2, x, y)=0,$$

can easily be seen. Taking the derivatives of the first order of both equations, we have

$$\frac{df}{dx}=0, \quad \frac{df}{dy}=0, \quad \frac{dg}{dx}=0, \quad \frac{dg}{dy}=0,$$

which, with $f=0$ and $g=0$, are six equations: and as regards z_2 and its derivatives, the quantities that occur in them are z_2, p_2,

q_2, r_2, s_2, t_2, in the usual notation. Six quantities cannot usually be eliminated between six equations; and therefore we cannot infer that z_1 is usually determined by an equation of the second order. But if $f=0$ and $g=0$ do not explicitly involve the dependent variables, then only five quantities occur for elimination: in that case, there survives a single equation of the second order.

In the more general case, when the dependent variables do occur explicitly, we associate with the preceding six equations the set

$$\frac{d^2f}{dx^2}=0,\quad \frac{d^2f}{dxdy}=0,\quad \frac{d^2f}{dy^2}=0,\quad \frac{d^2g}{dx^2}=0,\quad \frac{d^2g}{dxdy}=0,\quad \frac{d^2g}{dy^2}=0,$$

thus making twelve in all. There are ten quantities to be eliminated; and therefore two equations will survive after the elimination, being two equations of the third order satisfied by z_1.

Ex. Prove that, if there be m dependent variables and two independent variables, and if there be m partial differential equations of the first order involving the dependent variables explicitly, then usually the lowest order of differential equation satisfied by a single variable is $2m-1$, and that usually the number of equations of that order satisfied by the single variable is m.

We shall return to the discussion of this matter in Chapter XXI. Meanwhile, these results, as well as other considerations, indicate that we require the theory of equations of order higher than the first; accordingly, we proceed to the consideration of that theory.

CAMBRIDGE: PRINTED BY JOHN CLAY, M.A. AT THE UNIVERSITY PRESS.

www.ingramcontent.com/pod-product-compliance
Lightning Source LLC
LaVergne TN
LVHW011258110826
845149LV00001B/180
* 9 7 8 1 4 1 8 1 8 4 7 4 2 *